W9-BJB-102

Mathematica® for Physics

Robert L. Zimmerman

University of Oregon

Fredrick I. Olness

Southern Methodist University

Addison-Wesley Publishing Company
Reading, Massachusetts • Menlo Park, California
New York • Don Mills, Ontario
Wokingham, England • Amsterdam • Bonn
Sydney • Singapore • Tokyo • Madrid
San Juan • Milan • Paris

Library of Congress Cataloging-in-Publication Data

Zimmerman, Robert L.
Mathematica for physics / Robert L. Zimmerman, Fredrick I. Olness.
p. cm.
ISBN 0-201-53796-6
1. Mathematical physics. 2. Mathematica (Computer file)
I. Olness, Fredrick Olness. II. Title.
QC20.Z56 1994
530′.0285′53–dc20 94-2482
CIP

The programs and applications presented in this book have been included for their instructional value. They have been tested with care but are not guaranteed for any particular purpose. The publisher does not offer any warranties or representations, nor does it accept any liabilities with respect to the programs or applications.

Copyright © 1995 by Addison-Wesley Publishing Company, Inc.

All rights reserved. No part of this publication may be reproduced, stored in a retrieval system, or transmitted, in any form or by any means, electronic, mechanical, photocopying, recording, or otherwise, without the prior written permission of the publisher. Printed in the United States of America.

1 2 3 4 5 6 7 8 9 10-MA-9897969594

To Merrie,
and
to Gloria and Jonathan,
for support and encouragement.

Foreword

To be good at physics you no longer have to be good at calculating.

When I started doing physics in the early 1970s, it was generally thought that being able to do long calculations by hand was an essential skill. But as it happens I was never very good at such calculations. And so I looked for an alternative, and before long I found one: I discovered that I did not really need to do the calculations myself—I could get a computer to do them instead.

My physics papers from the late 1970s were full of elaborate calculations. And from seeing this many people came to the conclusion that I must be a great calculator. But in fact that was far from the truth, and what was actually going on was that I was doing all my calculations by computer.

Of course, in those days it required a considerable amount of programming effort to get a computer to do one's calculations—particularly when algebra and graphics were involved. But after a few years I realized that with new generations of computers and software engineering tools it would actually be possible to build a single software system that would be able to handle all the various kinds of calculations that one needed to do. And from this realization I came in the end to develop *Mathematica*.

So now that *Mathematica* exists, what does it mean for physics? In the six years since the first version of *Mathematica* became available, a vast amount of new physics has been done with it. Indeed, for example, if one looks today at any of the leading physics journals, one can tell that a large fraction of the calculations and pictures in them were done with *Mathematica*.

But it is not just at the level of research that *Mathematica* affects the way physics is done. Anyone who learns physics today can do so in a very different way because of *Mathematica*.

And that is the point of this book. What the authors have done is to take the topics from a mainstream physics course and show how each of them can be handled in a new way with *Mathematica*.

The results are impressive. Over and over again what was once a calculation too lengthy to be reproduced as part of a course now becomes a few lines of *Mathematica* input that can be executed in a matter of seconds. And instead of having to explain in painful detail the mechanics of the calculation, one can concentrate on the conceptual issues that underlie it.

One of the things that one sees in many of the examples in the book is the extent to which *Mathematica* has narrowed the gap between physics learning and physics research. For once a calculation has been set up in *Mathematica*, one can try the calculation in many different cases—not just ones already covered in textbooks, but also ones that may never have been tried before.

This book will no doubt be read by many students of physics and many professors. Students may wonder how many of the calculations it describes could ever have been done before *Mathematica*. Perhaps sometimes their professors will tell them tales from the heroic age of hand calculation. But mostly I hope that students and professors alike will be able to use the technology that we have built and material of the kind that is included in this book to learn and discover physics in a new and exciting way.

Stephen Wolfram
Creator of *Mathematica*

Introduction

Computer algebra software has already had an important impact on the way physics is taught and research is performed. While computers cannot replace thinking, they significantly enhance problem-solving abilities of the scientist and student by eliminating the tedious mathematics. As computers become more powerful and more readily available, the scope of problems solved in both research and teaching tremendously expands.

This book serves as a guide for using the computer algebra program *Mathematica* for physics research and teaching. The flexibility of *Mathematica* to manipulate analytical, numerical, and graphical expressions will further broaden the scope of problems that the student and researcher can solve.

Physics is not a spectator sport. The best way to demonstrate *Mathematica* is by solving a variety of physics problems chosen to illustrate its ability to display the output in many forms. A siginificant asset of *Mathematica* is the ease with which results can be visualized. This feature brings physics problems "alive" so that the reader can interact and experiment with the solutions. One can change the parameters and immediately observe the consequences, thereby gaining deeper insight into the physics of the solution. Instead of being throughly exhausted by the burdensome mathematics required to obtain the answer, *Mathematica* enables us to focus our attention on understanding the solution.

How to use this book:

This book is intended for undergraduate students, graduate students, and practicing physicists who want to learn new *Mathematica* techniques for solving a general class of physics problems. For the student, we expect this text to be a supplement to the standard course texts in Mechanics, Electrodynamics, Relativity, and Quantum Mechanics; the student should use this book to get ideas on how to use *Mathematica* to solve the problems assigned by the instructor. Since we cover the canonical problems from the core courses, the student can practice with our solutions, and then modify our solutions to solve the particular problems assigned. This should help the student move up the *Mathematica* learning curve quickly.

This book is also suitable for a course designed to teach the applications of *Mathematica* to physics.

As such, the design of this book is more like a reference book than a novel to be read cover to cover. Each problem is self-contained (up to user-defined functions discussed at the beginning of each chapter), so the reader can go immedately to the portion of the book that is relevant to their problem.

There are two sections of the book which we recommend all readers examine before trying the problems.

1) The section in this preface on Troubleshooting is important to skim so that the reader knows where to turn in case difficulties arise.
2) The first chapter contains useful information about style, notation, and short-cuts that we will make use of throughout the book.

Note that we assume the reader is reasonable familiar with *Mathematica* (at least at the level of the tutorial in the *Mathematica* manual), so we focus on the physics applications and not on rudimentary *Mathematica* techniques.

The book consists of two levels of material. The first three chapters are readily understandable by the undergraduate physics student. The latter chapters are intended for advanced undergraduate and graduate physics students, and cover a broader range of topics.

Each of the chapters 2 through 7 are divided into three parts:

i) an introduction to the physics and *Mathematica* commands;
ii) solved problems that cover the standard ideas and methods found in the discipline;
iii) unsolved exercises.

As the reader gains insight into the power of symbolic computations, they can easily extend the techniques demonstrated here to go beyond these examples and explore more difficult problems.

About the Electronic Supplement:

The *Mathematica* code for initialization and all user-defined functions is included in the electronic supplement so that you can begin working the examples immediately. We have also included some selected samples from this book. In the future, we plan to post tips and suggestions about using this book, extensions to other problems and related fields, and any bug fixes (should we encounter any).

The electronic supplement is available as item number 0206-862 from MathSource(TM). MathSource is an on-line archive of *Mathematica* related materials contributed by Wolfram Research and *Mathematica* users around the world. You can obtain this material from MathSource in three ways:

Electronic Mail:

To have materials sent to you atuomatically by electronic mail, send the single line e-mail message:

Communication with the authors:

Send 0206-862
to the mathsource server at
mathsource@wri.com
For more information about MathSource, send the message:
Help Intro

World Wide Web:

MathSource is available via World Wide Web at **http://mathsource.wri.com/**

Gopher:

MathSource is also available via Gopher at **mathsource.wri.com port=70**

Anonymous FTP:

MathSource is also available via anonymous FTP at **mathsource.wri.com**

Direct Dialup:

You can also direct dialup to MathSource with a modem to (217) 398-1898 (8N1: 8-bits, no-parity, 1-stop bit).

How to use the Electronic Supplement:

To spare the reader tedious typing, we have put all initializations for each chapter into a single initialization file. This enables the reader to start solving problems immediately without re-typing the initialization.

For example, if you wanted to solve Problem 6 in Chapter 7, all you need do is load in the initialization for Chapter 7 with at **Get** command

```
In[1]:=   <</maindir/subdir/.../ch7init.m
```

(The directory path above must be modified for your specific configuration. See your *Mathematica* installation manual, or your system manager.) This command will automatically load in all the necessary standard *Mathematica* packages, and define all the user-defined function needed for chapter 7. You can then proceed directly to problem 6 (or whatever problem you choose) and begin working.

Additionally, we have stored selected problems in sample Notebook files. For details on these files, refer to the Notebook files themselves, and the **Readme.txt** file.

Note that if you are working with a limited system and do not want to load all the standard *Mathematica* packages, you will have to edit the intialization files.

If you frequently work on a particular chapter, you can speed up the initialization process by creating a **Dump** file (if your system supports this feature) to save the initialization in a compiled form.

Communication with the authors:

We welcome any comments and suggestions regarding this book. You may contact us via e-mail or regular mail at:

Robert Zimmerman
Institute of Theoretical Science
University of Oregon
Eugene, OR 97403
bob@zim.uoregon.edu

Fredrick Olness
Department of Physics
Southern Methodist University
Dallas, TX 75275
olness@phyvms.physics.smu.edu

We welcome any bug reports, and will post such information on the MathSource server (should we encounter any). However, we are unable to offer any help debugging problems specific to the compatability of different *Mathematica* versions or hardware installations; for this type of assistance, we must refer you to the previous section on Troubleshooting, or to your local system manager.

Acknowledgments

It is our pleasure to acknowledge the people who have contributed to this project. We thank Fumitaka Umewaka whose masters thesis *Applications of Mathematica in Teaching Physics* at the University of Oregon initiated this project. We thank Ray Mikkelson of Macalester College for reading parts of the manuscript, and making constructive suggestions. We would like to thank Davison E. Soper of the University of Oregon for introducing us to *Mathematica*. We thank our many colleagues at the University of Oregon and Southern Methodist University, too numerous to name, who have supported the development of this book. We also extend our gratitude to the graduate students at these institutions who have tested solutions to several of the problems.

Fred Olness would like to thank the Lightner-Sams Foundation for a gift of computer equipment which helped make this project possible, and Southern Methodist University for an Instructional Technology Grant to support computer software. Fred Olness would also like to acknowledge the U.S. Department of Energy for support of my high energy physics research; many techniques developed in the course of my research were incorporated in this book.

At Addison-Wesley, we would like to acknowledge the invaluable contributions by Stuart Johnson (Physics Editor), Amy Willcutt, Jennifer Albanese, Nev Hanke, Eileen Hoff, and Laurie Petrycki.

At Wolfram Research, we thank Glenn Scholebo for elegantly converting the *Mathematica* notebooks into final typeset form, and Stephen Wolfram for contributing the foreword.

Troubleshooting

This manuscript was generated directly from a *Mathematica* notebook to minimize the possibility of introducing errors. The Tex commands were embedded directly into the *Mathematica* notebook, which was then converted to a Textures file with nb2tex. The final version of the *Mathematica* code was run on a SUN Sparc 2 with version 2.2; this book was developed using Macintosh, NeXT, and Sun platforms with *Mathematica* versions 1.2, 2.0, 2.1, and 2.2.

There are subtle differences between different versions and implementations of *Mathematica*. This means that it is likely that the reader will find that some of the examples presented in this book need to be slightly modified to adapt to your particular configuration. We have tested these examples extensively on different platforms and with different versions to ensure that they are robust. However, we list below the most common difficulties that the reader is likely to encounter.

In our examples, we have been careful to present enough intermediate output so that the reader can cross check their results for consistency. This allows the reader to isolate any differences that may arise, and find the cause quickly.

Initializations

If you would like your output to match this book as closely as possible, you can use the following initialization lines.

```
In[1]:=  (* Some "Mathematica for Physics" Initializations *)
         $DefaultFont={"Courier",8};
         Off[General::spell];
         Off[General::spell1];
         PageWidth->70;
```

Order of roots in **Solve**

A significant difference between *Mathematica* version 2.1 and 2.2 is the order of the solutions returned by **Solve** and **DSolve**. Throughout the book, you will note we are careful to select the desired root using the `[[i]]` notation. If you have difficulty with the problems, this is a first place to look. Compare the intermediate output displayed in the book with your results. Use this information to isolate the problem, and determine if it is due to the ordering of the roots of the **`Solve`** command.

How to debug **`Modules`** and user-defined functions

In Chapter 2, we discuss how to re-define a **`Module`** so that the variables are **`Global`** and can be examined. This allows the reader to step through each statement of the **`Module`** individually to try and isolate the error. This is a very useful debugging technique.

Commands that never return an answer

In cases where the *Mathematica* command takes a long time (over approximately 60 seconds on a Sparc 2) to return an answer, we have indicated this in the text. If you wait a long time, get no output, and suspect a problem, there are many ways to approach this problem. The most obvious is to break the procedure up into smaller steps. We often group commands together to shorten our solution. We must confess that when we initially solved the problem, we actually performed each step individually, examined the output, and afterwards decided the most efficient set of steps to use and display.

Avoid the **`Simplify`** command

In particular, if you are having trouble with lengthy expressions, try to avoid the **`Simplify`** command. Until you know the scope of your problem, you will find that the **`Expand`** and **`Together`** command are much more efficient at reducing the answer. Once you have massaged the result into something *Mathematica* can swallow in a single bite, then you may want to try and **`Simplify`** the result. In certain cases, the operation of a **`Together`** or **`Expand`** command before a **`Simplify`** command can make an order of magnitude difference in execution time.

Printing Problems

Complex PostScript pages that do not print

There is one printing problem we encounter rather frequently. If you are sending a very complex PostScript page to a laser printer, you may find that you are getting a PostScript error. Quite often, the problem is simply that the page description is simply too complex to fit into the memory of your printer. (Note, unfortunately the error messages returned by the PostScript printer generally will not point you in the right direction with this problem.)

Possible solutions include

1) insert page breaks so that no more than one complex graphic appears on a single page.

2) convert the graphics cells into bitmap format. This will allow them to print easily, but the resolution will be lost.
3) buy a more powerful printer with more memory.

We note that some of the pages in this text did not print on a standard laser printer with a reasonable amount of memory in the full PostScript resolution, so we were forced to use the bitmat graphics for the proofreading process.

In particular, the output from **`PlotGradientField`** and **`PlotGradientField3D`** seemed to generate PostScript errors most frequently.

Animation

The details of animating a series of graphics varies widely with different implementations of *Mathematica*. In this text we simply show the reader how to generate the sequence of graphics. The reader must refer to the computer specific user guide for the details of displaying the animation.

Clearing Variables

If *Mathematica* is yielding unusual results, a common cause is that there are variable definitions (possibly from a previous problem) which are confliciting with the assumed definitions for the present problem. The command **`Clear["Global`*"]`** will solve most of these problems. In fact, each chapter of this book was initially a single *Mathematica* notebook. Using the **`Clear["Global`*"]`** command, we were able to run each chapter as a single *Mathematica* session from start to finish. (Again, this helped us eliminate errors, and verify that our examples were correct.) While this command should solve most all problems of this type, it may be worth starting a new *Mathematica* session if mysterious problems still persist.

Limited Memory or CPU

If you are working with limited memory or CPU, there are a number of steps you can take to still be productive with *Mathematica*.

1) Load in only the packages necessary for the particular problem that you want to work on. These are listed at the beginning of each problem; in general, this list of packages is much shorter than the complete list for the chapter so it will save you memory and time.
2) When you are generating graphics, start the plot with a minimal number of points. Once you verify there are no errors and the Options for the graphics are correct, you can then increase the number of points and the resolution of the graphics.

Contents

Foreword v

Introduction vii

Troubleshooting xi

CHAPTER 1. **Getting Started** 1

1.1 Introduction 1
- 1.1.1 Computers as a Tool 1
- 1.1.2 Suggestions on Approaching the Exercises 2

1.2 Arithmetic and Algebra 2
- 1.2.1 Arithmetic and Notation 2
- 1.2.2 Algebra 3
- 1.2.3 Mapping Expressions 3
- 1.2.4 Rules 4
- 1.2.5 Conjugation 5
- 1.2.6 User-defined Complex Conjugate Rule 5
- 1.2.7 Algebraic Equations 5
- 1.2.8 Threading Expressions 7

1.3 Functions and Procedures 7
- 1.3.1 User-defined Functions 7
- 1.3.2 Discontinuous Functions 8
- 1.3.3 Nonanalytic Functions 8
- 1.3.4 Rules 9
- 1.3.5 Procedures 10

1.4 Miscellaneous 10
- 1.4.1 Packages 10
- 1.4.2 Contexts 11
- 1.4.3 Protecting Commands 12

1.5 Calculus 12

1.5.1 Integration 12
1.5.2 Analytic Solutions of Differential Equations 13
1.5.3 Changing Variables and Pure Functions 13
1.5.4 Numerical Solutions of Differential Equations 14
1.6 Graphics 15
1.6.1 Animated Plots 15
1.6.2 Vector Field Plots 16
1.6.3 Shadowing 16
1.6.4 Three-dimensional Graphics 18
1.6.5 Space Curve 18
1.7 Exercises 18

CHAPTER 2. General Physics 23

2.1 Introduction 23
2.2 General Physics 23
2.2.1 Newtonian Motion 23
2.2.2 Electricity, Magnetism, and Circuits 24
2.3 *Mathematica* Commands 24
2.3.1 Packages 24
2.3.2 User-defined Procedures 25
Procedure to find a least-squares fit to a set of data 25
Example: Slope of a Straight Line 25
Example: An Exponential 25
Procedure to find the electric potential for point charges 26
Example: Dipole 26
2.3.3 Protect User-defined Procedures 26
2.4 Problems 26
2.4.1 Projectile Motion in a Constant Gravitational Field 26
Problem 1: Escape Velocity 26
Problem 2: Projectile in a Uniform Gravitational Field 27
Problem 3: Projectile with Air Resistance 31
Problem 4: Rocket with Varying Mass 36
2.4.2 Projectile Motion in Rotating Reference Frames 42
Problem 1: Coriolis and Centrifugal Forces 42
Problem 2: Foucault Pendulum 47
2.4.3 Electricity and Magnetism 52
Problem 1: Charged Disk 52
Problem 2: Uniformly Charged Sphere 54
Problem 3: Electric Dipole 60
Problem 4: Magnetic Vector Potential for a Linear Current 63
2.4.4 Circuits 66
Problem 1: Series RC Circuit 66
Problem 2: Series RL Loop 68
Problem 3: RLC Loop 71

2.4.5 Modern Physics 75
Problem 1: The Bohr Atom 75
Problem 2: Relativistic Collision 77
2.5 Unsolved Problems 79

CHAPTER 3. Oscillating Systems 83
3.1 Introduction 83
3.1.1 Oscillations 83
Potentials 83
Phase Planes 83
Small Oscillations and Normal Modes 84
3.2 *Mathematica* Commands 84
3.2.1 Packages 84
3.2.2 User-defined Procedures 85
Series expansion for second-order equation 85
Example 1: Second order solution 85
Example 2: How the **diffSeriesOne** routine works 86
Phase plot for one-dimensional system 87
Example: Phase plots for harmonic motion 87
Time behavior of phase plot for a one-dimensional system 88
Example: Time-evolved phase plots for harmonic motion 88
doublePlot: Phase plots and time evolution for a one-dimensional system 89
Example: Phase plots and time evolution for harmonic motion 89
Fourier spectrum of a one-dimensional oscillating system 89
Example: Fast Fourier transform 90
Eigenvalues and eigenvectors for small oscillating systems 90
Example: Two coupled particles 91
Animation for linear motion 93
Example: Linear harmonic oscillator 93
3.2.3 Protect User-defined Procedures 94
3.3 Problems 94
3.3.1 Linear Oscillations 94
Problem 1: Analysis of Linear Oscillator 94
Problem 2: Solution of Linear Oscillator 97
Problem 3: Damped Linear Oscillator 99
Problem 4: Damped Harmonic Oscillator and Driving Forces 104
3.3.2 Nonlinear Oscillations 109
Problem 1: Duffing's Oscillator Equation 109
Problem 2: Forced Duffing Oscillator for Double-well Potential 114
Problem 3: van der Pole Oscillator and Limiting Cycles 118
Problem 4: Motion of a Damped, Forced Nonlinear Pendulum 121

3.3.3 Small Oscillations 124
Problem 1: Two Coupled Harmonic Oscillators 124
Problem 2: Three Coupled Harmonic Oscillators 130
Problem 3: Double Pendulum 135
3.4 Unsolved Problems 139

CHAPTER 4. Lagrangians and Hamiltonians 143

4.1 Introduction 143
4.1.1 Lagrange's Equations 143
Generalized Coordinates and Constraints 143
Lagrangian 144
Nonholonomic Constraints and Lagrangian Multipliers 144
4.1.2 Hamilton's and Hamilton-Jacobi Equations 144
Hamilton's Equations 144
Hamilton-Jacobi Technique 145
4.1.3 *Mathematica* Commands 146
Packages 146
User-defined Rules 147
HyperbolicToComplex and **ComplexToHyperbolic** 147
Example: **HyperbolicToComplex** and **ComplexToHyperbolic** 147
User-defined Procedures 147
Finding Lagrange's equations 147
Example: Lagrange's equation for a particle in a potential V[x] 148
Finding the canonical momentum, Hamiltonian, and equations of motion 148
Example 1: Hamilton's equations of motion in one dimension 149
Example 2: Hamilton's equations of motion in two dimensions 149
Example 3: How **Hamilton** works 150
Finding the canonical momentum, Hamilton's principal function, and Hamilton-Jacobi equations 151
Example 1: One-dimensional particle in a potential V[x] 151
Example 2: Two-dimensional particle in a potential V[x] 152
Example 3: How **HamiltonJacobi** works 152
Series expansion solution for second-order equation 153
Series expansion solution for two first-order equations 153
Example 1: Expansion of harmonic oscillator 154
Example 2: How **firstDiffSeries** works 154
First-order perturbation solution 155
Example 1: Perturbed harmonic oscillator 156
Example 2: Details of **firstOrderPert** 156
4.1.4 Protect User-defined Procedures 158

4.2 Problems 158
4.2.1 Lagrangian Problems 158
Problem 1: Atwood Machine 158
Problem 2: Bead Sliding on a Rotating Wire 161
Problem 3: Bead on a Rotating Hoop 165
Problem 4: Hoop Rolling on an Incline 171
Problem 5: Sphere Rolling on a Fixed Sphere 174
Problem 6: Mass Falling Through a Hole in a Table 178
4.2.2 Orbiting Bodies 183
Problem 1: Equivalent One-body Problem 183
Problem 2: Kepler Problem 188
Problem 3: Precessing Ellipse and Generalized Kepler Problem 192
Problem 4: Numerical Solution for Orbits with Central Forces 195
Problem 5: Quadripole Potential and Perturbative Solutions 199
4.2.3 Hamilton and Hamilton-Jacobi Problems 206
Problem 1: Harmonic Oscillator and Hamilton's Equations 206
Problem 2: Hamilton's Equations in Cylindrical and Spherical Coordinates 210
Problem 3: Spherical Pendulum and Hamilton's Equations 213
Problem 4: Harmonic Oscillator and Hamilton-Jacobi Equations 218
Problem 5: Kepler's Problem and Hamilton-Jacobi Equations 221
4.3 Unsolved Problems 225

CHAPTER 5. Electrostatics 229

5.1 Introduction 229
5.1.1 Electric Field and Potential 229
Electric field 229
Electrostatic potential 229
5.1.2 Laplace's Equation 230
Cartesian coordinates 230
Cylindrical coordinates 230
Spherical coordinates 230
5.1.3 *Mathematica* Commands 231
Packages 231
User-defined procedures 231
Operator: **`TrigToY`** 231
Example 232
Operator: **`TrigToP`** 232
Example 233
`Monopole` 233
Example 233
`PotentialExpansion` 234
Example: Asymptotic potential of two-point charges 234

VEPlot 234
Example: Equipotential surface and electric field of two-point charges 235
MultipoleSH 236
Example: Potential of non-axially symmetric charge density 236
MultipoleP 237
Example: Potential of an axially symmetric charge distribution 238
5.1.4 Protect User-defined Procedures 238
5.2 Problems 238
5.2.1 Point Charges, Multipoles, and Image Problems 238
Problem 1: Superposition of point charges 238
Problem 2: Point charges and grounded plane 242
Problem 3: Point charges and grounded sphere 245
Problem 4: Line charge and grounded plane 249
Problem 5: Multipole expansion of a charge distribution 252
5.2.2 Cartesian and Cylindrical Coordinates 258
Problem 1 : Separation of variables in Cartesian and cylindrical coordinates 258
Problem 2 : Potential and a rectangular groove 261
Problem 3: Rectangular conduit 265
Problem 4: Potential inside a rectangular box with five sides at zero potential 269
Problem 5: Conducting cylinder with a potential on the surface 274
5.2.3 Legendre Polynomials and Spherical Harmonics 278
Problem 1: A charged ring 278
Problem 2: Grounded sphere in an electric field 285
Problem 3: Sphere with an axially symmetric charge distribution 288
Problem 4: Sphere with a given axially symmetric potential 292
Problem 5: Sphere with upper hemisphere V and lower hemisphere $-V$ 295
5.3 Unsolved Problems 299

CHAPTER 6. Quantum Mechanics 303
6.1 Introduction 303
6.1.1 Foundations of Quantum Mechanics 303
Historical beginnings 303
Time-independent quantum mechanics 304
6.1.2 *Mathematica* Commands 304
Packages 304
User-defined procedures and rules 304
Change of variables procedure 304

Example: Change of variable 305
Series expansion solution for second-order equation 305
HyperbolicToComplex and **ComplexToHyperbolic** 306
Complex conjugate rule 306
Example: Complex exponential 306
User-defined solutions of differential equations 307
Solution: Hermite polynomials 307
Example: Hermite solution 307
Solution: Legendre polynomials 307
Example: Legendre solution 307
User-defined one-dimensional wave properties 308
One-dimensional wave function with constant potential 308
6.1.3 User-defined Three-dimensional Quantum Equations 308
Operator: **schrodinger** 308
Example: Spherical potential and spherical wave function 308
Operator: **hamiltonian** 309
Example: Harmonic oscillator 309
Operator: **flux** 309
Example: Plane wave flux 309
6.1.4 Protect User-defined Operators 310
6.2 Problems 310
6.2.1 One-dimensional Schrodinger's Equation 310
Problem 1: Particle bound in an infinite potential well 310
Problem 2: Particle bound in a finite potential well 314
Problem 3: Particle hitting a finite step potential 322
Problem 4: Particle propagating towards a rectangular potential 328
Problem 5: The one-dimensional harmonic oscillator 336
6.2.2 Three-dimensional Schrodinger's Equation 341
Problem 1: Three-dimensional harmonic oscillator in Cartesian coordinates 341
Problem 2: Schrodinger's equation for spherically symmetric potentials 344
Problem 3: Particle in an infinite, spherical box 349
Problem 4: Particle with negative energy in a finite, spherical box 353
Problem 5: The hydrogen atom in spherical coordinates 359
Problem 6: Separation in cylindrical and paraboloidal coordinates 364
6.3 Unsolved Problems 369
CHAPTER 7. **Relativity and Cosmology 371**
7.1 Introduction 371
7.1.1 Special Relativity 371
The two basic postulates of special relativity 372

Lorentz transformations 372
Covariant equations and tensors 373
Cartesian coordinates and "flat" spacetime 373
7.1.2 General Relativity and Cosmology 374
Spacetime metric 374
Field equations 374
Free-falling test particles and light trajectories 374
Robertson-Walker cosmology 375
7.1.3 *Mathematica* Commands 375
Packages 375
User-defined metric, boost, and velocity parameters 375
Metric 375
Rule for relativistic velocity parameters 376
Boost along the x-axis 376
User-defined geometric procedures 376
Christoffel symbols 376
Example: Christoffel symbols for a pseudo-Euclidean metric 377
Curvature tensor 377
Example: Curvature tensor for pseudo-Euclidean metric 378
Ricci tensor 378
Example: Ricci tensor for Godel metric 378
Killing's equations 379
Example: Killing vector equations in pseudo-Euclidean space 379
Einstein tensor 380
Example: Einstein tensor for wave metric 380
Geodesic equations 380
Example: Geodesics for a pseudo-Euclidean metric 381
User-defined metrics and Christoffel symbols 381
Schwarzschild metric 381
Kerr metric 382
Protect user-defined operators 384
7.2 Problems 384
7.2.1 Special Relativity Problems 384
Problem 1: Decay of a particle 384
Problem 2: Two-particle collision 385
Problem 3: Compton scattering 386
Problem 4: Moving mirror and generalized Snell's law 389
Problem 5: One-dimensional motion of a relativistic particle with constant acceleration 392
Problem 6: Two-dimensional motion of a relativistic particle in a uniform electric field 395
7.2.2 General Relativity and Cosmology 398
Problem 1: Schwarzschild solution in null coordinates 398

Problem 2: The horizons and surfaces of infinite redshift 400
Problem 3: Killing vectors and constants of motion 402
Problem 4: Potential analysis for timelike geodesics 405
Problem 5: Time it takes to fall into a black hole 408
Problem 6: Circular geodesics for the Schwarzschild metric 414
Problem 7: Field equations for Robertson-Walker cosmology 416
Problem 8: Zero-pressure cosmological models 421
Problem 9: The expansion and age of the standard model 424

7.3 Unsolved Problems 430

Index 433

CHAPTER ONE

Getting Started

1.1 Introduction

1.1.1 Computers as a Tool

The solutions for realistic physics problems are often hampered because the algebra is too complex for anyone but the dedicated researcher. Just as the calculator eliminated laborious numerical computations, symbolic software programs eliminate arduous algebraic computations. While computer power is no substitute for thinking, it spares the scientist from performing mundane mathematical steps and thereby frees time for creative thinking. The scientist is able to explore complex relationships among quantities, ask "What if...?", and obtain an immediate answer.

Because there are many ways to solve physics problems, the styles of the *Mathematica* solutions and commands presented in this book vary intentionally by problem. Furthermore, we emphasize that the solutions presented here are not necessarily the most efficient; while we do discuss writing efficient *Mathematica* code, we often sacrifice the most elegant or efficient solution in favor of one that is most easily understood from a teaching perspective. For example, the many user-defined procedures we use are written to automate repetitive tasks in an intuitive way. Had our goal been to write an efficient "black-box" program, the routines would have been less pedagogical, more complex to read, and encumbered with additional code to trap error conditions. Some of the unsolved examples prompt the reader to optimize the *Mathematica* code.

Each chapter has a short overview of the major physics and mathematics topics emphasized in the chapter. The problems are chosen to cover a broad range of physics problems and to illustrate a variety of *Mathematica* procedures. Unsolved exercises are included at the end of the chapter to reinforce the techniques developed in the examples and to suggest additional applications not covered.

Chapter one, "Getting Started," introduces the basic commands and notation that are extensively used in the remaining chapters. It does not include all capabilities and commands found in *Mathematica*. We assume that the reader is familiar with the most basic features as discussed in the *Mathematica* user's manual *Mathematica: A System for Doing Mathematics by Computer*, by Stephen Wolfram (Addison-Wesley, 1988). This book is intended to illustrate the advantages of *Mathematica* as applied to physics problems.

1.1.2 Suggestions on Approaching the Exercises

We suggest the reader work out all the problems in small parts. Only after the answer is reached in a straight-forward manner, should one attempt to "polish" the solutions into an elegant and optimized form. The reader should continually be looking for methods that are physically more intuitive, even though they may involve more steps. *Mathematica*'s advantage is that one can execute complicated problems in parts and obtain the intermediate results immediately, thereby minimizing the debugging necessary. The graphics capabilities of *Mathematica* are powerful, and the reader is encouraged to plot intermediate results to understand the physics behind the solutions.

Each problem is independent of the others in the chapter. One can execute any problem by

1. loading the required *Mathematica* packages. These packages are indicated at the beginning of each chapter. If you have limited CPU or memory, you may wish to load only those packages necessary for a single problem; this subset of packages is indicated at the beginning of each solution.
2. executing the user-defined setup commands located at the beginning of each chapter.

With this initialization, the reader can run any or all of the exercises.

We attempt to introduce some elements of programming style and consistency in our solutions. In general, we insert "**sol**" into the name of a **Solution**, "**dsol**" into the name of a **DSolution**, "**eq**" into the name of an **Equation**, and "**rule**" into the name of a **Rule**. We also prefer to use the postfix form for various operations. Although doing this may seem cumbersome at first, it greatly simplifies applying sequential operations. Furthermore, we generally avoid defining objects beginning with capital letters to minimize conflicts with internal *Mathematica* commands.

We now examine some of the more common commands that are used throughout this book. We also introduce tricks and other techniques that are useful for debugging and solving physics problems in general.

1.2 Arithmetic and Algebra

1.2.1 Arithmetic and Notation

Mathematica has several built-in constants such as **Pi** and **E**. To obtain the numerical value of **Pi**, simply type

```
In[1]:=  N[Pi]

Out[1]=  3.14159
```

We will often find it convenient to use the postfix form:

```
In[2]:=  Pi // N

Out[2]=  3.14159
```

If you want to specify a second argument for the postfix form, you must make a pure function using **#** and **&**:

```
In[3]:=  Pi // N[#,50]&

Out[3]=  3.1415926535897932384626433832795028841971693993751
```

The postfix form is useful because it allows repeated operations to be efficiently expressed. For example, a two-step procedure can be written:

```
In[4]:=  E^(I x) //Series[#,{x,0,3}]&  //Normal

Out[4]=
                 2
                x    I  3
         1 + I x - -- - - x
                2    6
```

1.2.2 Algebra

Mathematica can algebraically manipulate expressions. For example, the command **Expand** is used to expand the expression $(a + bx)^5$:

```
In[5]:=  eq1= Expand[(1+x^2)^5]

Out[5]=
              2       4       6      8    10
         1 + 5 x  + 10 x  + 10 x  + 5 x  + x
```

The command **Factor** or **Simplify** reduces **eq1** to its original form:

```
In[6]:=  eq1 // Simplify

Out[6]=
              2 5
         (1 + x )
```

We can also factor **eq1** over the complex numbers if the **GaussianIntegers** option of **Factor** is used:

```
In[7]:=  Factor[eq1,GaussianIntegers->True]

Out[7]=
                 5         5
         (-I + x)  (I + x)
```

To view the options allowed by a command type **Options[command]**, type

```
In[8]:=  Options[Factor]

Out[8]=  {GaussianIntegers -> False, Modulus -> 0, Trig -> False}
```

1.2.3 Mapping Expressions

The commands **Simplify** or **Expand** do not always reduce the expression to the desired form. If this is the case, then other commands and pattern-matching techniques must be used. For example, consider the expression

```
In[9]:=  eq1= a1 (b+c/2)^2 + a2 (d+e/3)^3 //Expand //Simplify

Out[9]=
                              2                              2        3
             2             a1 c          3        2      a2 d e    a2 e
         a1 b  + a1 b c + ----- + a2 d  + a2 d  e + ------- + -----
                             4                          3        27
```

Simplify does not return the expression to its original form. The expression is reduced to its original form using the commands **Collect** and **Map**. We first collect the **a1** and **a2** terms with **Collect**:

```
In[10]:= eq2= eq1 // Collect[#,{a1,a2}]&

Out[10]=
                      2                          2    3
             2       c          3    2        d e    e
a1 (b  + b c + --) + a2 (d  + d  e + ---- + --)
                     4                         3     27
```

Next, we **Map** the **Factor** command on each term separately

```
In[11]:= eq3= eq2 // Map[Factor,#,{2}]&

Out[11]=
                2                    3
a1 (2 b + c)      a2 (3 d + e)
-------------- + --------------
      4                 27
```

which is the original expression.

In the previous example, we used the **Map** function with an optional third argument to specify the level **({2})** at which the function was to be applied. You can visualize how this step works by replacing the real function **Factor** with a fake function **factor** (lower-case):

```
In[12]:= eq2 // Map[factor,#,{2}]&

Out[12]=
                                      2                                         2    3
                        2            c                          3    2       d e    e
factor[a1] factor[b  + b c + --] + factor[a2] factor[d  + d  e + ---- + --]
                                     4                                         3     27
```

It is now obvious at which level **({2})** of **Map** mapped **Factor**.

1.2.4 Rules

Rules are useful for making substitutions without making the definitions permanent. A simple example is to consider the expression $(x+z)^2$ and substitute $\mathrm{Sin}[ky]$ for z:

```
In[13]:= (x+z)^2 /. z->Sin[k y]

Out[13]=
                  2
(x + Sin[k y])
```

The substitution operator **/.** applied the rule once to $(x+z)^2$.

A list of rules can be applied to an expression in one step. Consider applying the list of rules **{z->Sin[k y], k->1}** to $(x+z)^2$. The substitution command **/.** will no longer work because **k** is found in the rule and the command is applied to $(x+z)^2$ only once. Repeated substitutions can be made using **//.**

```
In[14]:= (x+z)^2 //. {z->Sin[k y], k->1}

Out[14]=
                2
(x + Sin[y])
```

The replacement operator **//.** applies the list over and over again until the expression no longer changes.

1.2.5 Conjugation

A more complicated example in which the expression for the rule is not obvious is the rule for taking the complex conjugate of the expression **a + 2 I b**, where **a** and **b** are assumed to be real. The straightforward method is to substitute **I** for **-I**; unfortunately, this does not work. To see why, we examine how *Mathematica* handles **a +2 I b** by applying the command **FullForm**:

```
In[15]:= a + 2 I b // FullForm

Out[15]= Plus[a, Times[Complex[0, 2], b]]
```

This pattern suggests that a good rule for taking the complex conjugate is

```
In[16]:= conjugateRule= {Complex[re_,im_]->Complex[re,-im]};
```

Using the replacement operator **/.** to apply the rule, we get

```
In[17]:= a + 2 I b /. conjugateRule

Out[17]= a - 2 I b
```

This time it works. Be sure to use **/.** and not **//.** or you will have an endless loop.

1.2.6 User-defined Complex Conjugate Rule

For many physics problems, we will find that the above-defined **conjugateRule** is more useful than the built-in **Conjugate** function. For this reason, we find it convenient to create our own "homemade" **conjugate** function:

```
In[18]:= conjugate::usage =
         " A simple method of computing the conjugate of an object
          which is explicitly Complex";

         conjugateRule      = {Complex[re_,im_]:>Complex[re,-im]};
         conjugate[exp__] := exp /. conjugateRule;
```

Note that the use of the function (1) duplicates the way **Conjugate** is used and (2) ensures we do not encounter an endless loop.

Example: Complex Exponential

Here is a simple example of a case we will find useful in solving many physics problems. The curious reader is encouraged to compare this example with the built-in **Conjugate** function.

```
In[19]:=  I Exp[ I k x]  // conjugate

Out[19]=       -I k x
          -I E
```

1.2.7 Algebraic Equations

Consider the solution of the algebraic equation $ax^2 + bx + c == 0$. The most straightforward method for solving this equation is to use the command **Solve**:

```
In[20]:=  eq1=a x^2 + b x + c == 0;
          sol1= Solve[eq1 ,x] //Simplify

Out[20]=
                            2                                  2
                -b + Sqrt[b  - 4 a c]              -(b + Sqrt[b  - 4 a c])
          {{x -> ---------------------}, {x -> ------------------------}}
                         2 a                               2 a
```

The solution also follows from the commands **Roots** and **Reduce**.

The substitution operator **->** allows the solution to be substituted at a later time into any expression containing **x**. If only the second root is desired, it can be selected using **/.** and the double brackets **[[2]]**:

```
In[21]:=  x /. sol1[[2]]

Out[21]=                2
          -(b + Sqrt[b  - 4 a c])
          -----------------------
                   2 a
```

A similar command follows for taking the first solution in the list.

The roots can be verified by substituting **sol1** back into the original equation:

```
In[22]:=  eq1 /.sol1 //Simplify

Out[22]=  {True, True}
```

Let us consider the solution of two coupled equations:

```
In[23]:=  eq2= {a x+b y==c,
               a x-b y==d};
```

This time use **Reduce** instead of **Solve** to get the solution:

```
In[24]:=  eq3= Reduce[eq2,{x,y}]

Out[24]=
                                     c + d          c - d
          a != 0 && b != 0 && x == ----- && y == ----- ||
                                      2 a            2 b

                                                    c
          b != 0 && a == 0 && d == -c && y == - ||
                                                    b

                                                   c
          a != 0 && b == 0 && d == c && x == - || c == 0 && b == 0 && a == 0 && d == 0
                                                   a
```

Reduce tries to search for all possible solutions and therefore takes longer than **Solve**. To extract the desired part of the solution, we use syntax **[[...{}...]]**:

```
In[25]:=  eq3[[1,{2,3}]]

Out[25]=                  c + d
          b != 0 && x == -----
                           2 a
```

to obtain the second and third subpart of the first part. Caution, the order of **Reduce** and **Solve** is version dependent.

1.2.8 Threading Expressions

You can use **Thread** to relate the components of two separate lists. For example, to equate the components of two vectors, you write

```
In[26]:= vector1={x1,y1};
         vector2={x2,y2};
         Thread[vector1==vector2]

Out[26]= {x1 == x2, y1 == y2}
```

To equate the components of a vector to zero, simply equate the vector equal to a single zero and **Thread**:

```
In[27]:= Thread[vector1==0]

Out[27]= {x1 == 0, y1 == 0}
```

Similar properties hold for rules, and we can use this property to create a rule that will change from **Cartesian** to **Spherical** coordinates. First, load the package:

```
In[28]:= Needs["Calculus`VectorAnalysis`"];
```

The command **CoordinatesToCartesian** found in this package gives the relation between the two coordinate systems. A rule for converting from **Cartesian** to **Spherical** coordinates follows from using **Thread**:

```
In[29]:= Thread[{x,y,z} -> CoordinatesToCartesian[{r,theta,phi},Spherical] ]

Out[29]= {x -> r Cos[phi] Sin[theta], y -> r Sin[phi] Sin[theta], z -> r Cos[theta]}
```

Thread is also useful in threading algebraic procedures over equations. For example, given the equation

```
In[30]:= eq1= x+1/a == y/a;
```

and multiplying by **a**, we have the result

```
In[31]:= a eq1

Out[31]=    1         y
         a (- + x == -)
            a         a
```

This is not really what we wanted. Instead, we multiply both sides by **a** and use **Thread**:

```
In[32]:= Thread[a eq1,Equal] //ExpandAll

Out[32]= 1 + a x == y
```

1.3 Functions and Procedures

1.3.1 User-defined Functions

A trivial example of a user-defined function of two variables called **f[k,y]** is

```
In[1]:=  f[k_,y_] := N[BesselJ[k, y],2];
         f[1,2]

Out[1]=  0.58
```

The function **f[k,y]** is a **BesselJ[k, y]** evaluated to a precision of two digits. An equivalent definition using postfix notation is

```
In[2]:=  g[k_,y_] := BesselJ[k, y] //N[#,2]&;
         g[1,2]

Out[2]=  0.58
```

1.3.2 Discontinuous Functions

A function does not have to be continuous. Consider a step function that has values −1 for negative **x** and +1 for positive **x**. Two possible definitions for the step functions are

```
In[3]:=  step1[x_ ]:= -1  /;     x<= 0
         step1[x_ ]:=  1  /; 0 <x
```

and

```
In[4]:=  step2[x_ /;    x<=0]    := -1
         step2[x_ /; 0<x    ]    :=  1
```

In **step2**, the command **/;** allows **x** to be evaluated only when the condition on the right-hand side is true:

```
In[5]:=  Plot[step1[x],{x,-1,1}];
```

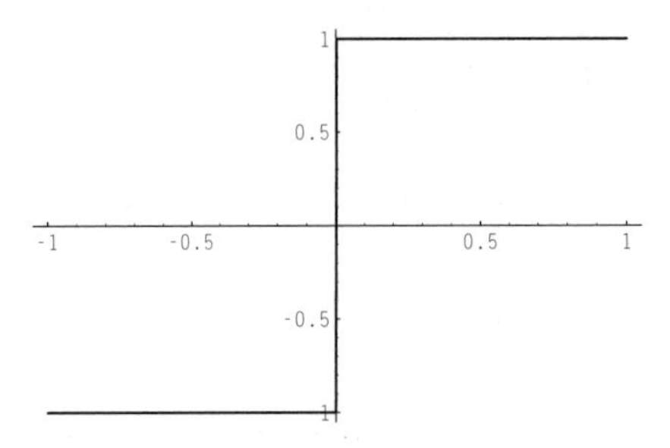

To see which definition produces a table of values the fastest, apply **Timing** to the two tables:

```
In[6]:=  {Timing[ Table[step1[x],{x,-2,  2,.0005}]]//First,
          Timing[ Table[step2[x],{x,-2,  2,.0005}]]//First}

Out[6]=  {8.15 Second, 6.93333 Second}
```

The definition given by **step2** is slightly faster.

1.3.3 Nonanalytic Functions

We can sometime get unexpected results when we try to perform operations on nonanalytic functions such as **Conjugate** and **Sign**. For example, consider

```
In[7]:=  D[ Conjugate[ f[y] ], y]

Out[7]=  Conjugate'[f[y]] f'[y]
```

We see that *Mathematica* attempts to treat **Conjugate** as an analytic function. The solution to such a problem is to teach *Mathematica* how to correctly treat the derivative of the **Conjugate** operator:

```
In[8]:=  Unprotect[Conjugate];
         D[ Conjugate[x___] ,y___] ^:= Conjugate[D[x,y]]
         Protect[Conjugate];
```

Note: we use **^:=** to assign the definition with **Conjugate** and not **D**.

Now we check to see that *Mathematica* gives us the answer we want:

```
In[9]:=  D[ Conjugate[ f[y] ], y]

Out[9]=  Conjugate[f'[y]]
```

1.3.4 Rules

Rules can be assigned to symbols to make them have certain properties. Let us assign rules to **i** to make it have the properties of the complex symbol i even though *Mathematica* already has this symbol defined by **I**. For **i** to be a complex symbol, it must have the properties $i^2 = -1$, $i^3 = -i$, and $i^4 = -i$. We use the command **/:** to assign these properties to **i**:

```
In[10]:= i/:i^2  =-1;
         i/:i^3  =-i;
         i/:i^4  = 1;
         i/:i^n_Integer :=i^Mod[n,4] /; n<0 || 4<n;
```

We make a table of values for **i^n** to verify that **i** has the correct properties:

```
In[11]:= Table[i^n,{n,-8,8}]

Out[11]= {1, i, -1, -i, 1, i, -1, -i, 1, i, -1, -i, 1, i, -1, -i, 1}
```

As a second example, we make $\mathrm{Cos}[n\pi] = (-1)^n$ and $\mathrm{Sin}[n\pi] = 0$ for integer **n**:

```
In[12]:= Unprotect[Cos,Sin];
         Cos/: Cos[n_ Pi] := (-1)^n /; IntegerQ[n];
         Sin/: Sin[n_ Pi] := 0      /; IntegerQ[n];
         Protect[Sin,Cos];
```

If we declare **n** an integer,

```
In[13]:= IntegerQ[n]^= True;
```

then

```
In[14]:= {Cos[n Pi], Sin[n Pi]}

Out[14]=      n
         {(-1) , 0}
```

1.3.5 Procedures

A procedure that involves several steps can be conveniently defined as a function. A two-step procedure that first forms a product and then expands the product is

```
In[15]:= prod[x_,n_] :=
           Module[{s,temp1,temp2},
                  temp1= Product[(x-s)^(s),{s,1,n,2}];
                  temp2= temp1 //ExpandAll;
                  Return[temp2]
           ];
         prod[x,2]

Out[15]= -1 + x
```

An example of a more complicated procedure is to define a function that will factor a polynomial when the roots are not integers. The following procedure finds the numerical value of the roots of a polynomial and then factors the equations over the numerical roots:

```
In[16]:= factor[equation_] :=
            Module[{temp1},
                   temp1= (x/.NSolve[equation,x]);
                   Return[ Times@@(x-temp1) == 0]
            ];
```

The first step described by **temp1** returns a list of numerical solutions for the polynomial equation. The second step puts the equation in a factor form:

```
In[17]:= factor[-1 + 3x - 3x^2 + x^3 - x^5 == 0 ] //N[#,3]&

Out[17]= (-0.591 - 0.211 I + x) (-0.591 + 0.211 I + x) (-0.354 - 1.1 I + x)

          (-0.354 + 1.1 I + x) (1.89 + x) == 0
```

1.4 Miscellaneous

1.4.1 Packages

We will make frequent use of the standard *Mathematica* packages. The best reference for these commands is the book, *Guide to Standard Mathematica Packages*, that came with your *Mathematica* software. Let us take the **Algebra`Trigonometry`** package as an example. First, we load the package:

```
In[1]:=  Needs["Algebra`Trigonometry`"]
```

To see quickly which functions the package has loaded, enter

```
In[2]:=  ?Algebra`Trigonometry`*

         ComplexToTrig TrigExpand    TrigFactor    TrigReduce    TrigToComplex
         TrigCanonical
```

To find out about a particular function, type

```
In[3]:=  ?ComplexToTrig

         ComplexToTrig[expr] writes complex exponentials as trigonometric functions
            of a real angle.
```

Should we want to see the entire package (for whatever reason), we enter (output suppressed)

```
In[4]:=  !!Algebra`Trigonometry`
```

This latter method is not recommended in general as it will fill your screen with the entire contents of the package and may take a long time to output over a network.

1.4.2 Contexts

We give only a very brief introduction to **Contexts** here. Refer to the *Mathematica* user's manual for further details. To see the complete list of **Contexts** currently defined, enter

```
In[5]:=  Contexts[]
Out[5]=  {Algebra`Trigonometry`, Algebra`Trigonometry`Private`,
           Calculus`VectorAnalysis`, Calculus`VectorAnalysis`Private`, DSolve`,
           EditPrivate`, Elliptic`Private`, FE`, Format`, Fourier`Private`, Global`,
           Graphics`Animation`, Graphics`Private`, HypergeometricPFQ`Private`, Inform`,
           Integrate`, InverseSeries`Private`, Limit`, NullSpace`, Obsolete`, Series`,
           Solve`, SpecialFunctions`Series`Private`, System`, System`ComplexExpand`,
           System`ComplexExpand`Private`, System`Private`}
```

Note that the list includes **Algebra`Trigonometry`**, since we have loaded this package. We can ask for the variable using

```
In[6]:=  Context[ComplexToTrig]
Out[6]=  Algebra`Trigonometry`
```

Because **ComplexToTrig** was defined by **Algebra`Trigonometry`**, its **Context** is **Algebra`Trigonometry`**. For a new variable **dummy**, the **Context** is

```
In[7]:=  Context[dummy]
Out[7]=  Global`
```

Essentially all variables that the user creates will be in the **Global`** context. We can use this information to our advantage in the following manner. Suppose we want to start a new problem, but we do not want any definitions created in the previous problem to conflict with the new problem. The solution is to execute the command

```
In[8]:=  Clear["Global`*"]
```

This has the advantage of clearing all variables defined by the user in the **Global`** context, while not removing any commands defined by our packages because they are in separate contexts such as **Algebra`Trigonometry`**. Should we want also to remove all the definitions created by the package **Algebra`Trigonometry`**, we can do this with the command

```
In[9]:=  Clear["Algebra`Trigonometry`*"]
```

Note: There are some subtle differences between the **Clear** and **Remove** commands. For our purposes, the **Clear** command will work best. Refer to the *Mathematica* user's manual for details.

1.4.3 Protecting Commands

The **Clear["Global`*"]** command works nicely for cleaning our slate. However, we may not want to erase the user-defined commands. We can solve this problem using **Protect** to keep these from being erased. For example, we define

```
In[10]:= dummy[x_]=x;
         Protect[dummy];
         Context[dummy]

Out[10]= Global`
```

Now we can clear all **Global** variables:

```
In[11]:= Clear["Global`*"]
```

without erasing our definition for **dummy**,

```
In[12]:= ?dummy
         Global`dummy

         Attributes[dummy] = {Protected}

         dummy[x_] = x
```

1.5 Calculus

1.5.1 Integration

Mathematica can integrate numerically and (in some cases) analytically. An example of an indefinite integral is

```
In[1]:=  eq1= Integrate[ a x/(b-c x^3),x] //Simplify
Out[1]=
                                  1/3
                               2 c    x
                           1 + ---------
                                  1/3                          1/3
                                 b                            c    x
         (a (-2 Sqrt[3] ArcTan[---------] - 2 Log[-1 + -------] +
                                Sqrt[3]                  1/3
                                                        b

                     1/3       2/3  2
                    c    x    c    x                1/3  2/3
            Log[1 + ------- + --------])) / (6 b   c    )
                      1/3       2/3
                     b         b
```

The original integrand is recovered if the derivative of **eq1** is taken:

```
In[2]:=  D[eq1,x] //Simplify

Out[2]=    a x
         --------
                3
         b - c x
```

1.5.2 Analytic Solutions of Differential Equations

Mathematica can numerically and (in some cases) analytically solve differential equations. Consider the solution for the harmonic oscillator equation,

```
In[3]:=  eq1= y''[t] + w0^2 y[t] == 0;
```

with the initial conditions

```
In[4]:=  initial={y[0]==y0, y'[0]==v0};
```

Appending the oscillator equation to the initial conditions, we get the list

```
In[5]:=  eq2= Append[initial,eq1]

Out[5]=                               2
         {y[0] == y0, y'[0] == v0, w0  y[t] + y''[t] == 0}
```

The analytic solution of **eq2** follows from applying **DSolve** to the list:

```
In[6]:=  eq3= DSolve[eq2,y[t],t ] //Flatten //ExpandAll

Out[6]=            I  -I t w0        I  I t w0
                   - E        v0     - E        v0     -I t w0          I t w0
                   2                 2                E          y0    E          y0
         {y[t] -> ----------------- - ----------------- + ---------------- + ----------------}
                         w0                 w0                  2                 2
```

The command **Flatten** removed one set of **{}** brackets. This result can be changed to a trigonometric function if we load the package

```
In[7]:=  Needs["Algebra`Trigonometry`"]
```

and apply the command **ComplexToTrig** to **eq3**:

```
In[8]:=  eq3 /.{Rule->Equal} //ComplexToTrig //ExpandAll

Out[8]=                              v0 Sin[t w0]
         {y[t] == y0 Cos[t w0] + --------------}
                                      w0
```

We have converted the rule to an equation with the substitution **{Rule->Equal}**.

1.5.3 Changing Variables and Pure Functions

It is often useful to express functions in their pure forms when changing variables in a differential equation. For example, consider the differential equation

```
In[9]:=  eq1 = a x[t] + b x'[t] + c x''[t] == 0;
```

and express the equation in terms of a new function **y[t]** defined by **x[t]=y[t]+2**. The straightforward substitution **x[t]->y[t]+2** does not generate what we want:

```
In[10]:= eq1 //.{x[t]->y[t]+2}

Out[10]= a (2 + y[t]) + b x'[t] + c x''[t] == 0
```

The substitution will work if we express the change in its pure form using **#** in place of the variable:

```
In[11]:= eq1 /.{x->((y[#]+2 )&)} //ExpandAll

Out[11]= 2 a + a y[t] + b y'[t] + c y''[t] == 0
```

Note the parenthesis outside the **&**.

Unfortunately, it is not always a trivial task to express a function in its pure form. A trick that can be used in more complicated cases is to use **DSolve** instead of **Solve**; for example,

```
In[12]:= dsol= DSolve[x[t]==y[t]+2,{x},t][[1]]

Out[12]= {x -> (2 + y[#1] & )}
```

Some versions may display this output using the **Function** command, it is functionally equivalent.

1.5.4 Numerical Solutions of Differential Equations

If *Mathematica* cannot analytically solve the differential equation, then **NDSolve** can be used to return a numerical solution. Consider the solution of the coupled differential equations,

```
In[13]:= eq1= {x''[t] == -x[t] - x[t] y[t]^2 ,
               y''[t] == -y[t] - x[t] y[t]^2 };
```

with the initial conditions,

```
In[14]:= initial= {x[0] == 2, x'[0] == 0,
                   y[0] == 1, y'[0] == 0};
```

The numerical solution of **eq1** for **t=0** to **20** follows from

```
In[15]:= ndsol=NDSolve[Join[eq1,initial],{x,y},{t,0,20}][[1]]

Out[15]= {x -> InterpolatingFunction[{0., 20.}, <>],

           y -> InterpolatingFunction[{0., 20.}, <>]}
```

The numerical data are expressed by an **InterpolatingFunction** whose values are found by interpolation. The graphs of **x[t]** and **y[t]** are

```
In[16]:= Plot[ Evaluate[{x[t],y[t]}/.ndsol]
              ,{t,0,20}
              ,PlotStyle->{  {Dashing[{    }]}
                            ,{Dashing[{.05}]} }
          ];
```

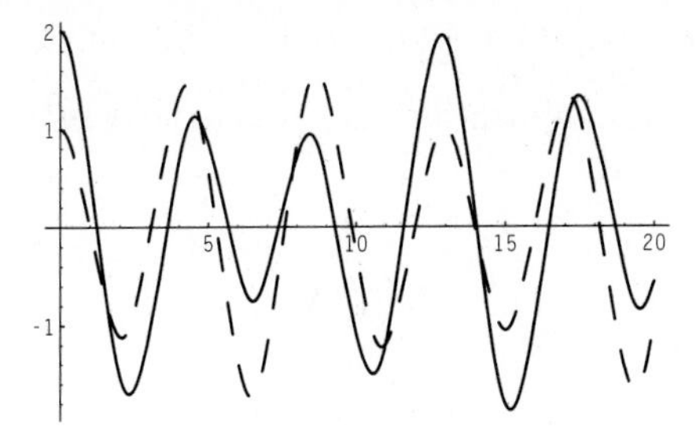

The solid curve is the **x[t]** value and the dashing curve is the **y[t]** value. We can also plot the solution in the **{x,y}** plane:

In[17]:=
```
ParametricPlot[Evaluate[{x[t],y[t]}/.ndsol] ,{t,0,20}];
```

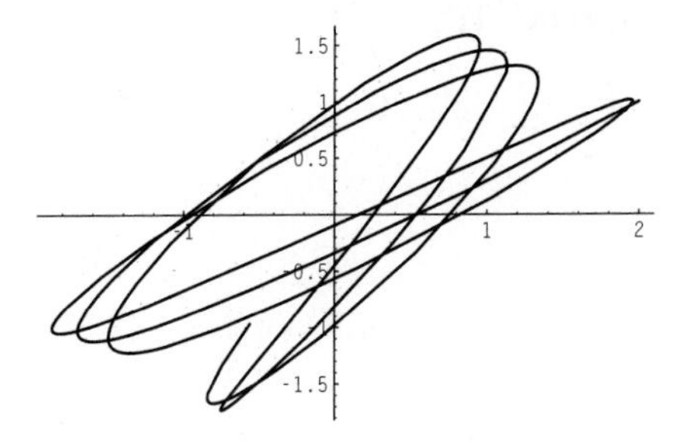

1.6 Graphics

1.6.1 Animated Plots

Mathematica's graphics capabilities are powerful and easy to use. (We have already used the commands **Plot** and **ParametricPlot**.) To animate a traveling wave, first construct a collection of traveling wave plots. Once the plots are generated, they can be displayed in sequence using the animation command in the Notebook front end or using the *Mathematica* animation command. (We display only two frames here.)

In[1]:=
```
v=1;
frames=2;
Do[ Plot[Cos[2 Pi(x-v t)],{x,-2,2},PlotRange->{-2,2}]
   ,{t,0,1-1/frames,1/frames}];
```

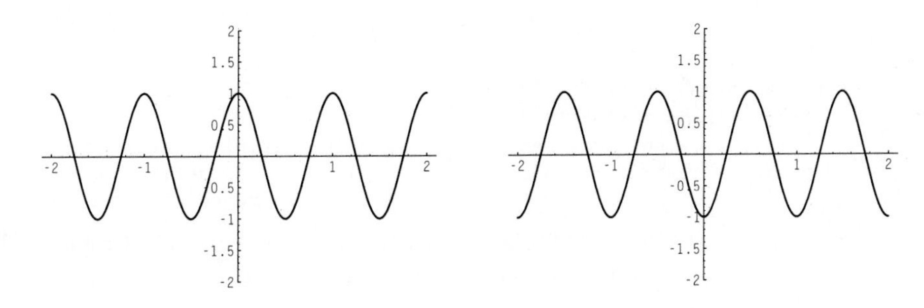

1.6.2 Vector Field Plots

A useful two-dimensional plot for physics applications is **PlotVectorField**. We define a two-dimensional vector field with the list

```
In[2]:=  vector= {x,y}/(x^2+y^2)^(3/2)

Out[2]=        x                y
         {--------------, --------------}
           2    2 3/2       2    2 3/2
         (x  + y )        (x  + y )
```

We then load the package:

```
In[3]:=  Needs["Graphics`PlotField`"];
```

The vector field plot follows from

```
In[4]:=  PlotVectorField[vector ,{x,-5,5,2} ,{y,-5,5,2}
             ,Graphics`PlotField`ScaleFunction->(1&) ];
```

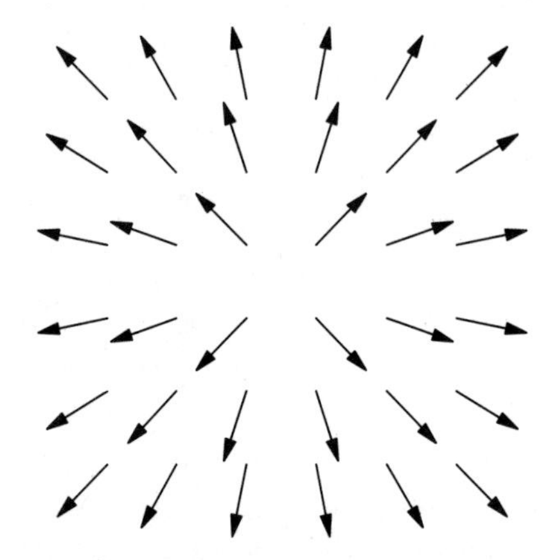

Note that we specify the complete context of the variable **ScaleFunction** using the syntax **Graphics`PlotField`ScaleFunction->(1&)** due to the shadowing of the **Graphics`PlotField3D`ScaleFunction** variable as discussed in the next section.

1.6.3 Shadowing

When you use multiple packages, there is always the possibility of shadowing; for example,

```
In[5]:=  Needs["Graphics`PlotField`"];
         Needs["Graphics`PlotField3D`"];

  Warning: Symbol ScaleFunction appears in multiple contexts
     {Graphics`PlotField3D`, Graphics`PlotField`}; definitions in context
     Graphics`PlotField3D` may shadow or be shadowed by other definitions.
```

We see that the **ScaleFunction** is assumed to be in the **Graphics`PlotField3D`** context rather than the **Graphics`PlotField`** context:

```
In[6]:=  Context[ScaleFunction]

Out[6]=  Graphics`PlotField3D`
```

This will cause a problem with the following example. Suppose we want to plot the gradient field for a point charge. This is singular at the origin, so we would like to rescale the vector lengths to a constant using **ScaleFunction**. The following example does not work because **ScaleFunction** is assumed to be in the three-dimensional context rather than the two-dimensional context.

```
In[7]:=  PlotGradientField[1/Sqrt[x^2+y^2]
               ,{x,-5,5,2}
               ,{y,-5,5,2}
               ,ScaleFunction->(1&)
            ];
```

The solution is to specify the full context we desire along with the variable. Now it does work:

```
In[8]:=  PlotGradientField[1/Sqrt[x^2+y^2]
               ,{x,-5,5,2}
               ,{y,-5,5,2}
               ,Graphics`PlotField`ScaleFunction->(1&)
            ];
```

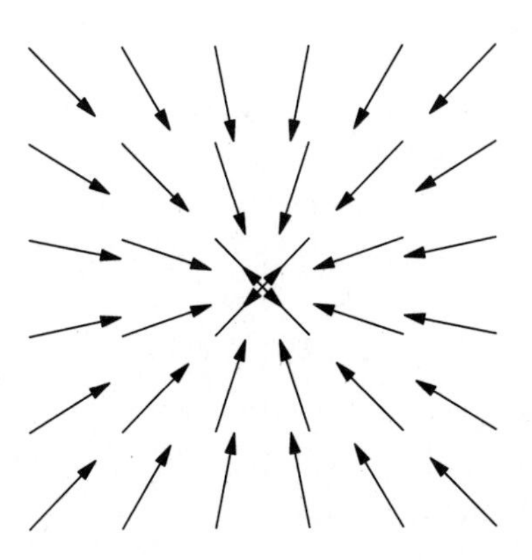

1.6.4 Three-dimensional Graphics

To illustrate the use of three-dimensional graphics, consider the plot of the **Log** function:

In[9]:=
```
Plot3D[ Log[x^2+y^2] ,{x,-20,20},{y,-20,20}
      ,PlotPoints->35
      ,AspectRatio->1
      ,Boxed->False
      ,Axes->None  ];
```

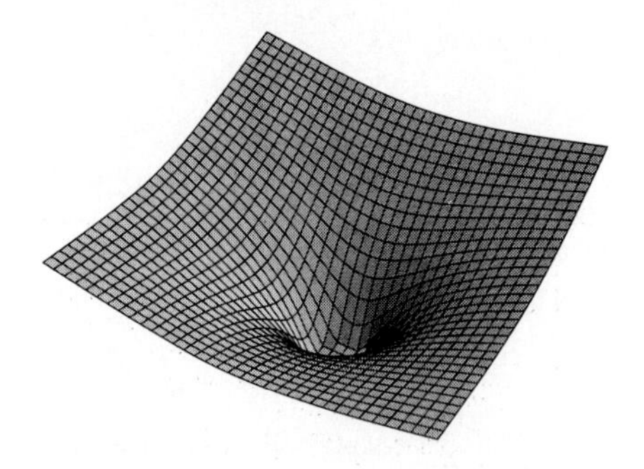

1.6.5 Space Curve

The last example that we include is a curve in space. Consider a space curve defined by the coordinates

In[10]:=
```
coordinates= {Cos[2t+Pi/4 ], 2 Cos[2 t ], Cos[3 t]};
```

The three-dimensional plot of this curve is

In[11]:=
```
ParametricPlot3D[ Evaluate[coordinates]
                ,{t,0,10 Pi}
                ,PlotPoints-> 300];
```

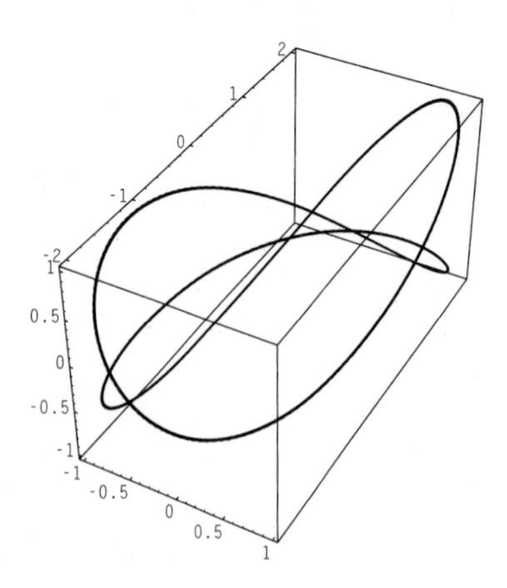

1.7 Exercises

Exercise 1.1: Tables of Zeros

Consider the functions **HermiteH[n,x]**, **BesselJ[n,x]**, **LegendreP[n,x]**, and **LaguerreL[n,x]**.

Part (a) For the first six integer values of **n**, make a table of zeros for these functions. Label the rows and columns and express the numerical values as rational numbers.

Part (b) Verify the results by plotting these functions.

Exercise 1.2: Substitutions

Consider the substitution rule, **rule= {a+b->x,c+d->y}**. Apply the substitution using **/.** and **//.** to **(a+b+c+d)**. Explain.

Exercise 1.3: Procedures

Consider the list of 30 numbers between 0 and 20, **list=Table[Random[Real,0,20],i,30]**. Write a procedure that will retain only those elements in the list having values between 6 and 15 and that will rank these elements from largest to smallest.

Exercise 1.4: Algebraic Simplification

Simplify the following expressions:

Part (a) Decompose $(x^2-1)/[(x-4)(x-5)]$ into its partial fractions.

Part (b) Show that $2+a^2/2-a^2\mathrm{Cos}[2x]/2-2a\,\mathrm{Sin}[x]$ can be written as a product of two factors that contain functions linear in $\mathrm{Sin}[x]$.

Part (c) Show that $1/[(1+x)(1+x^2)]$ can be written as partial fractions:

$$-\frac{(1+I)}{4(x-I)}-\frac{(1-I)}{4(x+I)}+\frac{1}{2(1+x)}.$$

Part (d) Show that $(\mathrm{Sin}[3x]-\mathrm{Cos}[4x])^2$ can be expressed as combinations of linear trigonometric functions.

Part (e) Reduce $\sqrt{1-\sqrt{(1-2x^2+x^4)}}$ to x.

Exercise 1.5: Algebraic Equations

Find all solutions for x in the following equations:

Part (a) $x^6-4x+1=0$

Part (b) $1+3x^2+3x^4+x^6=0$

Part (c) $(x-1)(x^2+1)e^{-x^2}=0$

Part (d) $a = \mathrm{Sin}[x - c]$

Part (e) $(ax^2 + bx + c)/(d + ex) = 0$

Part (f) $c\,\mathrm{Sin}[x]^2 - b\,\mathrm{Sin}[x] + a = 0$

Find the solutions using as many of the commands **`NSolve`**, **`Solve`**, **`Reduce`**, **`Eliminate`**, **`NRoots`**, and **`FindRoot`** as you can.

Exercise 1.6: Algebraic Equations

Consider the two equations $x^2 = a^2 + 2cab + b^2$ and $a^2 = b^2 + x^2 - (2bx)/\sqrt{1-y^2}$. Solve for y and express it in terms of a, b, and c.

Exercise 1.7: Coupled Equations

Solve the following coupled equations for x, y, and z:

$$\begin{aligned} x - ay^2 + 3z &= 34 \\ x + y - z &= -1 \\ -x + 2y + 3z &= 4. \end{aligned}$$

Exercise 1.8: Integration and Simplification

Integrate the following functions:

$$\frac{\sqrt{1-(x/b)^2}}{1-(x/p)^2}$$

and

$$\frac{1-x^2}{1-x^2+x^4}.$$

Take the derivatives of the results and reduce the expressions to their original forms.

Exercise 1.9: Integration

Part (a) Integrate the function $\mathrm{Cos}[kr\mathrm{Cos}[x]]$ from $x = 0$ to 2π. *(Hint: Change variables,* $x = \mathrm{ArcCos}[z]$.*)*

Part (b) Write a procedure that will change the variable of integration and then integrate.

Exercise 1.10: Differential Equation

Consider the differential equation $mz''[t] == -mg - bz'[t]$.

Part (a) Find the solutions for $z'[t]$ and $z[t]$.

Part (b) Express the results in the form of pure functions.

Exercise 1.11: Two Coupled Differential Equations

Consider the two coupled differential equations $x''[t] == -ax[t] + by'[t]$ and $y''[t] == -ay[t] - bx'[t]$. Solve these equations for $x[t]$ and $y[t]$. *(Hint: Try forming one complex equation for the variable* $u = x + Iy$.*)*

Exercise 1.12: Superposition of Waves

Consider two oscillations of slightly different frequencies: $\mathrm{Cos}[(\omega + \delta)t] + \mathrm{Cos}[\omega t]$. Show that the expression can be written as $2\mathrm{Cos}[t\delta/2]\mathrm{Cos}[t(\omega + \delta/2)]$ and plot the results.

Exercise 1.13: Animation of Traveling Waves

Consider the two traveling waves moving in opposite directions, $\mathrm{Cos}[2\pi(x - vt)]$ and $\mathrm{Cos}[2\pi(x + vt)]$. Animate the traveling waves.

CHAPTER TWO

General Physics

2.1 Introduction

One important advantage of *Mathematica* is that the physicist is freed from time-consuming algebra and can use this extra time to explore the problems in greater depth. The researcher can scrutinize the *Mathematica* commands to find alternate derivations and can change the parameters to understand their effects on the solution.

Mathematica will tell you if there is an error in the syntax of an input statement. If you need help with the syntax, use the help facility (**?**) or find the command in the *Mathematica* user's manual. The more problems you solve, the better you will learn the strengths and weaknesses of *Mathematica* as it applies to physics.

A brief view of the *Mathematica* language was presented in Chapter 1. In Chapter 2, we apply *Mathematica* to simple problems encountered in undergraduate physics. We take problems from classical mechanics, electricity and magnetism, circuits, and modern physics. The simplicity of the problems allows the reader to concentrate on the *Mathematica* commands; the more difficult problems are introduced in subsequent chapters.

This chapter is divided into three sections:

1. Introduction
2. Problems
3. Exercises.

In the first section, we give a brief overview of the physics, the packages that are used in this chapter, and some examples of user-defined procedures. In the second section, a few remarks and suggestions follow the statement of each problem, and then the solution is given. Suggested exercises are found in the third section.

2.2 General Physics

2.2.1 Newtonian Motion

Basic to any presentation of mechanics is Newton's second law, which states that an external force acting on a particle will cause the momentum of the particle to change:

$$\vec{F} = \frac{d\vec{p}}{dt},$$

where $\vec{p} = m\vec{v}$ is the momentum. The force and momentum are vector quantities and are represented by lists in *Mathematica*. The second law represents a set of three, second-order differential equations for three-dimensional problems. As these are second-order equations, their solutions require the initial values of the position and velocity vectors.

The analysis of the equations of motion simplifies if the force is conservative. The force is conservative if it can be written as

$$\vec{F} = -\vec{\nabla} V,$$

where V is the scalar potential. In a conservative force field, the total energy is conserved.

In many cases, it is useful to discuss the motion of a particle, or system of particles, relative to a rotating (non-inertial) coordinate system. Newton's laws of motion must be rewritten to express the modification due to the non-inertial frame of reference. Newton's second law in a rotating coordinate system is

$$m\vec{a} = \vec{F} - m(\vec{\omega}' \times \vec{r}) - 2m\,(\vec{\omega} \times \vec{v}) - \vec{v} \times (\vec{\omega} \times \vec{r}),$$

where $\vec{\omega}$ is the angular velocity of the moving system and $\vec{a}$ is the acceleration relative to the moving coordinates. If the angular velocity $\vec{\omega}$ is assumed to be constant, the second term on the right-hand side (proportional to $\vec{\omega}'$) vanishes. The last two terms are called the **coriolis** and **centripetal forces**, respectively.

2.2.2 Electricity, Magnetism, and Circuits

The electric potential at position r created by a point particle at the origin of the coordinates and with charge q is $U = q/r$. The electric field E is related to the potential U by

$$E = -\nabla U.$$

The infinitesimal magnetic vector potential dA at positon r created by a filamentary current $I\,dl$ at r' is

$$dA = \frac{I\,dl'}{|r - r'|},$$

where dl' is a differential length along the current. Integrating along the current gives the total magnetic vector potential. The magnetic induction B is related to the vector potential by $B = \nabla \times A$.

In the case of a series circuit that contains an electromotive force $V[t]$, inductance L, capacitance C, and resistance R, the basic equation for the charge $q[t]$ is

$$V[t] = L\,q''[t] + R\,q'[t] + q[t]/C.$$

The charge $q[t]$ is related to the current $I[t]$ by $q'[t] = I[t]$.

2.3 *Mathematica* Commands

2.3.1 Packages

There are several packages that are used in this chapter. First, for convenience we turn off the spell checker

```
In[1]:= Off[General::spell ];
        Off[General::spell1];
```

This chapter needs the following packages. The packages should be activated now to avoid shadowing by user-defined commands. (*Note*: There is some shadowing among these packages.) A list of the packages is

```
In[2]:=  Needs["Algebra`SymbolicSum`"];
         Needs["Algebra`Trigonometry`"] ;
         Needs["Calculus`FourierTransform`"];
         Needs["Calculus`LaplaceTransform`"];
         Needs["Calculus`VectorAnalysis`"];
         Needs["Graphics`ParametricPlot3D`"];
         Needs["Graphics`PlotField`"];
         Needs["Graphics`PlotField3D`"];
         Needs["Miscellaneous`Units`"];
         Needs["Miscellaneous`PhysicalConstants`"];
```

2.3.2 User-defined Procedures

Procedure to find a least-squares fit to a set of data

The standard function **Fit** is quite useful for fitting a set of data; however, the form of the functions that it can handle is restricted. Here we define a function **leastSquares**, which is similar but which allows a fit for an arbitrary parameter in the exponent, for example. (*Cf.*, Problem 3: Projectile with air resistance.)

We will not give a detailed description of how this function works. The best way to understand it is to take the following simple example and examine the intermediate steps.

```
In[3]:=  leastSquares::usage=" leastSquares[data,function,var ,start]
         This procedure will fit a set of data points to a function
         with unknown constants. The constants are determined by finding
         the values that give the least squared fit to the data.";

         leastSquares[p_List,f_,var_ ,start___]:=
           Module[{temp1,temp2 },
             temp1=Apply[Plus, Map[(#[[2]]-(f/.var->#[[1]]))^2&,p]];
             temp2=FindMinimum[Evaluate[temp1//Expand],start];
             Return[{f//.temp2[[2]],temp2[[2]]}] ];
```

Example: Slope of a Straight Line

Here is a simple example that both **leastSquares** and **Fit** can do:

```
In[4]:=  leastSquares[{{0,0},{1,2},{2,4},{3,6}},a x ,x ,{a,1/2}]

Out[4]=  {2. x, {a -> 2.}}

In[5]:=  Fit[{{0,0},{1,2},{2,4},{3,6}},  {1,x},x] //Chop

Out[5]=  2. x
```

Example: An Exponential

Here is a simple example with a functional form that **Fit** cannot do directly:

```
In[6]:=  leastSquares[{{0,1},{1,10},{2,10^2},{3,10^3}}
                ,E^(a x) ,x ,{a,1/2}]
```

```
Out[6]=    2.30259 x
         {E           , {a -> 2.30259}}
```

Procedure to find the electric potential for point charges

Using the superpostition principle for the scalar potential, we find it useful to define the potential for a monopole of charge q, located at the point $\{x, y, z\}$. Using this function, we can then build up complex multipole systems:

```
In[7]:=  monopole::usage =" monopole[q,{x,y,z}]
         This procedure gives the potential for a point charge q
         located at the point {x,y,z}.";

         monopole[q_, p0_:{0,0,0}, p1_:{x,y,z}]=
            q/Sqrt[(p0-p1).(p0-p1)]

Out[7]=                  q
         ----------------------------------
         Sqrt[(p0 - p1) . (p0 - p1)]
```

Example: Dipole

Using this simple function, it is easy to build up a dipole or more-complex charge configurations:

```
In[8]:=  dipole=( monopole[+q,{0,0,+d/2}]
                + monopole[-q,{0,0,-d/2}] )

Out[8]=                   q                                  q
         -(-------------------------------) + ----------------------------
                 2    2    -d      2                2    2    d     2
          Sqrt[x  + y  + (-- - z) ]           Sqrt[x  + y  + (- - z) ]
                            2                                 2
```

2.3.3 Protect User-defined Procedures

```
In[9]:=  Protect[leastSquares,monopole];
```

2.4 Problems

2.4.1 Projectile Motion in a Constant Gravitational Field

Problem 1: Escape Velocity

From the conservation of energy, show that the escape speed for a particle shot from Earth's surface is about 6.95 m/sec. Neglect air resistance.

Remarks and Outline This simple problem illustrates the use of algebra and the use of unit conversions. The solution of the problem follows from equating the energy at Earth's surface with the energy at $r = \infty$.

Required Packages

```
In[1]:=  Needs["Miscellaneous`Units`"];
         Needs["Miscellaneous`PhysicalConstants`"];
```

Solution

```
In[2]:=  Clear["Global`*"];
```

The kinetic energy **T**, potential energy **V**, and total energy **energy** of the particle at Earth's surface is

```
In[3]:=  T= (1/2) m v^2;
         V= -G M m/r;
         energy=T+V

Out[3]=
                        2
           G m M     m v
         -(-----) + ----
             r        2
```

If a particle just escapes from Earth's gravitational field, the energy at infinity is zero. The conservation of energy gives the equation **energy=0**, which can be solved for **v**, where **v** is the escape speed:

```
In[4]:=  sol= Solve[energy==0,v][[2]]

Out[4]=
               Sqrt[2] Sqrt[G] Sqrt[M]
         {v -> -----------------------}
                       Sqrt[r]
```

(Be sure to take the positive solution here.) Using the information in the package on physical constants,
Miscellaneous`PhysicalConstants`, the numerical value of the escape speed is

```
In[5]:=  v0= ( v /.sol
                //.{G->GravitationalConstant,
                    r->EarthRadius,
                    M->EarthMass,
                    Newton->Kilogram Meter/Second^2}
                //N
                //PowerExpand
            )

Out[5]=  11182. Meter
         ------------
            Second
```

This speed can be expressed in **Miles/Second** using the package **Miscellaneous`Units`**:

```
In[6]:=  Convert[v0, Mile/Second ]

Out[6]=  6.94818 Mile
         ------------
            Second
```

Problem 2: Projectile in a Uniform Gravitational Field

Consider a particle of mass m moving in a constant gravitational field.

Part (a) Solve the equations of motion for the trajectories that have the same initial and final position but different final times.

Part (b) Plot the trajectories for several different final times.

Part (c) Solve the equations of motion for trajectories with a given initial speed and angle. Find the position of maximum height.

Part (**d**) Plot the trajectories in Part (c) for different initial angles.

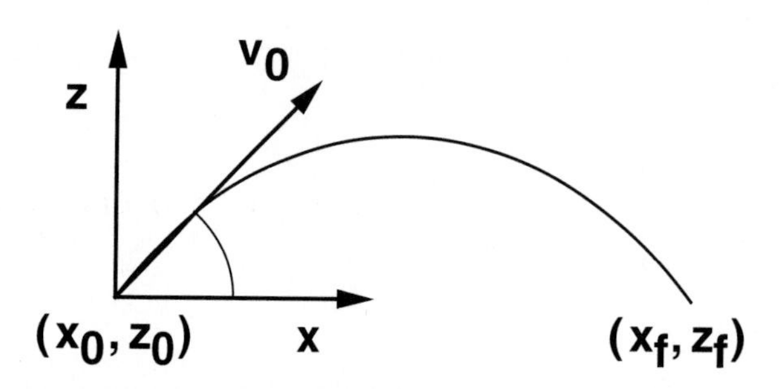

Remarks and outline This solution illustrates the use of **DSolve** and various graphics options. There are several ways to plot the results and you are encouraged to explore other graphics options. **DSolve** gives the analytic solution for the equations of motion in a constant gravitational field. The proper choice of boundary conditions gives the desired solution.

Required packages *There are no required packages.*

Solution

```
In[7]:= Clear["Global`*"];
```

Part (**a**) The equations of motion for a particle of mass **m** moving in a constant gravitational field are

```
In[8]:= eq1 ={m x''[t] == 0 ,
               m z''[t] == -m g };
```

where **z** is the projectile height and **x** is the horizontal distance. If the position at time **t=0** is **{x,z}={0,0}** and the final position at time **t=tf** is **{xf,zf}**, the boundary conditions are

```
In[9]:= initial1={x[0]==0, x[tf]==xf,
                  z[0]==0, z[tf]==zf};
```

Joining **eq1** with **initial1** and using **DSolve**, you get

```
In[10]:= dsol1=( DSolve[Join[eq1,initial1],{x[t],z[t]},t][[1]]
                //Simplify)

Out[10]=
                                       2
                t xf               -(g t )    g t tf    t zf
         {x[t] -> ----, z[t] -> -------- + ------ + ----}
                 tf                  2          2        tf
```

Part (**b**) Plot the trajectories for three different final times **tf={5,10,15}** and for the values of the parameters given by

```
In[11]:= values1={xf->300, zf->0, g->9.8};
```

The coordinates as a function of **t** for these values become

```
In[12]:= coord1[tf_]= {x[t],z[t]} /.dsol1 /.values1

Out[12]=  300 t          2
         {-----, -4.9 t   + 4.9 t tf}
           tf
```

A graph of the continuous trajectories follows immediately from **ListPlot**. Choose times from 0 to **tf** in steps of 0.5 sec.

```
In[13]:= Clear[plot1];
         plot1[tf_]:=
           ListPlot[
              Evaluate[ Table[coord1[tf],{t,0,tf,0.5}] ]
             ,PlotStyle->PointSize[0.02]
             ,GridLines->Automatic
           ]
```

plot1[tf] will generate one plot. Then make a table of plots with the desired times:

```
In[14]:= plotarray2=Table[plot1[tf],{tf,5,15,5}];
```

Finally, show the table of plots (intermediate plot output suppressed):

```
In[15]:= Show[plotarray2];
```

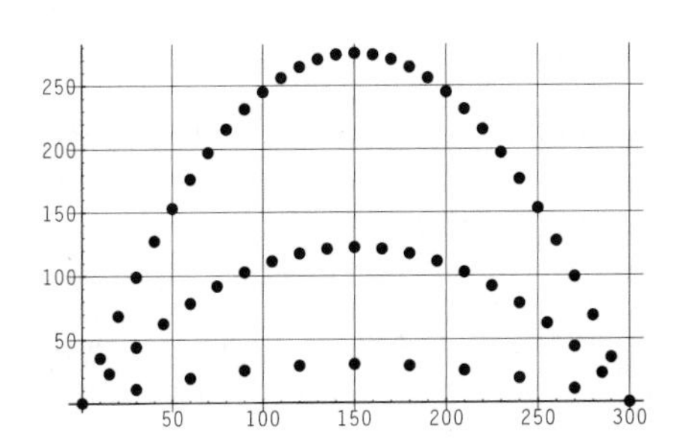

Part (c) Solve the equations of motion with initial speed **v0** and initial angle **alpha**:

```
In[16]:= initial2={  x[0]==0 ,x'[0]==v0 Cos[alpha],
                     z[0]==0, z'[0]==v0 Sin[alpha]};
```

The solution of **eq1** with these initial conditions is

```
In[17]:= dsol2=DSolve[Join[eq1,initial2],{x[t],z[t]},t][[1]]

Out[17]=
                                                  2
                                           -(g t )
         {x[t] -> t v0 Cos[alpha], z[t] -> -------- + t v0 Sin[alpha]}
                                              2
```

The time it takes to reach the highest point follows from taking the time derivative of **z[t]**, setting the results to zero, and solving for **t**:

```
In[18]:=  eq2= D[ z[t] /. dsol2, t]==0;
          tsol= Solve[eq2,t][[1]]

Out[18]=
                 v0 Sin[alpha]
          {t -> -------------}
                      g
```

The position of maximum height is obtained by evaluating **{x,z}** at this time:

```
In[19]:=  {x[t],z[t]} /.dsol2 /.tsol //Simplify

Out[19]=
             2                    2            2
           v0  Sin[2 alpha]     v0  Sin[alpha]
          {----------------, ----------------}
                2 g                 2 g
```

Part (d) Plot the three trajectories that have **alpha** equal to $\{\pi/8, 2\pi/8, 3\pi/8\}$. You should plot the positions of the trajectories at discrete time points. The plot follows immediately from the command **ListPlot**. For the choice of parameters, choose

```
In[20]:=  values2={g->9.8,v0->100};
```

with the following coordinates of the trajectory:

```
In[21]:=  coord2[alpha_]= {x[t],z[t]} /.dsol2 /.values2

Out[21]=
                                     2
          {100 t Cos[alpha], -4.9 t  + 100 t Sin[alpha]}
```

A graph of the continuous trajectories follows immediately from **ListPlot**. Choose times from 0 to 20 seconds in steps of 1 sec. (*Note*: All but the first two lines of **plot2** are simply to enhance the plot; these may be ignored.)

```
In[22]:=  Clear[plot2];
          plot2[alpha_]:=
            ListPlot[Evaluate[
              Table[coord2[alpha],{t,0,20,1}] ]
               ,PlotStyle->PointSize[0.02]
               ,Frame -> True
               ,PlotLabel->(" Same Initial and Final Positions " )
               ,GridLines->Automatic
               ,FrameLabel->{"x[t]","z[t]","Distance","Height"}
            ]
```

plot2[alpha] will generate one plot. Then you make a table of plots with the desired **alpha** (intermediate output suppressed):

```
In[23]:=  plotarray2=Table[plot2[alpha],{alpha,Pi/8, 3 Pi/8, Pi/8}];
```

Finally, we show the table of plots:

```
In[24]:=  Show[plotarray2
                ,PlotRange->{{0,1000},{0,500}}
                ,RotateLabel->False];
```

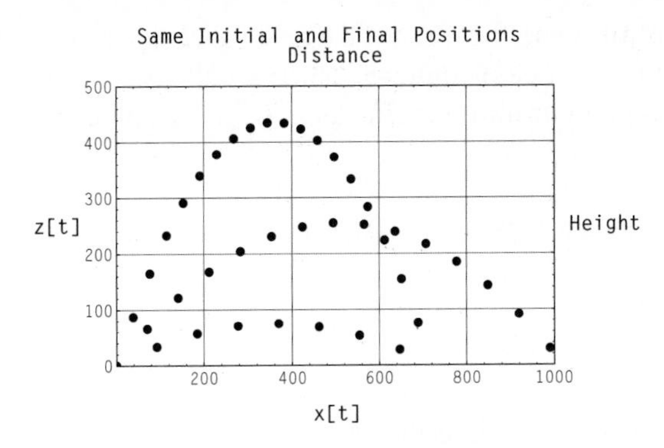

Note that alpha $= \pi/4$ gives the largest range.

Problem 3: Projectile with Air Resistance

A particle of mass m is acted on by a constant pull of gravity with acceleration g and air resistance. The particle is constrained to move only along the direction of the gravitational force and the force of the air drag is assumed to be $-bv$.

Part (a) Solve the equations of motion for the particle's position and velocity. Find the terminal velocity. Take the limit of the solution as $b \to 0$ (i.e., no air resistance).

Part (b) Plot the speed for $b \neq 0$ and $b = 0$. Label the graphs and show the terminal velocity.

Part (c) Plot the particle's position at equal intervals of time for $b \neq 0$ and $b = 0$. Label the graphs.

Part (d) Construct an artificial table of data points for $\{t, z[t]\}$ with an experimental error in $z[t]$ of 10%. Fit the data to a polynomial in t. Compare the results to the series expansion of the exact answer and plot the results.

Part (e) Determine the coefficient of drag b from an artificial table of data points constructed with an error of 10%.

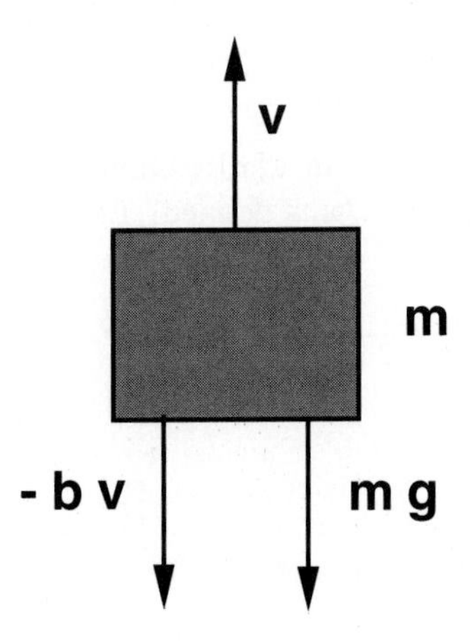

Remarks and outline This problem illustrates the use of the commands **Limit**, **Series**, **Fit**, and the user-defined function **leastSquares**. **DSolve** will give the analytic solution for the equations of motion in a constant gravitational field. Graphics options are used to embellish the plots.

Required packages *There are no required packages.*

Solution

```
In[25]:=  Clear["Global`*"];
```

Part (a) The motion of the projectile is described by the equation

```
In[26]:=  m z''[t] ==  -m g -b z'[t];
```

where **-m g** is the gravitational term, **-b z'[t]** is the air resistance term, and **b** is greater than or equal to zero. The solution follows immediately from **DSolve**. Alternatively, you can solve **z[t]** and **z'[t]** by applying **DSolve** to the two first-order equations:

```
In[27]:=  eq1 =  { m v'[t] == -m g -b v[t],
                   z'[t] == v[t]            }

Out[27]=  {m v'[t] == -(g m) - b v[t], z'[t] == v[t]}
```

Assume the initial position and velocity are **z0** and **v0**, respectively:

```
In[28]:=  initial={ z[0]==z0, z'[0]==v0};
```

The solution of **eq1** follows from **DSolve**:

```
In[29]:=  dsol=( DSolve[Join[eq1,initial], {v,z},t]
                 //Flatten
                 //Simplify );
          {v[t],z[t]}/.dsol

Out[29]=
                 (b t)/m                        2
           g m - E          g m + b v0   g m       g m t    m v0    m (g m + b v0)
          {--------------------------, ---- - ----- + ----- - -------------- + z0}
                   (b t)/m                2       b       b      2  (b t)/m
                b E                      b                      b  E
```

The terminal velocity is obtained by letting **t**$\rightarrow\infty$ in **v[t]**. Obviously, **b**> 0 for a sensible solution, so the positive quantity **large=b t/m** is defined. Expressing **v[t]** in terms of **large**,

```
In[30]:=  vLimit= v[t] /.dsol /.{t-> large m/b} //ExpandAll

Out[30]=
            g m        g m          v0
          -(---) + ---------- + -------
             b         large      large
                    b E          E
```

and taking the limit as **large->Infinity**, you get

```
In[31]:=  vLimit=Limit[vLimit,large->Infinity]
```

```
Out[31]=    g m
          -(---)
             b
```

The solution reduces to the standard free-fall solution as **b**→ 0, and **z0[t]** and **v0[t]** are defined to be these limiting values:

```
In[32]:= {z0[t],v0[t]}=Limit[{z[t],v[t]}/.dsol    ,b->0]

Out[32]=          2
           -(g t )
          {-------- + t v0 + z0, -(g t) + v0}
              2
```

Part (b) Plot the height **z[t]** versus **t** for **b**= $\{0, 1/2, 1\}$ and for the following values of the parameters:

```
In[33]:= values1={g->9.8, m->1, z0->0, v0->0};
```

The coordinates to plot are then a function of **b**:

```
In[34]:= coordinates1[b_]={t,z[t]}/.dsol/.values1

Out[34]=      9.8      9.8       9.8 t
         {t, ---- - --------- - -----}
               2      2  b t      b
              b      b  E
```

Take care of limiting case where **b**=0 by hand to ensure that you avoid a divide-by-zero:

```
In[35]:= coordinates1[0]={t,z0[t]}/.values1;
```

For convenience, define a function to make a plot for arbitrary values of **b**:

```
In[36]:= Clear[plot1];
         plot1[b_]:=
           ParametricPlot[Evaluate[coordinates1[b]],{t,0,10}]
```

Collect a group of plots into a table,

```
In[37]:= plotarray1=Table[plot1[b],{b,0,1,1/2}];
```

and then add some graphics options (intermediate plots suppressed):

```
In[38]:= Show[plotarray1

         ,Graphics[{Text[FontForm["b=0   ",{"CourierB",10}],{6,-200}]
                   ,Text[FontForm["b=1/2",{"CourierB",10}],{8,-150}]
                   ,Text[FontForm["b=1   ",{"CourierB",10}],{9,-50}]
                    }]
           ,Frame -> True
           ,GridLines->Automatic
           ,FrameLabel-> {"t","z"
                         ,"Effect of Air Resistance","z"}
           ];
```

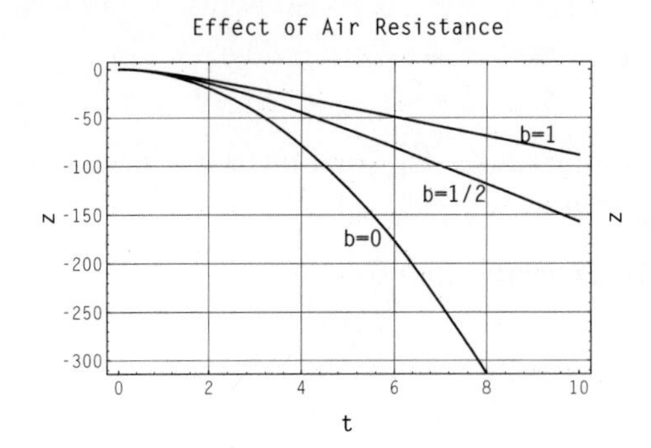

Part (c) Similarly, plot the velocity **v[t]** versus **t** for **b**= $\{0, 1/2, 1\}$. The coordinates to plot are then a function of **b**:

```
In[39]:= coordinates2[b_]={t,v[t]}/.dsol/.values1 //ExpandAll
```

```
Out[39]=      -9.8      9.8
         {t, ----- + -------}
               b        b t
                     b E
```

Take care of limiting case where **b=0** by hand to ensure that we avoid a divide-by-zero:

```
In[40]:= coordinates2[0]={t,v0[t]}/.values1;
```

The coordinates for the terminal velocity are

```
In[41]:= coordinates3[b_]={t,vLimit}/.dsol/.values1 //ExpandAll;
         coordinates3[0]={t,0};
```

For convenience, define a function to make a plot for arbitrary values of **b** of both the velocity and the terminal velocity:

```
In[42]:= Clear[plot2];
         plot2[b_]:=
           ParametricPlot[Evaluate[{coordinates2[b]
                                   ,coordinates3[b]}]
              ,{t,0,10}]
```

Collect a group of plots into a table (intermediate output suppressed),

```
In[43]:= plotarray2=Table[plot2[b],{b,0,1,1/2}];
```

```
In[44]:= Show[plotarray2

         ,Graphics[{Text[FontForm["b=0  ",{"CourierB",10}],{4,-25}]
                   ,Text[FontForm["b=1/2",{"CourierB",10}],{7,-15}]
                   ,Text[FontForm["b=1  ",{"CourierB",10}],{8,-5}]
                   }]
           ,PlotRange->{0,-30}
           ,Frame -> True
           ,GridLines->None
           ,RotateLabel->False
           ,FrameLabel-> {"t","v"
                       ,"Effect of Air Resistance","v"}
           ];
```

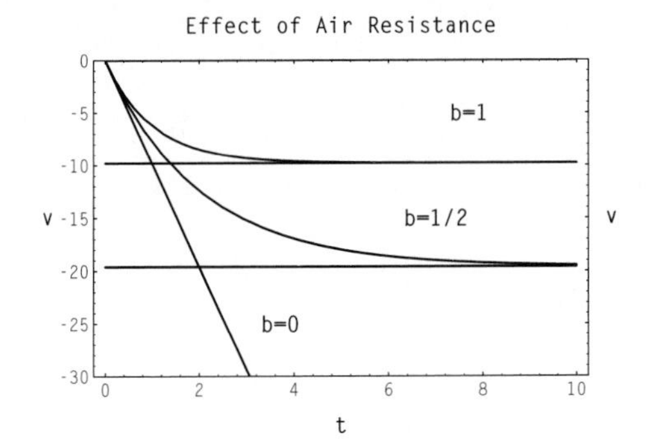

Part (d) Using **b**→ 1/2, construct an artificial table of data points for **{t,z[t]}** with an experimental error in **z[t]** of 10%. For the following values of the parameters given by **values1**, the artificial data table is

```
In[45]:= dataPoint[t_]:=( {t,z[t] (1.0 + Random[Real,{-0.1,0.1}]) }
                        //.Join[dsol,values1] //N );

         dataTable[b_]=Table[dataPoint[t],{t,0,20,1}];

         b0=0.5;
         data=dataTable[b0];
         data //Short

Out[45]= {{0, 0.}, {1., -3.96731}, {2., <<7>>2}, <<17>>, {20., -382.379}}
```

A fit to a polynomial in **t** with exponents that vary from two to five follows from applying **Fit** to the data:

```
In[46]:= fit[t_]=Rationalize[ Fit[data,Table[t^i,{i,2,5}],t] ,10^-4]

Out[46]=
              2        3      4
         -505 t    23 t      t
         ------- + ------ - ---
           162      117     211
```

Rationalize was used to reduce the coefficients to simple fractions. Note that **fit[t]** (lowercase) is now a function. Compare this fit to the expansion of the exact solution:

```
In[47]:= exact[t_,b_]=  z[t]/.dsol/.values1

Out[47]= 9.8      9.8        9.8 t
         --- - -------- - ------
          2       2 b t      b
         b     b  E
```

Overlay the plots:

```
In[48]:= dataPlot=ListPlot[dataTable[0.5]
                  ,PlotStyle->PointSize[0.02] ];
         fitPlot=Plot[fit[t],{t,0,20}];
         exactPlot=Plot[exact[t,0.5],{t,0,20}];
```

```
In[49]:= Show[dataPlot, fitPlot, exactPlot

             ,Frame -> True
             ,GridLines->Automatic
             ,FrameLabel-> {"t","z"
                         ,"Experimental Data & Air Resistance","z"}
         ];
```

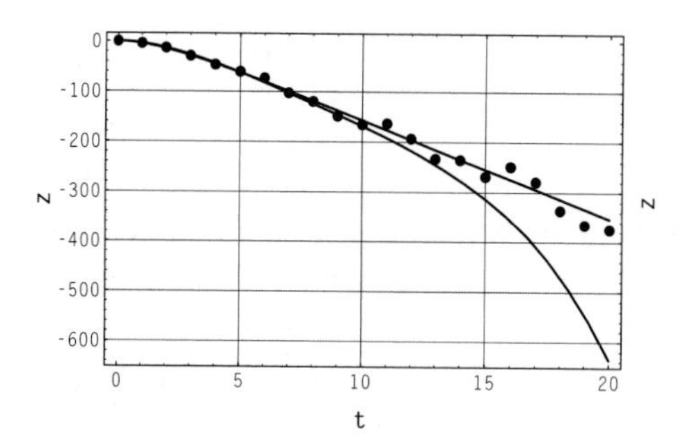

Part (e) Determine the drag coefficient by fitting the analytic solution

```
In[50]:= trialfunction=z[t]/.dsol/.values1
Out[50]= 9.8       9.8        9.8 t
         --  -  --------  -  -----
          2       2 b t       b
         b       b E
```

to a **leastSquares** fit. Consider the 20 experimental data points in **data**. This table of points corresponds to a drag coefficient of **b=1/2** and an error in measurement of 10%. Applying **leastSquares** to **data**, you get

```
In[51]:= fit2= leastSquares[data ,trialfunction, t ,{b,1}]
Out[51]=
                      41.6227
         {41.6227 - ------------ - 20.1966 t, {b -> 0.48523}}
                     0.48523 t
                    E
```

This answer is very close to the exact answer of **b**=0.5.

Problem 4: Rocket with Varying Mass

A rocket is fired vertically upward with a mass loss of $m'[t]$ due to exhaust. At time t, it has velocity v relative to Earth (inertial frame). The exhaust leaves the rocket with velocity u relative to Earth (inertial frame). Let v_e be the exhaust speed relative to the rocket, or $u = (v - v_e)$. The sign of v_e has been chosen to be positive, so the exhaust velocity is directed toward Earth ($v_e > 0$). The change in mass dm is a loss, so it has a negative value ($dm < 0$).

Part (a) Show that the rocket's equation of motion is

$$m\,\frac{dv}{dt} = -v_e\,\frac{dm}{dt} - mg.$$

Part (b) Assume a constant mass loss and use **DSolve** to find x and v. Find the height of the rocket when all the mass of the rocket has been used as fuel.

Part (c) Plot the distance versus height for different exhaust speeds.

Part (d) Solve the rocket equation by assuming it can be written as a power series in t.

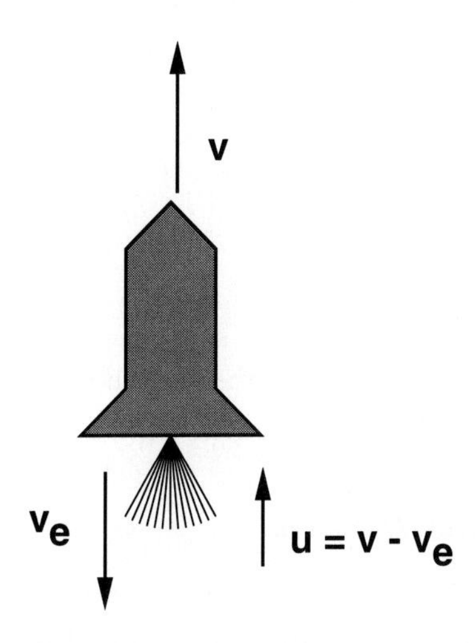

Remarks and outline This problem illustrates the use of power series in solving differential equations and the **Sum** command. The equation of motion follows from finding the momentum change of the rocket and using Newton's second law. The solution of the differential equation follows from **DSolve**, and the results can be plotted using **ParametricPlot**. The solution also follows from assuming a power series expansion in t. The series solution reduces to solving a set of algebraic equations. The exact solution is recovered after summing the series to infinity. Though this second method is not as straightforward as **DSolve**, it does illustrate the use of power-series solutions.

Required packages

```
In[52]:= Needs["Algebra`SymbolicSum`"];
```

Solution

```
In[53]:= Clear["Global`*"];
```

Part (a) Let **v** and **m** be the velocity and mass of the rocket at **t** and **u** the velocity of the exhaust gas relative to Earth. Let **ve** be the exhaust velocity relative to the rocket or **u=(v-ve)**. At time **t**, the mass is **m**, and at time **t+dt**, the rocket mass is **(m+dm)** and the exhaust mass is **(-dm)**. The momentum of the rocket-exhaust system at times **t** and **t+dt** are

```
In[54]:= pi= m[t] v;
         pf= (m[t]+dm)(v+dv) + (-dm)(v-ve);
```

In the expression for **pf**, the first term represents the rocket's momentum with the new mass value **(m[t]+dm)** and new velocity **(v+dv)**. The second term represents the momentum of the mass exhaust **(-dm)** moving with velocity **u=(v-ve)** in Earth's inertial frame. The differential relations are

```
In[55]:= dv= x''[t] dt;
         dm=  m'[t] dt;
```

The rate of momentum change is

```
In[56]:= dp =(pf-pi)/dt  //ExpandAll

Out[56]= ve m'[t] + m[t] x''[t] + dt m'[t] x''[t]
```

Set the rate of momentum change equal to the external force $F = -mg$ using Newton's second law and take the limit **dt->0**:

```
In[57]:= eq1= -m[t] g  == dp  /.{dt->0}

Out[57]= -(g m[t]) == ve m'[t] + m[t] x''[t]
```

where **m'[t]** is the rate of decrease of rocket mass. **eq1** is the desired result.

Part (b) If we assume **m'[t] = -rate** and **m[0]=m0**, then the time dependence of the mass is

```
In[58]:= dsol1= DSolve[{m'[t]==-rate,m[0]==m0},m,t] //Flatten

Out[58]= {m -> Function[t, m0 - rate t]}
```

The **dsol1** is expressed in pure form so that the substitution will also work for the derivatives of **m**. Use **dsol1** to eliminate **m[t]** and **m'[t]** in **eq1**:

```
In[59]:= eq2= eq1 /.dsol1 //Simplify

Out[59]= g (-m0 + rate t) == -(rate ve) + (m0 - rate t) x''[t]
```

and solve for **x''[t]**:

```
In[60]:= eq3= Solve[eq2,x''[t] ][[1,1]] /.{Rule->Equal} //Simplify

Out[60]=
                   g m0 - g rate t - rate ve
         x''[t] == -------------------------
                        -m0 + rate t
```

The substitution **Rule->Equal** changes the rule to an equation. Applying **DSolve** to **eq3** with the initial conditions **x[0]==0** and **x'[0]==0**, the solution for **x[t]** follows:

```
In[61]:= dsol2= DSolve[{eq3,x[0]==0,x'[0]==0},x,t][[1]];
         x[t] /.dsol2

Out[61]=
             2
         -(g t )   m0 ve Log[m0]
         ------- - ------------- + t ve (1 + Log[m0]) +
            2          rate

           m0 ve Log[m0 - rate t]
           ---------------------- - t ve Log[m0 - rate t]
                    rate
```

The results can be further simplified if you define the log rules:

```
In[62]:= logRules={c_. Log[a_] + c_. Log[b_]:> c Log[a b],
                   c_. Log[a_] - c_. Log[b_]:> c Log[a/b] };
```

Applying the **logRules** to **x[t]**, you get

```
In[63]:=  position=
           ((x[t] /.dsol2
                 //.logRules
                 //Simplify
              ) //.logRules
                 /.{Log[m0/(m0-rate t)]->-Log[1-(rate t)/m0]}
                 //Collect[#,Log[1- rate t/m0]]&
              )
```

```
Out[63]=      2
         -(g t )              m0 ve                     rate t
         ------- + t ve + (------- - t ve) Log[1 - ------]
            2                  rate                      m0
```

The speed follows from the time derivative of the position:

```
In[64]:= speed=   x'[t] /.dsol2 //.logRules //Simplify
```

```
Out[64]= -(g t) + ve Log[m0] - ve Log[m0 - rate t]
```

Notice that all the rocket mass is consumed as fuel after a time **t->m0/rate**. The rocket's height when this occurs is

```
In[65]:= height= Limit[position, t->m0/rate] //.logRules //Simplify
```

```
Out[65]= m0 (-(g m0) + 2 rate ve)
         ------------------------
                        2
                2 rate
```

Part (c) Use the command **ParametricPlot** to plot the position versus speed for different values of the exhaust speed **ve**. Choose the parameters:

```
In[66]:= values1={ m0->100, rate->10, g->9.8};
```

and let the parametric parameter **t** vary between **t=0** and **t=tf**, where

```
In[67]:= tf= (m0/rate) /.values1
```

```
Out[67]= 10
```

The time **t=m0/rate** is the time it takes for the rocket's mass to be consumed. Define a plot command that will graph **{position,speed}** for a given value of **ve**:

```
In[68]:= Clear[plot1];
         plot1[veIn_]:=
         ParametricPlot[
             Evaluate[{position,speed}/.values1/.{ve->veIn} ]
               ,{t,0,tf(1.0-10^-6)}]
```

The **tf(1.0-10^-6)** prevents the expression from being indeterminate. Note that we use the rule **{ve->veIn}** to replace **ve** in the expressions of **{position,speed}**. The value of **veIn** is input for the **plot1** function. Next, use the **Graphics** primitive to place a point at the end of the curve:

```
In[69]:= Clear[endPoint];
         endPoint[veIn_]:=
         Graphics[ { PointSize[0.05]
                    ,Point[( {position,speed}
                             //.{t->tf(1.0-10^-6)}
                             /.values1
                             /.{ve->veIn} )
                          ]
                    }
                 ]
```

Superimposing the graph for **ve**= {100, 200, 300, 400} and adding options, you get

```
In[70]:= Show[  Table[{plot1[ve],endPoint[ve]},{ve,100,400,100}]

             ,PlotRange->All
             ,Frame -> True
             ,FrameLabel->{"position","speed","distance"," "}
             ,RotateLabel->False
             ,GridLines->Automatic ];
```

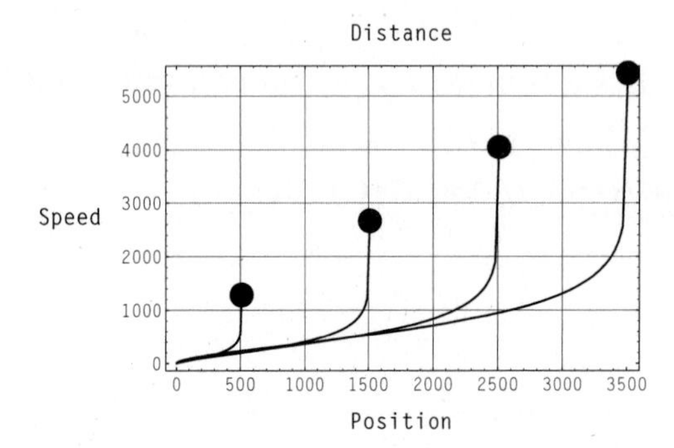

Part (d) Expand **eq3** in a time series and solve for the unknown coefficients of the expansion. Retain terms up to fourth order but use **order** as a variable parameter, so it is easy to increase this. We assume the initial conditions are

```
In[71]:= initial={x[0]->0, x'[0]->0};
         order=4;
```

Moving the right-hand side of **eq3** to the left-hand side and expanding, you get

```
In[72]:= eq4= ( (eq3[[1]] - eq3[[2]]  + O[t]^order)
             /.initial
             //Simplify
            )

Out[72]=
                                            2
           rate ve                      rate  ve       (3)
     (g - ------- + x''[0]) + (-(-----------) + x    [0]) t +
             m0                            2
                                         m0

              3            (4)                     4           (5)
          rate  ve      x    [0]    2          rate  ve      x    [0]   3        4
     (-(----------) + ---------) t  + (-(----------) + ---------) t  + O[t]
               3            2                   4            6
             m0                               m0
```

The solution of the differential equation reduces to algebraic equations if you set the coefficients in front of the powers of **t** to zero. The variables that must be solved for are

```
In[73]:= variables=Table[ D[x[t],{t,n}]   ,{n,0,order+1}] /.{t->0}
Out[73]=
                           (3)      (4)      (5)
{x[0], x'[0], x''[0], x      [0], x    [0], x    [0]}
```

The solution of these equations follow from

```
In[74]:= sol1= Solve[eq4==0,variables][[1]]
Out[74]=
                                                      2
                g m0 - rate ve      (3)          rate   ve
{x''[0] -> -(-----------------), x    [0] -> -----------,
                     m0                             2
                                                  m0

               3                         4
  (4)        2 rate   ve    (5)        6 rate   ve
 x    [0] -> -----------, x    [0] -> -----------}
                  3                         4
                m0                        m0
```

The series expansion solution for **x[t]** is

```
In[75]:= series1= x[t] + O[t]^order /.sol1 /.initial //ExpandAll
Out[75]=
                                 2    3
 -g     rate ve    2    rate   ve t            4
(-- + ---------) t  + ------------- + O[t]
 2      2 m0                  2
                          6 m0
```

You can sum this expansion to get an exact solution by noticing that this series can be written in the form of a sum with terms:

```
In[76]:= term[n_]= ve (rate/m0)^n t^(n+1) /((n+1) n);
```

Simply crosscheck the series:

```
In[77]:= series2= -(1/2) g t^2 + Sum[term[n],{n,1,order-2}];
         series1==series2 //Normal //ExpandAll
Out[77]= True
```

Sum the series to infinity to get the exact solution using the package **Algebra`SymbolicSum`**:

```
In[78]:= exact2=( -(1/2) g t^2 + Sum[term[n],{n,1,Infinity}]
                   //Collect[#,Log[1- rate t/m0]]&
               )
Out[78]=
      2
-(g t )                m0 ve                        rate t
-------- + t ve + (------- - t ve) Log[1 - -------]
   2                   rate                           m0
```

which agrees with the previous solution given by **position**:

```
In[79]:= (( (exact2-position)
             //ExpandAll
             //.logRules
             //Simplify
          ) //.logRules
             //Simplify
         )
Out[79]= 0
```

2.4.2 Projectile Motion in Rotating Reference Frames

Problem 1: Coriolis and Centrifugal Forces

Assume Earth to be a sphere rotating about the z-axis with angular velocity $\vec{\omega}_0$. Its rotation and gravity are assumed to be constant.

Part (a) Find the equations of motion for a projectile as measured by an observer on Earth's surface. Solve the equations by assuming a series expansion in t (time) and compare the coriolis and centrifugal effects.

Part (b) Solve for the exact answer at the pole and plot the trajectory.

Part (c) Solve for the exact answer at the equator and plot the trajectory.

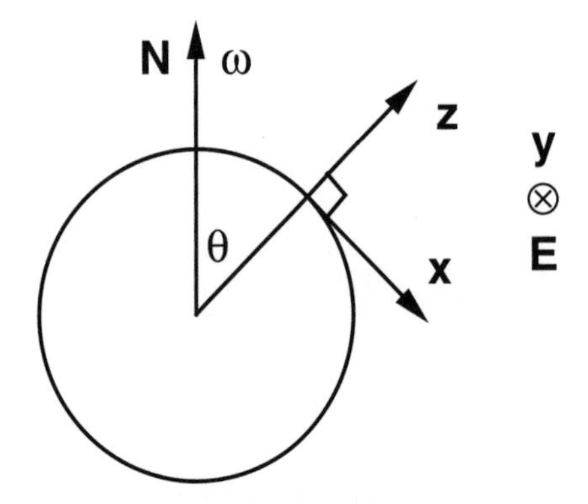

Inertial Reference Frame

Remarks and outline This problem illustrates the use of vector products and the series expansion of a differential equation. The equations of motion follow from Newton's second law as expressed in a rotating coordinate system

$$\vec{F}_{body} = \vec{F}_{inertial} - m(\vec{\omega}' \times \vec{r}) - 2m(\vec{\omega} \times \vec{v}) - m\vec{\omega} \times (\vec{\omega} \times \vec{r}).$$

The general solution can be expressed as a power series in t. The results are then plotted with **ParametricPlot3D**. The equation can be solved with **DSolve** for the special case in which the motion is at the pole or at the equator.

Required packages

```
In[80]:= Needs["Algebra`Trigonometry`"] ;
         Needs["Calculus`VectorAnalysis`"];
         Needs["Graphics`ParametricPlot3D`"];
```

Solution

```
In[81]:= Clear["Global`*"];
```

Part (a) Assume Earth to be a sphere rotating about the z-axis with constant angular velocity **w0**. Choose a coordinate system on Earth with the z-axis along the vertical, the

x-axis pointing south, and the y-axis pointing east. Earth's rotation vector and the force of gravity in this rotating coordinate system are given by

```
In[82]:= w[t_]= {-w0 Sin[theta], 0, w0 Cos[theta]};
         fInertial= {0,0,-m g};
```

where the angle **theta** is the colatitude and $90° -$ **theta** is the latitude. Here, write **w[t]**, since this is in general a function of **t**; for the present problem, **w[t]** is a constant and **w'[t]**=0. The position vector is

```
In[83]:= r[t_]:= {x[t],y[t],z[t]};
```

The force vector ($\vec{F}_{body}$) in the body frame is proportional to **m** times the acceleration:

```
In[84]:= leftHandSide=  m D[r[t],{t,2}];
```

and the force terms on the right-hand side of the equation are

$$(\vec{F}_{body} = \vec{F}_{inertial} - m(\vec{\omega}' \times \vec{r}) - 2m(\vec{\omega} \times \vec{v}) - m\vec{\omega} \times (\vec{\omega} \times \vec{r})).$$

```
In[85]:=  rightHandSide =
            ((
              + fInertial
              -          m CrossProduct[w'[t],r[t]]
              - mCor 2 m CrossProduct[w[t], D[r[t],t] ]
              - mCen   m CrossProduct[w[t],(CrossProduct[w[t],r[t]])]
             )//ExpandAll
            );
```

Note that $\vec{F}_{inertial}$ is $-mg$. An extra factor has been placed in front of the coriolis (**mCor**) and centrifugal (**mCen**) terms in order to identify their effects. The solution follows from setting **mCor** and **mCen** equal to one. The command **CrossProduct** is found in the package **Calculus`VectorAnalysis`**.

The equations of motion follow from equating the left-hand and right-hand sides. Display one of the three equations:

```
In[86]:= eq1=Thread[leftHandSide==rightHandSide];
         eq1[[1]]

Out[86]=
                              2            2
         m x''[t] == m mCen w0  Cos[theta]  x[t] +

                         2
            m mCen w0  Cos[theta] Sin[theta] z[t] +

            2 m mCor w0 Cos[theta] y'[t]
```

Construct a time series solution for arbitrary **theta** by moving all the terms to the left-hand side of the equation and expanding to order **nOrder** in the variable **t**. (*Note:* Use **nOrder**=1 here only to keep the expressions printable. It is trivial to redo the calculation for higher orders.)

```
In[87]:= nOrder=1;
         right=Series[rightHandSide,{t,0,nOrder}]//Normal;
         left =Series[leftHandSide ,{t,0,nOrder}]//Normal

Out[87]=
                               (3)                          (3)
         {m x''[0] + m t x    [0], m y''[0] + m t y    [0],

                                (3)
           m z''[0] + m t z    [0]}
```

Obtain a set of algebraic equations by setting the coefficients in front of the various powers of **t** in **right** and **left** equal to each other.

```
In[88]:= eqs=Thread[
         (CoefficientList[left ,{t}]//Flatten)
         ==
         (CoefficientList[right,{t}]//Flatten)
         ];
```

Use the following initial conditions for the position and velocity:

```
In[89]:= initial={ x[0]==0,     y[0]==0,     z[0]==0,
                   x'[0]==vx0, y'[0]==vy0, z'[0]==vz0};
```

with the following variables:

```
In[90]:= vars=(Table[ D[{x[t],y[t],z[t]},{t,i}],{i,0,nOrder+2}]
                   /.t->0
                   //Flatten
              )

Out[90]= {x[0], y[0], z[0], x'[0], y'[0], z'[0], x''[0], y''[0], z''[0],

           (3)        (3)        (3)
          x   [0], y    [0], z    [0]}
```

Solving the set of equations gives the solution:

```
In[91]:= sol= Solve[Join[eqs,initial],vars][[1]] //Simplify;
```

The general solution for $\{x[t], y[t], z[t]\}$ comes from this solution:

```
In[92]:= genSolution=
         (( {x[t],y[t],z[t]}
                 // Map[Series[#,{t,0,nOrder+2}]&,#]&
                 //Normal
              ) //.sol
                 //Expand
                 //Collect[#,{t}]&
                 //Map[Simplify,#,{2}]&
              )

Out[92]=
                       2
         {t vx0 + mCor t  vy0 w0 Cos[theta] +

                             2   3   2
            ((mCen - 4 mCor ) t  w0  Cos[theta]

               (vx0 Cos[theta] + vz0 Sin[theta])) / 6,

                  3                               2
           t vy0 + (t  w0 (mCen vy0 w0 - 4 mCor  vy0 w0 +

                2 g mCor Sin[theta])) / 6 -

                     2
            mCor t  w0 (vx0 Cos[theta] + vz0 Sin[theta]),

                                      2   3   2
           t vz0 + ((mCen - 4 mCor ) t  w0  Sin[theta]

               (vx0 Cos[theta] + vz0 Sin[theta])) / 6 +
```

```
    2  -g
   t  (-- + mCor vy0 w0 Sin[theta])}
       2
```

It is clear how the coriolis and centrifugal terms affect various powers in **t**.

Part (b) Solve the equations of motion for the exact solutions at the North Pole (**theta**=0) with the initial conditions

```
In[93]:= initial={ x[0]==x0,   y[0]==y0,   z[0]==z0,
                   x'[0]==vx0, y'[0]==vy0, z'[0]==vz0};
```

The equations at this pole are

```
In[94]:= eq2= Join[eq1,initial] /.{theta->0, mCen->1, mCor->1}

Out[94]=                  2
         {m x''[t] == m w0  x[t] + 2 m w0 y'[t],

                            2
           m y''[t] == m w0  y[t] - 2 m w0 x'[t], m z''[t] == -(g m),

           x[0] == x0, y[0] == y0, z[0] == z0, x'[0] == vx0, y'[0] == vy0,

           z'[0] == vz0}
```

and the solution at this pole is

```
In[95]:= dsol2=( DSolve[eq2, {x[t],y[t],z[t]},t][[1]]
                 //ComplexToTrig
                 //Simplify
               )

Out[95]= {x[t] -> t vx0 Cos[t w0] + x0 Cos[t w0] - t w0 y0 Cos[t w0] +

            t vy0 Sin[t w0] + t w0 x0 Sin[t w0] + y0 Sin[t w0],

          y[t] -> t vy0 Cos[t w0] + t w0 x0 Cos[t w0] + y0 Cos[t w0] -

            t vx0 Sin[t w0] - x0 Sin[t w0] + t w0 y0 Sin[t w0],

                          2
                    -(g t )
          z[t] -> -------- + t vz0 + z0}
                      2
```

Plot the motion at the pole for the parameters,

```
In[96]:= values={ g->9.8, w0->1/100,
                  x0->0,  y0->0,  z0->500,
                  vx0->1, vy0->0, vz0->0
                };
```

Use **ParametricPlot3D** to plot the general space curve $\{x, y, z\}$. The solution at the North Pole (**dsol2**) is (output suppressed)

```
In[97]:= p1=ParametricPlot3D[
            Evaluate[{x[t],y[t],z[t]}//.dsol2//.values]
              ,{t,0,10}
              ,BoxRatios->{1,1,1}];
```

Also include two other comparison curves on the same graph. First, the trajectory's projection onto the z=0 plane is (output suppressed)

```
In[98]:=  p2=ParametricPlot3D[
              Evaluate[{x[t],y[t],0}//.dsol2//.values]
                ,{t,0,10}
                ,BoxRatios->{1,1,1}];
```

Second, the trajectory for **w0**=0 is (output suppressed)

```
In[99]:=  p3=ParametricPlot3D[
              Evaluate[{x[t],y[t],z[t]}//.dsol2 /.{w0->0} //.values]
                ,{t,0,10}
                ,BoxRatios->{1,1,1}];
```

Combining these three curves and adding various options, you get

```
In[100]:= Show[ p1,p2,p3
             ,PlotLabel->FontForm["Coriolis Deflection at North-Pole"
                                                  ,{"CourierB",10}]
             ,FaceGrids->{{0,1,0},{-1,0,0}}
             ,BoxRatios->{1,1,1}
             ,ViewPoint->{1,-1,0.7}
             ,AxesLabel->{"x-south","y-east","z-vertical"}];
```

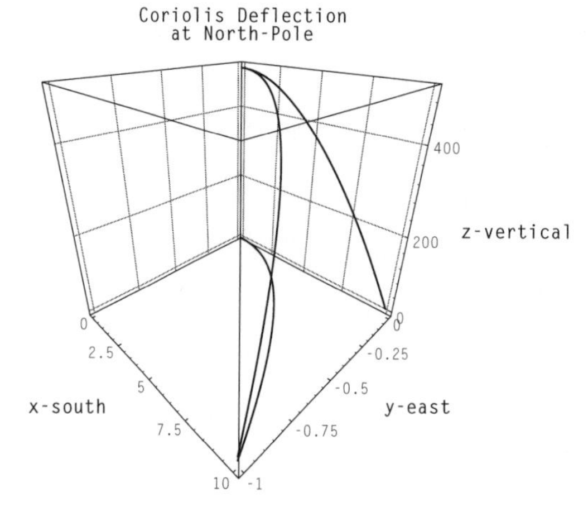

Part (c) Solve the equations of motion for the exact solutions at the equator (**theta**=$\pi/2$) with the same initial conditions as in Part (b).

```
In[101]:= eq3= Join[eq1,initial] /.{theta->Pi/2, mCen->1, mCor->1};
```

The solution at the equator is

```
In[102]:= dsol3=( DSolve[eq3, {x[t],y[t],z[t]},t][[1]]
                    //ComplexToTrig
                    //Simplify
                );
```

Plot the general space curve $\{x, y, z\}$ on the same three-dimensional graph using the set of **values** defined in Part (b). The solution at the equator is (output suppressed)

```
In[103]:= p1=ParametricPlot3D[
              Evaluate[{x[t],y[t],z[t]}//.dsol3//.values]
                ,{t,0,10}
                ,BoxRatios->{1,1,1}];
```

Also include two other comparison curves on the same graph. First, the projection of the trajectory onto the z=0 plane is (output suppressed)

```
In[104]:= p2=ParametricPlot3D[
              Evaluate[{x[t],y[t],0}//.dsol3//.values]
                ,{t,0,10}
                ,BoxRatios->{1,1,1}];
```

Second, the trajectory for **w0**=0 (need to use **Limit** for **w0->0**) is

```
In[105]:= Limit[{x[t],y[t],z[t]}//.dsol3 ,w0->0]
Out[105]=
                                          2
                                    -(g t )
          {t vx0 + x0, t vy0 + y0, --------- + t vz0 + z0}
                                       2
```

The trajectory for **w0**=0 is (output suppressed)

```
In[106]:= p3=ParametricPlot3D[
              Limit[ {x[t],y[t],z[t]}//.dsol3,w0->0] //.values //Evaluate
                ,{t,0,10}
                ,BoxRatios->{1,1,1}];
```

Combining these three curves and adding various options, you get

```
In[107]:= Show[ p1,p2,p3
            ,PlotLabel->FontForm["Coriolis Deflection at Equator"
                                                    ,{"CourierB",10}]
            ,FaceGrids->{{0,-1,0},{-1,0,0},{0,1,0},{0,0,1}}
            ,BoxRatios->{1,1,1}
            ,ViewPoint->{1,0,0}
            ,AxesLabel->{"x-south","y-east","z-vertical"}];
```

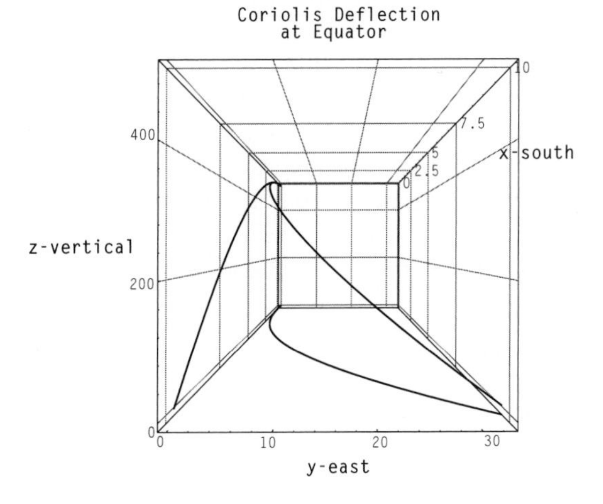

Problem 2: Foucault Pendulum

Assume Earth to be a sphere rotating about the z-axis with angular velocity ω_0. Its rotation and gravity are assumed to be constant. Choose a coordinate system with the z-axis along the vertical, the x-axis pointing south, and the y-axis pointing east.

Part (a) Find the equations of motion for a pendulum as measured by an observer on Earth's surface. Simplify the equations by assuming (1) the motion is confined to the horizontal plane, (2) ω_0^2 terms are neglected, and (3) $\text{Sin}[\theta]$ terms are neglected (small oscillations).

Part (b) Solve the equations of motion for $x[t]$ and $y[t]$ for arbitrary initial conditions. Evaluate the solution in Part (a) with the initial conditions $x[0] = 0$, $x'[0] = 0$, $y[0] = y0$, and $y'[0] = 0$ and with the approximation $\omega_0 \text{Cos}[\theta] < \sqrt{g/L}$.

Part (c) Plot the solution found in Part (b) over one period $\tau = 2\pi/\omega_0$.

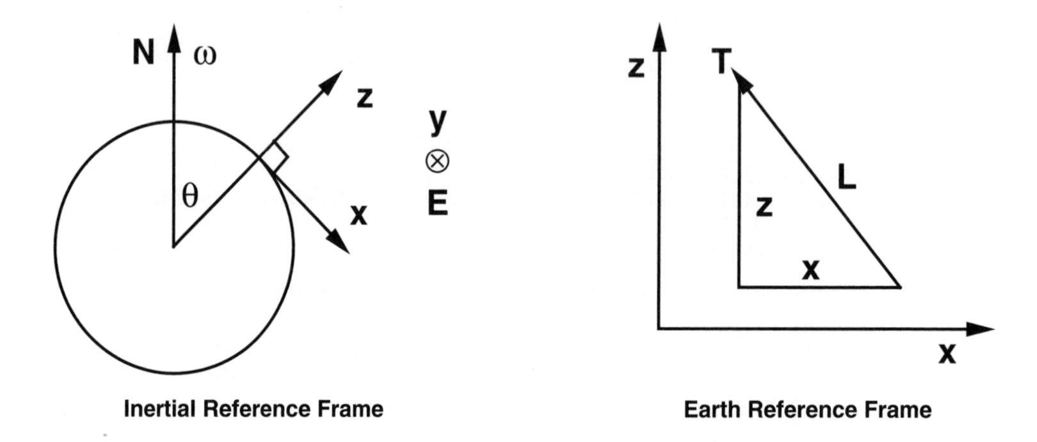

Remarks and outline These equations are complicated and must be simplified by assuming small oscillations. The solution of the resulting equations are solved by reducing the coupled equations to one complex equation. **DSolve** can be used to evaluate the complex equation and to get a simple analytic expression for the x and y variables. The x and y variables are then plotted in Part (c). It is useful to note that in the figure, the components of the tension for a pendulum of length L are given by $T_x = (x/L)T$, $T_y = (y/L)T$, $T_z = (1 - z/L)T$ in Earth's reference frame (noninertial).

The graph of the solution done in Part (c) also follows from the original equations of motion without any simplifications if the command **NDSolve** is used to get a numerical answer. Try it.

Required packages

```
In[108]:= Needs["Algebra`Trigonometry`"] ;
          Needs["Calculus`VectorAnalysis`"];
```

Solution

```
In[109]:= Clear["Global`*"];
```

Part (a) Earth's rotational vector is

```
In[110]:= w[t_] = {-w0 Sin[theta],0, w0 Cos[theta]};
```

where the angle **theta** is the colatitude and 90° – **theta** is the latitude. In general, **w[t]** can depend on time; here it is taken to be constant. The force of gravity and the force of tension due to the string are, respectively,

```
In[111]:= force1 = {0,0,-m g};
          force2 = {-tension x[t]/L,
                    -tension y[t]/L,
                    -tension z[t]/L + tension};
          force=force1+force2;
```

The coordinates of the pendulum bob in Earth's reference frame are

```
In[112]:= r[t_]:= {x[t],y[t],z[t]};
```

The equations of motion in the Earth's reference frame follow from

$$\vec{F}_{body} = \vec{F}_{inertial} - m(\vec{\omega}' \times \vec{r}) - 2m(\vec{\omega} \times \vec{v}) - m\vec{\omega} \times (\vec{\omega} \times \vec{r}).$$

From Newton's second law, the left-hand side is just $ma[t] = mr''[t]$:

```
In[113]:=  leftHandSide=  m D[r[t],{t,2}];
```

while the right-hand side is given by

```
In[114]:=  rightHandSide =
             (( + force
              -     m CrossProduct[w'[t],r[t]]
              -  2 m CrossProduct[w[t], D[r[t],t] ]
              -     m CrossProduct[w[t],(CrossProduct[w[t],r[t]])]
                                            )//ExpandAll  );
```

The equations of motion are then

```
In[115]:= eq1=Thread[rightHandSide==leftHandSide];
          Short[eq1,2]

Out[115]=     tension x[t]            2              2
          {-(-----------------) + m w0   Cos[theta]   x[t] + <<1>> +
                    L

              2 m w0 Cos[theta] y'[t] == m x''[t], <<2>>}
```

The assumption that the motion is confined to the horizonal plane means

```
In[116]:= constraint1=Thread[Table[D[z[t],{t,n}],{n,0,2}]->0]

Out[116]= {z[t] -> 0, z'[t] -> 0, z''[t] -> 0}
```

Also assume **w0^2->0** and **Sin[theta]->0**:

```
In[117]:= constraint2={w0^2->0,Sin[theta]->0};
```

The complete set of constraints is

```
In[118]:= constraint=Join[constraint1,constraint2];
```

Applying **constraint** to **eq1**, the equations of motion become

```
In[119]:= eq2=eq1 /.constraint //Simplify

Out[119]=    tension x[t]
          {-(------------) + 2 m w0 Cos[theta] y'[t] == m x''[t],
                  L

              tension y[t]
            -(------------) - 2 m w0 Cos[theta] x'[t] == m y''[t],
                   L

            -(g m) + tension == 0}
```

Eliminating **tension** from **eq2**, you get

```
In[120]:= {eq3x,eq3y}= ( Solve[eq2,{x''[t],y''[t]},tension]
                    /.Rule->Equal
                    //Flatten )

Out[120]=            -(g m x[t]) + 2 L m w0 Cos[theta] y'[t]
          {x''[t] == ---------------------------------------,
                                    L m

                       -(g m y[t]) - 2 L m w0 Cos[theta] x'[t]
            y''[t] == ---------------------------------------}
                                     L m
```

where **{eq3x,eq3y}** has been used to identify the two equations. Note that these equations are coupled in **x** and **y**.

Part (b) These equations can be treated as one complex equation if you multiply the second equation (**eq3y**) by **I**:

```
In[121]:= eq4y= I eq3y //Thread[#,Equal]&

Out[121]=              I (-(g m y[t]) - 2 L m w0 Cos[theta] x'[t])
          I y''[t] == -------------------------------------------
                                        L m
```

and add it to the first equation **eq3x**:

```
In[122]:= eq5= eq3x+eq4y //Thread[#,Equal]&

Out[122]=                        I (-(g m y[t]) - 2 L m w0 Cos[theta] x'[t])
          x''[t] + I y''[t] == ------------------------------------------- +
                                               L m

            -(g m x[t]) + 2 L m w0 Cos[theta] y'[t]
            ---------------------------------------
                            L m
```

eq5 reduces to one complex equation for the variable **u=x+I y**:

```
In[123]:= eq6=  eq5 /.{x->(( u[#]-I y[#])&)} //ExpandAll

Out[123]=            g u[t]
          u''[t] == -(------) - 2 I w0 Cos[theta] u'[t]
                       L
```

Using initial conditions

```
In[124]:= initial={u[0]==I amp, u'[0]==0};
```

the solution for **eq6** follows from applying **DSolve**:

```
In[125]:= dsol=( DSolve[Join[{eq6},initial],{u[t]},t][[1]]
                //.constraint
                //ExpandAll
                //ComplexToTrig
                //Simplify
            )

Out[125]=
                                  (Sqrt[-4 g L] t)/L
          {u[t] -> (2 I g + 2 I E                     g -

               Sqrt[-4 g L] w0 Cos[theta] +

                (Sqrt[-4 g L] t)/L
               E                    Sqrt[-4 g L] w0 Cos[theta])

                                                 I
                                                 - amp Sin[t w0 Cos[theta]]
               amp Cos[t w0 Cos[theta]]          4
             (------------------------------ - -----------------------------)}
                (Sqrt[-4 g L] t)/(2 L)          (Sqrt[-4 g L] t)/(2 L)
             4 E                        g     E                        g
```

Part (c) Examine the precession at three angles of **theta**: $\theta = 0$ for the North Pole, $\theta = \pi/2$ for the Equator, and $\theta = \pi$ for the South Pole. Choose the following parameters:

```
In[126]:= values={ g->9.8, w0->1/10, L->1, amp->1};

In[127]:= ParametricPlot[
            Evaluate[{Re[u[t]],Im[u[t]]}
                  /.dsol //.values/.{theta->0}]
            ,{t, 0,10}
            ,PlotLabel->"North Pole"
            ,AxesLabel->{"x-axis, south","y-axis, east"}];
```

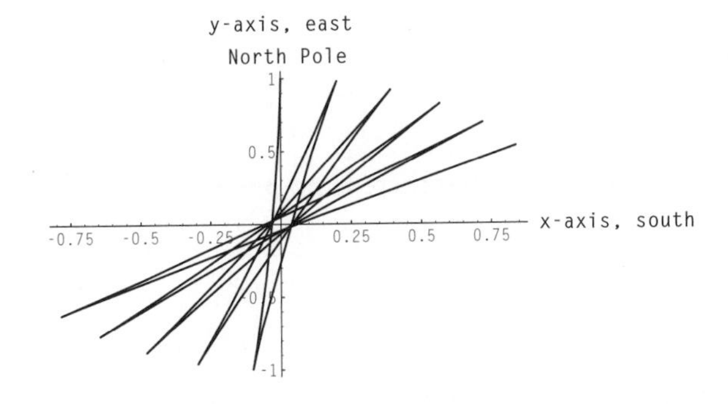

At the North Pole, set **theta->0**. Note that the precession is in a clockwise direction:

```
In[128]:= ParametricPlot[
    Evaluate[{Re[u[t]],Im[u[t]]}
          /.dsol //.values/.{theta->Pi}]
    ,{t, 0,10}
    ,PlotLabel->"South Pole"
    ,AxesLabel->{"x-axis, south","y-axis, east"}];
```

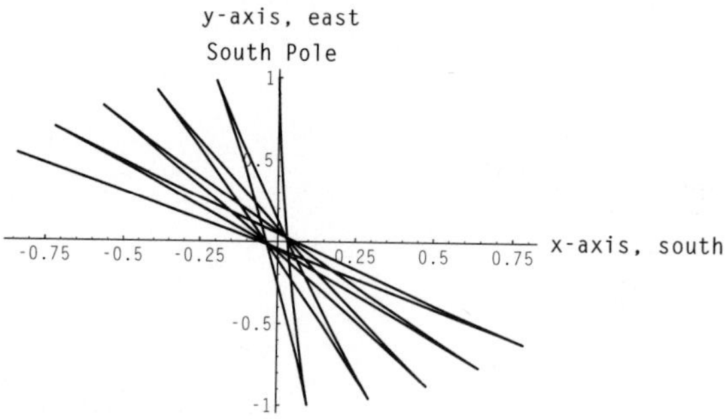

At the South Pole, set **theta->Pi**. Note that the precession is in a counter-clockwise direction:

```
In[129]:= ParametricPlot[
    Evaluate[{Re[u[t]],Im[u[t]]}
          /.dsol //.values/.{theta->Pi/2}]
    ,{t, 0,10}
    ,PlotLabel->"Equator"
    ,AxesLabel->{"x-axis, south","y-axis, east"}
    ,Axes->False];
```

At the Equator, set **theta->Pi/2**. Note that there is no precession.

2.4.3 Electricity and Magnetism

Problem 1: Charged Disk

A disk of radius R is uniformly charged with a surface charge given by density $\sigma = Q/(\pi R^2)$, where Q is the total charge.

Part (a) Find the potential and electric field along the axis of the disk.

Part (**b**) Take the limit $R \to \infty$ to get the field for a uniformly charged plane.

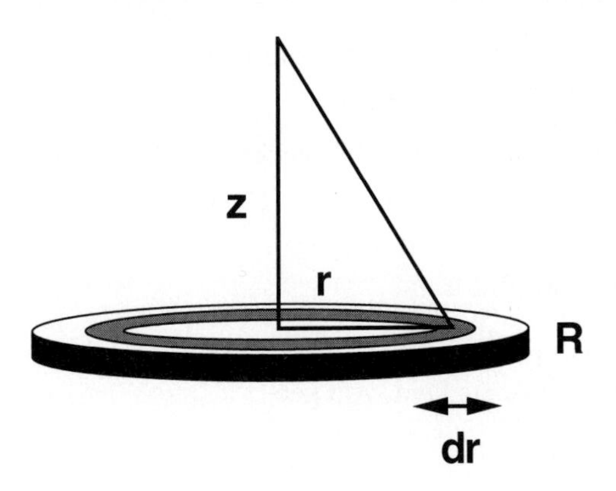

Remarks and outline This simple problem illustrates the use of the command **Grad** found in the package **Calculus`VectorAnalysis`** and also uses the command **Limit**.

Required packages

```
In[130]:= Needs["Calculus`VectorAnalysis`"];
```

Solution

```
In[131]:= Clear["Global`*"];
```

Part (**a**) Consider an infinitesimal, charged ring of thickness **dr**, charge **dq**, and radius **r** centered around the origin of the disk. The potential at position **z** along the z-axis produced by this ring is

```
In[132]:= dV = dq/Sqrt[(z^2+r^2)];
```

The charge on the ring is related to the surface charge by

```
In[133]:= dq = 2 Pi r dr density;
```

where **density** is the surface charge density. Because the distance from the ring to the field point on the z-axis is the same for all points on the ring, the potential from the differential ring is

```
In[134]:= dV
Out[134]= 2 density dr Pi r
          ------------------
                 2    2
            Sqrt[r  + z ]
```

To get the total potential on the axis of the disk, integrate **dV/dr** from **r**=0 to **R**:

```
In[135]:= diskPotential= ( Integrate[dV/dr,{r,0,R}]
                            //PowerExpand
                            //Simplify  )
```

```
Out[135]=
                                  2    2
          2 density Pi (-z + Sqrt[R  + z ])
```

The electric field on the axis is obtained from the gradient of the potential (**Grad** is found in the package **Calculus`VectorAnalysis`**):

```
In[136]:= Edisk= -Grad[diskPotential,Cartesian[x,y,z]] //PowerExpand //Simplify

Out[136]=
                                             z
          {0, 0, -2 density Pi (-1 + ---------------)}
                                             2    2
                                      Sqrt[R  + z ]
```

Part (b) To get the electric field for a plane, let the radius of the disk go to infinity. Taking the limit of the z-component of the electric field, you get

```
In[137]:= Eplane =  Limit[Edisk,R->Infinity]

Out[137]= {0, 0, 2 density Pi}
```

It follows from the symmetry of a plane that the electric field will have only a z-component.

Problem 2: Uniformly Charged Sphere

Consider a sphere of uniform charge density and radius R.

Part (a) Find the electric field for $r > R$ and $r < R$ and plot the results.

Part (b) Find the potential for $r > R$ and $r < R$ and plot the results.

Part (c) Make a two-dimensional plot of the equipotential surfaces and the electric field on the same graph.

Part (d) Make a three-dimensional plot of the potential surfaces and electric field.

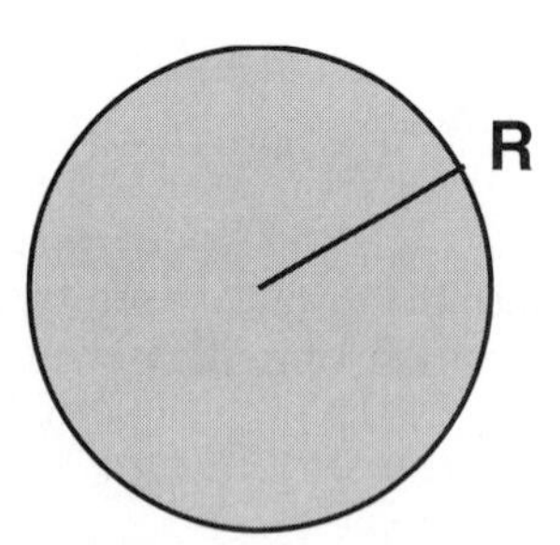

Remarks and outline This problem illustrates the use of discontinuous functions and the combination of different kinds of graphics. The electric field has only a radial component and its value follows from Gauss's law. The potential follows from integrating the electric field from $r = \infty$ to the field point. The various kinds of graphics are combined with the command **Show**.

Required packages

```
In[138]:= Needs["Graphics`ParametricPlot3D`"];
          Needs["Graphics`PlotField`"];
          Needs["Graphics`PlotField3D`"];
```

Solution

```
In[139]:= Clear["Global`*"];
```

Part (a) By symmetry, the only component of the electric field is the radial component. Gauss's law simplifies here to $EA = Q_{enclosed}$, where A is the surface area of the Gaussian surface. When applied to a spherical surface of radius **r**, this gives two equations, depending on whether **r<R** or **r>R**:

```
In[140]:= eq1 = { 4 Pi r^2 Eoutside == totalCharge    (*r>R*),
                  4 Pi r^2 Einside  == enclosedCharge (*r<R*)};
```

We use the notation **(*...*)** to embed a comment into the equation. The total charge and the charge enclosed by the Gaussian sphere are

```
In[141]:= totalCharge    = 4 Pi R^3/3 density;
          enclosedCharge = 4 Pi r^3/3 density;
```

where **density** is the charge volume density of the sphere. Solving **eq1** for the radial component of the electric field, you get

```
In[142]:= sol= Solve[eq1,{Einside,Eoutside}][[1]]
Out[142]=
                                                  3
                density r                density R
          {Einside -> ---------, Eoutside -> ----------}
                    3                          2
                                            3 r
```

where **Eoutside** is the electric field for **r>R** and **Einside** is the electric field for **r<R**. The general radial electric field for all **r** is

```
In[143]:= efield[rr_,RR_]:=(Eoutside/.sol/.{r->rr,R->RR}) /; rr > RR;
          efield[rr_,RR_]:=(Einside /.sol/.{r->rr,R->RR}) /; rr <=RR;
```

Two notes on *Mathematica* here. First, use the dummy variables **rr** and **RR** as arguments of the **efield** function. The values of **r** and **R** are then replaced by the input values of **rr** and **RR**, respectively. Second, use the notation **/;rr>RR** to restrict the first definition of **efield** to the case in which **rr>RR**; the second definition will apply only when **rr<=RR**. This is a useful method to define a function in a piecewise manner.

The plot of the radial component of the electric field (for **density**=1) follows from (output suppressed)

```
In[144]:= p1=Plot[efield[r,1]/.density->1,{r,10^-6,3}];
```

For those interested in more sophisticated plots, shade the region inside and outside the sphere with

```
In[145]:= medium1={GrayLevel[.7],Polygon[{{1,.51},{1,0},{3,0},{3,.51}}]};
          medium2={GrayLevel[.5],Polygon[{{0,.51},{0,0},{1,0},{1,.51}}]};
```

Putting the plots together, you have

```
In[146]:= Show[ {Graphics[{medium1,medium2}]
                 ,Graphics[{ Text["Inside ",{.5,.3}],
                             Text["Outside",{2,.3}] }]
                 ,p1
                 } ,AxesLabel-> {"Distance ", "Electric Field"}
                   ,Axes->True
              ];
```

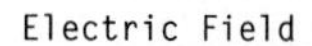
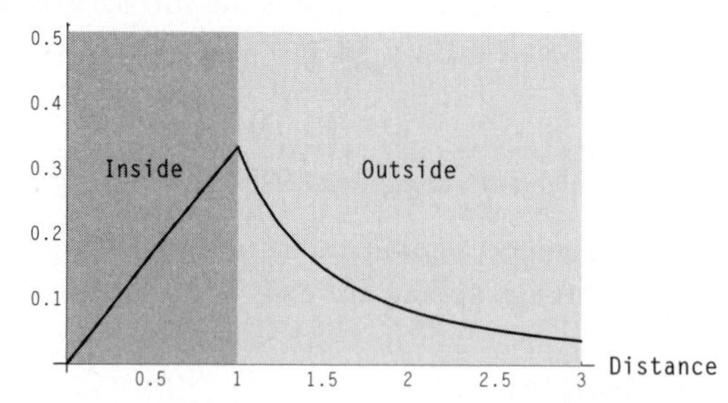

Note that the plot **p1** is listed last so that it is shown on top of the shaded polygons.

Part (b) The potential follows from integrating the radial component of the electric field from infinity to **r**: $V[r] = -\int_{\infty}^{r} \vec{E}_{outside} \cdot d\vec{r}$. For **r** >**R**, you get

```
In[147]:= potoutside[r_,R_]=
          (  -Integrate[ Eoutside/.sol, {r,rMax,r}]
                // Limit[#,rMax->Infinity]&
          )

Out[147]=            3
          density R
          -----------
             3 r
```

For the case **r** <**R**, the region of integration must be done in two parts. The first part requires an integration of **Eoutside** from $r = [\infty, R]$. This term is given by **potoutside[R,R]**. The second term is an integration of **Einside** from $r = [R, r]$. The potential for **r**<**R** is a combination of these two terms and it follows that $V[r] = V[R] - \int_{R}^{r} \vec{E}_{inside} \cdot d\vec{r}$. Therefore

```
In[148]:= potinside[r_,R_]=
          (potoutside[R,R]  - Integrate[Einside/.sol,{r,R,r}]
             )//Simplify

Out[148]=                2      2
          density (-r  + 3 R )
          --------------------
                   6
```

The potential for all **r** is

```
In[149]:= potential[r_,R_]:= If[r>R,potoutside[r,R]
                                  ,potinside[ r,R]];
```

Here, use the **If** construction to define potential in a piecewise manner. The plot of the potential (**density**=1, **R**=1) is (output suppressed)

```
In[150]:= p2=Plot[potential[r,1]/.density->1,{r,10^-6,3}];
```

Overlaying the plot with the shaded **Polygons**, you have

```
In[151]:= Show[ {Graphics[{medium1,medium2}]
          ,Graphics[{ Text["Inside ",{.5,.3}],
                      Text["Outside",{2,.3}] }]
          ,p2
          } ,AxesLabel-> {"Distance ", "Potential Field"}
            ,Axes->True
       ];
```

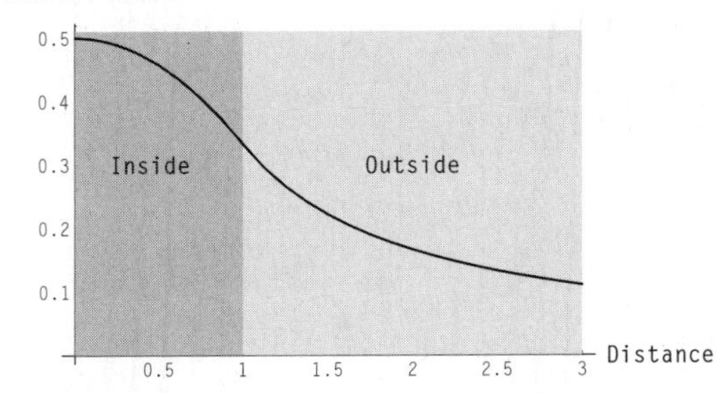

Part (c) We use **ContourPlot** and **PlotGradientField** to make a two-dimensional plot of the equipotential surfaces and the electric vector field. The potential in Cartesian coordinates is

```
In[152]:= Clear[pot];
          pot[x_,y_,z_:0,R_:1,density_:1]:=
           If[(x^2+y^2+z^2)<R^2
                 ,(density (-(x^2+y^2+z^2) + 3 R^2))/6
                 ,(density R^3)/(3 Sqrt[x^2+y^2+z^2])
             ]
```

An **If** statement has been used to choose between the inside and outside solution found in Part (b). In Cartesian coordinates, we have replaced $r = \sqrt{x^2+y^2+z^2}$. The default parameters are **z** =0, **R** =1, and **density**=1. The graphics for the equipotential surfaces follow from the function **potPlot**:

```
In[153]:= potPlot=
          ContourPlot[ pot[x,y] ,{x,-2,2},{y,-2,2}
                   ,ContourShading->False
                   ,ContourSmoothing->True];
```

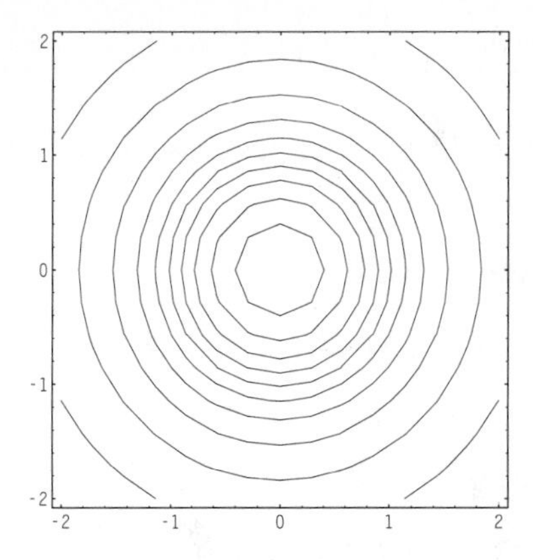

The graphics for the electric field follow from **ePlot** (**PlotGradientField** is found in the package **Graphics`PlotField`**):

```
In[154]:= ePlot=
          PlotGradientField[-pot[x,y] ,{x,-2,2},{y,-2,2}];
```

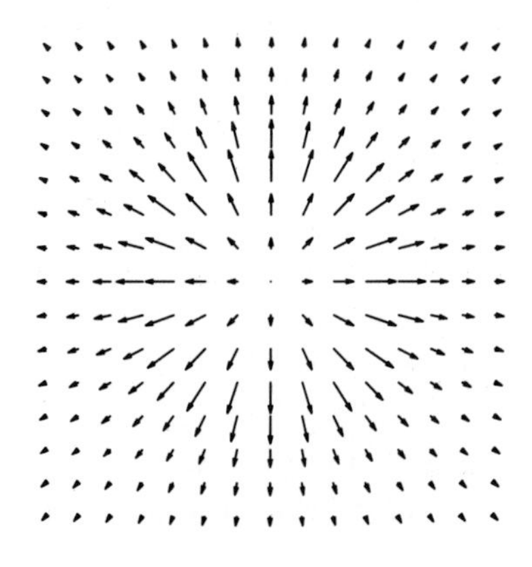

Combining the electric field and equipotential surfaces, you get

```
In[155]:= Show[potPlot,ePlot];
```

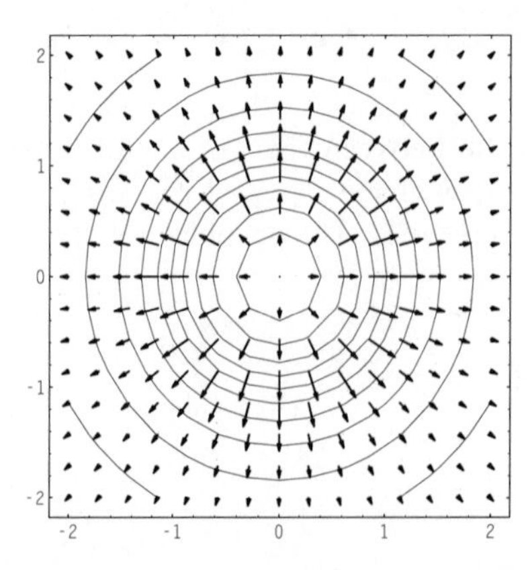

Part (**d**) The three-dimensional electric field follows from

```
In[156]:= ePlot3D=
          PlotGradientField3D[-pot[x,y,z] ,{x,0,2},{y,0,2},{z,0,2}
                ,VectorHeads->True];
```

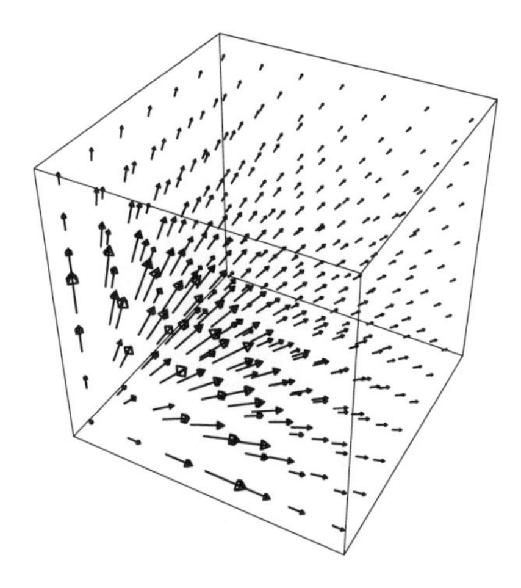

The equipotential surfaces are calculated in two steps. First, define radial position vector:

```
In[157]:= point[r_,R_:1]:= {r Sin[theta] Cos[phi]
                           ,r Sin[theta] Sin[phi]
                           ,r Cos[theta]}
```

The equipotential surfaces are formed by rotating the position vectors to form spherical surfaces:

```
In[158]:= potPlot3D[r_]:=
          ParametricPlot3D[  Evaluate[ point[r] ]
                      ,{theta,0,Pi/2},{phi,0, Pi/2}
                      ,PlotPoints->12  ]
```

Set the equipotential surfaces at r $= \{0.5, 1.0, 1.5, 2.0\}$. Displaying the superposition of the electric field and equipotential surfaces, you get

```
In[159]:= Show[ePlot3D
            ,Table[potPlot3D[r],{r,1/2,2,1/2}]
            ,PlotRange->{0,1} ];
```

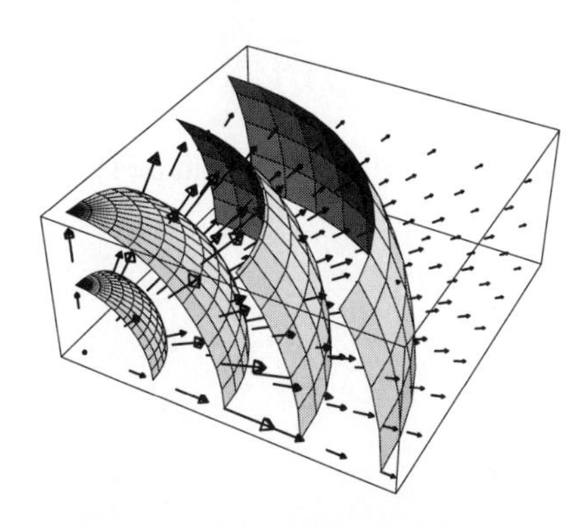

Problem 3: Electric Dipole

Consider a positive and a negative charge of the same magnitude separated by a distance d.

Part (a) Find the potential and plot the equipotential lines in the $y = 0$ plane. Make a three-dimensional plot of the electric field.

Part (b) Express the potential in spherical coordinates and expand in d/r, keeping only the leading-order term. Calculate the electric field.

Part (c) Animate the electric field of a rotating dipole.

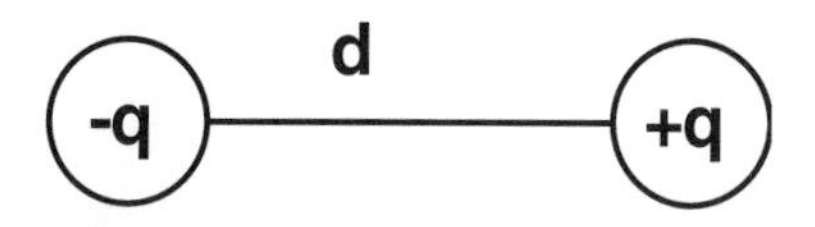

Remarks and outline This problem illustrates the use of commands found in the packages **Calculus`VectorAnalysis`**, **Graphics`PlotField`**, and **Graphics`PlotField3D`**. The dipole potential is the potential produced by two point charges and is calculated from the user-defined function **monopole**. The electric field follows from the negative of the gradient of the potential, $E = -\nabla V$.

Required packages

```
In[160]:= Needs["Calculus`VectorAnalysis`"];
          Needs["Graphics`PlotField`"];
          Needs["Graphics`PlotField3D`"];
```

Solution

```
In[161]:= Clear["Global`*"];
```

Part (a) The coulomb potential for a point charge is given by the user-defined function **monopole**. A dipole is formed from the superposition of two monopoles. Assume a positive charge is located along the positive z-axis at a distance **d/2** from the origin and a negative charge is located along the negative z-axis at a distance **d/2** from the origin. The superposition of these two charges gives the dipole potential:

```
In[162]:= dipole= ( monopole[-q,{0,0,-d/2}]
                   + monopole[+q,{0,0,+d/2}] )

Out[162]=
                    q                                    q
          -(----------------------------) + ----------------------------
                   2    2   -d      2                2    2    d      2
           Sqrt[x  + y  + (-- - z) ]         Sqrt[x  + y  + (- - z) ]
                             2                                2
```

The plot of the equipotential lines in the plane **y**=0 (for **q**=1 and **d**=1) is

```
In[163]:= ContourPlot[Evaluate[dipole/.{q->1,d->1,y->0}]
              ,{x,-2,2},{z,-2,2}
              ,PlotPoints->45
              ,Contours ->11
              ,Axes->True
              ,AxesLabel ->{"x-axis","z-axis"}];
```

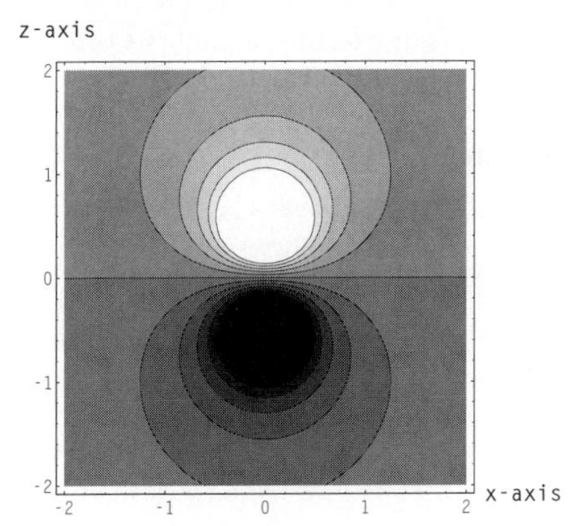

The plot of the electric field follows from **PlotGradientField3D** (which uses the package **Graphics`PlotField3D`**):

```
In[164]:= PlotGradientField3D[ Evaluate[ -dipole/.{q->1, d->1}]
            ,{x,-0.5,0.5},{y,-0.5,0.5},{z,-1.0,1.0}
            ,PlotPoints->6
            ,VectorHeads->True
            ,Graphics`PlotField3D`ScaleFunction->(1&)];
```

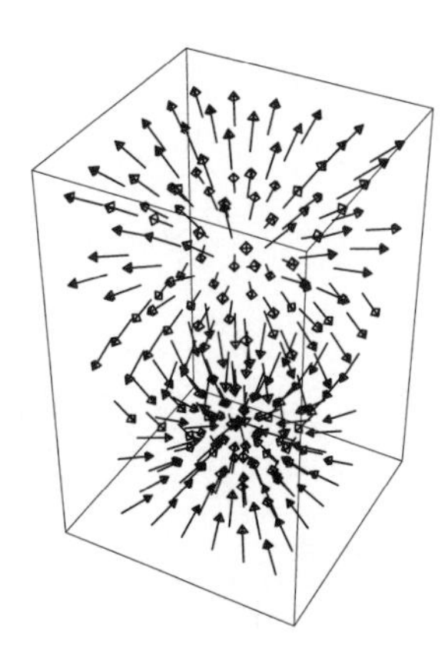

Note that **Graphics`PlotField3D`ScaleFunction->(1&)** is used. This scales the length of all vectors by the function **(1&)**; that is, they will all have length one. This is useful when you are near singularities. Otherwise, most of the vectors will appear to have zero length, except for those near the singularity. Also note that we specify the complete context for **Graphics`PlotField3D`ScaleFunction**. We do this in case the variable **Graphics`PlotField3D`ScaleFunction** is shadowed by **Graphics`PlotField`ScaleFunction**, which may have been defined previously. If you suspect a problem with shadowing, you can use the **Context** command to see which context is implied for the specific variable.

Part (b) To express the potential in **Spherical** coordinates, define the spherical position vector (**CoordinatesToCartesian** command found in the package **Calculus`VectorAnalysis**

```
In[165]:= x2rRule=
          Thread[{x,y,z}->
            CoordinatesToCartesian[{r,theta,phi},Spherical]]

Out[165]= {x -> r Cos[phi] Sin[theta], y -> r Sin[phi] Sin[theta],

            z -> r Cos[theta]}
```

It follows from **dipole** that the potential in spherical coordinates is

```
In[166]:= potential=dipole /.x2rRule //Simplify

Out[166]=               q                                  q
          ---------------------------------- - ----------------------------------
                  2                                    2
                 d     2                              d     2
          Sqrt[-- + r  - d r Cos[theta]]       Sqrt[-- + r  + d r Cos[theta]]
                 4                                    4
```

Expanding to lowest order in $1/r \sim 0$ (or equivalently, $r \sim \infty$), the potential becomes

```
In[167]:= dipoler= (Series[potential,{r,Infinity,2}]
                    ) //Normal //Simplify

Out[167]= d q Cos[theta]
          --------------
                2
               r
```

The electric field follows from the negative of the gradient of the potential, $E = -\nabla V$:

```
In[168]:= electricfield= -Grad[dipoler,Spherical[r,theta,phi] ]

Out[168]=  2 d q Cos[theta]  d q Sin[theta]
          {----------------, --------------, 0}
                  3                3
                 r                r
```

Part (c) To animate a rotating dipole, locate the charges in the $\{x, z\}$ plane and rotate the dipole around the y-axis. The expression for a dipole with unit charge and separation one in the $\{x, z\}$ plane is

```
In[169]:= rotdipole=( monopole[+1, 1/2{+Sin[theta],0,+Cos[theta]}]
                    + monopole[-1, 1/2{-Sin[theta],0,-Cos[theta]}] );
```

The animation follows from a table of dipoles that have different values of **theta**. (The details of the animation depend on your specific computer interface.) Only one frame is displayed here.

```
In[170]:= frames=2;
          Do[ PlotGradientField[Evaluate[-rotdipole/.{y->0}]
                  ,{x,-2,2},{z,-2, 2}
                  ,PlotPoints->14
                  ,VectorHeads->True
                  ,Graphics`PlotField`ScaleFunction->(1&)]
          ,{theta,2 Pi/frames,2 Pi,2 Pi/frames}]
```

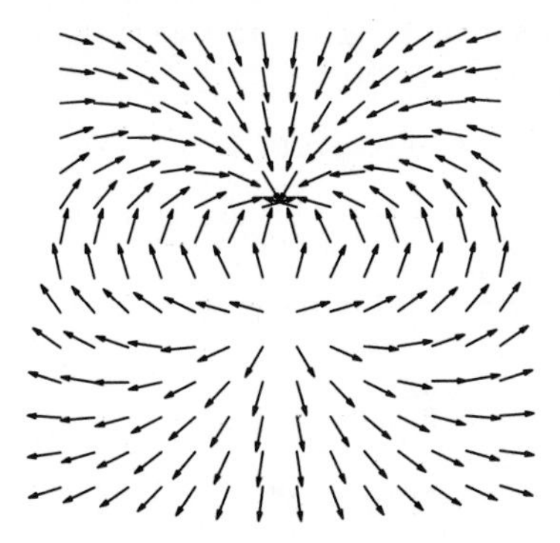

Problem 4: Magnetic Vector Potential for a Linear Current

Consider a long straight wire of length $2L$ and with a current I_0.

Part (a) Determine the magnetic vector potential along a line perpendicular to the midpoint of the wire. Assume the L is sufficiently large so that end effects can be neglected.

Part (b) Calculate the B field from $\vec{B} = \vec{\nabla} \times \vec{A}$ and take the limit as L goes to infinity.

Part (c) Plot the B field of the infinite wire in two and three dimensions.

Part (d) Calculate the B field of an infinite wire in cylindrical coordinates.

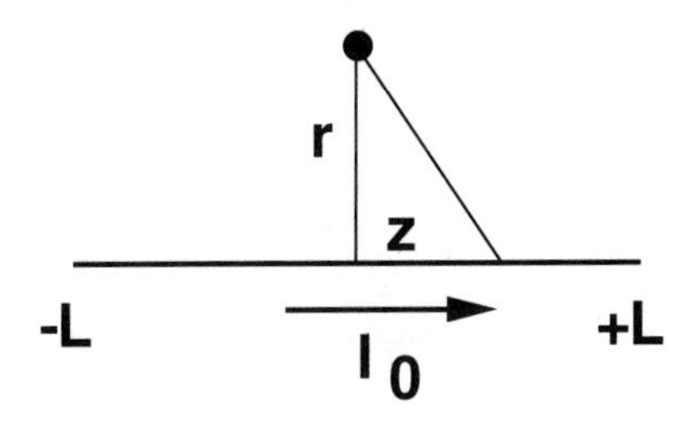

Remarks and outline This problem illustrates the use of command **Curl** in both Cartesian and cylindrical coordinates. The **Curl** command is found in the package **Calculus'VectorAnalysis'**. The vector potential follows from integrating over the length of the wire. The curl of the vector potential gives the **B** field.

Required packages

```
In[171]:= Needs["Calculus'VectorAnalysis'"];
          Needs["Graphics'PlotField'"];
          Needs["Graphics'PlotField3D'"];
```

Solution

```
In[172]:= Clear["Global'*"];
```

Part (a) Choose the origin of the coordinates at the midpoint of the wire and let the z-axis be along the direction of the wire. It follows from symmetry that the vector potential along the line perpendicular to the midpoint of the wire has only a z-component. The

z-component of the vector potential is obtained at a distance **r** from integrating over the current from **-L** to **L**, $A_z = \int_{-L}^{+L} dz\, I_0/\sqrt{z^2+r^2}$:

```
In[173]:= Az[r_]= Integrate[I0/Sqrt[z^2+r^2],{z,-L,L}] //Simplify
```

```
Out[173]=
                            2    2                          2    2
          I0 (-Log[-L + Sqrt[L  + r ]] + Log[L + Sqrt[L  + r ]])
```

The magnetic vector potential in **Cartesian** coordinates $\{x, y, z\}$ is

```
In[174]:= A[x_,y_,z_]= {0,0,Az[Sqrt[x^2+y^2]]}
```

```
Out[174]=
                                     2    2    2
          {0, 0, I0 (-Log[-L + Sqrt[L  + x  + y ]] +

                            2    2    2
              Log[L + Sqrt[L  + x  + y ]])}
```

Part (b) The magnetic field follows from taking $\vec{\nabla} \times \vec{A}$ (from the package **Calculus`VectorAnalysis`**):

```
In[175]:= B[x_,y_,z_]= (  Curl[A[x,y,z],Cartesian[x,y,z]]
                          //Simplify  )
```

```
Out[175]=
                      -2 I0 L y                          2 I0 L x
          {-------------------------------, -------------------------------, 0}
             2    2          2    2    2       2    2          2    2    2
           (x  + y ) Sqrt[L  + x  + y ]     (x  + y ) Sqrt[L  + x  + y ]
```

The **B** field in the limit **L**$\to \infty$ becomes

```
In[176]:= BLimit[x_,y_,z_]= Limit[B[x,y,z],L->Infinity]
```

```
Out[176]=  -2 I0 y   2 I0 x
          {-------, -------, 0}
            2    2   2    2
           x  + y   x  + y
```

Part (c) For **I0=1**, the three-dimensional plot of the magnetic field is

```
In[177]:= p1=PlotVectorField3D[BLimit[x,y,z]/.I0->1
              ,{x,-1,1},{y,-1,1},{z,-1,1}
              ,PlotPoints->6
              ,VectorHeads->True
              ,ViewPoint->{1.4, -1.4, 3}
              ,Graphics`PlotField3D`ScaleFunction->(1&)];
```

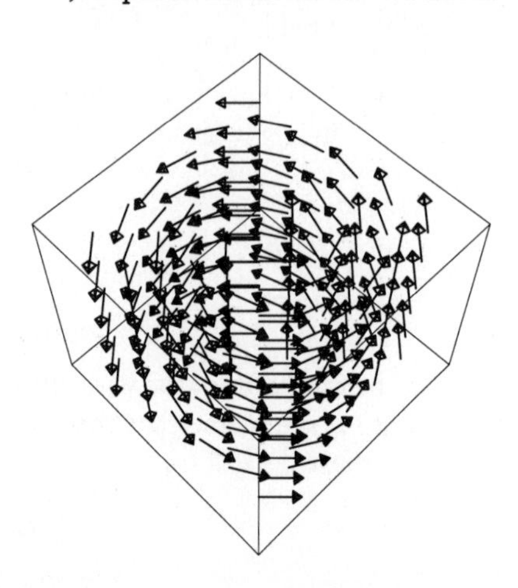

From a different **ViewPoint**, you can more clearly see the pattern of this vector field.

```
In[178]:= Show[p1, ViewPoint->{2,2,16}];
```

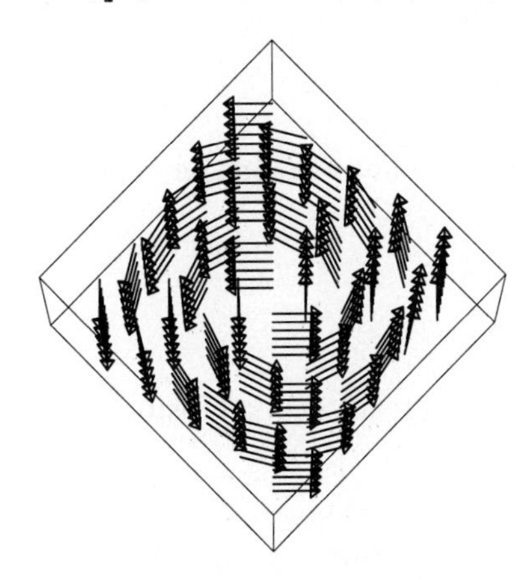

The two-dimensional plot of the magnetic field in the $\{x, y\}$ plane is obtained if you **Drop** the z-component:

```
In[179]:= PlotVectorField[Drop[BLimit[x,y,z],-1]/.I0->1
              ,{x,-3,3},{y,-3,3}
              ,PlotPoints->20
              ,VectorHeads->True
              ,Graphics`PlotField`ScaleFunction->(1&)];
```

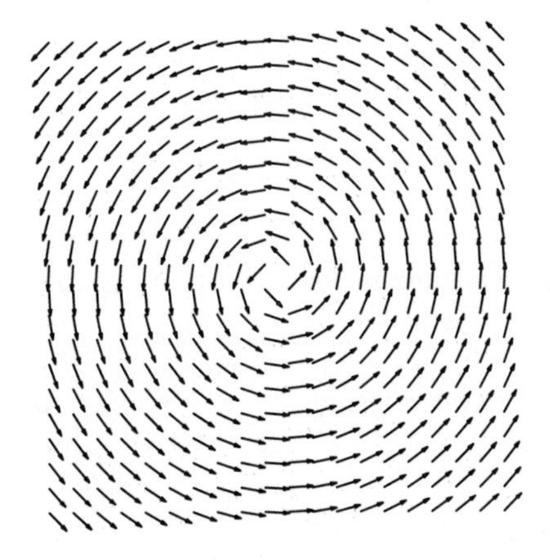

Part (d) The magnetic field in cylindrical coordinates follows from taking the **Curl** of the vector potential in **Cylindrical** coordinates. The vector potential expressed in cylindrical coordinates $\{r, theta, z\}$ is

```
In[180]:= Acylind[r_]= {0,0, Az[r]};
```

Taking the **Curl** of **Acylind** in cylindrical coordinates, and taking the limit as $\text{L} \to \infty$, you get

```
In[181]:= Bcylind[r_] = ( Curl[Acylind[r],Cylindrical]
                          //Simplify
                          //Limit[#,L->Infinity]& )

Out[181]=     2 IO
          {0, ----, 0}
               r
```

You will find, as expected, that **Bcylind** had only a **theta** component.

2.4.4 Circuits

Problem 1: Series RC Circuit

Consider a circuit consisting of a time-dependent emf $V[t]$, a resistor R, and a capacitor C connected in series.

Part (**a**) Solve for the charge $q[t]$ with an arbitrary $V[t]$.

Part (**b**) Evaluate the general solution found in Part (a) when $V[t] = V_0 \operatorname{Cos}[\omega t]$ and $q[0] = q_0$.

Part (**c**) Make a two-dimensional plot of the charge as a function of t. Make a three-dimensional plot of the charge as a function of t and C.

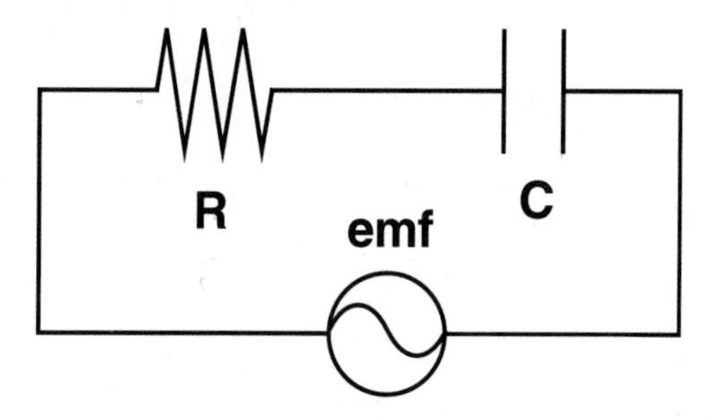

Remarks and outline Part (a) shows how **DSolve** returns a symbolic solution for the first-order RC differential equation.

Required packages *There are no required packages.*

Solution

```
In[182]:= Clear["Global`*"];
```

Part (**a**) The charge **q[t]** in the RC circuit obeys the differential equation

```
In[183]:= eq1=    V[t] == R q'[t] + q[t]/C;
```

Solving **eq1** for **q[t]** with an arbitrary voltage **V[t]**, you get

```
In[184]:= dsol= DSolve[eq1,q[t],t][[1]]
```

```
Out[184]=              C[1]
          {q[t] -> ———————————— +
                    t/(C R)
                   E

                         DSolve`t/(C R)
             Integrate[E                 V[DSolve`t], {DSolve`t, 0, t}]
             ————————————————————————————————————————————————————————————}
                                      t/(C R)
                                     E        R
```

Part (b) **dsol** for the potential **V0 Cos[w t]** becomes

```
In[185]:= V[t_]= V0 Cos[w t];
          q[t_]= q[t] /. dsol //Simplify;
```

Expressing **C[1]** in terms of the initial conditions **q[0]=q0**, you get

```
In[186]:= c1Sol=Solve[q[0]==q0,C[1]][[1]]

Out[186]= {C[1] -> q0}
```

and the solution for the charge becomes

```
In[187]:= q[t_]= q[t] /.c1Sol //Simplify

Out[187]=                          t/(C R)                   t/(C R)
             q0        C V0 (-1 + E        Cos[t w] + C E          R w Sin[t w])
          ——————— + ———————————————————————————————————————————————————————————
          t/(C R)                       t/(C R)      2  2  2
          E                             E        (1 + C  R  w )
```

Part (c) The plot of the charge as a function of time for the parameters of **values1** is

```
In[188]:= values1={q0->1, V0->1, C->1, R->1, w->6};

In[189]:= Plot[ Evaluate[ q[t]/.values1]
               ,{t,0,6}
               ,AxesLabel->{"time","charge"}
               ,PlotLabel->FontForm["RC Loop",{"CourierB",14}] ];
```

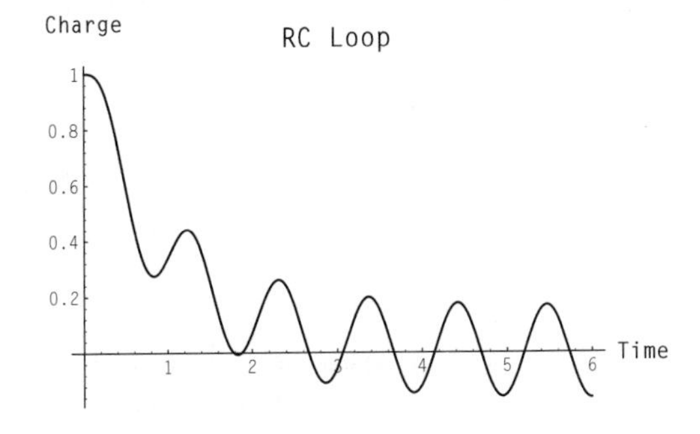

The **FontForm** command was used to specify the font of the **PlotLabel**.

A three-dimensional plot of the charge as a function of the variables **t** and **C** is

```
In[190]:= values2={q0->1, V0->1, R->1, w->6};
```

```
In[191]:= Plot3D[ Evaluate[ q[t]/.values2]
          ,{t,0,6}
          ,{C,0.01,0.25}
          ,AxesLabel->{"Time","Capacitance","Charge"}
          ,FaceGrids->{{0,-1,0}}
          ,ViewPoint->{.3,-1,.3}
          ,PlotPoints->35];
```

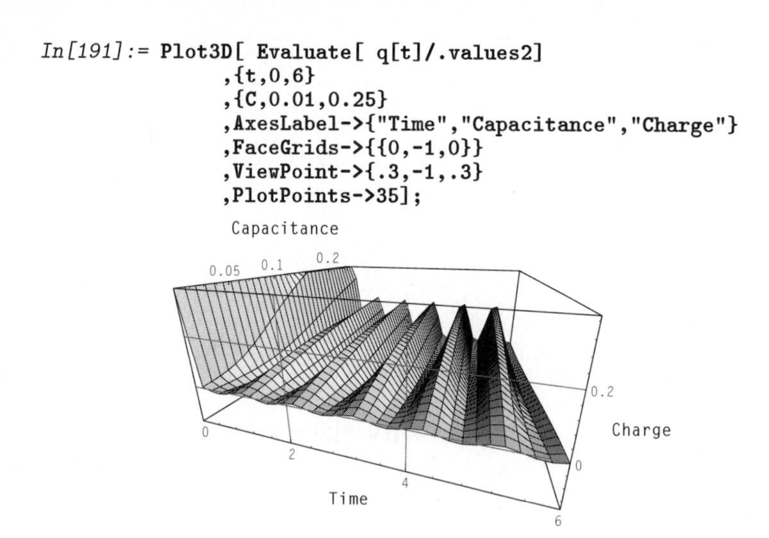

Problem 2: Series RL Loop

Consider a circuit consisting of a time-dependent emf $V[t]$, a resistor R, and an inductor L connected in series.

Part (a) Solve for $q[t]$ for arbitrary $V[t]$.

Part (b) Solve for the charge and current when $V[t] = V_0\text{Cos}[\omega t]$.

Part (c) Make a two-dimensional plot of the charge and current as a function of t for the initial condition $q[0] = 0$.

Part (d) Take the **Fourier** transform of the data formed by the solution of the charge found in Part (b).

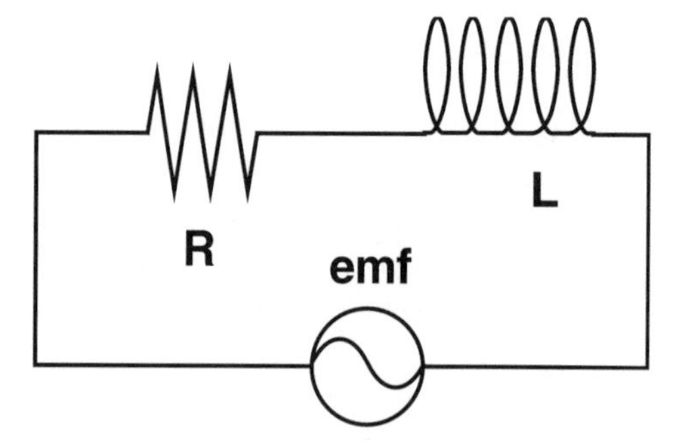

Remarks and outline This problem shows how **DSolve** returns a symbolic solution for the second-order RL equation with constant coefficients. In Part (d), you will see how the data from the oscillating system can be analyzed with a **Fourier** transform of the data.

Required packages

```
In[192]:= Needs["Calculus`FourierTransform`"];
```

Solution

```
In[193]:= Clear["Global`*"];
```

Part (a) The time dependence of the charge is described by the equation

```
In[194]:= eq1=  V[t] == L q''[t] + R q'[t]

Out[194]= V[t] == R q'[t] + L q''[t]
```

Applying **DSolve** to **eq1** for an arbitrary **V[t]** and arbitrary initial conditions, you get

```
In[195]:= initial={q[0]== q0, q'[0]==i0};

          dsol=  ( DSolve[{eq1}~Join~initial,q[t],t]
                   //Flatten
                   //Simplify )

Out[195]=
                         i0 L        i0 L
          {q[t] -> q0 + ------ - ------------ +
                          R        (R t)/L
                                  E        R

                    (R (DSolve`t - t))/L
            Integrate[(1 - E                    ) V[DSolve`t],

              {DSolve`t, 0, t}]
            ------------------------------------------------------
                                          R

            }
```

Note that the infix form of **Join** was used.

Part (b) When the voltage is given by **V0 Cos[w t]**, the charge becomes

```
In[196]:= V[t_]=V0 Cos[w t];

          q[t_]= (q[t] /.dsol //Simplify)

Out[196]=
                i0 L      i0 L
          q0 + ------ - ---------- +
                 R       (R t)/L
                        E        R

                     (R t)/L                   (R t)/L
           V0 (L w - E         L w Cos[t w] + E          R Sin[t w])
           ----------------------------------------------------------
                           (R t)/L   2       2  3
                          E         (R  w + L  w )
```

The current follows from the derivative of the charge:

```
In[197]:= current[t_]=D[q[t] ,t] //Simplify

Out[197]=
                2                2  2     (R t)/L
          (i0 R  - R V0 + i0 L  w  + E          R V0 Cos[t w] +

             (R t)/L                          (R t)/L    2     2  2
            E          L V0 w Sin[t w]) / (E           (R  + L  w ))
```

Part (c) For the values of the parameters,

```
In[198]:=  values={R->0.1, L->2, V0->1, w->1, i0->0.1, q0->0};
```

the graphics for the current and charge are

```
In[199]:= Plot[
            Evaluate[{q[t],current[t]}/.values]
                ,{t,0,20}
                ,PlotStyle->{ {Dashing[{0.06,0.00}],Thickness[0.008]}
                             ,{Dashing[{0.03,0.03}],Thickness[0.008]}
                            }
                ,Frame -> True
                ,FrameLabel->{{"Time"},{"Charge & Current"}}
                ,GridLines->Automatic ];
```

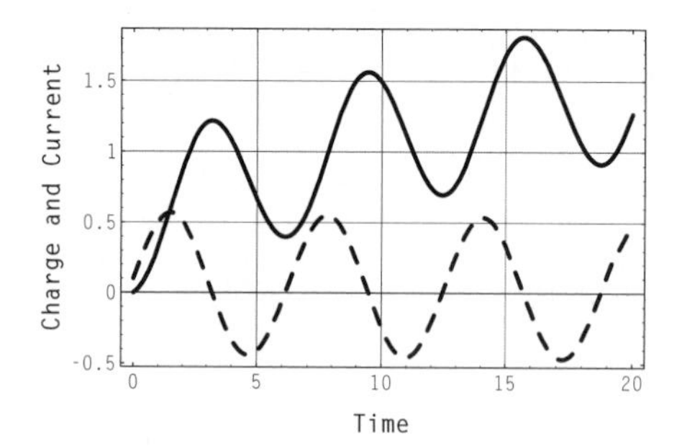

Here we see that the current (dashed line) oscillates in a regular fashion, but the charge through the circuit (solid line) is slowly accumulating.

Part (d) The **Fourier** transform and inverse Fourier transform are basic tools for analyzing oscillating systems. To Fourier-analyze the time dependence of the charge, generate a 300-item list of **q[t]** for the parameters given by **values**. The Fourier data is

```
In[200]:= fourierdata =
            Table[q[t] Random[Real,{0.9,1.1}] /.values //N
                  ,{t,0, 100 Pi,100 Pi/300}];
          fourierdata //Short

Out[200]= {0, 0.367964, 0.887446, 1.26495, <<295>>, 1.83369, 1.46875}

In[201]:= ListPlot[ fourierdata, PlotJoined -> True];
```

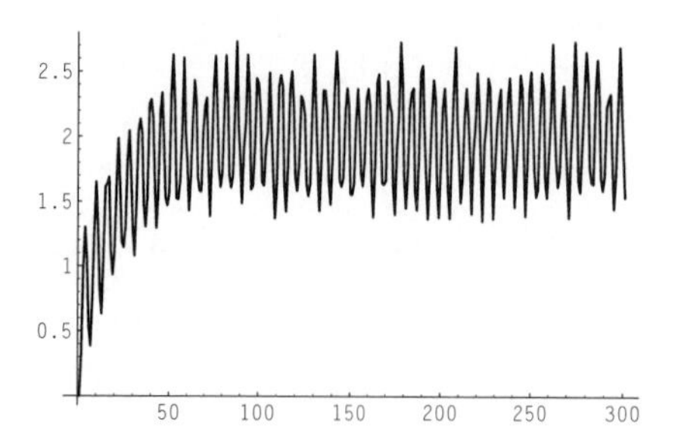

A random error function has been included to make the data table appear more like the results of an actual measurement. If the data is of the form Sin[t], the time coordinate will be a peak in the Fourier transform of the data at 50 and 300 – 50. The **Fourier** transform of the data is

```
In[202]:= ListPlot[ Abs[ Fourier[fourierdata] ]
                ,PlotJoined->True
                ,PlotRange->{0,5}];
```

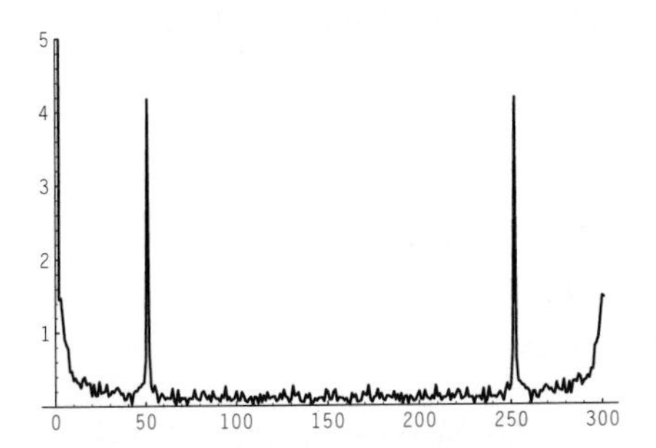

This graph shows a strong peak at 50 and another at $300 - 50$ due to aliasing; this result is expected from the Fourier transform of the data generated.

Problem 3: RLC Loop

Consider an RLC circuit consisting of a time-dependent emf $V[t]$, a resistor R, an inductor L, and a capacitor C connected in series.

Part (a) Find the **LaplaceTransform** of $q[t]$ for an arbitrary $V[t]$.

Part (b) Solve for the charge and current when $V[t] = 0$.

Part (c) Make a two-dimensional plot of the charge and current as a function of time. Make a three-dimensional plot of the charge as a function of t and L.

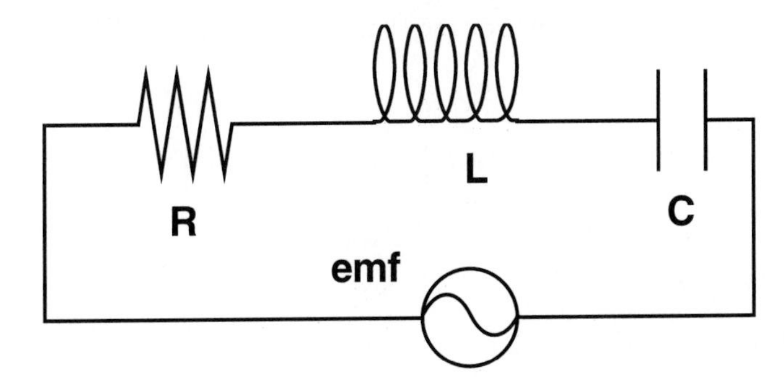

Remarks and outline Although **DSolve** will solve the RLC differential equation, you can use **LaplaceTransform** to illustrate how this can be used to solve certain differential equations.

Required packages

```
In[203]:= Needs["Algebra`Trigonometry`"] ;
          Needs["Calculus`LaplaceTransform`"];
```

Solution

```
In[204]:= Clear["Global`*"];
```

Part (**a**) The differential equation for the charge **q[t]** in an RLC circuit is

```
In[205]:= eq1= L q''[t]+ R q'[t] + q[t]/C == V[t]

Out[205]= q[t]
          ---- + R q'[t] + L q''[t] == V[t]
           C
```

Taking the Laplace transform of **eq1**, you get (this takes a while)

```
In[206]:= eq2= LaplaceTransform[eq1,t,s] //Simplify

Out[206]= LaplaceTransform[q[t], t, s]
          ---------------------------- + R s LaplaceTransform[q[t], t, s] +
                       C

             2
          L s  LaplaceTransform[q[t], t, s] - R q[0] - L s q[0] - L q'[0] ==

          LaplaceTransform[V[t], t, s]
```

It follows from the algebraic solution of **eq2** that the **LaplaceTransform** of **q[t]** is:

```
In[207]:= sol= Solve[eq2, LaplaceTransform[q[t],t,s] ][[1]]

Out[207]= {LaplaceTransform[q[t], t, s] ->

           C (LaplaceTransform[V[t], t, s] + R q[0] + L s q[0] + L q'[0])
           --------------------------------------------------------------}
                                                 2
                                1 + C R s + C L s
```

Part (**b**) Consider the explicit solution for the charge when **V[t]**=0. The Laplace transform follows from **sol**:

```
In[208]:= sol0 =(sol/.V[t]->0)

Out[208]=
                                         C (R q[0] + L s q[0] + L q'[0])
          {LaplaceTransform[q[t], t, s] -> -------------------------------}
                                                                2
                                                1 + C R s + C L s
```

For the initial conditions

```
In[209]:= initial={q[0]->q0,q'[0]->i0};
```

The solution for **q[t]** follows from taking the inverse Laplace transform:

```
In[210]:= charge[t_]=
            (((InverseLaplaceTransform[
                        LaplaceTransform[q[t],t,s]
                        /.sol0 /.initial ,s,t]
                 //Simplify
               ) //PowerExpand
                 //ExpandAll
               ) //Simplify
               );
          charge[t] //Short[#,2]&
```

```
Out[210]=
                                                2
     ((-(Sqrt[C] R) + Sqrt[-4 L + C R ]) t)/(2 Sqrt[C] L)
    E                                                          q0
    ---------------------------------------------------------------- + <<4>> +
                                  2

                         2
      Sqrt[C] q0 R Sqrt[-4 L + C R ]
      ------------------------------
                            2
           2 <<1>> (4 L - C R )
```

The expression for the charge is rather cumbersome but can be simplified if you define the frequency $-w_0^2 = (-4L + CR^2)/(4L^2C)$. This notation is implemented by the rule

```
In[211]:= freqrule= {  Sqrt[-4 L+ C R^2]->     2 I w0 Sqrt[C] L,
                      1/Sqrt[-4 L+ C R^2]-> 1/(2 I w0 Sqrt[C] L),
                            (-4 L+ C R^2)->      -4 w0^2 C L^2,
                      1/    (-4 L+ C R^2)-> 1/( -4 w0^2 C L^2)
                       };
```

We also have the inverse rule:

```
In[212]:= wrule=Solve[Sqrt[-4 L+ C R^2] == 2 I w0 Sqrt[C] L,w0][[1]]

Out[212]=
                -I               2
                -- Sqrt[-4 L + C R ]
                2
       {w0 -> ----------------------}
                    Sqrt[C] L
```

With this change in notation, the expression for the charge simplifies to

```
In[213]:=  charge0=  ((charge[t] //.freqrule
                                  //ComplexToTrig
                                  //Simplify
                                ) //.freqrule
                                  //Simplify
                                )

Out[213]= 2 L q0 w0 Cos[t w0] + 2 i0 L Sin[t w0] + q0 R Sin[t w0]
          ------------------------------------------------------
                               (R t)/(2 L)
                            2 E             L w0
```

The current follows from the time derivative of the charge:

```
In[214]:= current0= ( D[charge0,t]
                      //Collect[#,{q0,i0}]&
                      //Map[Simplify,#,{2}]& )

Out[214]=
                 2      2   2
          -(q0 (R  + 4 L  w0 ) Sin[t w0])    i0 (2 L w0 Cos[t w0] - R Sin[t w0])
          ------------------------------- + -----------------------------------
                 (R t)/(2 L)  2                       (R t)/(2 L)
              4 E            L  w0                2 E             L w0
```

Part (c) Consider the parameters

```
In[215]:= values= {R->1 ,V0->1, C->1, L->4, q0->1, i0->0};
```

for the plots of the charge and current. The labels for the graphs follow from (output suppressed)

```
In[216]:= text=
          Show[Graphics[
           {Text[FontForm[Charge ,{"CourierB",10}],{5,0.8}],
            Text[FontForm[Current,{"CourierB",10}],{8,0.3}]}]];
```

Adding the labels to the curves using the option **Epilog**, the plot of the charge and current is

```
In[217]:= Plot[Evaluate[{charge0,current0} /.wrule /.values]
                ,{t,.001,20}
                ,PlotStyle->{
                      {Dashing[{0.01,0.01}],Thickness[0.01 ]},
                      {Dashing[{0.02,0.02}],Thickness[0.008]} }
                ,Frame -> True
                ,FrameLabel->{{"Time"},{"Charge & Current"}  }
                ,GridLines->Automatic
                ,Epilog->text[[1]]
             ];
```

To study the effects of inductance on the charge, consider a three-dimensional plot of the charge as a function of the variables time and inductance. For the parameters

```
In[218]:= values2= {R->1 ,V0->1, C->1, q0->1, i0->1};
```

the plot of the charge is

```
In[219]:= p1=Plot3D[Evaluate[charge0 /.wrule /.values2]
               ,{t,0.01,30},{L,1,4}
               ,ViewPoint->{1,+3,1}
               ,FaceGrids->{{0,0,1},{0,-1,0}}
               ,AxesLabel->{"Time","Inductance","Charge"}
             ,PlotPoints->30
            ];
```

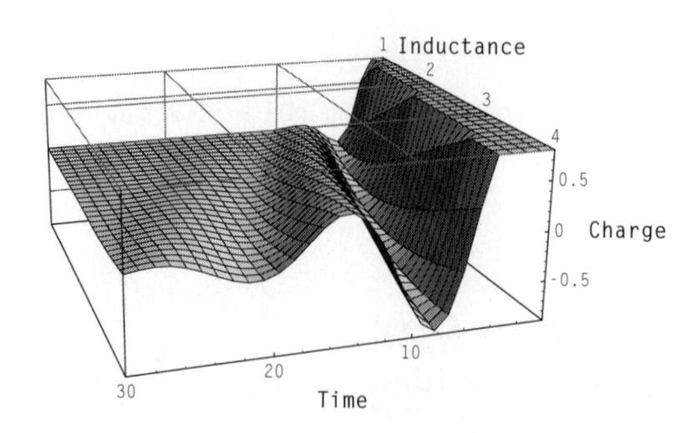

2.4.5 Modern Physics

Problem 1: The Bohr Atom

Consider an electron that is attracted to the nucleus with a force Ze/r^2 and revolves around the nucleus in a circular orbit.

Part (a) Using Bohr's method of quantization, derive the equation for the energy levels in terms of n, where n is the nth Bohr orbit. Make a plot of the energy levels.

Part (b) Derive the frequency of the radiation associated with a transition from the Bohr orbit n_2 to n_1.

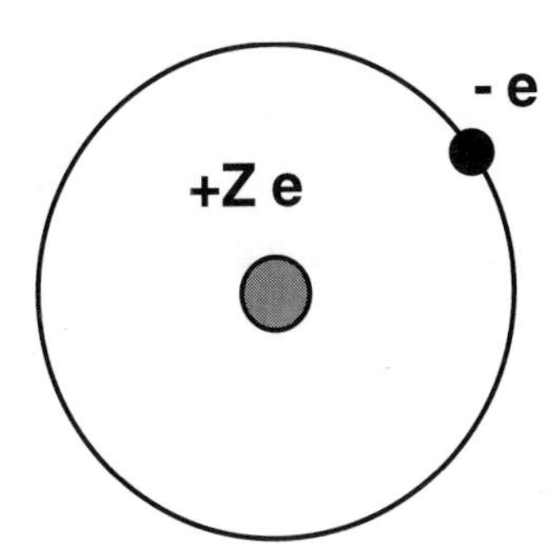

Remarks and outline One equation for the parameters follows from equating the centrifugal force to the coulomb force; a second equation follows from the Bohr quantization rule. Solving these two equations gives the quantized expressions for v and r. The energy is the sum of the potential and kinetic energies.

Required packages

```
In[220]:= Needs["Miscellaneous`Units`"];
          Needs["Miscellaneous`PhysicalConstants`"];
```

Solution

```
In[221]:= Clear["Global`*"];
```

Part (a) We balance the coulomb force against the centrifugal force for a circular orbit of radius r:

```
In[222]:= coulomb     = Z e^2/r^2;
          centrifugal = m v^2/r;

          eq1=  coulomb == centrifugal

Out[222]=  2           2
          e  Z      m v
          ----- == ----
            2        r
           r
```

The angular momentum is defined by $r \times p = mvr$:

```
In[223]:= angularMomentum=m v r;
```

Quantizing the angular momentum in units of **hbar** ($\hbar$) results in

```
In[224]:= eq2=  angularMomentum == n hbar

Out[224]= m r v == hbar n
```

where n = $\{1, 2, \ldots\}$. You can get an equation for **{v,r}** by solving the algebraic equations **eq1** and **eq2**:

```
In[225]:= rule1=Solve[{eq1,eq2},{v,r}][[1]]

Out[225]=
                 2  2              2
             hbar  n              e  Z
          {r -> --------, v -> --------}
                  2             hbar n
                 e  m Z
```

The total energy of the nth orbit is the sum of the kinetic and potential energies:

```
In[226]:= kineticEnergy   = 1/2 m v^2;
          potentialEnergy = -Z e^2/r;

          totalEnergy = kineticEnergy + potentialEnergy

Out[226]=   2    2
         m v    e  Z
         ---- - ----
          2      r
```

Expressing the energy in terms of **n**, it follows that

```
In[227]:= energy[n_]= totalEnergy //.rule1

Out[227]=   4     2
          -(e  m Z )
          ----------
                2  2
          2 hbar  n
```

To make a plot of the energy, choose the specific values:

```
In[228]:= values={
             e->ElectronCharge
            ,m->ElectronMass
            ,Z->1/(4 Pi epsilon0)
            ,epsilon0-> 8.854 10^-12 Coulomb^2/(Newton Meter^2)
            ,hbar->PlanckConstantReduced};
```

A $1/(4\pi\epsilon_0)$ has been artificially inserted into **Z** to express the result in MKS units. Next, define a function that gives the energy in **ElectronVolt** units:

```
In[229]:= energyEV[n_]:=
              Convert[energy[n] //.values,ElectronVolt];
```

and check to see that the ionization energy for the first hydrogen orbit is the usual value:

```
In[230]:= energyEV[1]

Out[230]= -13.6063 ElectronVolt
```

Pictorially, the energy values of the hydrogen atom in units of ElectronVolts are:

```
In[231]:= Plot[Evaluate[Table[energyEV[n] ,{n,1,10}]/ElectronVolt]
            ,{x,0,1}
            ,Axes->{False,True}
            ,PlotRange->All];
```

This shows the energy levels in a simple, graphical way. (We will solve for the full quantum wavefunctions in Chapter 6, "Quantum Mechanics".)

Part (b) The angular frequency of the radiation associated with the transition from orbit **ni** to **nf** follows from Bohr's postulate, $\hbar\omega = \Delta E$:

```
In[232]:= eq3=  hbar omega == energy[ni] - energy[nf]

Out[232]=
                         4   2         4   2
                        e  m Z        e  m Z
          hbar omega == ----------- - -----------
                               2   2         2   2
                        2 hbar   nf   2 hbar   ni
```

You get an expression for **omega** by dividing both sides of **eq3** by **hbar**:

```
In[233]:= Thread[eq3/hbar,Equal]//ExpandAll

Out[233]=
                    4   2         4   2
                   e  m Z        e  m Z
          omega == ----------- - -----------
                          3   2         3   2
                   2 hbar   nf   2 hbar   ni
```

Problem 2: Relativistic Collision

A particle at rest with m_2 is hit by a second relativistic particle of m_1 and speed v_1. Upon collision, they coalesce into one particle of rest mass m_3 moving with speed v_3 relative to the observer. Find m_3 and v_3 for the composite particle in terms of m_1, m_2, and γ_1, where $\gamma_1 = 1/\sqrt{1 - v_1^2}$.

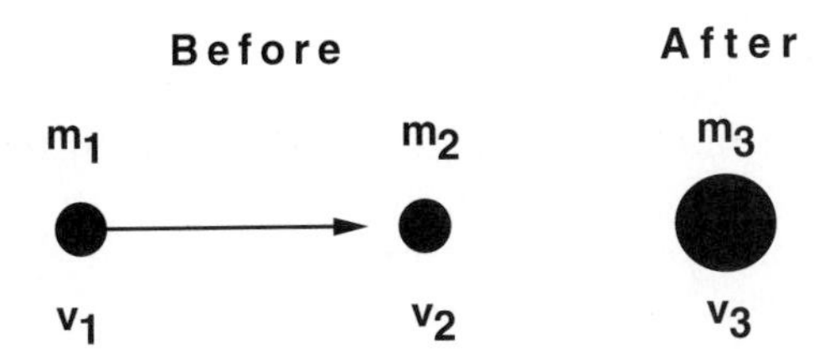

Remarks and outline This problem illustrates the Minkowski metric tensor and the use of vectors and tensors. The results follow from the conservation of momentum. The algebra is simplified by working with scalar invariants, which are valid in all Lorentz frames.

Required packages *There are no required packages.*

Solution

```
In[234]:= Clear["Global`*"];
```

Part (a) The two four-momentums of the initial particles are

```
In[235]:= p1=m1 gam1 {1,0,0,v1};
          p2=m2      {1,0,0,0 };
```

Here the momenta have been written in four-vector notation: $p^\mu = \{E, p_x, p_y, p_z\}$. The energy of m_1 is (**m1 gam1**), and the momentum of m_1 is (**m1 gam1 v1**) in the z-direction. (Natural units have been used such that $c \equiv 1$.) That $\gamma = 1/\sqrt{1-v^2}$ will be demonstrated shortly.

The four-momentum of the final particle is

```
In[236]:= p3=m3 gam3 {1,0,0,v3};
```

The metric tensor is

```
In[237]:= eta=DiagonalMatrix[1,-1,-1,-1];
```

You can solve for **gam1** and **gam3** by using the mass-shell condition $p_1^2 = m_1^2$ and $p_3^2 = m_3^2$:

```
In[238]:= gamrule=Join[ Solve[{p1.eta.p1 == m1^2},gam1][[2]]
                       ,Solve[{p3.eta.p3 == m3^2},gam3][[2]]
                      ]

Out[238]=
                    I                          I
          {gam1 -> ----------------, gam3 -> ----------------}
                                2                          2
                    Sqrt[-1 + v1 ]             Sqrt[-1 + v3 ]
```

You can put this in the usual form with some simplification:

```
In[239]:= iRule={1/Sqrt[-1+v_^2]:>1/(I Sqrt[1-v^2])};
          gamrule= gamrule //.iRule

Out[239]=
                    1                         1
          {gam1 -> ---------------, gam3 -> ---------------}
                               2                         2
                    Sqrt[1 - v1 ]             Sqrt[1 - v3 ]
```

Note that because you are not in a Euclidian space, you must compute the dot product of two four-vectors with the proper metric. You may find it of interest to separately examine **p1.eta.p1**.

Having found **gam1** and **gam3**, you can find the inverse relation simply by solving for **v1** and **v3**. Both relations will be useful.

```
In[240]:= vrule=Solve[{p1.eta.p1 == m1^2
                      ,p3.eta.p3 == m3^2},{v1,v3}][[4]]

Out[240]=
                             2                          2
                Sqrt[-1 + gam1 ]           Sqrt[-1 + gam3 ]
          {v1 -> ----------------, v3 -> ----------------}
                      gam1                       gam3
```

(Be sure to select the positive root here.) Conservation of four-momentum gives you a relation between the vectors that is described by the equation $p_1 + p_2 == p_3$:

```
In[241]:= eq1= Thread[p1+p2==p3] /.gamrule //Simplify

Out[241]=
                       m1                m3
          {m2 + ------------- == -------------, True, True,
                             2                2
                Sqrt[1 - v1 ]    Sqrt[1 - v3 ]

              m1 v1             m3 v3
          ------------- == -------------}
                       2                2
          Sqrt[1 - v1 ]    Sqrt[1 - v3 ]
```

You can easily solve for **m3**:

```
In[242]:= sol1= (( Solve[ eq1[[1]],{m3}][[1]]
                    /.vrule
                    //PowerExpand
                    //Simplify
                  ) //PowerExpand
                    //Simplify
                  )

Out[242]=
                 gam1 m1 + m2
          {m3 -> ------------}
                     gam3
```

and then use this relation to find **v3**:

```
In[243]:= sol2= ( Solve[ eq1[[4]]/.sol1 /.gamrule ,{v3}][[1]]
                  /.{Sqrt[1-v1^2] -> 1/gam1}
                  //Simplify
                )

Out[243]=
                 gam1 m1 v1
          {v3 -> ------------}
                 gam1 m1 + m2
```

A quick check shows the four-momentum constraint is satisfied.

```
In[244]:= p1+p2==p3 //.Join[sol1,sol2]

Out[244]= True
```

2.5 Unsolved Problems

Exercise 2.1: Exploding projectiles

A projectile is launched at 20 m/s at an angle of 30° with the horizontal. In the course of its flight, it explodes, breaking into two fragments, one of which has twice the mass of the other. The two fragments land simultaneously. The lighter one lands 20 m from the launch point in the direction in which the projectile was fired.

Part (a) Where does the other fragment land?

Part (b) Plot the motion of the two fragments and the motion of the center of mass point.

Part (c) Write a function that will plot the motion of the two fragments and the motion of the center of mass when the initial velocity, mass of fragments, and position of one of the fragments is given.

Exercise 2.2: Intersecting trajectories

A hunter with a blowgun wishes to shoot a monkey hanging from a branch. The hunter aims right at the monkey and fires the dart. At the same instant, the monkey lets go of the branch and drops to the ground.

Part (a) Show that the monkey will be hit regardless of the angle of the dart to the monkey and the dart's initial velocity so long as it is great enough to travel the horizontal distance to the tree before the dart hits the ground.

Part (b) Plot the points for the positions of the dart and the monkey when those positions are separated by equal time intervals. Choose the time intervals so that the last point is the intersection of the dart and monkey.

Part (c) Do the plot for several different initial values of the dart and for several different heights of the monkey.

Exercise 2.3: Reflecting trajectories

A boy stands a distance x_0 from a vertical wall and throws a ball. The ball leaves the boy's hand at a height y_0 above the ground with initial velocity v_0. When the ball hits the wall, its horizontal component of velocity is reversed and its vertical component remains unchanged.

Part (a) Where does the ball hit the ground?

Part (b) Plot the motion of the ball for several different initial velocities.

Exercise 2.4: Principle of Galilean invariance

Two balls are released simultaneously from a platform above Earth's surface: One is merely dropped from rest, while the other is launched with an initial horizontal velocity v_0. Graph the stroboscopic images of the two balls at successive instants of time and show that they fall downward in unison, reaching equal heights at the same time. Consider several different values of v_0.

Exercise 2.5: Elastic collision

A ball moving with nonrelativistic velocity v_i makes an off-center, perfectly elastic collision with another ball of equal mass initially at rest. The incoming ball is deflected at an angle θ from its original direction of motion.

Part (a) Find the velocities and trajectories of the balls after the collision.

Part (b) Plot the trajectories and center of mass point for different values of v_i and θ.

Exercise 2.6: Quadrupole

Consider three charges located on the z-axis. One negative charge of magnitude q is located at $+s$; the other, of the same magnitude, is at $-s$. A positive charge of magnitude $2q$ is located at the origin. The field point is at $\vec{r}$, and it is assumed that the separation s of the charges is small compared to the distance r.

Part (a) Find the potential and electric fields to leading order in s/r.

Part (b) Plot the equipotential surfaces and the electric field lines.

Exercise 2.7: Fast Fourier transform

Consider the function

$$\mathrm{Sin}[2\pi t]\,(1+\frac{1}{5}\,\mathrm{Sin}[6\pi t]+\frac{1}{10}\,\mathrm{Sin}[8\pi t]).$$

Generate a table of data points from this function with random noise added. Take the Fourier transform of the table and plot the results.

Exercise 2.8: B field of a spinning disk

A thin disk of charge density s, radius R, and thickness $t < R$ rotates with an angular velocity ω about the z-axis. Find the B field on the axis of the uniformly rotating disk and plot the results.

Exercise 2.9: Fit to data

Consider the following data table:

```
In[1]:=  data =
         Table[{x, 5 Sin[3 Pi x]+Random[Real,{-.2,.2}]}
                 //N //Chop[#,10^-3]&
           ,{x,0,0.5 ,0.05}];
```

Part (a) Fit the data to a power series in x and plot the fit and data on the same graph.

Part (b) Use **leastSquares** to fit the data to $A\mathrm{Sin}[Bx]$ and plot the fit and data on the same graph.

CHAPTER THREE

Oscillating Systems

3.1 Introduction

In this chapter, we consider oscillating systems. Patterns of movement that repeat over and over again are common in celestial mechanics, mechanical systems, quantum mechanics, and electrodynamics. The solutions given in this chapter are not unique; you are encouraged either to change the *Mathematica* commands and find procedures that will further illuminate the physics or to find other commands that will make the calculations faster. The speed of the calculation can be obtained from the command **`Timing`**. This chapter is divided into three sections:

1. Introduction
2. Problems
3. Unsolved Problems

The problems section is subdivided into problems on linear oscillators, nonlinear oscillators, and small oscillations.

3.1.1 Oscillations

Potentials

The equations describing oscillating systems are usually complicated. In many cases, they can be analyzed with potential or phase space diagrams. Potential diagrams can be used to analyze oscillatory motion and to find positions of equilibrium. The system is in equilibrium when the net force acting on it vanishes. If the system is initially at the equilibrium position with zero initial velocity, then the system will remain in equilibrium indefinitely. The equilibrium is stable if a small disturbance of the system from equilibrium results only in small bounded motion about the rest position. The equilibrium is unstable if an infinitesimal disturbance eventually produces unbounded motion. The potential has an extremum at both of these equilibrium positions. When the extremum of the potential is a minimum, the equilibrium is stable; when it is a maximum, the equilibrium is unstable.

Phase Planes

In some cases, it is advantageous to study the motion in terms of its phase plane. For example, consider a system having one degree of freedom that is described by a differential equation of the form

$$x''[t] + f(x'[t], t) == 0.$$

If we define $v[t] = x'[t]$, then this equation can be expressed as two first order-equations:

$$v'[t] = -f(x, v, t)$$

and

$$x'[t] = v[t].$$

The $\{x, v\}$ plane is called the phase plane, and the changing state of the system makes a curve in the phase plane called a trajectory. The trajectory displays the dependence between the position and velocity. Equilibrium occurs when the velocity v and force $f(x, v, t)$ are simultaneously equal to zero and correspond to points in the phase plane, where $f(x, v, t) = 0$ and $v = 0$.

Small Oscillations and Normal Modes

Linear systems having many degrees of freedom such as coupled oscillators are most conveniently described by finding the normal modes of the system. Consider linear systems in which the oscillatory motion is described by a potential and the motion takes place in the neighborhood of an equilibrium point. The motion can be described by coordinates $q_i[t]$ $(i = 1, 2, 3, \ldots n)$, where the q_i's are measured from the equilibrium position. If the forces are linear functions of the displacements, the expression for the potential V will be quadratic in q_i:

$$V = \frac{1}{2}\, V_{ij}\, q_i[t]\, q_j[t]$$

and the kinetic energy T will be quadratic in the derivatives of the coordinates:

$$T = \frac{1}{2}\, T_{ij}\, q_i'[t]\, q_j'[t].$$

The coefficients V_{ij} and T_{ij} form the potential and kinetic energy matrices, respectively.

The equations of motion for this system of coupled linear oscillators can be expressed in terms of the potential and kinetic energy matrices:

$$T_{ij} q_j''[t] + V_{ij} q_j[t] == 0.$$

Only for certain initial conditions will the system oscillate at one natural frequency. When it does, we speak of a normal mode of oscillation. Let the parameters $\omega_1, \omega_2, \omega_3, \ldots$ be the normal frequencies of oscillation. For each ω_i, there is a corresponding normal mode of oscillation in the coordinates $\{q_{i1}, q_{i2}, \ldots\}$. All of the particles in each normal mode vibrate with the same frequency. A complete solution of the motion involves a superposition of normal modes.

3.2 *Mathematica* Commands

3.2.1 Packages

Several packages are used in this chapter. First turn off the spell checker:

```
In[1]:= Off[General::spell ];
        Off[General::spell1];
```

This chapter needs the following packages:

In[2]:=
```
Needs["Algebra`SymbolicSum`"];
Needs["Algebra`Trigonometry`"];
Needs["Calculus`FourierTransform`"];
Needs["Calculus`LaplaceTransform`"];
Needs["Graphics`FilledPlot`"];
Needs["Graphics`Animation`"];
Needs["Graphics`Graphics3D`"];
```

3.2.2 User-defined Procedures

It is useful to create user-defined procedures for those calculations that are used repeatedly. We will not document all the procedures exhaustively; to understand them in detail, go through the steps one at a time. (We have illustrated how to do this for some selected procedures in this chapter.) Try to make the procedures more time efficient, add default conditions, and add options to the procedures. Some of these procedures will fail in special cases; write conditional statements to handle these cases. Again, we emphasize that our goal is to solve the physics, not write an all-purpose foolproof package.

Series expansion for second-order equation

You will often find it useful to solve differential equations using a series expansion. This function performs the necessary operations automatically:

In[3]:=
```
diffSeriesOne::usage="
diffSeriesOne[eq,z,t,order,{initialList}]
eq is an equation, z is the function, t is the independent
variable, order is the order of the Series solution, and
{initialList} is a list of initial conditions such as
{z[0]->z0}.
This procedure assumes the solution can be written as
a power series.";
```

In[4]:=
```
Clear[diffSeriesOne];
diffSeriesOne[eqin_,z_,t_,order_,initList__List]:=
 Module[{eq,zSer,eSer,eqs,vars,sol,zSol},
   eq=eqin[[1]]-eqin[[2]]==0;
   zSer= Series[z[t],{t,0,order}];
   eSer= eq/.(z[t]->zSer)//ExpandAll;
   eqs = Thread[CoefficientList[eSer[[1]],t]==0];
   vars=(Table[D[z[t],{t,j}],{j,2,order+2}]/.t->0);
   sol = Solve[eqs,vars][[1]];
   zSol= zSer/.sol/.initList;
   Return[ zSol]
 ]
```

Example 1: Second order solution

Consider the differential equation

In[5]:=
```
eq1= y''[t] + w^2 y[t]^2 + a y'[t]^2 + y[t]^3  == 0;
```

In[6]:=
```
diffSeriesOne[eq1,y,t,2,{y[0]->y0,y'[0]->vy0}]
```

Out[6]=

```
                     2     2   2     3   2
           (-(a vy0 ) - w  y0  - y0 ) t          3
y0 + vy0 t + ---------------------------------- + O[t]
                            2
```

Example 2: How the **diffSeriesOne** routine works

To understand how **diffSeriesOne** works, examine the intermediate steps. To do this, clear and re-execute a slightly modified command, **diffSeriesOne2**. This time, however, delete all the local variables in the **Module** and then re-execute the example:

In[7]:=

```
Clear[diffSeriesOne2];
diffSeriesOne2[eqin_,z_,t_,order_,initList__List]:=
 Module[{},
   eq=eqin[[1]]-eqin[[2]]==0;
   zSer= Series[z[t],{t,0,order}];
   eSer= eq/.(z[t]->zSer)//ExpandAll;
   eqs = Thread[CoefficientList[eSer[[1]],t]==0];
   vars=(Table[D[z[t],{t,j}],{j,2,order+2}]/.t->0);
   sol = Solve[eqs,vars][[1]];
   zSol= zSer/.sol/.initList;
   Return[ zSol]
 ]

 diffSeriesOne2[eq1,y,t,2,{y[0]->y0,y'[0]->vy0}]
```

Out[7]=

```
                     2     2   2     3   2
           (-(a vy0 ) - w  y0  - y0 ) t          3
y0 + vy0 t + ---------------------------------- + O[t]
                            2
```

(*Note*: **Module[{}, ...]** has been written above.) Now you can access the intermediate results. Obviously, **eq** simply takes the two parts of the input equation and puts them on the left-hand side of the equation. **zSer** is the series expansion of **z[t]** to the appropriate order:

In[8]:=

```
zSer
```

Out[8]=

```
                         2
                y''[0] t         3
y[0] + y'[0] t + --------- + O[t]
                    2
```

eSer is the differential equation with **zSer** substituted. Next set individual powers of **t** to zero and solve the resulting equations. To do this, use **CoefficientList** on **eSer[[1]]** (this is why all the terms are put on the left-hand side of the equation in step 1). Using the **Thread** command, you then have your set of **order+1** equations:

In[9]:=

```
eqs
```

Out[9]=

```
  2        2        3              2
{w  y[0]  + y[0]  + a y'[0]  + y''[0] == 0,

    2                       2                                  (3)
  2 w  y[0] y'[0] + 3 y[0]  y'[0] + 2 a y'[0] y''[0] + y    [0] == 0,
```

```
                                                                  2
 2          2               2    2                    3 y[0]  y''[0]
w  y'[0]  + 3 y[0] y'[0]  + w  y[0] y''[0] +  ---------------  +
                                                         2

                                (4)
         2              (3)    y    [0]
 a y''[0]  + a y'[0] y    [0] + ----------  == 0}
                                    2
```

The independent variables to solve for are stored in **vars**:

```
In[10]:= vars

Out[10]=          (3)       (4)
         {y''[0], y   [0], y   [0]}
```

The solution is given in **sol**. Then substitute this solution, along with the initial conditions, into the series expression for the function **zSer**, producing the returned solution:

```
In[11]:= zSol= zSer/.sol/.{y[0]->y0,y'[0]->vy0}

Out[11]=                      2     2  2     3  2
                    (-(a vy0 ) - w  y0  - y0 ) t         3
         y0 + vy0 t + ---------------------------- + O[t]
                                  2
```

Phase plot for one-dimensional system

This routine is simply a specialized **ParametricPlot**; however, you will often find it useful for an oscillating system:

```
In[12]:= phasePlot::usage="phasePlot[q,p,{tList}]
         This procedure will plot the phase trajectory {q,p}.
         {tlist} is of the form {t,tMin,tMax}.";

In[13]:= Clear[phasePlot];
         phasePlot[q_,p_,tList_List]:=
           ParametricPlot[{p,q},tList
             ,PlotRange->All
             ,Axes->False
             ,PlotLabel->"Phase Trajectory"
             ,Frame->True
             ,FrameLabel->{"p-axis","x"}]
```

Example: Phase plots for harmonic motion

Consider the harmonic functions $q = \mathrm{Sin}[2\pi a t]$ and $p = D[q,t]$. The phase diagram is

```
In[14]:= phasePlot[Sin[2 Pi t],D[Sin[2 Pi t],t],{t,0,3}];
```

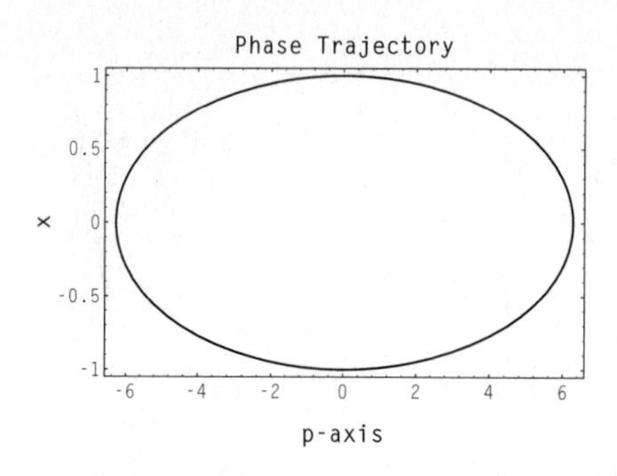

Time behavior of phase plot for a one-dimensional system

This routine is simply a specialized **ParametricPlot3D**; however, you will often find it useful for an oscillating system:

In[15]:=
```
timePhasePlot::usage=" timePhasePlot[q,p,{tList}]
This procedure will plot the time evolution of the trajectory.
{tlist} is of the form {t,tMin,tMax}.";
```

In[16]:=
```
Clear[timePhasePlot];
timePhasePlot[q_,p_,tList_List]:=
 ParametricPlot3D[{p,q,tList[[1]]},tList
    ,PlotRange->All
    ,PlotLabel->"Evolution of Phase"
    ,AxesLabel->{"p-axis","x-axis","time"}
    ,PlotPoints->300
    ,BoxRatios->{1,1,1}]
```

Example: Time-evolved phase plots for harmonic motion

Consider the harmonic functions $q = \mathrm{Sin}[2\pi a t]$ and $p = D[q, t]$. The time-evolved phase diagram is

In[17]:=
```
timePhasePlot[Sin[2 Pi t],D[Sin[2 Pi t],t],{t,0,3}];
```

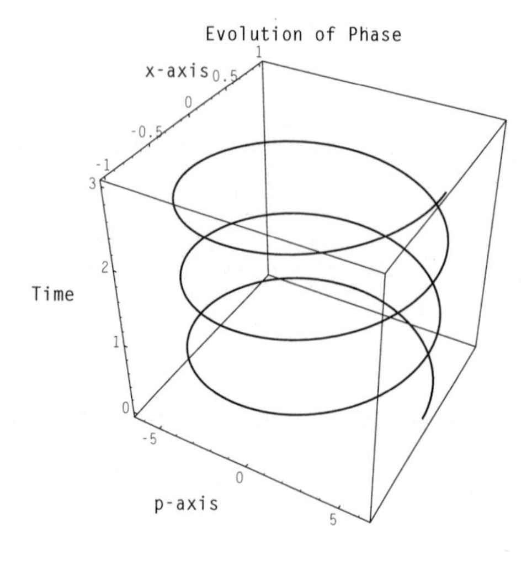

doublePlot: Phase plots and time evolution for a one-dimensional system

This routine combines **phasePlot** and **timePhasePlot** into a **GraphicsArray**:

```
In[18]:= doublePlot::usage=" doublePlot[q,p,{tList}]
         Combines phasePlot and timePhasePlot.";
```

```
In[19]:= Clear[doublePlot];
         doublePlot[q_,p_,tList_List]:=
         Show[GraphicsArray[{     phasePlot[q,p,tList]
                             ,timePhasePlot[q,p,tList]
                             }]
             ];
```

Example: Phase plots and time evolution for harmonic motion

Consider the harmonic functions $q = \text{Sin}[2\pi at]$ and $p = D[q,t]$. The phase plot and time evolution are (intermediate output suppressed)

```
In[20]:= doublePlot[Sin[2 Pi t],D[Sin[2 Pi t],t],{t,0,3}];
```

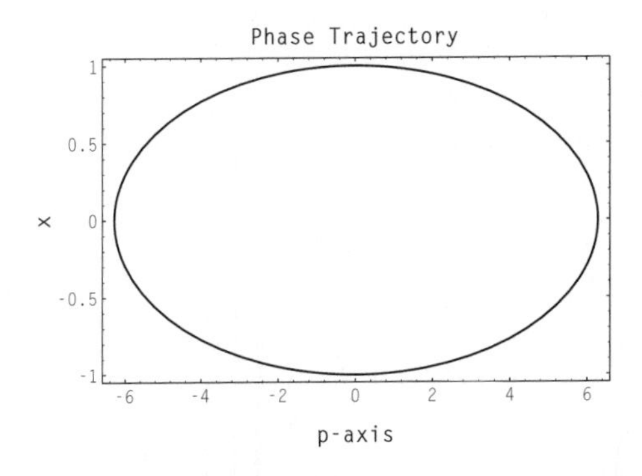

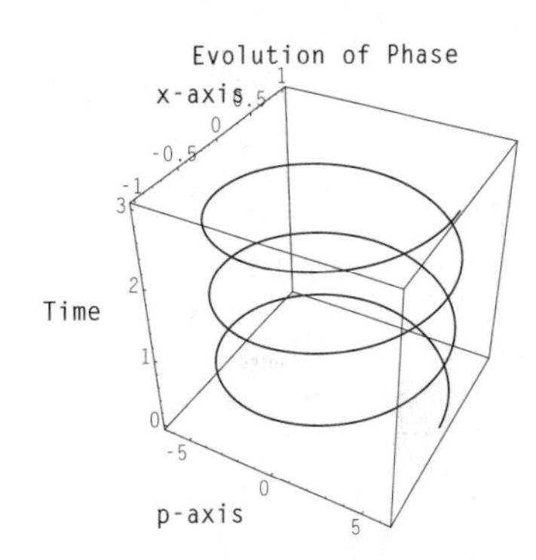

Fourier spectrum of a one-dimensional oscillating system

This function uses the **Fourier** command to produce a plot of the Fourier transform:

```
In[21]:= fftSpectrum::usage="
         fftSpectrum[f, t, tMax]
         This procedure will plot the fast Fourier transform of
         the function f. The time interval is from 0 to tMax.
         The number of steps is fixed at 512.";
```

```
In[22]:= fftSpectrum[f_,t_,tMax_] :=
           Module[{dt,fData,ffData,plot},
              dt    = tMax/(512-1)//N;
              fData = Table[f,{t,0,tMax,dt}]//N;
              ffData= Abs[Fourier[fData]] ;
              plot  = ListPlot[ffData,PlotJoined->True]
           ]
```

Example: Fast Fourier transform

Consider the following example:

```
In[23]:=  eq1=Sin[2Pi x](1-1/5 Sin[2Pi x/2]+ 1/10 Sin[2Pi x/4]);
```

You can plot all the elements of the transform on one graph:

```
In[24]:=  p3=fftSpectrum[eq1,x,100];
```

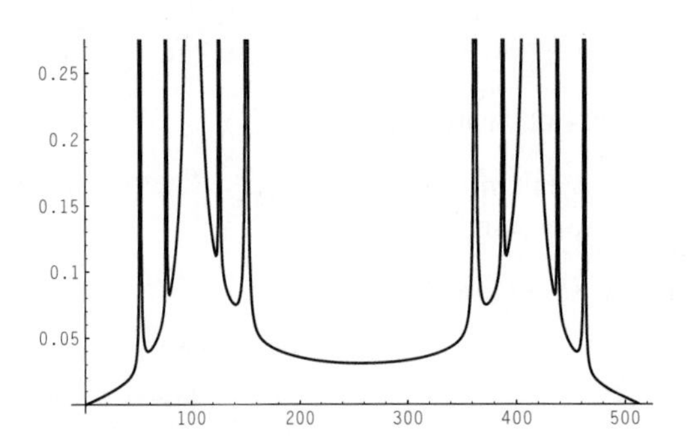

and set **Options** in the **Show** command to focus on a particular range of the plot:

```
In[25]:=  Show[p3,PlotRange->{{0,200},All}];
```

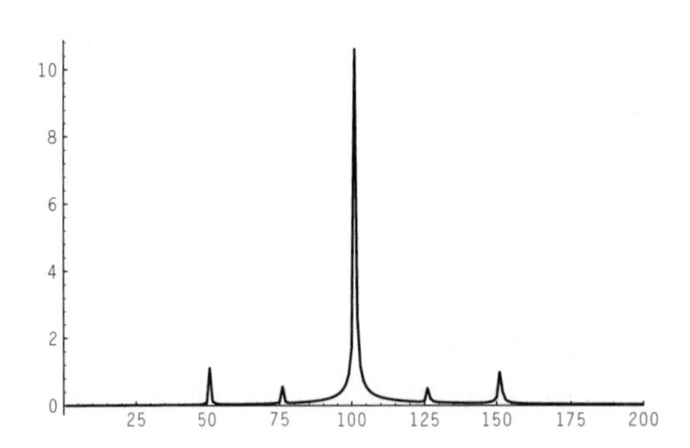

Eigenvalues and eigenvectors for small oscillating systems

You will often want to solve for the eigenvalues and eigenvectors for systems in the limit of small oscillations. This function computes the answer and returns the solution in a useful format:

```
In[26]:=  smallOsc::usage="
          smallOsc[T,V,{q1[t],q2[t],...},{q10,q20,...},t]
          Given the expressions for the kinetic and potential energies
```

```
of the system, this procedure will find the eigenfrequencies
and eigenvectors. Enter the kinetic energy T, potential V,
coordinates, equilibrium positions, and time variable. The
first list in the output are the eigenfrequencies and the
remaining lists are the eigenvectors. ";
```

```
In[27]:= Clear[smallOsc];
smallOsc[T_,V_ ,coordList__,equlibList__,t_:t]:=
  Module[{q,p,dim,equlibRule,Vij,Tij,matrix,w2Sol,
              freq,vector,eqs,vSol},
    q=coordList;
    p=D[coordList,t];
    dim=Length[q];
    equlibRule=Thread[coordList-> equlibList];
    Vij=Outer[D[V,#1,#2]&,q,q];
    Tij=Outer[D[T,#1,#2]&,p,p];
    matrix= Vij-w2 Tij /.equlibRule;
    w2Sol= Solve[ Det[matrix]==0,w2]//Simplify;
    freq=  w2/.w2Sol;
    vector=Array[a,dim];
    eqs= Thread[0==matrix.vector]//ExpandAll//Simplify ;
    vSol=
    Table[
       Solve[
          Join[eqs/.w2Sol[[i]],{vector.vector==1}]
      ,vector][[2]]
    ,{i,1,dim}];
    Return[{freq,vSol}]]
```

Example: Two coupled particles

Again, to see how this works, we recommend you examine the intermediate results. Do this by eliminating the local variables by executing a related function **smallOsc2** with the modification:

```
In[28]:= smallOsc2[T_,V_ ,coordList__,equlibList__,t_:t]:=
  Module[{},
    q=coordList;
    p=D[coordList,t];
    dim=Length[q];
    equlibRule=Thread[coordList-> equlibList];
    Vij=Outer[D[V,#1,#2]&,q,q];
    Tij=Outer[D[T,#1,#2]&,p,p];
    matrix= Vij-w2 Tij /.equlibRule;
    w2Sol= Solve[ Det[matrix]==0,w2]//Simplify;
    freq=  w2/.w2Sol;
    vector=Array[a,dim];
    eqs= Thread[0==matrix.vector]//ExpandAll//Simplify ;
    vSol=
    Table[
       Solve[
          Join[eqs/.w2Sol[[i]],{vector.vector==1}]
      ,vector][[2]]
    ,{i,1,dim}];
    Return[{freq,vSol}]]
```

Here the local variables have been removed from the statement **Module[...]**.

Consider two equal masses that are described by the kinetic and potential energies:

```
In[29]:= T=(m  /2) (q1'[t]^2)+(m /2)( q2'[t]^2 );
         V= (k1/2)(q2[t]-q1[t])^2 ;
```

The eigenvalues and eigenfunctions for small oscillations are

```
In[30]:= smallOsc2[ T,V,{q1[t],q2[t]},{0,0},t]

Out[30]=
           2 k1                  1                  1
         {{0, ----}, {{a[1] -> -------, a[2] -> -------},
            m                 Sqrt[2]            Sqrt[2]

                      1                      1
           {a[1] -> -------, a[2] -> -(-------)}}}
                    Sqrt[2]               Sqrt[2]
```

Examining the internal variables, it is obvious that **q** is the coordinate list, **p** is the velocity (momentum) list, and **dim** is the dimension of normal modes. **equilbList** is the list of equilibrium positions, and you convert this into **equlibRule** so that you can make the appropriate replacements:

```
In[31]:= equlibRule

Out[31]= {q1[t] -> 0, q2[t] -> 0}
```

Vij and **Tij** are the potential and kinetic energy tensors, respectively. Note the use of **Outer** to compute the derivative of these matrices with respect to the appropriate coordinates. The matrix is formed from $V_{ij} - \omega^2 T_{ij}$, and the solutions of the **Det** of this matrix yield the eigenfrequencies:

```
In[32]:= w2Sol

Out[32]=
                                2 k1
         {{w2 -> 0}, {w2 -> ----}}
                                 m
```

To find the eigenvectors, solve the equation **matrix.vector==0** with various **w2** values substituted in with **w2Sol**. For example, the eigenvector associated with the first eigenfrequency is

```
In[33]:= Solve[
         Join[eqs/.w2Sol[[1]],{vector.vector==1}]
             ,vector]

Out[33]=
                       1                       1
         {{a[1] -> -(-------), a[2] -> -(-------)},
                     Sqrt[2]                 Sqrt[2]

                      1                 1
           {a[1] -> -------, a[2] -> -------}}
                    Sqrt[2]           Sqrt[2]
```

Note that you get two (dependent) solutions. Select the second of these with **[[2]]**. Using the **Table** command, collect all eigenvectors associated with all eigenfrequencies. Finally **Return** the list of eigenfrequencies and eigenvectors:

```
In[34]:= {freq,vSol}
```

```
Out[34]=        2 k1                 1                 1
         {{0, ────}, {{a[1] -> ───────, a[2] -> ───────},
               m                   Sqrt[2]           Sqrt[2]

                    1                    1
          {a[1] -> ───────, a[2] -> -(───────)}}}
                  Sqrt[2]              Sqrt[2]
```

Also note that the eigensystem was solved by hand (rather than by using a built-in function) to ensure the order of eigenfrequencies and eigenvectors matches.

Animation for linear motion

A valuable *Mathematica* feature is the capability to animate a sequence of graphics. The **oscMovie** function will automate this process for some simple one-dimensional cases:

```
In[35]:=  oscMovie::usage="oscMovie[f,tMax,frames,{range}:Automatic]
          Given a periodic function f of variable t, this procedure
          animates the motion in 1-dimension.
          tMax is the final value of t, the range
          determines the graphics window, and frames is the number
          of frames. The frames are stored in the array plot[i].";
```

```
In[36]:= Clear[oscMovie];
         oscMovie[f_,tMax_,frames_,range_:Automatic]:=
         Module[{t,i},
         Do[
          t=i tMax/frames;
          plot[i]=
            Show[Graphics[{PointSize[.05],Point[{f[t]//N,0}]}
                          ,PlotRange->{range,{-1,1}}]
              ,Axes->True]
         ,{i,1,frames}]
         ];
```

Example: Linear harmonic oscillator

Consider a harmonic function and apply **oscMovie**:

```
In[37]:= oscMovie[ Sin,2 Pi,1,{-1,1}]
```

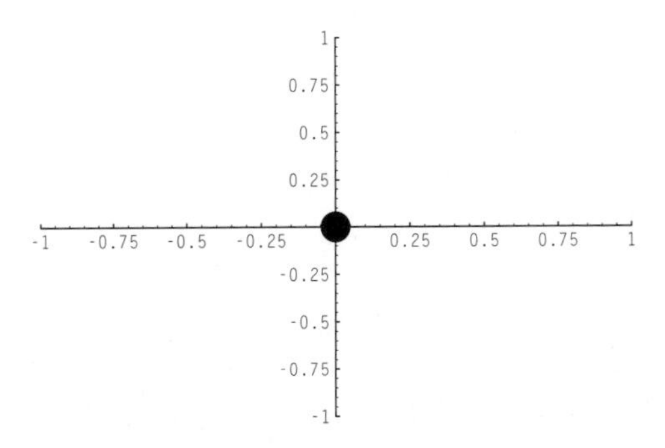

Note: For non-Notebook front ends, the command package **Graphics`Animation`** will define **ShowAnimation**.

```
In[38]:= ShowAnimation[Array[plot,2]]
```

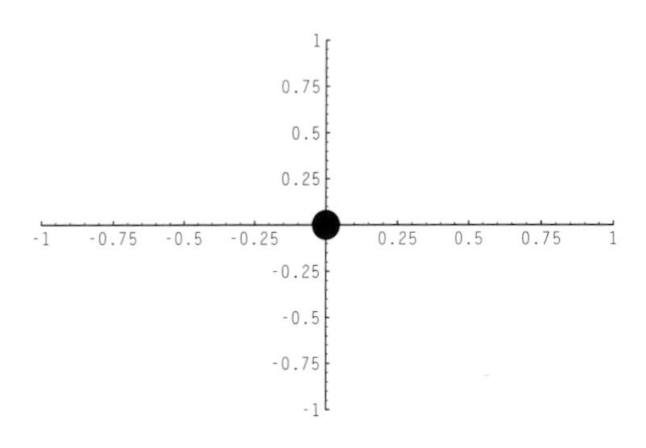

The details of animating the graphs depend on your specific computer configuration.

3.2.3 Protect User-defined Procedures

```
In[39]:= Protect[diffSeriesOne,phasePlot,timePhasePlot,doublePlot
         ,fftSpectrum,smallOsc,oscMovie];
```

3.3 Problems

3.3.1 Linear Oscillations

Problem 1: Analysis of Linear Oscillator

Consider a linear oscillator described by the equation

$$mx''[t] + kx[t] = 0.$$

Part (a) Reduce the linear equation to a first-order differential equation and define its kinetic and potential energies.

Part (b) Plot the potential function and describe the system's behavior.

Part (c) Plot the phase diagram using **ContourPlot**, **DensityPlot**, and **ShadowPlot3D**.

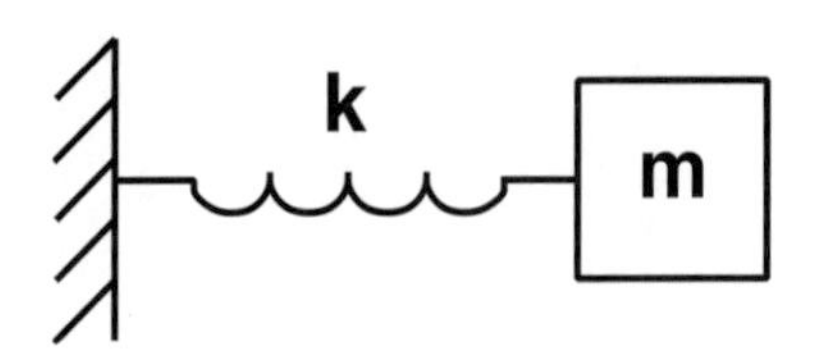

Remarks and outline Potential and phase diagrams are very useful in understanding the behavior of dynamic systems. An expression for the conserved energy follows from multiplying the equation of motion by $x'[t]$ and then integrating. The potential follows from the energy expression. The system's behavior can be visualized by examining a plot of the potential. The phase trajectories also follow from the energy relation. There are several different ways the phase plots can be graphed. Three possible graphics commands that can be used to display the phase plots are illustrated here.

Required packages

```
In[1]:=  Needs["Graphics`FilledPlot`"];
         Needs["Graphics`Graphics3D`"];
```

Solution

```
In[2]:=  Clear["Global`*"];
```

Part (**a**) The oscillator equation is

```
In[3]:=  Fke=m x''[t];
         Fpe=k x[t];
```

To reduce the problem to a first integral, multiply by `x'[t]` and integrate to obtain the kinetic (`T`), potential (`V`), and total **energy** of the system: $E = \int F\,dx = \int F\,(dx/dt)\,dt$:

```
In[4]:=  T=Integrate[x'[t] Fke,t];
         V=Integrate[x'[t] Fpe,t];
         energy =T+V

Out[4]=       2           2
         k x[t]     m x'[t]
         ------- + --------
            2          2
```

Part (**b**) The graph of the potential function follows from applying `Plot` to `V`. To illustrate an alternate way, use the command `FilledPlot`. The graph for `k=1` and `m=1` is

```
In[5]:=  FilledPlot[V/.{x[t]->x,k->1,m->1},{x,-1,1}
             ,Frame->True
             ,PlotLabel->"V"
             ,FrameLabel->{"x-axis","Potential"}
             ,RotateLabel->False
           ];
```

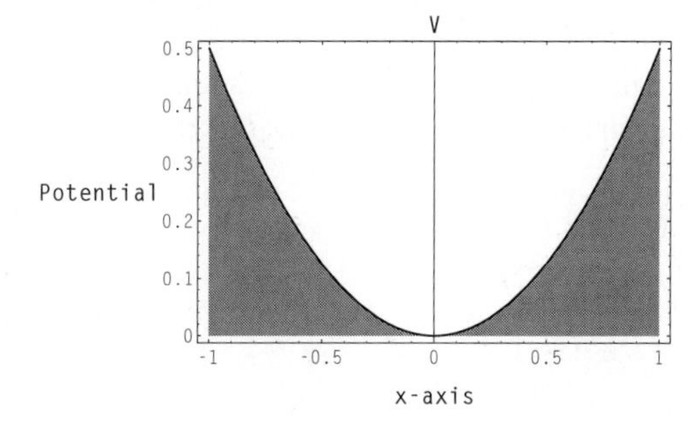

The motion is confined by the potential curve and is bounded for all initial conditions. There is one stable equilibrium position at **x=0**.

Part (c) Different ways are available to make a phase plot for this linear oscillator. In the first, you use **ContourPlot**. Alternatively, you could use **DensityPlot**, although by setting **ContourShading** to **True**, you can achieve the equivalent result:

```
In[6]:=  rules={m->1, k->1,x[t]->x,x'[t]->v};
         energy=energy /.rules
```

```
Out[6]=   2    2
         v    x
         -- + --
         2    2
```

```
In[7]:=  ContourPlot[energy,{x,-1,1},{v,-1,1}
                ,ContourShading->True
                ,Axes->True
                ,AxesLabel->{"x[t]","v[t]"}];
```

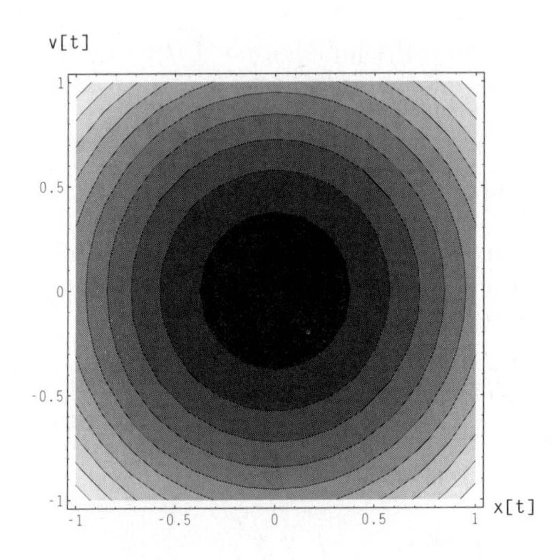

The phase plane is filled with nested circles. The point at $\{x[t] = 0, x'[t] = 0\}$ represents the position of equilibrium and corresponds to zero energy. Periodic motion results from all other initial conditions.

A third way you can graph the phase plane is to plot the energy surface as a function of $\{x[t], x'[t]\}$ and then project the surface onto the plane $\{x[t], x'[t]\}$:

```
In[8]:=  ShadowPlot3D[energy,{x, -1, 1},{v,-1,1}
            ,PlotLabel->"Energy Surface"
            ,Axes->True
            ,AxesLabel->{"x[t]","v[t]","energy"}
            ,ShadowPosition->1];
```

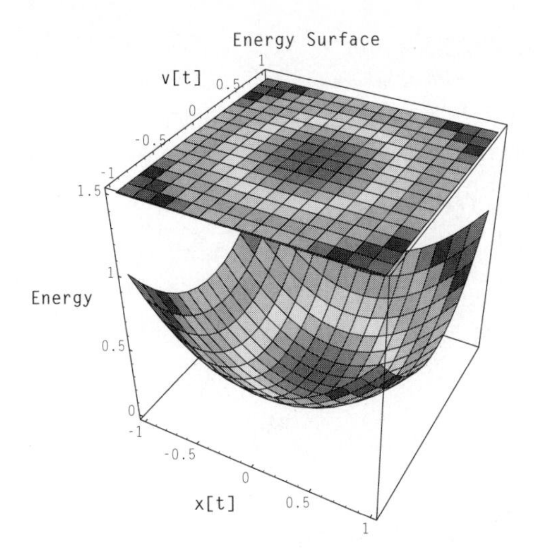

Problem 2: Solution of Linear Oscillator

Consider a linear oscillator described by the equation

$$x''[t] + w_0^2 x[t] = 0.$$

Part (a) Use **DSolve** to solve for the motion. Express the solution in terms of a phase and amplitude and of the initial conditions $x[0]$ and $x'[0]$.

Part (b) **Solve** the equations of motion by assuming the solution can be expressed as a power series in t. **Sum** the series to get the solution in closed form.

Part (c) Use **LaplaceTransform** to solve the equations.

Remarks and outline **DSolve** expresses the solution in terms of two constants. By a proper choice of constants, you can have the solution represented in terms of a phase and amplitude or in terms of the initial conditions $x[0]$ and $x'[0]$. The solution also follows from assuming a power series expansion and summing the series. The series solution follows from using the user-defined procedure defined by the command **diffSeriesOne**. The third method follows from taking the **LaplaceTransform** of the differential equation. These last two methods are not practical for this simple example but are useful for more complex differential equations.

Required packages

```
In[9]:= Needs["Algebra`SymbolicSum`"];
        Needs["Algebra`Trigonometry`"];
        Needs["Calculus`LaplaceTransform`"];
```

Solution

```
In[10]:= Clear["Global`*"];
```

Part (a) The linear oscillator equation is

```
In[11]:=  eq1= x''[t] + w^2 x[t] ==0;
```

The solution follows from **DSolve**:

```
In[12]:= dsol1= DSolve[{eq1},x[t],t][[1]] //PowerExpand

Out[12]=
                 -I t w           I t w
{x[t] -> E             C[1] + E        C[2]}
```

dsol1 can be expressed in terms of a phase **del** and amplitude **A** by defining the constants **C[1]** and **C[2]** by the rule

```
In[13]:= cRule= {C[1]->A/ 2 Exp[-I del]
               ,C[2]->A/2  Exp[ I del]};
```

dsol1 now becomes

```
In[14]:= dsol2=  dsol1 /.cRule //ComplexToTrig //Simplify

Out[14]= {x[t] -> A Cos[del + t w]}
```

The analytic solution can also be expressed in terms of the **initial** conditions:

```
In[15]:= initial={x[0]==x0, x'[0]==v0};
```

Solving **eq1** with these conditions, you get

```
In[16]:= dsol3= ( DSolve[Join[{eq1},initial] ,x[t],t]
                    //ComplexToTrig
                    //PowerExpand
                    //Simplify
                    //Flatten
               )

Out[16]=
                                    v0 Sin[t w]
{x[t] -> x0 Cos[t w] +  ------------}
                                         w
```

Part (b) The series expansion solution of **eq1** follows from the user-defined function **diffSeriesOne**. For more information about **diffSeriesOne**, enter **?diffSeriesOne**. Applying **diffSeriesOne** function to **eq1** and evaluating to order 8, you get

```
In[17]:= dsol4= (diffSeriesOne[eq1,x,t,9,{x[0]->x0,x'[0]->v0}]
               //Collect[#,{x0,v0}]& )

Out[17]=
                 3  2      5  4      7  6      9  8
                t  w      t  w      t  w      t  w
v0 (t - ------- + ------- - ------- + -------) +
                  6        120       5040     362880

             2  2      4  4      6  6      8  8
            t  w      t  w      t  w      t  w
   (1 - ------- + ------- - ------- + -------) x0
              2        24        720      40320
```

Terms are kept to order 8 and grouped according to their initial conditions. Notice that this series solution can be expressed in terms of two sums:

```
In[18]:= termX[n_]= (-1)^n (t w)^(2n)/(2n)!;
         termV[n_]= (-1)^n (t w)^(2n+1)/(w (2n+1)!);
```

as is verified by the finite sum:

```
In[19]:= test= ( Sum[x0 termX[n] + v0 termV[n] ,{n,0,4}]
                 //Collect[#,{x0,v0}]& );
         test==dsol4

Out[19]= True
```

To get the exact answer, **Sum** these two terms to infinity:

```
In[20]:= exactSum=
           (Sum[x0 termX[n] + v0 termV[n] ,{n,0,Infinity}]
            //PowerExpand //Cancel //Apart )

Out[20]=
                     v0 Sin[t w]
         x0 Cos[t w] + -----------
                          w
```

Part (c) Next solve **eq1** using **LaplaceTransform**. Take the **LaplaceTransform** of **eq1**:

```
In[21]:= lapTx= LaplaceTransform[eq1,t,s]

Out[21]=  2                                         2
         s  LaplaceTransform[x[t], t, s] + w  LaplaceTransform[x[t], t, s] -

           s x[0] - x'[0] == 0
```

and solve for **LaplaceTransform[x[t],t,s]**:

```
In[22]:= lapSol= Solve[lapTx,LaplaceTransform[x[t],t,s]][[1]]

Out[22]=
                                                  -(s x[0]) - x'[0]
         {LaplaceTransform[x[t], t, s] -> -(--------------------)}
                                                     2    2
                                                    s  + w
```

Taking the inverse, you get the solution

```
In[23]:= ( InverseLaplaceTransform[
               LaplaceTransform[x[t],t,s]/.lapSol
           ,s,t]
          //PowerExpand )

Out[23]=
                           Sin[t w] x'[0]
         Cos[t w] x[0] + --------------
                                w
```

Problem 3: Damped Linear Oscillator

Consider a damped linear oscillator, where the dissipation is proportional to the velocity. The equation of motion for the damped oscillator is

$$x''[t] + \gamma x'[t] + \omega_0^2 x[t] = 0.$$

The damping is characterized by the coefficient γ.

Part (a) Solve the equation with the initial conditions $x[0] = x_0$, $x'[0] = v_0$. Make a plot of the solution as a function of the variables t and $2\omega = \sqrt{\gamma^2 - 4\omega_0^2}$.

Part (b) For the over-damped case (ω real), plot the solution.

Part (c) For the critically-damped case ($\omega = 0$), find and plot the solution.

Part (d) For the under-damped case (ω imaginary), reduce the solution to trigonometric form. Plot the solution. Also use the user-defined routines to plot the **phasePlot** and **timePhasePlot** using **doublePlot**.

Part (e) Overlay the two-dimensional plots and then comment.

Remarks and outline There are three characteristic types of motion, depending on the values of ω_0 and γ:

1. Over-damped case, when $\gamma^2 > 4\omega_0^2$
2. Critically-damped case, when $\gamma^2 = 4\omega_0^2$
3. Under-damped cases, when $\gamma^2 < 4\omega_0^2$

The general solution follows from **DSolve** and the solution can be plotted using **Plot3D**. The expression for the solution simplifies if the results are expressed in terms of ω, where $\omega^2 = \gamma^2/4 - \omega_0^2$. The expression for the phase trajectories follows from the user-defined functions **timePhasePlot** and **phasePlot**.

Required packages

```
In[24]:= Needs["Algebra`Trigonometry`"];
```

Solution

```
In[25]:= Clear["Global`*"];
```

Part (a) Set $\omega \equiv$ **w** and $\gamma \equiv$ **gam** and consider the damped oscillator equation:

```
In[26]:= eq1= x''[t]+ gam x'[t] + w0^2 x[t] ==0;
```

and apply the command **DSolve**:

```
In[27]:= dsol= DSolve[{eq1, x[0]==x0,x'[0]==v0},x[t],t][[1]];
```

Before displaying the result, you will find it convenient to define $w = \sqrt{gam^2/4 - w_0^2}$ with the substitutions

```
In[28]:= wRule={    (gam^2 - 4*w0^2)^(1/2) ->  (2w)
             , 1/(gam^2 - 4*w0^2)^(1/2) ->1/(2w)};
```

We then find

```
In[29]:= dsol= (dsol //Simplify) //.wRule;
         x[t_]= x[t] /.dsol

Out[29]=
                                       (t (-gam + 2 w))/2
         -2 v0 - gam x0 + 2 w x0      E                    (2 v0 + gam x0 + 2 w x0)
         ----------------------- + --------------------------------------------------
              (t (gam + 2 w))/2                            4 w
         4 E                  w
```

It is obvious from **dsol** that there are three cases, depending on whether $2\omega = \sqrt{\gamma^2 - 4\omega_0^2}$ is real ($\gamma^2 > 4\omega_0^2$), zero ($\gamma^2 = 4\omega_0^2$), or imaginary ($\gamma^2 < 4\omega_0^2$). All three behaviors are illustrated in a single plot with the command **Plot3D**. Choose the following initial values:

```
In[30]:=  values={x0->1, v0->0, gam->1};
```

Note also that the replacement $\omega \to \sqrt{\omega^2}$ is made so that **w** can be made complex by setting **w2** to a negative value. By varying **w2** over negative, zero, and positive values, you obtain all three types of solutions noted above.

```
In[31]:=  p1=Plot3D[ Evaluate[ x[t] //.values /.{w->Sqrt[w2]}]
                ,{w2,25,-150},{t,0,1}
                ,PlotPoints->25
                ,PlotRange->{-1,2}
                ,BoxRatios->{1,1,1}
                ,ViewPoint->{-4, 3, 2}
                ,AxesLabel->{"w2","t","x[t]"}];
```

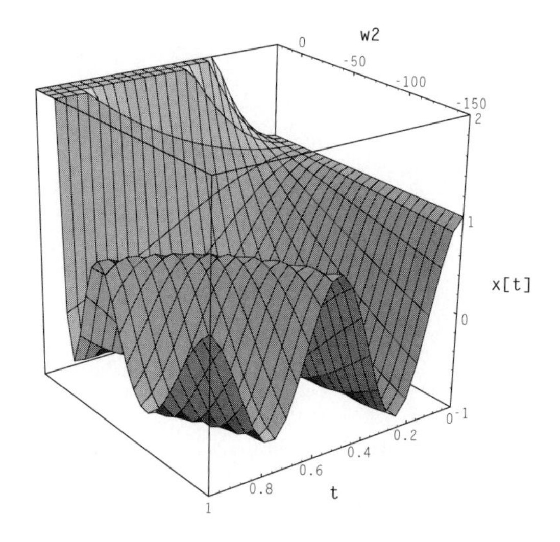

In the following plot, it is clear that as **w2** becomes large and negative, the solution **x[t]** becomes a damped oscillating function. **w2=0** corresponds to the critical damping case. For **0<2w<gam**, the solution is a decaying exponential. For **gam<2w**, the solution is an increasing exponential.

Part (b) Consider the over-damped case in which **w** is real: ($\gamma^2 > 4\omega_0^2$). In this case, **x[t]** is purely exponential. The exponential can be increasing ($2\omega > \gamma$) or decreasing ($0 < 2\omega < \gamma$). Both cases are illustrated next.

```
In[32]:=  p1=Plot[Evaluate[x[t] //.values /.{w->+0.6}],{t,0,10}];
```

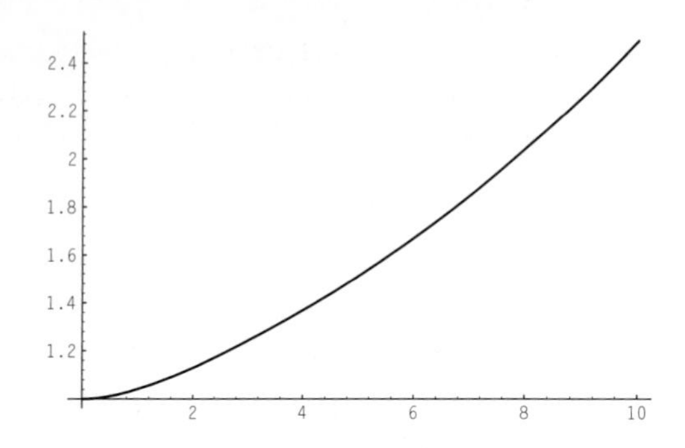

In[33]:= `p2=Plot[Evaluate[x[t] //.values /.{w->+0.4}],{t,0,10}];`

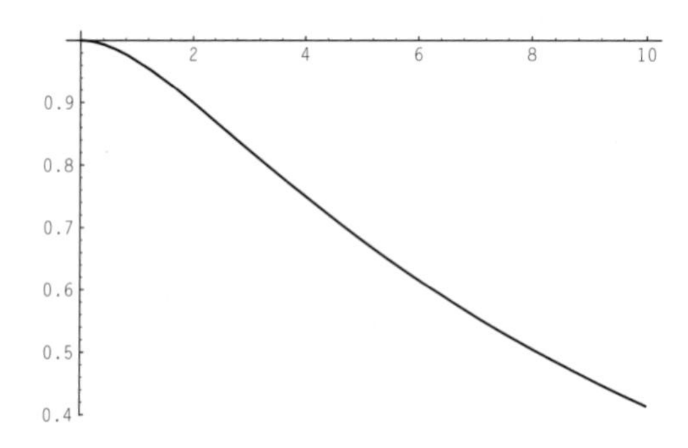

Part (c) Consider the critically-damped case in which **w=0** ($\gamma^2 = 4\omega_0^2$). In this case, the solution for **x[t]** becomes simply

In[34]:= `Limit[x[t],w->0]`

```
Out[34]= 2 t v0 + 2 x0 + gam t x0
         ------------------------
                (gam t)/2
             2 E
```

In[35]:= `p3=Plot[Evaluate[Limit[x[t] //.values, w->0]],{t,0,10}];`

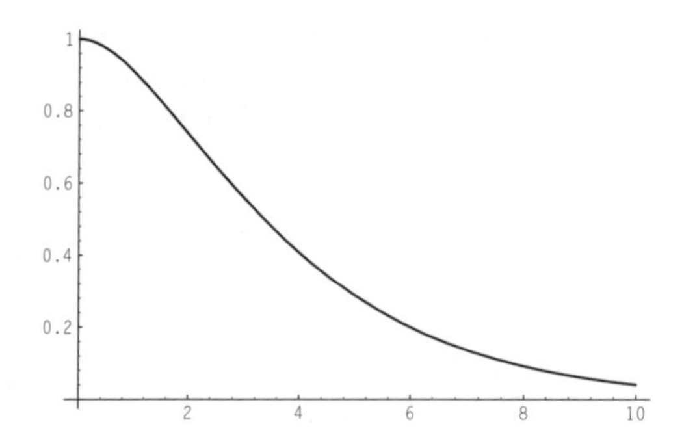

Part (d) Consider the under-damped case in which **w** is imaginary ($\gamma^2 < 4\omega_0^2$). The notation is simplified by replacing **w** by an explicitly imaginary quantity **I wRe** and then converting the complex exponentials to trigonometric functions:

```
In[36]:= ( x[t] /.{w-> I wRe}
                //ComplexToTrig
                //Simplify
         )

Out[36]= 2 wRe x0 Cos[t wRe] + 2 v0 Sin[t wRe] + gam x0 Sin[t wRe]
         -----------------------------------------------------------
                                (gam t)/2
                             2 E          wRe
```

Clearly, this solution oscillates with frequency **wRe** and is decaying with exponential $e^{-\gamma t/2}$. This behavior is showed in the following plot:

```
In[37]:= p4=Plot[Evaluate[x[t] //.values /.{w-> I 10}],{t,0,10}];
```

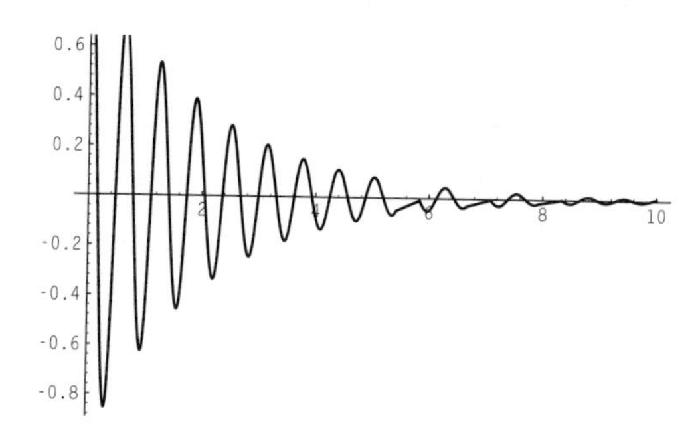

It is useful to plot the phase trajectory and the time evolution. This is easily done using the user-defined function **doublePlot** (output suppressed):

```
In[38]:= q= x[t] //.values/.{w-> I 10};
         p=x'[t] //.values/.{w-> I 10};

         doublePlot[q,p,{t,0.01,5}];
```

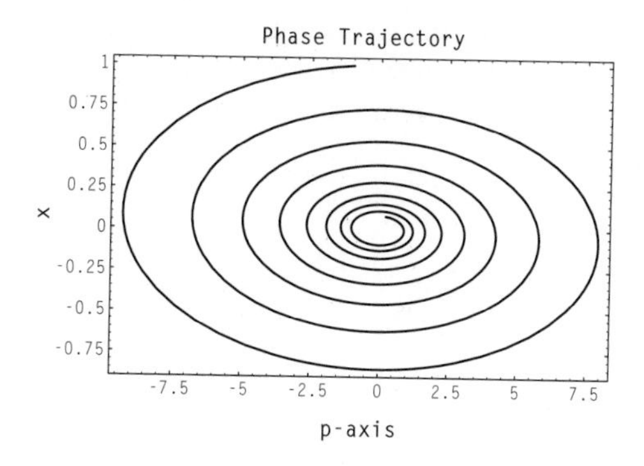

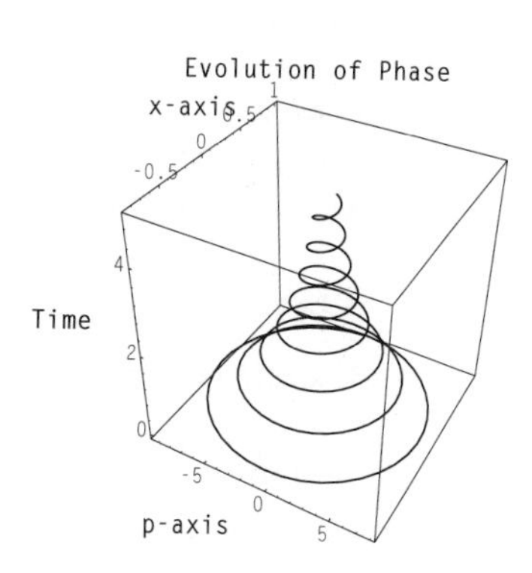

Part (e) Finally, overlay all the two-dimensional plots for purposes of comparison:

```
In[39]:= Show[p1,p2,p3,p4];
```

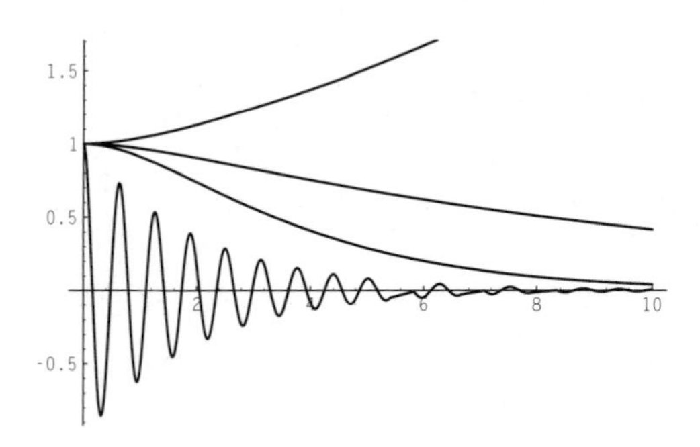

Problem 4: Damped Harmonic Oscillator and Driving Forces

Consider the motion of an under-damped harmonic oscillator with a harmonic forcing term. The motion is described by the equation

$$x''[t] + \gamma x'[t] + \omega_0^2 x[t] == Q_0 \mathrm{Cos}[\omega_d t],$$

where $4\omega_0^2 > \gamma$. The solution of this equation is the superposition of a homogeneous solution plus a particular solution.

Part (a) Solve for the homogeneous solution.

Part (b) Assume the particular solution is of the form $A\mathrm{Cos}[\omega_d t + \delta_f]$ and solve for the amplitude A and phase δ_f. Write the general form of the solution.

Part (c) Plot the solution as a function of ω_0 and t. Plot the amplitude A as a function of ω_0 and γ. Notice the resonance behavior at $\omega_0 = \omega_d$.

Part (d) Pick parameters to illustrate the transient and steady-state behaviors and plot $x[t]$ and its phase trajectory.

Remarks and outline Substituting $A\mathrm{Cos}[\omega_d t + \delta_f]$ into the differential equation reduces the problem to an algebraic equation with $\mathrm{Sin}[\omega_d t]$ and $\mathrm{Cos}[\omega_d t]$ terms. Setting the coefficients in front of the $\mathrm{Sin}[\omega_d t]$ and $\mathrm{Cos}[\omega_d t]$ terms to zero gives two equations that can be solved for δ_f and A. The solution can be plotted with the command **Plot3D**. The plot of the phase trajectory follows from the user-defined procedures **phasePlot** and **timePhasePlot**.

Required packages

```
In[40]:= Needs["Algebra`Trigonometry`"];
```

Solution

```
In[41]:= Clear["Global`*"];
```

Set $\omega \equiv$ **w** and $\gamma \equiv$ **gam**. The damped harmonic oscillator with a forcing term is

```
In[42]:= eq1= x''[t] +gam x'[t] +w0^2 x[t] ==Q0 Cos[wd t];
```

Part (a) Solve the homogeneous equation by setting the forcing term to zero. This reduces to Problem 3. Following these steps, you can quickly obtain

```
In[43]:= dsol= DSolve[{eq1/.{Q0->0}, x[0]==x0,x'[0]==v0},x[t],t][[1]];

         wRule={    (gam^2 - 4*w0^2)^(1/2) ->  (2w)
               , 1/(gam^2 - 4*w0^2)^(1/2) ->1/(2w)};

         dsol=dsol //.wRule //Simplify;

         homoSol=( x[t] /.dsol
                        /.{w -> I wRe}
                        //ComplexToTrig
                        //Simplify
                  )

Out[43]= 2 wRe x0 Cos[t wRe] + 2 v0 Sin[t wRe] + gam x0 Sin[t wRe]
         ------------------------------------------------------------
                                (gam t)/2
                             2 E          wRe
```

Because we will be interested in oscillating solutions, **w** has been taken to be pure imaginary and **w-> I wRe** has been substituted.

Part (b) Determine the particular solution by guessing a function of the form

```
In[44]:= guess= {x->  (A Cos[wd # + df]&)};
```

and solve for **A** and **df**. By substituting **guess** into **eq1** you get an algebraic equation:

```
In[45]:= eq2= (eq1/.guess)//TrigReduce

Out[45]= -(A gam wd (Cos[t wd] Sin[df] + Cos[df] Sin[t wd])) -

              2
           A wd  (Cos[df] Cos[t wd] - Sin[df] Sin[t wd]) +

              2
           A w0  (Cos[df] Cos[t wd] - Sin[df] Sin[t wd]) == Q0 Cos[t wd]
```

Two independent equations are obtained from **eq2** by considering the orthogonal **Sin[t wd]** and **Cos[t wd]** dependence:

```
In[46]:= eq3=( {eq2/.{Sin[_ t]:>0}
              ,eq2/.{Cos[_ t]:>0}}
                    /.{Sin[df]-> dfTan Cos[df]}
             )

Out[46]=                                                  2
         {-(A dfTan gam wd Cos[df] Cos[t wd]) - A wd  Cos[df] Cos[t wd] +

               2
            A w0  Cos[df] Cos[t wd] == Q0 Cos[t wd],

                                                               2
           -(A gam wd Cos[df] Sin[t wd]) + A dfTan wd  Cos[df] Sin[t wd] -

                      2
            A dfTan w0  Cos[df] Sin[t wd] == 0}
```

Notice the replacement **Sin[df]-> dfTan Cos[df]**. This substitution allows you to solve for **dfTan** (= Tan[*df*]) without the complication of a transcendental equation. The solution is

```
In[47]:= sol1=Solve[eq3,{A,dfTan}][[1]] //Simplify

Out[47]=
                                          2     2
                 gam wd              Q0 (-wd  + w0 ) Sec[df]
{dfTan -> ---------------, A -> -------------------------------------}
              2     2             2   2     4       2   2     4
            wd   - w0          gam  wd  + wd  - 2 wd  w0  + w0
```

This is almost what is needed, except **A** is implicitly defined in terms of **Sec[df]**. You can use the solution for **dfTan** to find **A** explicitly:

```
In[48]:= dfRule={df->ArcTan[dfTan]};
         secRule={Sec[ArcTan[x_]]:>Sqrt[1+x^2]};

         answerA= ( A //.Join[sol1,dfRule,secRule]
                      //.sol1
                      //Together
                      //PowerExpand
                      //Cancel
                  )

Out[48]=
                            Q0
-(---------------------------------------------------)
              2   2     4       2   2     4
    Sqrt[gam  wd  + wd  - 2 wd  w0  + w0 ]
```

Having evaluated **A**, next construct the particular solution. Joining the relevant rules into **sol**, you get

```
In[49]:= sol=Join[{A->answerA},{sol1[[1]]},dfRule]

Out[49]=
                            Q0                                            gam wd
{A -> -(---------------------------------------------------), dfTan -> ------------,
                    2   2     4       2   2     4                          2     2
          Sqrt[gam  wd  + wd  - 2 wd  w0  + w0 ]                         wd  - w0

  df -> ArcTan[dfTan]}
```

This rule yields the particular solution:

```
In[50]:= partSol= x[t] //.guess //.sol

Out[50]=
                                     gam wd
        Q0 Cos[t wd + ArcTan[-----------------]]
                                   2      2
                                 wd   - w0
-(---------------------------------------------------)
              2   2     4       2   2     4
    Sqrt[gam  wd  + wd  - 2 wd  w0  + w0 ]
```

The general solution **genSol** is then a superposition of the particular solution **partSol** and the homogeneous solution **homoSol**:

```
In[51]:= genSol= partSol + homoSol
```

```
Out[51]=
                                  gam wd
         Q0 Cos[t wd + ArcTan[―――――――――]]
                                2     2
                              wd  - w0
  -(―――――――――――――――――――――――――――――――――――――――――――) +
            2   2     4       2   2     4
      Sqrt[gam  wd  + wd  - 2 wd  w0  + w0 ]

      2 wRe x0 Cos[t wRe] + 2 v0 Sin[t wRe] + gam x0 Sin[t wRe]
      ―――――――――――――――――――――――――――――――――――――――――――――――――――――――――
                             (gam t)/2
                          2 E          wRe
```

Part (c) To plot the solution as a function of **w0** and **t**, let

```
In[52]:= values={A->3,  Q0->1, wd->1
                ,x0->0, v0->1
                ,wRe->Sqrt[w0^2-gam^2/4]};
```

Applying the command **Plot3D** to **genSol** with these values, you get

```
In[53]:= Plot3D[ genSol //.values //.{gam->0.1} //Evaluate
                ,{t,0,30}
                ,{w0,0.5,1.5}
                ,PlotPoints->40
                ,AxesLabel->{"Time","w0","x[t]"}
                ,FaceGrids->{{0,1,0},{0,-1,0}}
                ,ViewPoint->{1,.4,.6}];
```

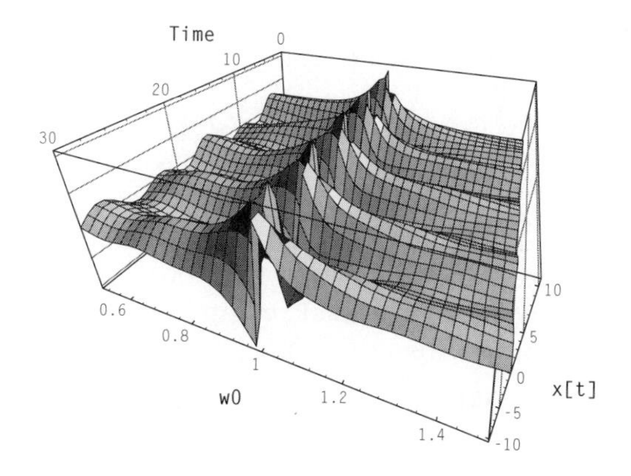

Notice the resonance behavior at **w0=wd=1**. The oscillations are not simple harmonic but contain transient effects. The resonance behavior follows from the form of **A** (plot **-A** for convenience):

```
In[54]:= Plot3D[  -A //.sol //.values //Evaluate
                ,{gam,0,4}
                ,{w0,0,2}
                ,PlotPoints->25
                ,PlotRange->{0,5}
                ,AxesLabel->{"gam","w0","Amplitude"}
                ,FaceGrids->{{-1,0,0} }
                ,ViewPoint->{-1,0.4,0.4} ];
```

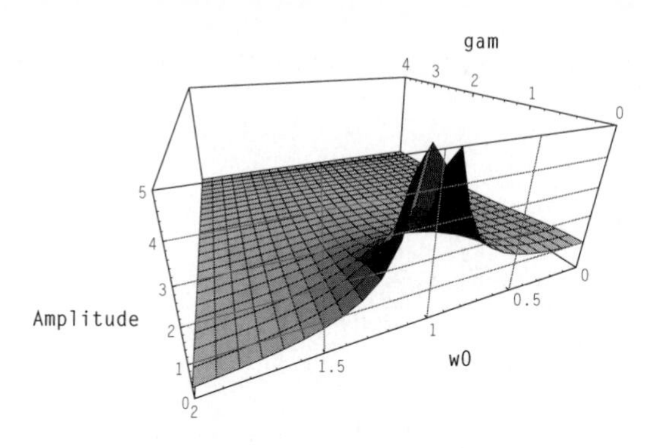

The resonance occurs at **w0=wd=1**, and its height diverges as **gam->0**. The divergence is clipped, since **PlotRange** is set to $\{0,5\}$.

Part (d) The transient and steady-state behavior of **genSol** are illustrated by

In[55]:=
```
Plot[genSol //.values //.{gam->0.1, w0->2} //Evaluate
      ,{t,0,80}];
```

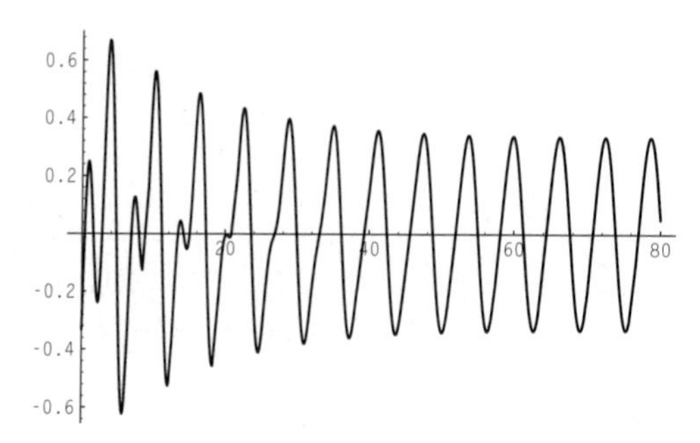

For **t**<50, both the transient and steady-state solutions are important; for **t**>50, the solution is dominated by the particular (steady-state) solution. The phase trajectory follows from the user-defined commands **timePhasePlot** and **phasePlot**, which are combined in **doublePlot** (output suppressed):

In[56]:=
```
q=genSol //.values //.{gam->0.1, w0->2};
p=D[q,t];
doublePlot[q,p,{t,0,80}];
```

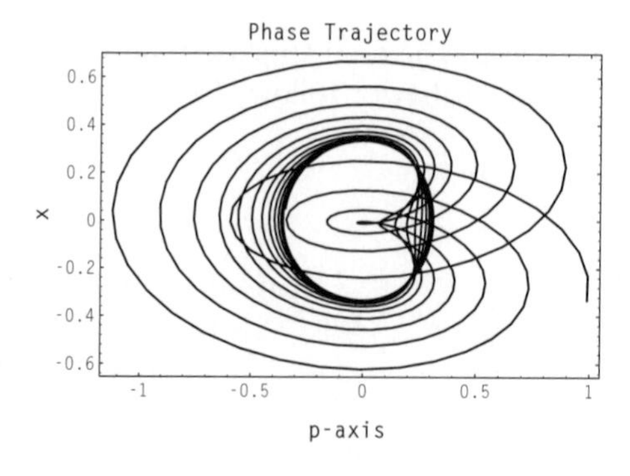

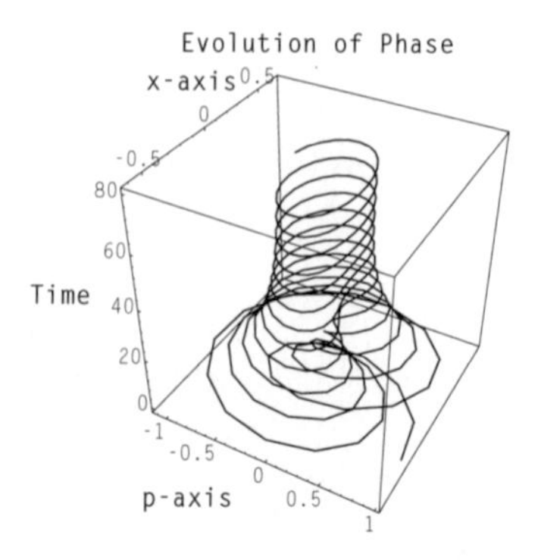

The first plot illustrates how the transient effect dies off with time; the second plot shows the asymptotic phase trajectory.

3.3.2 Nonlinear Oscillations

Problem 1: Duffing's Oscillator Equation

Consider a system described by the nonlinear equation of motion,

$$q''[t] - aq[t] + bq^3[t] = 0.$$

Assume $b > 0$ and let a be either positive or negative. This equation is called the **Duffing equation**.

Part (a) Illustrate the double-well and single-well behavior of the potential by plotting the potential as a function of a and q. Pick values of a that correspond to the two kinds of behavior and plot the potential as a function of q. Plot the phase diagram for the single- and double-well behavior.

Part (b) Numerically solve for the trajectories and plot $q[t]$ and $q'[t]$.

Part (c) Plot the phase trajectories for the single- and double-well behavior.

Part (d) Animate one of the solutions.

Remarks and outline An expression for energy follows from multiplying the equation of motion by $q'[t]$ and then integrating. Potential and phase diagrams are graphed using the commands **Plot**, **Plot3D**, and **ContourPlot**. The nonlinear equation does not have a closed form for the solution, so an interpolating function is calculated using **NDSolve**. The phase trajectories follow from the user-defined function **phasePlot**. The animation follows from the user-defined function **oscMovie**.

Required packages *There are no required packages.*

Solution

```
In[57]:= Clear["Global`*"];
```

Part (a) The oscillations are described by the forces

```
In[58]:= Fke  =  q''[t];
         Fpe  = -a q[t];
         Fpert=  b q[t]^3;

         force= Fke + Fpe + Fpert

Out[58]=
                              3
         -(a q[t]) + b q[t]  + q''[t]
```

in which have been separated the kinetic energy force **Fke**, the linear potential force **Fpe**, and the nonlinear force **Fpert**, which we will consider a perturbation on the linear potential. To find the energy, multiply **force** by **q'[t]** and **Integrate** with respect to **t**: $E = \int F\,dq = \int F\,(dq/dt)dt$:

In[59]:=
```
T =Integrate[ q'[t] Fke   ,t];
V0=Integrate[ q'[t] Fpe   ,t];
V1=Integrate[ q'[t] Fpert ,t];

energy =  T + V0 + V1
```

Out[59]=
```
            2          4        2
-(a q[t] )    b q[t]     q'[t]
----------- + ------- + ------
     2           4         2
```

To graph the potential, fix **b** and let **q** and **a** be variables. Next apply **Plot3D** to the total potential **V0+V1** and vary **q** and **a**:

In[60]:=
```
Plot3D[V0+V1 //.{q[t]->q, b->0.05} //Evaluate
        ,{q,-15,15}
        ,{a,-1,4}
        ,AxesLabel->{"q","a","Potential"}
        ,BoxRatios->{1,1,1}
        ,PlotRange->{-100,100}
        ,ViewPoint->{0,3,2}
      ];
```

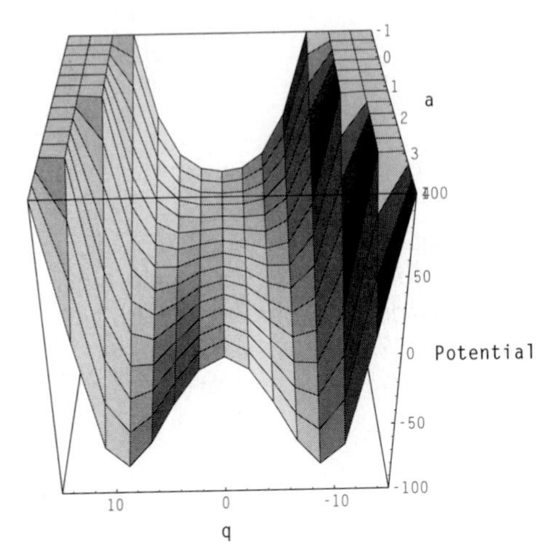

The three-dimensional plot illustrates how the shape of the potential changes with **a**, from a two-well potential (**a**>0) to a one-well potential (**a**<0). A qualitative change in the dynamics is produced when the potential makes this change. This value **a=0** is called a **bifurcation value**, and this particular bifurcation point is known as a pitchfork. Graph the two characteristic types of potentials by choosing values for **a** and **b** (output suppressed):

In[61]:=
```
plot1=
   Plot[V0+V1 //.{q[t]->q, b->0.05, a->-1}
     ,{q,-15,15}
     ,PlotLabel->"One Well Potential"];
```

In[62]:=
```
plot2=
   Plot[V0+V1 //.{q[t]->q, b->0.05, a->+4}
     ,{q,-15,15}
     ,PlotLabel->"Two Well Potential"];
```

```
In[63]:= Show[ GraphicsArray[{plot1,plot2}] ];
```

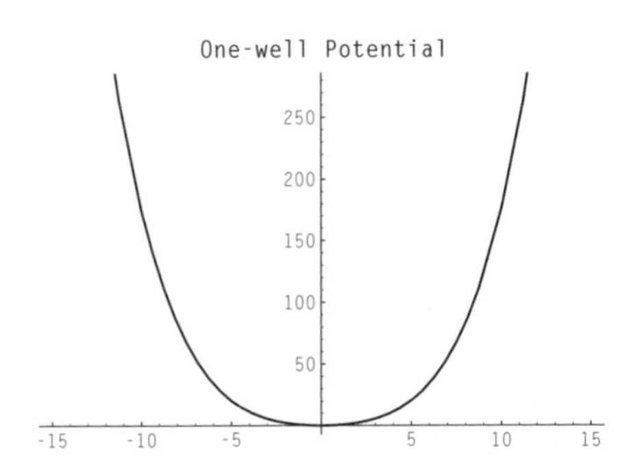

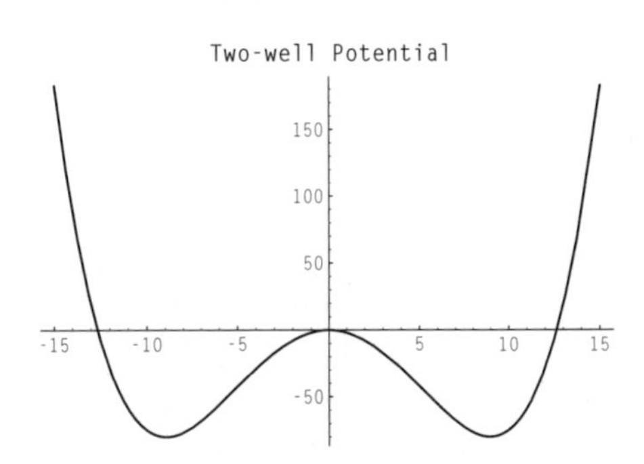

The two-well potential (**a**>0) has three equilibrium positions: Two are stable, while that at the origin is unstable. There is only one stable equilibrium position for the potential with **a**<0. The phase diagrams for these two characteristic potentials follow from a **ContourPlot** of the energy. Make the replacements:

```
In[64]:= values={q[t]->q, q'[t]->v, b->0.05};
```

For the single-well potential, you have (output suppressed)

```
In[65]:= cplot1=
           ContourPlot[energy //.values //.{a->-1} //Evaluate
              ,{q,-15,15}
              ,{v,-15,15}
              ,Contours->8
              ,PlotPoints->20];
```

For the double-well potential, you have (output suppressed)

```
In[66]:= cplot2=
           ContourPlot[energy //.values //.{a->+4} //Evaluate
              ,{q,-15,15}
              ,{v,-15,15}
              ,Contours->8
              ,PlotPoints->20];
```

Combine the plots to simplify comparison:

```
In[67]:= Show[ GraphicsArray[{cplot1,cplot2}] ];
```

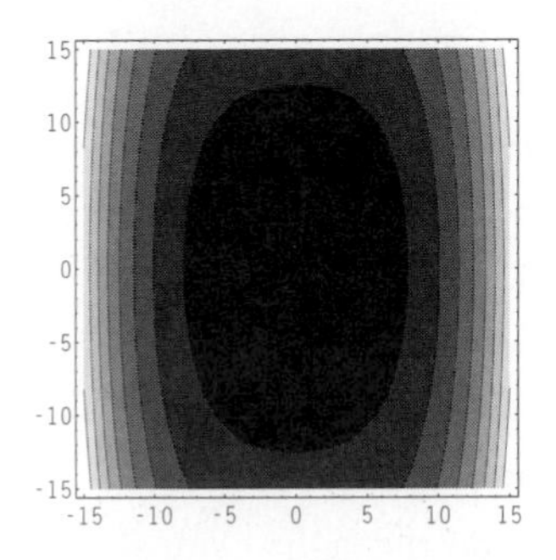

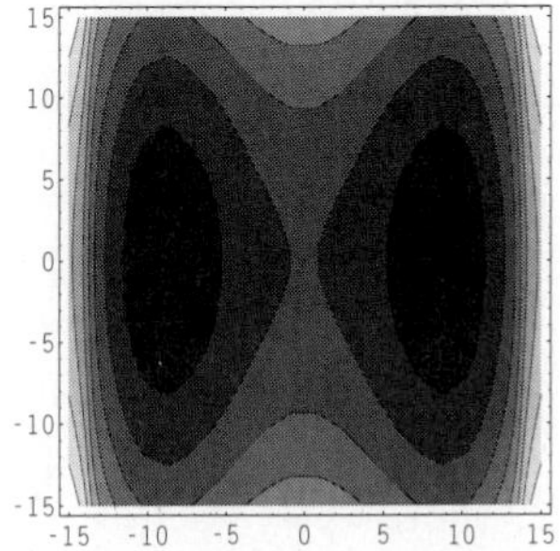

Part (b) The solution of the nonlinear equation must be done numerically. Express this as two first-ordered equations:

```
In[68]:= eq1= {(force /.{q''[t]->p'[t]})==0
              ,q'[t]==p[t]}

Out[68]=                       3
         {-(a q[t]) + b q[t]  + p'[t] == 0, q'[t] == p[t]}
```

and apply **NDSolve**, with the initial conditions

```
In[69]:= initial= {p[0]==0.001, q[0]==0};
         eq2=Join[eq1,initial];
```

Applying **NDSolve** to **eq2** for the case of a single-well potential (again, choosing **b=0.05**), you get

```
In[70]:= ndsol1=NDSolve[eq2 //.{b->0.05, a->-1}
                 ,{q[t],p[t]},{t,0,30}][[1]]

Out[70]= {q[t] -> InterpolatingFunction[{0., 30.}, <>][t],

           p[t] -> InterpolatingFunction[{0., 30.}, <>][t]}
```

The solution is returned in terms of an **InterpolatingFunction**. You will find it convenient to define **q1** and **p1** as the **q** and **p** solutions of the single-well case.

```
In[71]:= q1[t_]=q[t] //.ndsol1;
         p1[t_]=p[t] //.ndsol1;
```

Likewise, applying **NDSolve** to **eq2** for the case of a double-well potential (again choosing **b=0.05**), you get

```
In[72]:= ndsol2=NDSolve[eq2 //.{b->0.05, a->+4}
                 ,{q[t],p[t]},{t,0,30}][[1]]

Out[72]= {q[t] -> InterpolatingFunction[{0., 30.}, <>][t],

           p[t] -> InterpolatingFunction[{0., 30.}, <>][t]}

In[73]:= q2[t_]=q[t] //.ndsol2;
         p2[t_]=p[t] //.ndsol2;
```

where **q2** and **p2** are the **q** and **p** solutions of the double-well case.

Part (c) Plot **q** and **p** versus **t** for the single-well potential (output suppressed):

```
In[74]:= Show[
           GraphicsArray[
             { Plot[q1[t],{t,0,30}]
              ,Plot[p1[t],{t,0,30}] }
           ]
         ];
```

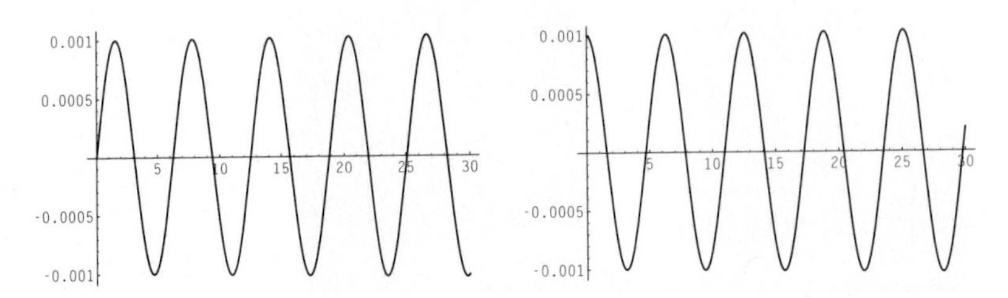

The graph on the left-hand side is for **q1**; that on the right-hand side is for **p1**. Note that the solution is simple oscillatory. For the double-well potential, the **q2** and **p2** plots versus **t** are (output suppressed)

```
In[75]:=  Show[
            GraphicsArray[
              { Plot[q2[t],{t,0,30}]
               ,Plot[p2[t],{t,0,30}] }
            ]
          ];
```

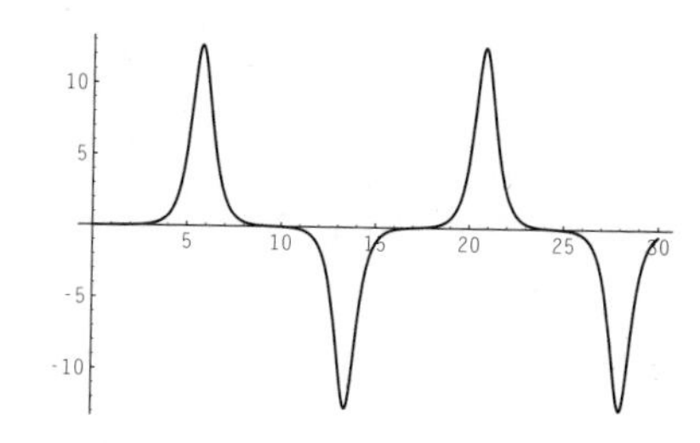

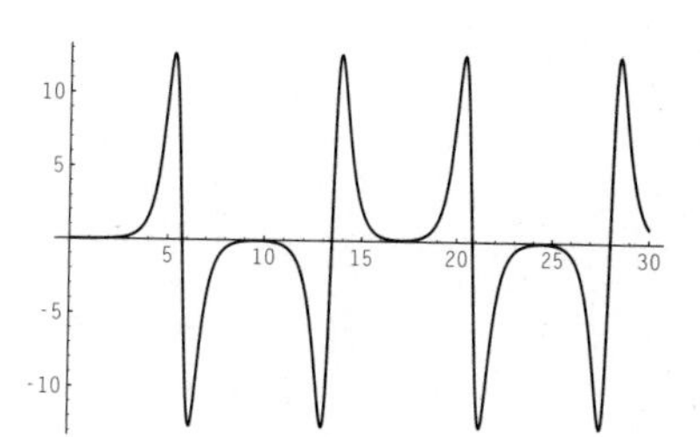

The left-hand graph is for **q2**; the right-hand graph is for **p2**. Note that the solution is a complex oscillating function.

Next plot the phase trajectory and the time evolution in phase space. This is done efficiently with **doublePlot**. First, for the single-well case, the result is (output suppressed)

```
In[76]:=  doublePlot[ q1[t] ,p1[t] ,{t,0,30}];
```

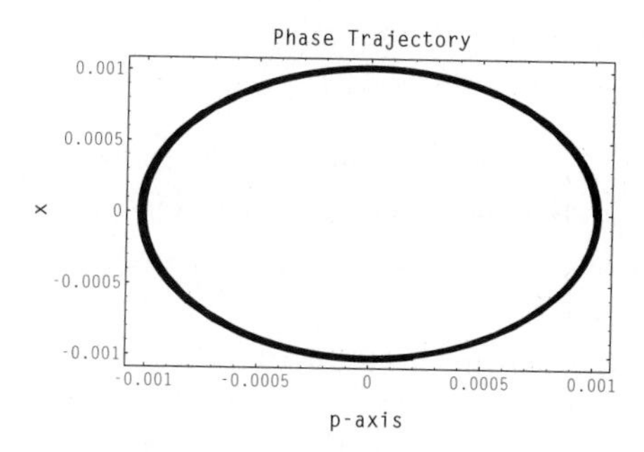

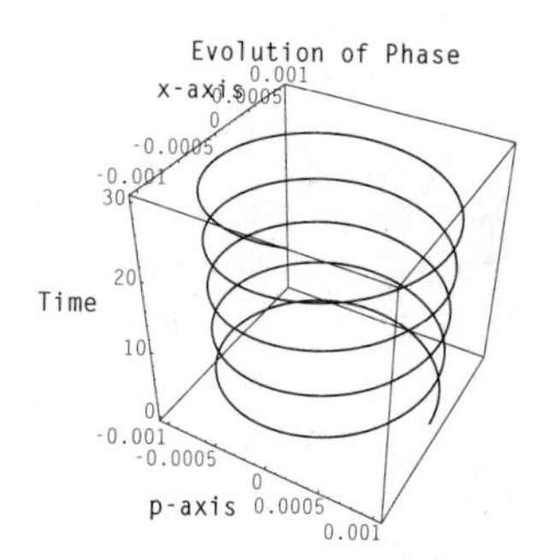

For the double-well case, the result is (output suppressed)

```
In[77]:=  doublePlot[ q2[t] ,p2[t] ,{t,0,30}];
```

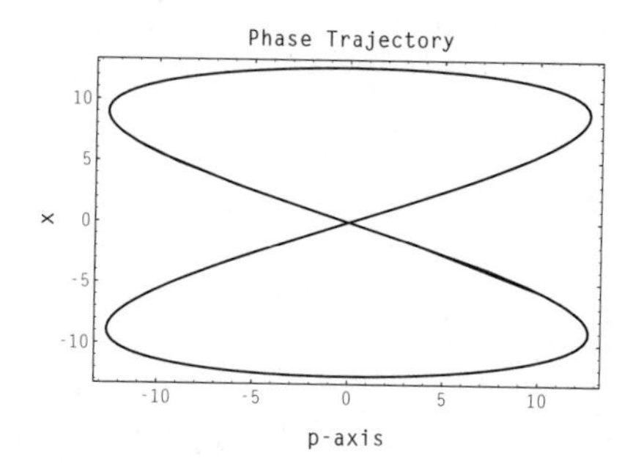

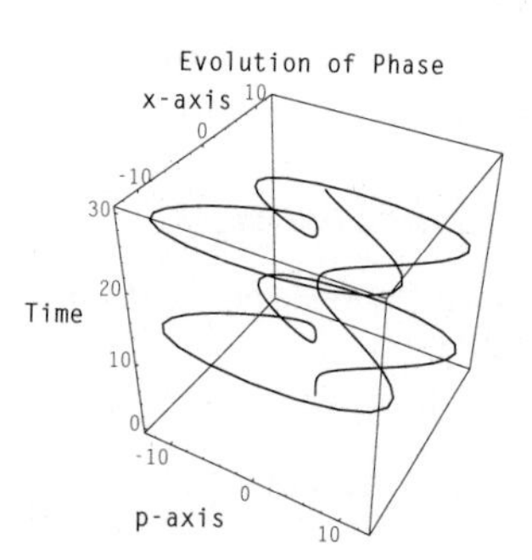

Part (d) The animation of the motion follows from the user-defined function **oscMovie**. For details about this command, enter **?oscMovie**.

Consider the animation for motion in the double-well potential. The variable **q2** corresponds to the linear motion of this system. To generate the graphics frames, apply **oscMovie**:

```
In[78]:= frames=2;
         oscMovie[q2, 30, frames,{-15,15}];
```

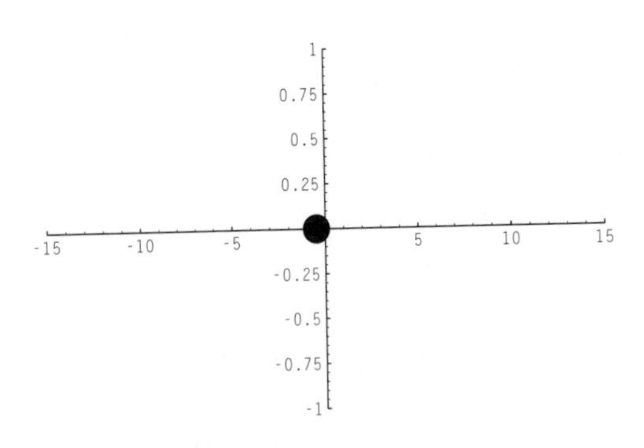

Problem 2: Forced Duffing Oscillator for Double-well Potential

Consider the two-well Duffing oscillator ($a > 0$) with a damping term and a harmonic forcing term. Assume the motion is described by the equation

$$q''[t] == aq[t] - bq^3[t] - \gamma q'[t] + Q_0 \mathrm{Cos}[\omega t]$$

with the initial conditions $q[0] = 0$ and $p[0] = 0.001$. Assume $a = 0.4$, $b = 0.5$, $\gamma = 0.2$, and $\omega = 1/8$ and consider the two trajectories defined by (1) $Q_0 = 0$ and (2) $Q_0 = 0.1$.

Part (a) Numerically solve the equations of motion for these two cases and plot $q[t]$ and $p[t]$.

Part (b) Graph the phase trajectory and spectrum for these two cases.

Remarks and outline The interpolating function for the differential equation is calculated using **NDSolve**.The damping term causes the solution to spiral towards the bottom of one of the potential wells. The forcing term produces chaotic-like behavior. The phase trajectories follow from the user-defined function **phasePlot** and **timePhasePlot**. These are combined with **doublePlot**. The Fourier spectrum follows from the user-defined function **ffspectrum**. All of these graphs illustrate the chaotic behavior produced by the forcing function.

Required packages

```
In[79]:= Needs["Calculus`FourierTransform`"];
```

Solution

```
In[80]:= Clear["Global`*"];
```

Part (a) The forced Duffing equation

```
In[81]:= eq1= q''[t] == a q[t] -  b q[t]^3 - gam q'[t] + Q0 Cos[w t];
```

can be written as two first-order differential equations:

```
In[82]:= eq2= eq1 //.{q''[t]->p'[t], q'[t]->p[t]};
         eq3= q'[t]==p[t];
         eq4={eq2,eq3}

Out[82]=                                                     3
         {p'[t] == Q0 Cos[t w] - gam p[t] + a q[t] - b q[t] , q'[t] == p[t]}
```

The initial conditions are assumed to be

```
In[83]:= initial={ p[0]==0.001 ,q[0]==0};
```

and the parameters have the following values:

```
In[84]:= values={a->0.4, b->0.5, w->1/8, gam->0.02};
```

To start with, set **Q0->0**. This corresponds to a solution with damping, but without a forcing function. The solutions for **q** and **p** follow from applying **NDSolve** to **eq4**:

```
In[85]:= ndsol1=
         NDSolve[Join[eq4,initial] //.values //.{Q0->0}
            ,{q[t],p[t]}
            ,{t,0,120}
            ,MaxSteps->700 ][[1]]

Out[85]= {q[t] -> InterpolatingFunction[{0., 120.}, <>][t],

           p[t] -> InterpolatingFunction[{0., 120.}, <>][t]}
```

The solution is returned in terms of an **InterpolatingFunction**. For convenience, define **q1** and **p1** as the **q** and **p** solutions of the **Q0=0** case:

```
In[86]:= q1[t_]=q[t] //.ndsol1;
         p1[t_]=p[t] //.ndsol1;
```

Likewise, the solution for **Q0->0.1** follows from applying **NDSolve** to **eq4** with the appropriate substitution:

```
In[87]:= ndsol2=
         NDSolve[Join[eq4,initial] //.values //.{Q0->0.1}
            ,{q[t],p[t]}
            ,{t,0,120}
            ,MaxSteps->700 ][[1]]

Out[87]= {q[t] -> InterpolatingFunction[{0., 120.}, <>][t],

           p[t] -> InterpolatingFunction[{0., 120.}, <>][t]}
```

The solution is returned in terms of an **InterpolatingFunction**. For convenience, define **q2** and **p2** as the **q** and **p** solutions of the **Q0=0.1** case:

```
In[88]:= q2[t_]=q[t] //.ndsol2;
         p2[t_]=p[t] //.ndsol2;
```

The plots for **q[t]** and **p[t]** follow from the command **Plot**. To facilitate comparison, combine these plots into a **GraphicsArray**. First the **Q0=0** case:

In[89]:=
```
Show[
  GraphicsArray[
    {Plot[q1[t],{t,0,120}]
    ,Plot[p1[t],{t,0,120}]
    }
  ]
];
```

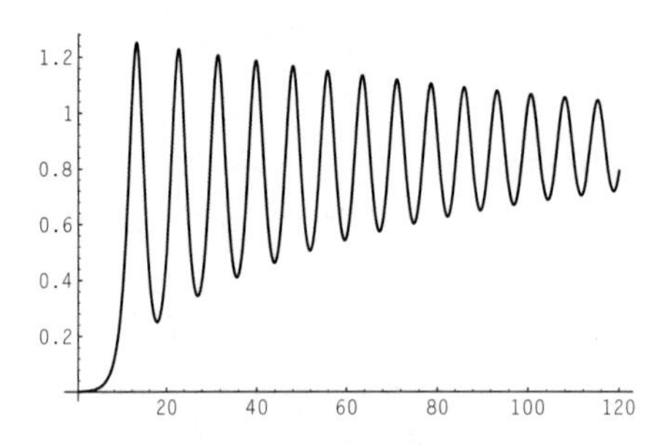

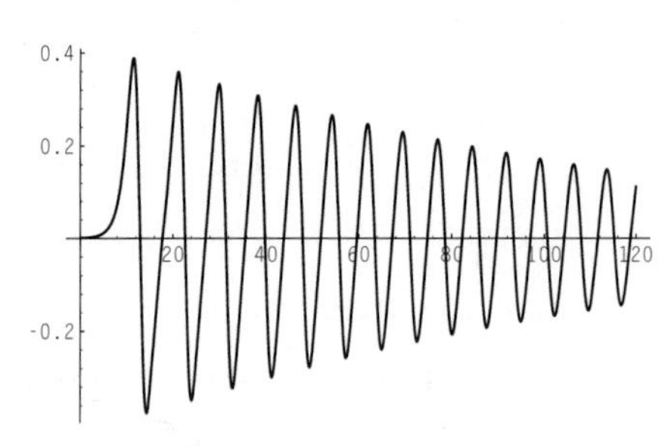

Note that this appears to be a damped oscillating system, as it should. Next the **Q0=0.1** case:

In[90]:=
```
Show[
  GraphicsArray[
    {Plot[q2[t],{t,0,120}]
    ,Plot[p2[t],{t,0,120}]
    }
  ]
];
```

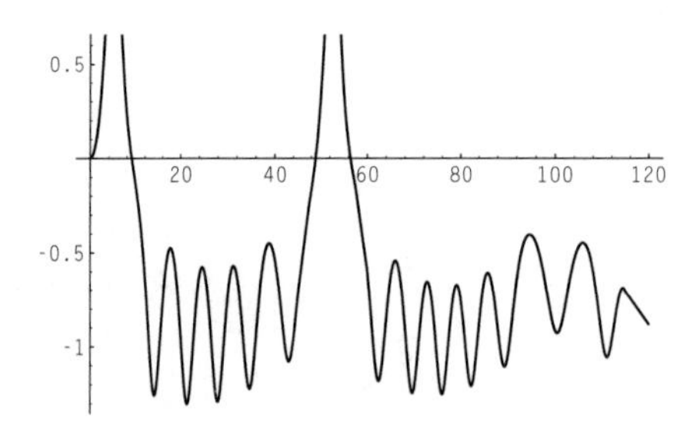

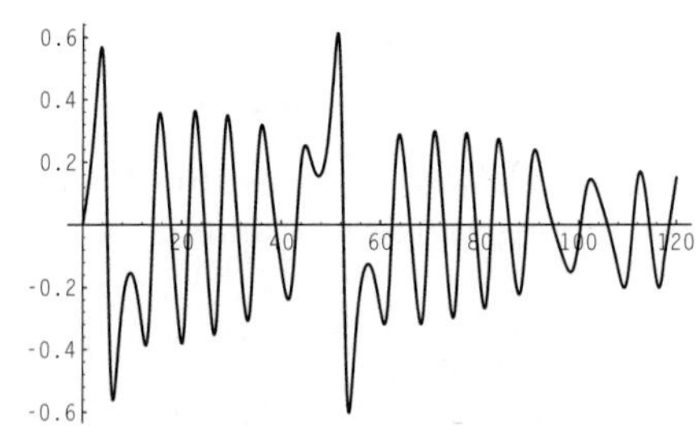

Note that the forced solution has a more irregular behavior.

Part (b) The phase trajectories follow from the user-defined procedures **phasePlot** and **timePhasePlot**. These are combined into **doublePlot**. The trajectory for nonforced motion (**Q0=0**) is

In[91]:=
```
doublePlot[q1[t],p1[t],{t,0,120}];
```

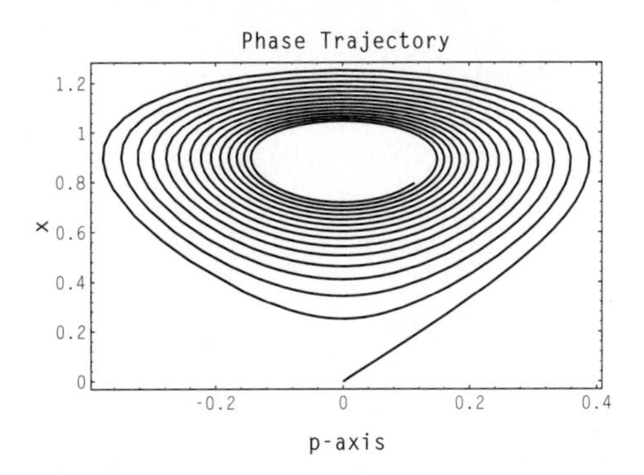

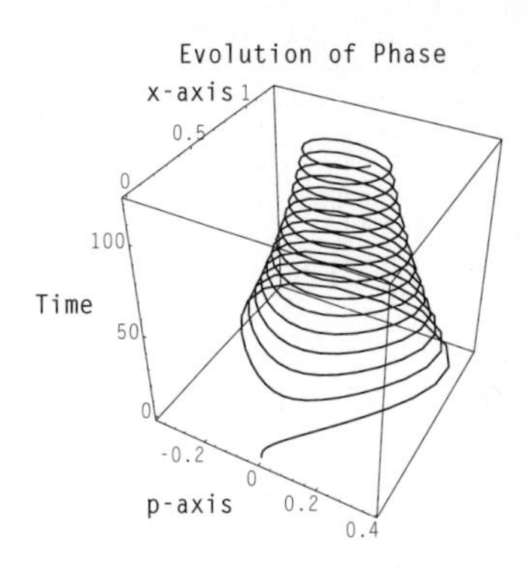

The trajectory is a spiral to the bottom of the potential well.
For the forced motion (**Q0=0.1**),

```
In[92]:=  doublePlot[q2[t],p2[t],{t,0,120}];
```

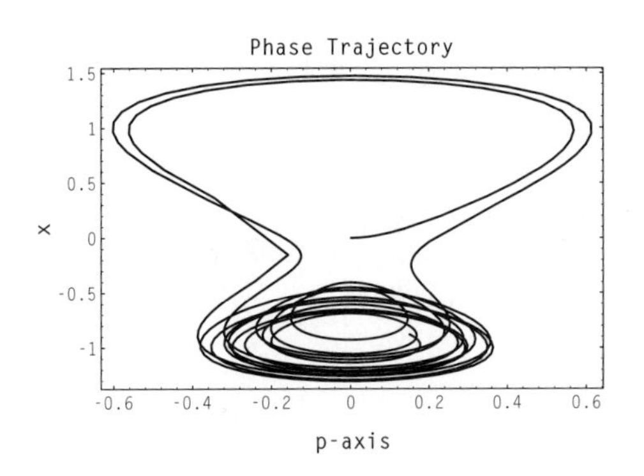

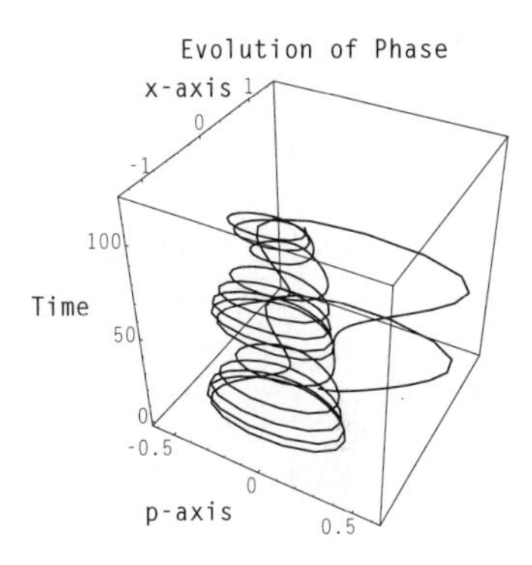

This phase trajectory illustrates the chaotic-like behavior as the solution jumps across the potential barrier at the origin.

The Fourier spectrum follows from the user-defined function **fftSpectrum**. Applying **fftSpectrum** to **q[t]** for the nonforced and forced cases, you get (output suppressed)

```
In[93]:=  plot1=Show[fftSpectrum[q1[t],t,120]
                     ,PlotRange->{{0,50},All}];

In[94]:=  plot2=Show[fftSpectrum[q2[t],t,120]
                     ,PlotRange->{{0,50},All}];

In[95]:=  Show[GraphicsArray[{plot1,plot2}]];
```

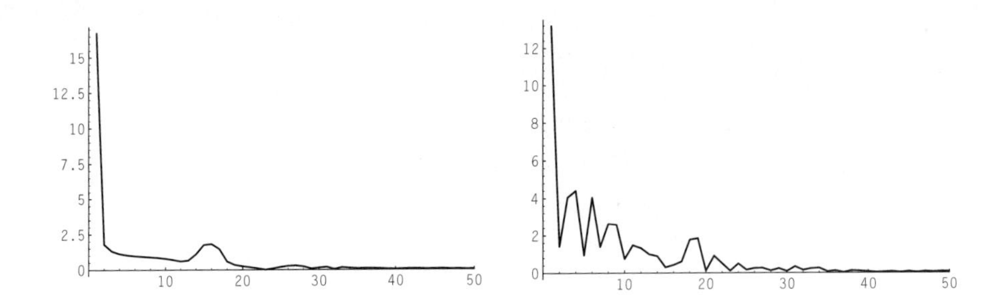

The left-hand graph is for the nonforced motion; the right-hand graph is for the forced motion. The chaotic-like behavior of the forced case is obvious from the number of peaks in the Fourier spectrum.

Problem 3: van der Pole Oscillator and Limiting Cycles

Consider an oscillator described by the coupled nonlinear equation of motion:

$$y'[t] = a(1 - x^2[t])\,y[t] - \omega_0^2 x[t]$$
$$x'[t] = y[t].$$

This equation is called the **van der Pole oscillator equation** and is an example of a system with a limiting cycle. Consider the parameters $a = 2$ and $\omega_0 = 1$ and two trajectories singled out by the initial conditions $\{x[0] = 0.06, y[0] = 0\}$ and $\{x[0] = 0, y[0] = 4\}$.

Part (a) Find the numerical solutions for these two different cases and plot the trajectories.

Part (b) Compare their phase trajectories.

Part (c) Animate one of the solutions.

Remarks and outline The solution will approach a limiting cycle independent of the initial conditions. The two initial conditions in the problem have been chosen so that one trajectory starts inside the limiting cycle and spirals outward and the other starts outside the trajectory and spirals inward. The interpolating function for the differential equation is calculated using **NDSolve**. The phase trajectories follow from the user-defined function **doublePlot**. The animation follows from the user-defined function **oscMovie**.

Required packages *There are no required packages.*

Solution

```
In[96]:= Clear["Global`*"];
```

Part (a) Consider the van der Pole equation

```
In[97]:= eq1= {y'[t] ==  a (1-x[t]^2) y[t] - w0^2 x[t]
              ,x'[t] ==  y[t]};
```

with the parameters

```
In[98]:= values= {a->2, w0->1};
```

Single out two trajectories based on the initial conditions

```
In[99]:=  initial1={x[0]== 0.06, y[0]==0};
          initial2={x[0]== 0,    y[0]==4};
```

The solutions of **eq1** for **initial1** conditions are

```
In[100]:= ndsol1=NDSolve[Join[eq1,initial1] //.values
                     ,{x[t],y[t]}
                     ,{t,0,25}
                     ,MaxSteps->700][[1]]

Out[100]= {x[t] -> InterpolatingFunction[{0., 25.}, <>][t],

    y[t] -> InterpolatingFunction[{0., 25.}, <>][t]}
```

The solution is returned in terms of an **InterpolatingFunction**. For convenience, define **x1** and **y1** as the **x** and **y** solutions of the **initial1** case:

```
In[101]:= x1[t_]=x[t] //.ndsol1;
          y1[t_]=y[t] //.ndsol1;
```

Likewise, the solutions of **eq1** for **initial2** conditions are

```
In[102]:= ndsol2=NDSolve[Join[eq1,initial2] //.values
                     ,{x[t],y[t]}
                     ,{t,0,25}
                     ,MaxSteps->700][[1]]

Out[102]= {x[t] -> InterpolatingFunction[{0., 25.}, <>][t],

    y[t] -> InterpolatingFunction[{0., 25.}, <>][t]}

In[103]:= x2[t_]=x[t] //.ndsol2;
          y2[t_]=y[t] //.ndsol2;
```

The plots of **x[t]** and **y[t]** for the first solution are

```
In[104]:= Show[
            GraphicsArray[
              {Plot[x1[t],{t,0,25}],Plot[y1[t],{t,0,25}]}
            ]
          ];
```

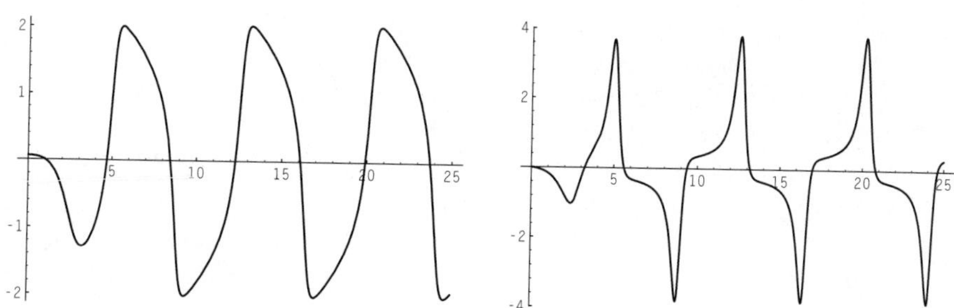

The plots for the second solution are

```
In[105]:= Show[
            GraphicsArray[
              {Plot[x2[t],{t,0,25}],Plot[y2[t],{t,0,25}]}
            ]
          ];
```

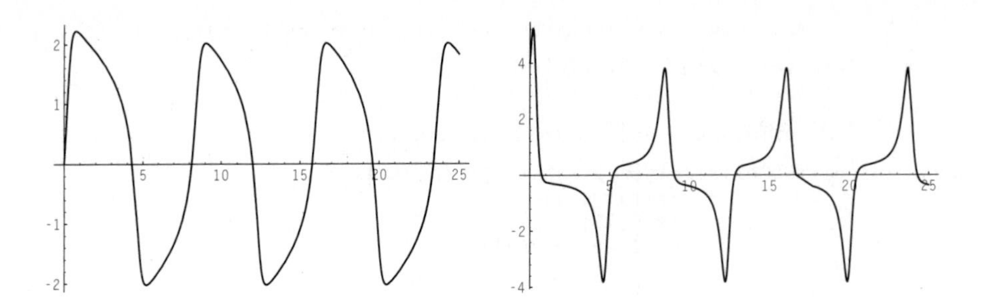

Part (b) The phase trajectories for these cases follow from the user-defined function **doublePlot**. For the **initial1** conditions, you will find

```
In[106]:= doublePlot[x1[t],y1[t],{t,0,25}];
```

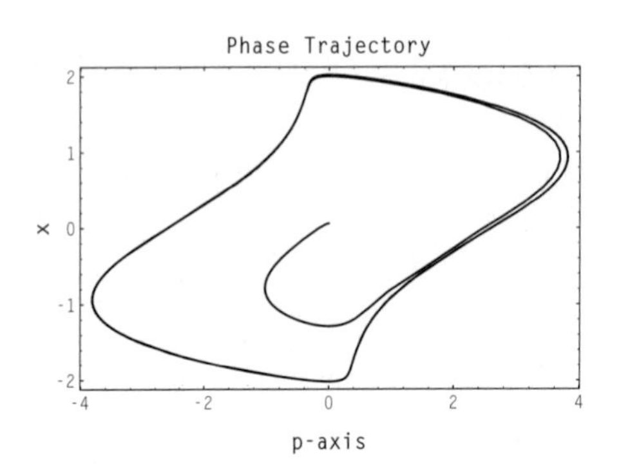

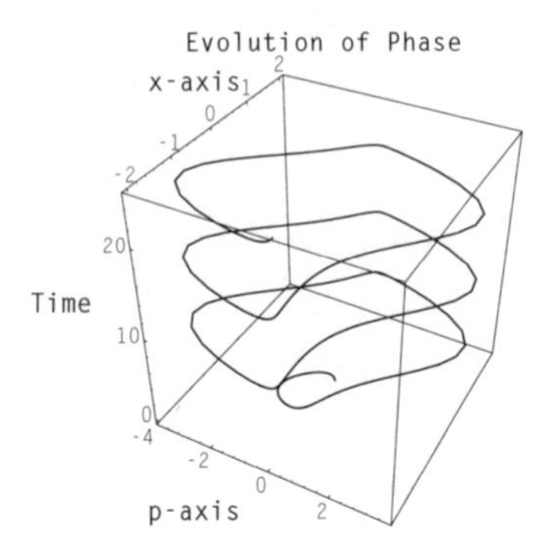

For the **initial2** conditions, you will find

```
In[107]:= doublePlot[x2[t],y2[t],{t,0,25}];
```

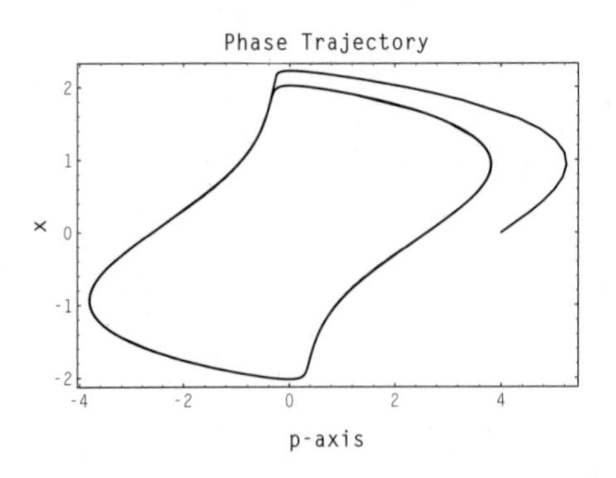

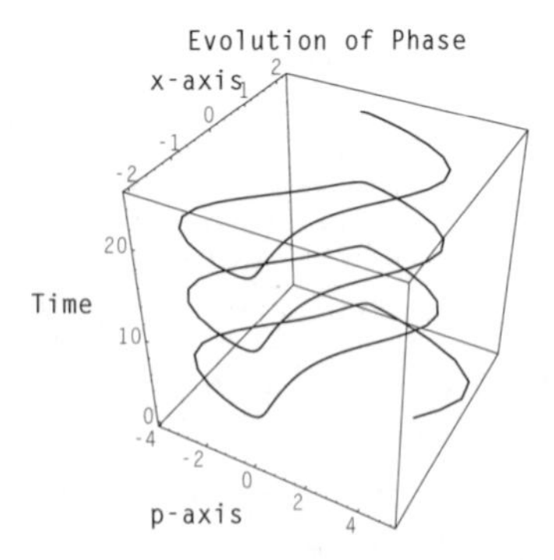

The trajectories spiral towards a limiting cycle independent of the initial conditions. Careful examination of the **ParametricPlot3D** (right-hand side) will show that the

initial1 conditions spiral outward, while the **initial2** conditions spiral inward.

Part (c) The animation follows from **oscMovie**:

```
In[108]:= oscMovie[x1,25,3,{-2,2}];
```

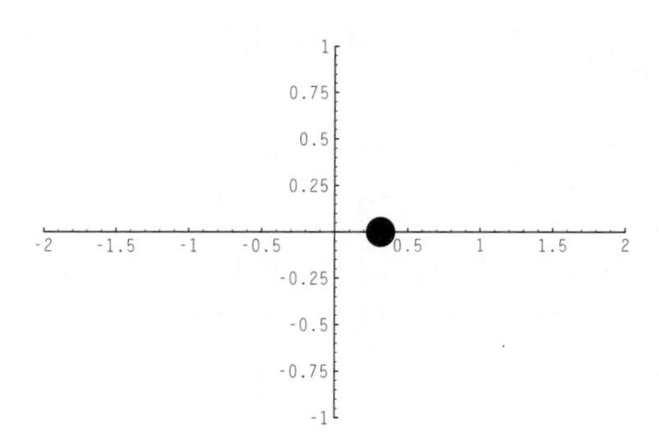

Problem 4: Motion of a Damped, Forced Nonlinear Pendulum

The nonlinear equation for a pendulum that is damped and has a forcing function is

$$\theta''[t] + \gamma\theta'[t] + \omega_0^2\mathrm{Sin}[\theta[t]] = F_0\mathrm{Cos}[\omega_d t]$$

and the linear form of this equation is

$$\theta''[t] + \gamma\theta'[t] + \omega_0^2\theta[t] = F_0\mathrm{Cos}[\omega_d t].$$

Assume the initial conditions are $\theta[0] = 0.9\pi$ and $\theta'[0] = 0$.

Part (a) Illustrate the nonlinear effects by letting $\omega_0 = 1$, $\omega_d = 1$, $\gamma = 0.01$, and $F_0 = 0.2$.

Part (b) Graph the solutions for the linear and nonlinear equations.

Part (c) Graph the frequency spectrum for these two solutions.

Part (d) Plot the phase trajectories for these two solutions.

Remarks and outline This problem illustrates how nonlinear effects can radically alter the behavior of a solution. The interpolating function for the differential equations are calculated using **NDSolve**. The phase trajectories follow from the user-defined function **doublePlot** and the Fourier spectrum follows from the user-defined function **ffspectrum**.

Required packages

```
In[109]:= Needs["Calculus`FourierTransform`"];
```

Solution

```
In[110]:= Clear["Global`*"];
```

Part (a) The equation of motion for a pendulum that is damped and has a forcing function, is

```
In[111]:= eq1=theta''[t] + gam theta'[t]+ w0^2 Sin[theta[t]]==f0 Cos[wd t]

Out[111]=   2
          w0  Sin[theta[t]] + gam theta'[t] + theta''[t] == f0 Cos[t wd]
```

You can rewrite this as a pair of first-order differential equations:

```
In[112]:= eq2= eq1 /.{theta'[t]->dtheta[t], theta''[t]->dtheta'[t]};
          eq3={eq2, theta'[t]==dtheta[t]}

Out[112]=                   2
          {gam dtheta[t] + w0  Sin[theta[t]] + dtheta'[t] == f0 Cos[t wd],

            theta'[t] == dtheta[t]}
```

The linear differential equation is simply the leading-order term in the expansion of **theta[t]**. This essentially replaces **Sin[theta[t]]** by **theta[t]**:

```
In[113]:= linearEq= eq3 //Map[Series[#,{theta[t],0,1}]&,#,{3}]& //Normal

Out[113]=                   2
          {gam dtheta[t] + w0  theta[t] + dtheta'[t] == f0 Cos[t wd],

            theta'[t] == dtheta[t]}
```

We consider the parameters

```
In[114]:= values={w0->1, wd->1, gam->0.01, f0->0.2};
```

and the initial conditions

```
In[115]:= initial={theta[0]==0.9 Pi, dtheta[0]==0};
```

Numerically solving the nonlinear equations, you get

```
In[116]:= ndsol1=NDSolve[Join[eq3,initial] //.values //Evaluate
                        ,{theta[t],dtheta[t]}
                        ,{t,0,100}
                        ,MaxSteps->1000][[1]]

Out[116]= {theta[t] -> InterpolatingFunction[{0., 100.}, <>][t],

            dtheta[t] -> InterpolatingFunction[{0., 100.}, <>][t]}
```

The solution is returned in terms of an **InterpolatingFunction**. For convenience, define **theta1** and **dtheta1** as the **theta** and **dtheta** solutions of the nonlinear case:

```
In[117]:=  theta1[t_]= theta[t] //.ndsol1;
          dtheta1[t_]=dtheta[t] //.ndsol1;
```

Numerically solving the linear equations, you get

```
In[118]:= ndsol2=NDSolve[Join[linearEq,initial] //.values //Evaluate
                        ,{theta[t],dtheta[t]}
                        ,{t,0,100}
                        ,MaxSteps->1000][[1]]

Out[118]= {theta[t] -> InterpolatingFunction[{0., 100.}, <>][t],

            dtheta[t] -> InterpolatingFunction[{0., 100.}, <>][t]}
```

Here **theta2** and **dtheta2** are the **theta** and **dtheta** solutions of the linear case:

```
In[119]:=  theta2[t_]= theta[t] //.ndsol2;
           dtheta2[t_]=dtheta[t] //.ndsol2;
```

Part (**b**) Compare the plots of the following nonlinear and linear cases:

```
In[120]:= Plot[{theta1[t],theta2[t]}
                ,{t,0,100}
                ,PlotStyle->{{Thickness[0.010]},{Thickness[0.005]}}];
```

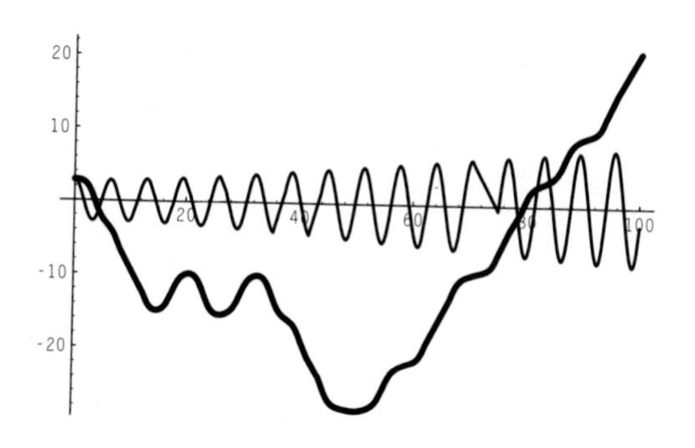

The first solution (thick line) corresponds to the nonlinear oscillator; the second solution (thin line) is the linear oscillator. The difference is obvious.

Part (**c**) The Fourier spectrum of the two solutions follows from the user-defined function **fftSpectrum**:

```
In[121]:= fplot1= Show[ fftSpectrum[theta1[t], t, 100]
                       ,PlotRange->{{0,30},{0,25}}];

In[122]:= fplot2= Show[ fftSpectrum[theta2[t], t, 100]
                       ,PlotRange->{{0,30},{0,25}}];

In[123]:=  Show[GraphicsArray[{fplot1,fplot2}]];
```

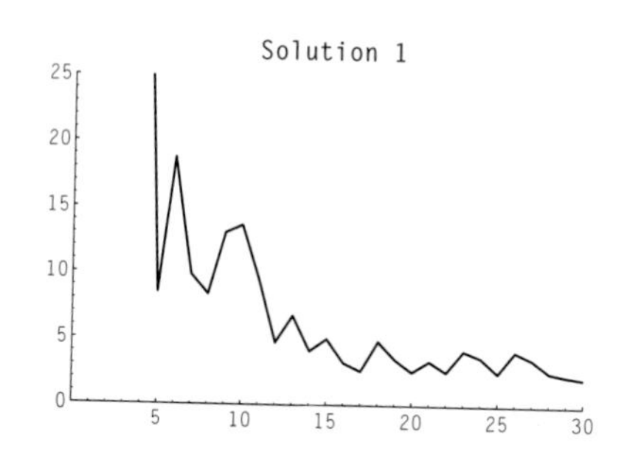

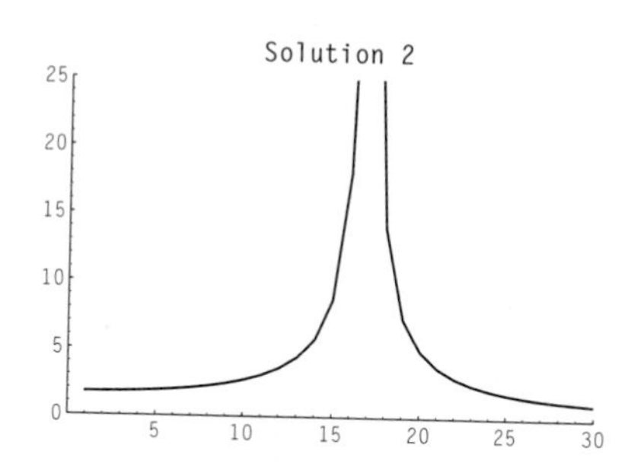

The first solution corresponds to the nonlinear oscillator; the second solution is the linear oscillator.

Part (**d**) The phase trajectories follow from the user-defined function **doublePlot**. First, for the nonlinear case,

```
In[124]:= doublePlot[theta1[t],dtheta1[t],{t,0,100}];
```

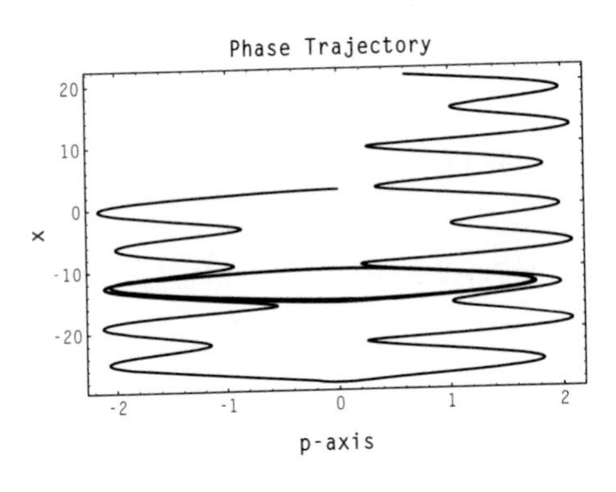

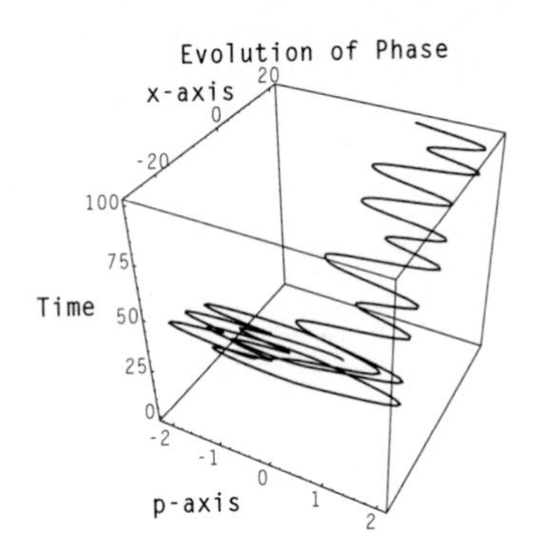

For the linear case,

```
In[125]:= doublePlot[theta2[t],dtheta2[t],{t,0,100}];
```

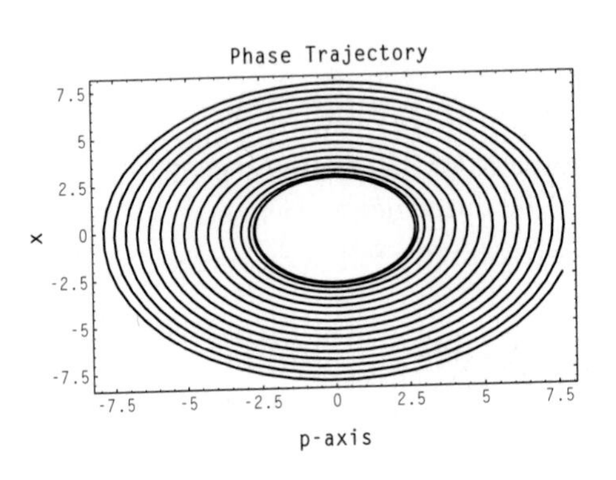

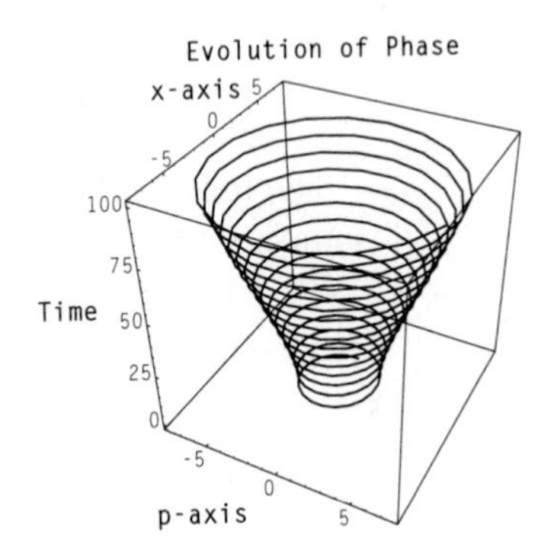

3.3.3 Small Oscillations

Problem 1: Two Coupled Harmonic Oscillators

Consider the vibrations of two particles of mass m connected to each other by a spring of stiffness k_1. Each particle is also connected to the wall by two more springs with equal spring stiffness k_2. Let $x[1]$ and $x[2]$ be the displacements from the equilibrium configuration for the two particles, respectively. Consider small vibrations along the line of the particles.

Part (a) Derive the algebraic equations for the eigenfrequencies and eigenvectors. Solve for the eigenfrequencies and eigenvectors using the command **Solve**.

Part (b) Use the command **Reduce** to solve the algebraic equations for the eigenfrequencies and eigenvectors.

Part (c) Use the user-defined function **smallOsc** to solve for the eigenfrequencies and eigenvectors.

Part (d) Write the general solution and plot the motion for the normal modes and for a mixture of normal modes.

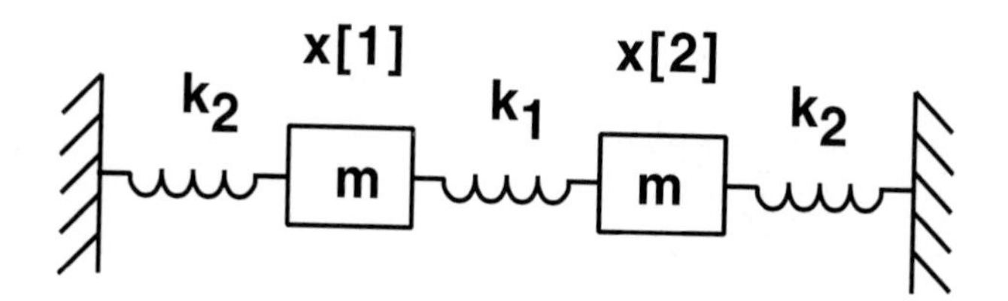

Remarks and outline The algebraic equations follow from expanding the kinetic and potential energies of the system about the point of equilibrium. The terms for the eigenvector matrix follow from the expansion. The algebraic equations for the eigen-frequencies and eigenvalues can be solved using the commands **Solve** and **Reduce** and the user-defined command **smallOsc**. The solution also follows from the commands **Eigenvalue** and **Eigenvector**, or the command **Eigensystem** (this latter approach is left as an exercise at the end of the chapter).

Required packages *There are no required packages.*

Solution

```
In[126]:= Clear["Global`*"];
```

Part (a) The kinetic and potential energies for the coupled oscillators are

```
In[127]:= T= (m/2) ( (x[1])'[t]^2 + (x[2])'[t]^2 );

          V=( (k1/2)( x[2][t] - x[1][t])^2
             + k2/2   x[1][t]^2
             + k2/2   x[2][t]^2
            );
```

where the coordinates and momenta

```
In[128]:= q={x[1][t], x[2][t]};
          p=D[q,t]

Out[128]= {(x[1])'[t], (x[2])'[t]}
```

are the displacements from the equilibrium configuration.

Note here the use of the **x[i][t]** notation. This allows you to use the **i** index to label the coordinates and momenta, while still taking advantage of the derivative shorthand notation **(x[i])'[t]**:

The **Tij** and **Vij** matrices for small oscillations follow from the coefficients in front of the quadratic terms in **x'[i][t]** and **x[i][t]**:

```
In[129]:= Tij=Outer[D[T,#1,#2]&,p,p];
          Vij=Outer[D[V,#1,#2]&,q,q];
```

The eigenvector **matrix** is

```
In[130]:= matrix = Vij - w2 Tij;
          matrix //MatrixForm
```

```
Out[130]= k1 + k2 - m w2    -k1

          -k1               k1 + k2 - m w2
```

where **w2** is the square of the eigenfrequency. The eigenfrequencies follow from setting the determinate of matrix equal to zero and solving for **w2**:

```
In[131]:= w2Sol= Solve[ Det[matrix]==0, w2] //Simplify

Out[131]=
                   2 k1 + k2            k2
          {{w2 -> ---------}, {w2 -> --}}
                       m                 m
```

The square root of **w2** gives the two eigenfrequencies. The eigenvector $\{a1, a2\}$ for the first eigenfrequency, **w2=k/m**, follows from solving the algebraic equations ($M_{ij}v_j = 0$):

```
In[132]:= vector= Array[a,2]

Out[132]= {a[1], a[2]}

In[133]:= eq= Thread[0==matrix.vector]

Out[133]= {0 == (k1 + k2 - m w2) a[1] - k1 a[2],

            0 == -(k1 a[1]) + (k1 + k2 - m w2) a[2]}
```

To obtain the appropriate eigenvector, you must substitute the appropriate eigenvalue for **w2**. Here you solve for the first eigenvector by substituting with **w2Sol[[1]]**:

```
In[134]:= eq1 = eq /.w2Sol[[1]] //ExpandAll //Simplify

Out[134]= {0 == -(k1 (a[1] + a[2])), 0 == -(k1 (a[1] + a[2]))}
```

The normalization equation is

```
In[135]:= eqNorm={vector.vector==1};
```

Solving **eq1** for the eigenvector (i.e., the components of **vector**), you have

```
In[136]:= ev1rule=Solve[Join[eq1,eqNorm],vector][[2]]

Out[136]=
                     1                    1
          {a[1] -> -------, a[2] -> -(-------)}
                   Sqrt[2]              Sqrt[2]
```

The eigenvector for the second normal frequency follows by substituting with **w2Sol[[2]]**:

```
In[137]:= eq2 = eq /.w2Sol[[2]] //ExpandAll //Simplify

Out[137]= {0 == k1 (a[1] - a[2]), 0 == k1 (-a[1] + a[2])}

In[138]:= ev2rule=Solve[Join[eq2,eqNorm],vector][[2]]

Out[138]=
                     1                  1
          {a[1] -> -------, a[2] -> -------}
                   Sqrt[2]            Sqrt[2]
```

Part (b) Instead of first solving for the eigenfrequencies and then eigenvectors, **Reduce** does both calculations simultaneously. Applying **Reduce** to the matrix equation,

```
In[139]:= eq3= Thread[0== matrix.vector]

Out[139]= {0 == (k1 + k2 - m w2) a[1] - k1 a[2],

           0 == -(k1 a[1]) + (k1 + k2 - m w2) a[2]}
```

and reducing for **{w2,a1,a2}**, you get

```
In[140]:= sol= Reduce[{eq3,eqNorm,k1 !=0,m !=0}//Flatten
                ,{w2,a[1],a[2]}]

Out[140]=
                                     k2                  1
          k1 != 0 && m != 0 && w2 == -- && a[1] == -(-------) &&
                                     m                Sqrt[2]

                        1                                        k2
            a[2] == -(-------) || k1 != 0 && m != 0 && w2 == -- &&
                      Sqrt[2]                                    m

                       1                 1
            a[1] == ------- && a[2] == ------- ||
                    Sqrt[2]            Sqrt[2]

                                        2 k1 + k2               1
            k1 != 0 && m != 0 && w2 == --------- && a[1] == -(-------) &&
                                            m                Sqrt[2]

                       1                                      2 k1 + k2
            a[2] == ------- || k1 != 0 && m != 0 && w2 == --------- &&
                    Sqrt[2]                                       m

                       1                    1
            a[1] == ------- && a[2] == -(-------)
                    Sqrt[2]              Sqrt[2]
```

where you have added the auxiliary conditions **k1!=0** and **m!=0** to eliminate trivial solutions.

Part (c) The eigenfrequencies and eigenvectors follow immediately from the user-defined function **smallOsc**. Applying **smallOsc** to **T** and **V**, you get

```
In[141]:=  eq4= smallOsc[ T,V,{x[1][t],x[2][t]},{0,0},t]

Out[141]=   2 k1 + k2  k2                  1                      1
          {{---------, --}, {{a[1] -> -------, a[2] -> -(-------)}},
                m      m               Sqrt[2]              Sqrt[2]

                      1                  1
            {a[1] -> -------, a[2] -> -------}}}
                     Sqrt[2]           Sqrt[2]
```

which agrees with the results derived in Parts (a) and (b). **smallOsc** returns two lists. The first list is the set of eigenvalues; the second is the set of eigenvectors, in the same order.

Part (d) The general solution is a linear combination of terms of the form $\mathrm{Cos}[\omega_1 t + \delta_1]$ and $\mathrm{Cos}[\omega_2 t + \delta_2]$, where ω_1 and ω_2 are the two eigenfrequencies. Here, identify ω_1 with **freq1** and ω_2 with **freq2**. By defining the **eigenmatrix**:

```
In[142]:=  eigenmatrix= vector //.{ev1rule,ev2rule}
```

```
Out[142]=      1            1              1          1
          {{-------, -(-------)}, {-------,  -------}}
            Sqrt[2]     Sqrt[2]     Sqrt[2]  Sqrt[2]
```

the general solution is

```
In[143]:= xRule={x1[t],x2[t]} ->
              ( Transpose[eigenmatrix]. {c1 Cos[ freq1 t + phase1]
                                        ,c2 Cos[ freq2 t + phase2]}
               )//Thread //Simplify

Out[143]=
                c1 Cos[phase1 + freq1 t] + c2 Cos[phase2 + freq2 t]
          {x1[t] -> ---------------------------------------------------,
                                        Sqrt[2]

                 -(c1 Cos[phase1 + freq1 t]) + c2 Cos[phase2 + freq2 t]
           x2[t] -> ------------------------------------------------------}
                                        Sqrt[2]
```

where **c1** and **c2** are the relative amplitudes of the two normal modes. The eigenfrequencies, **freq1** and **freq2**, are

```
In[144]:= freqrule= Thread[{freq1,freq2} -> (Sqrt[w2] /.w2Sol) ]

Out[144]=
                         2 k1 + k2                      k2
          {freq1 -> Sqrt[---------], freq2 -> Sqrt[--]}
                             m                          m
```

The two normal modes correspond to the special cases **c2=0** and **c1=0**. Consider the value of the parameters given by

```
In[145]:=  values= {k1->1, k2->1, m->1, phase1->0, phase2->0};
```

For simplicity, we collect all the appropriate rules into **allRules**:

```
In[146]:= allRules=Join[xRule,freqrule,values];
```

Next, examine the first normal mode $\{c1 = 1, c2 = 0\}$. Use **Plot** to plot the motion of the two coordinates **{x1[t],x2[t]}** on the same graph separated by a distance of ± 1. Then use the command **ParametricPlot** to map the trajectory of this system in the **{x1[t],x2[t]}** plane. (Write out each plot command separately here. In the next example, you will define a function to do this step.)

```
In[147]:= plot1a=
          Plot[{x1[t]+1, x2[t]-1} //.allRules //.{c1->1, c2->0} //Evaluate
               ,{t,0,20}
               ,PlotStyle->{{Thickness[0.010]},{Thickness[0.005]}}
            ];

In[148]:= plot1b= ParametricPlot[
            {x1[t], x2[t]} //.allRules //.{c1->1, c2->0} //Evaluate
           ,{t,0,20}];

In[149]:= Show[GraphicsArray[{plot1a,plot1b}]];
```

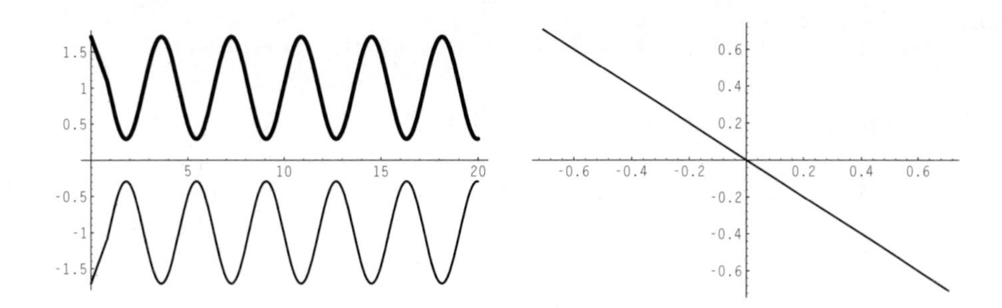

The thick line corresponds to **x[1]**; the thinner line to **x[2]**. From both plots, you can see that the motion is simple harmonic and the objects are out of phase.

Next, examine the second normal mode $\{c1 = 0, c2 = 1\}$:

```
In[150]:= plot2a=
          Plot[{x1[t]+1, x2[t]-1} //.allRules //.{c1->0, c2->1} //Evaluate
                ,{t,0,20}
                ,PlotStyle->{{Thickness[0.010]},{Thickness[0.005]}}
             ];
```

```
In[151]:= plot2b= ParametricPlot[
             {x1[t], x2[t]} //.allRules //.{c1->0, c2->1} //Evaluate
            ,{t,0,20}];
```

```
In[152]:= Show[GraphicsArray[{plot2a,plot2b}]];
```

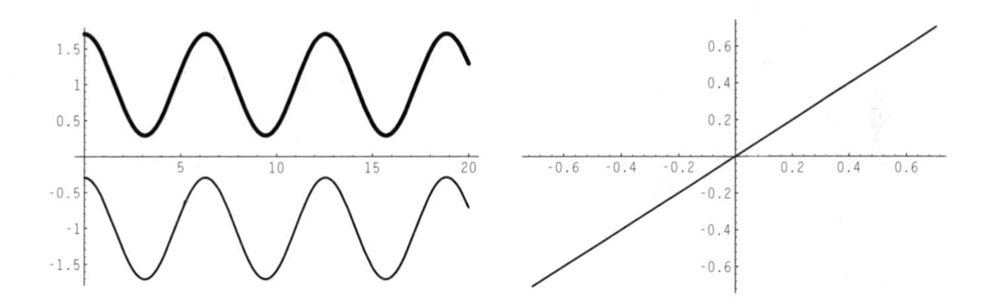

Comparing the second normal mode to the first, note that the second has a lower eigenfrequency. Also note that the objects are in phase.

Finally, examine a mixed mode $\{c1 = 1, c2 = 1\}$:

```
In[153]:= plot3a=
          Plot[{x1[t]+1, x2[t]-1} //.allRules //.{c1->1, c2->1} //Evaluate
                ,{t,0,20}
                ,PlotStyle->{{Thickness[0.010]},{Thickness[0.005]}}
             ];
```

```
In[154]:= plot3b= ParametricPlot[
            {x1[t], x2[t]} //.allRules //.{c1->1, c2->1} //Evaluate
            ,{t,0,20}];

In[155]:= Show[GraphicsArray[{plot3a,plot3b}]];
```

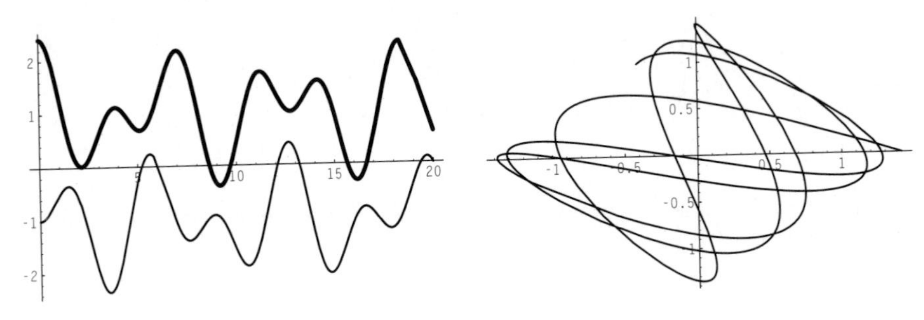

Problem 2: Three Coupled Harmonic Oscillators

Consider the vibrations of three particles coupled by springs. Two particles of mass m are symmetrically located on each side of a third particle of mass M. All three particles are on a straight line. Consider only vibrations along the line of the particles. The three particles are joined by two springs with force constants k. Let $x[1]$, $x[2]$, and $x[3]$ be the displacements from the equilibrium configuration for the three particles.

Part (a) Find the kinetic and potential energies for the system. Solve for the eigenfrequencies and eigenvectors.

Part (b) Write the general solution and plot the motion for the normal modes and for a mixture of normal modes.

Part (c) Animate one of the solutions.

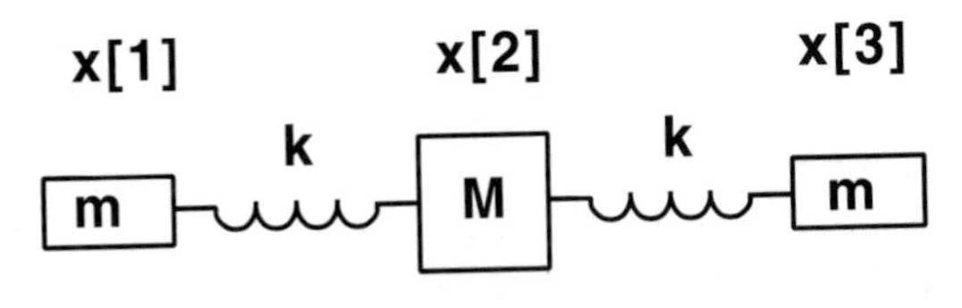

Remarks and outline The eigenfrequencies and eigenvectors follow from the user-defined command **smallOsc**. The general solution is a linear combination of the normal modes.

Required packages *There are no required packages.*

Solution

```
In[156]:= Clear["Global`*"];
```

Part (a) The kinetic and potential energies are

```
In[157]:= T=(     (m/2)( (x[1])'[t]^2 + (x[3])'[t]^2 )
           + (M/2)( (x[2])'[t]^2 )
          );

          V=( (k/2)( (x[2][t]-x[1][t])^2
                   + (x[3][t]-x[2][t])^2 )
          );
```

where the coordinates are the displacements from the equilibrium positions. Again use the notation **x[i][t]** so that you can easily index the coordinates. The eigenfrequencies and eigenvectors follow from the user-defined function **smallOsc**:

```
In[158]:= solution= smallOsc[T,V,{x[1][t],x[2][t],x[3][t]},{0,0,0},t]

Out[158]=
              k   k   2 k                1                    1
          {{0, -, - + ---}, {{a[1] -> -------, a[3] -> -------,
              m   m    M               Sqrt[3]              Sqrt[3]

                     1                     1                       1
           a[2] -> -------}, {a[1] -> -------, a[3] -> -(-------),
                   Sqrt[3]               Sqrt[2]               Sqrt[2]

                                               M
           a[2] -> 0}, {a[1] -> -----------------------------,
                                                      2    2
                                   Sqrt[2] Sqrt[2 m  + M ]

                                M                                 Sqrt[2] m
           a[3] -> -----------------------------, a[2] -> -(----------------)}}}
                                       2    2                           2    2
                   Sqrt[2] Sqrt[2 m  + M ]                   Sqrt[2 m  + M ]
```

The first list is the square of the eigenfrequencies, **w2**; the second contains the eigenvectors. The first root, **w2=0** with eigenvector $\{1/\sqrt{3}, 1/\sqrt{3}, 1/\sqrt{3}\}$, is not really an oscillation but represents a translatory motion of all three masses.

Part (b) The general oscillatory motion follows from a superposition of the normal modes. Form the **eigenmatrix**:

```
In[159]:= eigenmatrix= Array[a,3] /. solution[[2]]

Out[159]=
             1         1         1            1                  1
          {{-------, -------, -------}, {-------, 0, -(-------)},
            Sqrt[3]  Sqrt[3]  Sqrt[3]     Sqrt[2]            Sqrt[2]

                       M                       Sqrt[2] m
           {-----------------------------, -(----------------),
                                  2    2                2    2
             Sqrt[2] Sqrt[2 m  + M ]          Sqrt[2 m  + M ]

                        M
            -----------------------------}}
                                  2    2
             Sqrt[2] Sqrt[2 m  + M ]
```

(Note that to efficiently generate the eigenmatrix, the form of the eigenvectors returned by **smallOsc** has been used.) The general solution follows from

```
In[160]:= xRule=
          {x1[t],x2[t],x3[t]} ->
            ( Transpose[eigenmatrix]. { c1 t            + phase1,
                                       c2 Cos[freq2 t + phase2],
                                       c3 Cos[freq3 t + phase3]}
            )//Thread
Out[160]=
                 phase1 + c1 t   c2 Cos[phase2 + freq2 t]
          {x1[t] -> ------------- + ------------------------ +
                    Sqrt[3]              Sqrt[2]

             c3 M Cos[phase3 + freq3 t]
             --------------------------,
                             2    2
             Sqrt[2] Sqrt[2 m  + M ]

                  phase1 + c1 t   Sqrt[2] c3 m Cos[phase3 + freq3 t]
           x2[t] -> ------------- - ----------------------------------,
                    Sqrt[3]                    2    2
                                        Sqrt[2 m  + M ]

                  phase1 + c1 t   c2 Cos[phase2 + freq2 t]
           x3[t] -> ------------- - ------------------------ +
                    Sqrt[3]              Sqrt[2]

             c3 M Cos[phase3 + freq3 t]
             --------------------------}
                             2    2
             Sqrt[2] Sqrt[2 m  + M ]
```

where the frequencies are given by

```
In[161]:= freqrule= Thread[{freq1,freq2,freq3} -> Sqrt[ solution[[1]] ] ]
Out[161]=
                                   k                  k   2 k
          {freq1 -> 0, freq2 -> Sqrt[-], freq3 -> Sqrt[- + ---]}
                                   m                  m    M
```

(Note that the **c1** corresponds to a zero frequency pure translation.) The normal modes correspond to {c1 = 1, c2 = 0, c3 = 0,}, {c1 = 0, c2 = 1, c3 = 0,}, and {c1 = 0, c2 = 0, c3 = 1,}.

Plot the motion of the three coordinates with the parameters given by

```
In[162]:= values={k ->1, m->1, M->2, phase1->0, phase2->0, phase3->0};
```

Then collect all relevant rules into **allRules**:

```
In[163]:= allRules=Join[xRule,freqrule,values];
```

In contrast to the previous example, define functions to make the plots. First, **xPlot** plots $\{x[1], x[2], x[3]\}$ as a function of time (space the trajectories by ± 2 units):

```
In[164]:= Clear[xPlot];
          xPlot[C1_,C2_,C3_]:=
          Plot[ ( {x1[t]+2, x2[t], x3[t]-2} //.allRules
                                            //.{c1->C1, c2->C2, c3->C3}
                                            //Evaluate   )
                ,{t,0,20}
                ,PlotStyle->{{Thickness[0.012]}
                            ,{Thickness[0.008]}
                            ,{Thickness[0.004]}}
              ];
```

Next **xpPlot** plots $\{x[1], x[2], x[3]\}$ as a three-dimensional parametric plot:

```
In[165]:= Clear[xpPlot];
          xpPlot[C1_,C2_,C3_]:=
            ParametricPlot3D[
               ( {x1[t], x2[t], x3[t]} //.allRules
                                               //.{c1->C1, c2->C2, c3->C3}
                                               //Evaluate   )
                ,{t,0,20}
                ,BoxRatios->{1,1,1}];
```

We choose to **Show** these two plots in a **GraphicsArray** so that we can compare the two different interpretations of the trajectories.

First, display the first normal mode {c1 = 1, c2 = 0, c3 = 0}. Note that this corresponds to pure translation of all coordinates, which is why the eigenfrequency was zero. The parametric plot is also trivial for this case.

```
In[166]:= Show[GraphicsArray[{xPlot[1,0,0],xpPlot[1,0,0]}]];
```

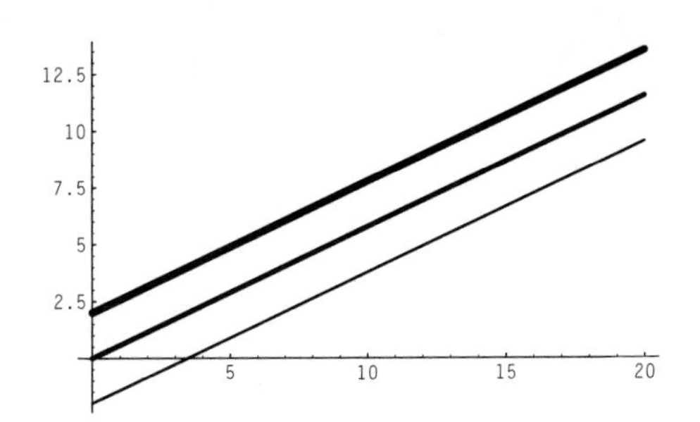

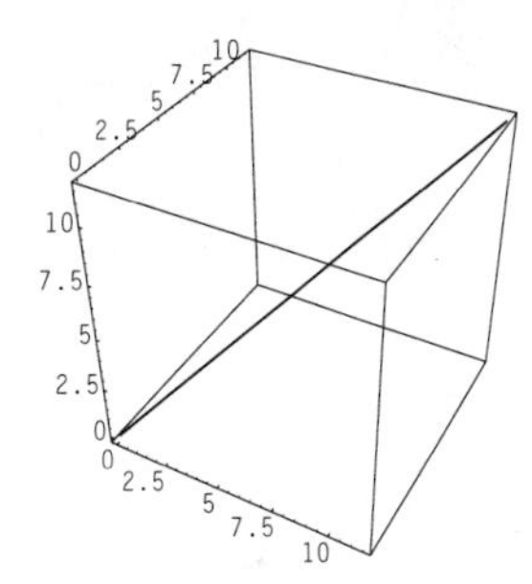

Next, display the second normal mode {c1 = 0, c2 = 1, c3 = 0}. Note masses 1 and 3 are out of phase, while mass 2 is stationary with respect to the others.

```
In[167]:= Show[GraphicsArray[{xPlot[0,1,0],xpPlot[0,1,0]}]];
```

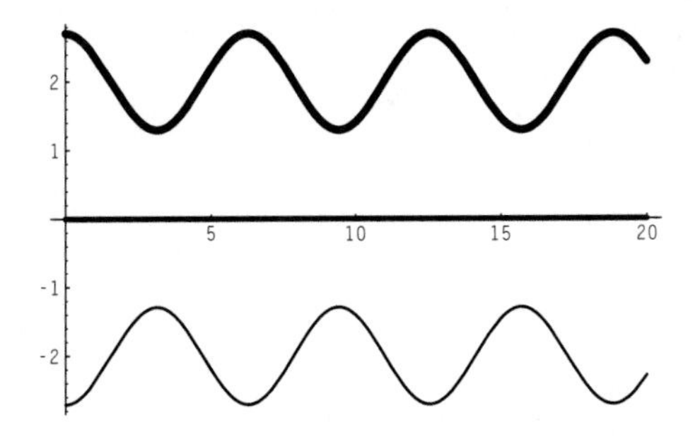

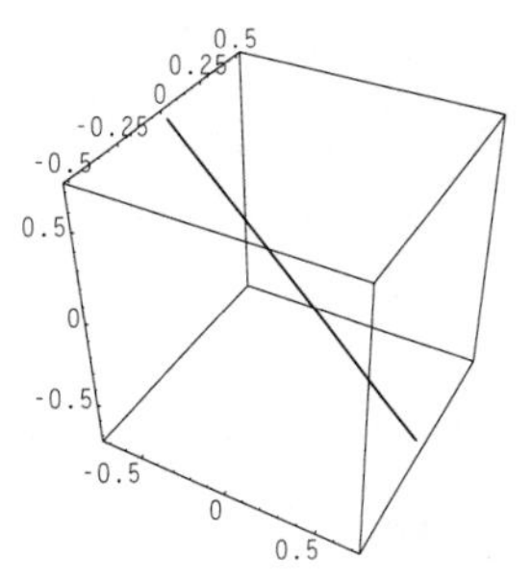

Here, display the third normal mode {c1 = 0, c2 = 0, c3 = 1}. Note masses 1 and 3 are in phase, while mass 2 is out of phase.

```
In[168]:= Show[GraphicsArray[{xPlot[0,0,1],xpPlot[0,0,1]}]];
```

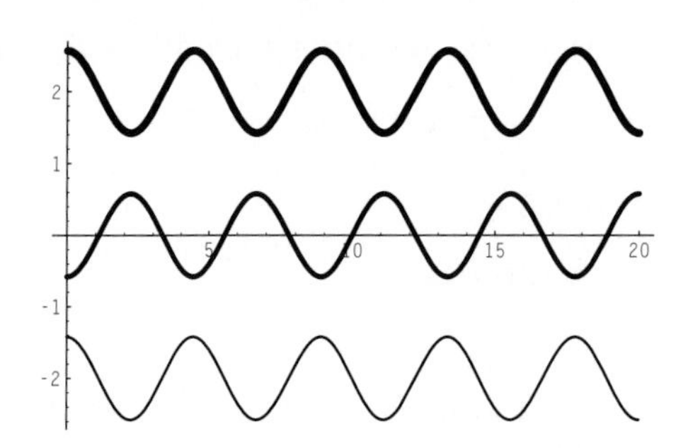

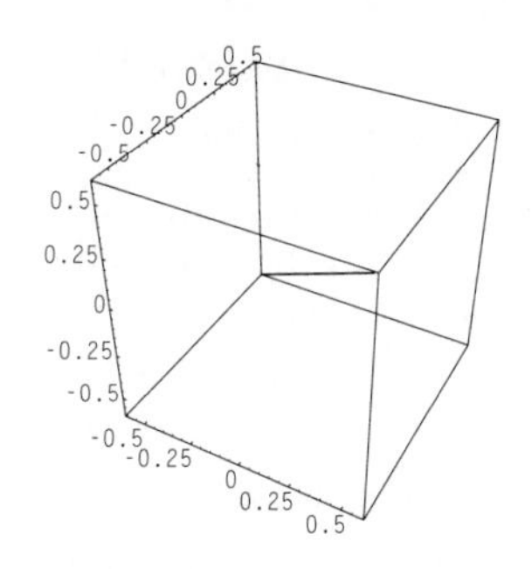

Finally, display a mixed mode $\{c1 = 0, c2 = 1, c3 = 1\}$ (keep $c1 = 0$, since this mode is trivial):

In[169]:=
```
Show[GraphicsArray[{xPlot[0,1,1],xpPlot[0,1,1]}]];
```

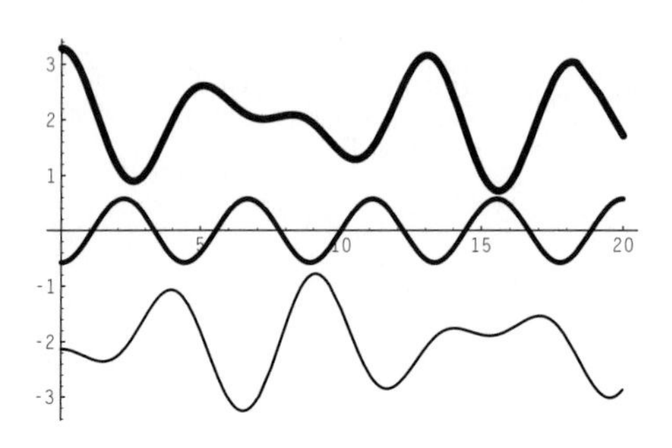

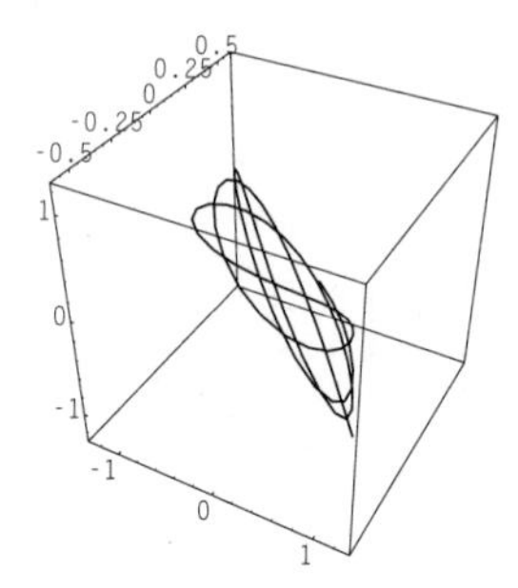

Part (c) To animate the solution, use **ListPlot** inside a **Do** loop to generate the individual graphics frames. Here, take the mixed mode $\{c1 = 0, c2 = 1, c3 = 1\}$.

In[170]:=
```
Do[
  ListPlot[( {{x1[t]-2,0},{x2[t],0},{x3[t]+2,0}}
                 //.allRules
                 //.{c1->0, c2->1, c3->1}
                 //Evaluate )
            ,PlotStyle->{PointSize[0.05]}
            ,PlotRange->{{-4,4},{-1,1}}
         ]
,{t,1,2,0.1}];
```

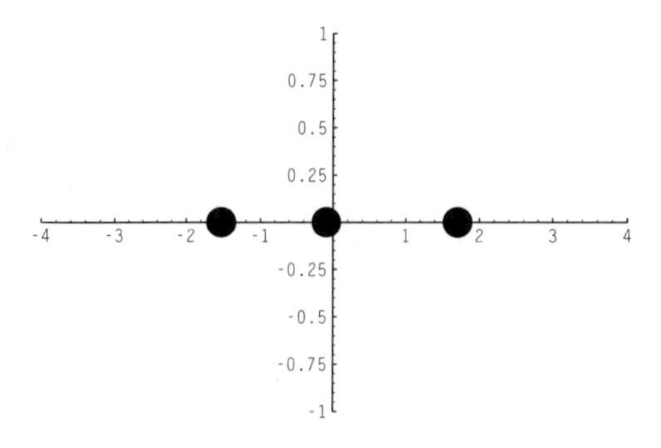

Problem 3: Double Pendulum

Consider a double pendulum consisting of a massless rigid rod. The first segment of the rod has one end fixed and the other supporting a particle of mass m_1. The length of this segment is L_1. Attached to m_1 is another segment of length L_2. At the other end of the second segment is a another particle of mass m_2. Assume the motion is confined to a plane.

Part (a) Find the normal modes for small oscillations.

Part (b) Plot the normal modes and mixed modes for the motion of the mass points for one cycle.

Part (c) Animate one of these solutions.

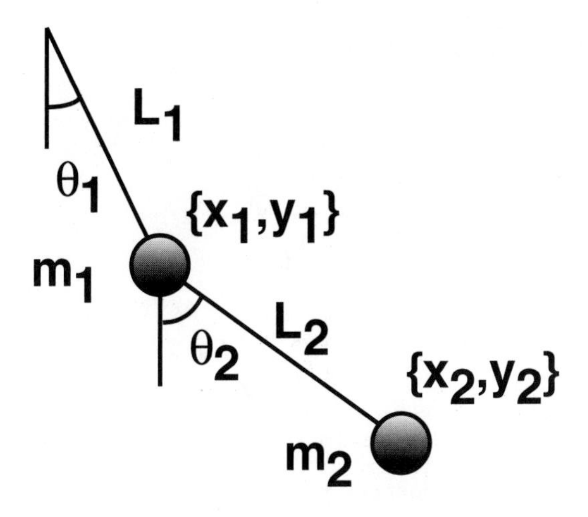

Remarks and outline The eigenfrequencies and eigenvectors follow from the user-defined command **smallOsc**. The general solution is a linear combination of the normal modes.

Required packages *There are no required packages.*

Solution

```
In[171]:= Clear["Global`*"];
```

Part (a) The coordinates of the masses are $\{x1, y1\}$ and $\{x2, y2\}$. Let **theta1[t]** and **theta2[t]** be the angles between the vertical line formed by the first and second rods, respectively. Expressing the mass points in polar coordinates, you have

```
In[172]:= x1=  L1 Sin[theta1[#]]&;
          y1=  L1 Cos[theta1[#]]&;

          x2=   x1[#] + L2 Sin[theta2[#]]&;
          y2=   y1[#] + L2 Cos[theta2[#]]&;
```

Use pure functions so that these definitions also apply to derivatives. For simplicity, let **L1=L2=len**:

```
In[173]:= L1=L2=len;
```

The kinetic and potential energies for the two masses are

```
In[174]:= T1= (m1/2) (x1'[t]^2+y1'[t]^2);
          T2= (m2/2) (x2'[t]^2+y2'[t]^2);
          V1= -m1 g y1[t];
          V2= -m2 g y2[t];
          T=T1+T2;
          V=V1+V2;
```

The eigenfrequencies and eigenvectors follow from the user-defined command **smallOsc**:

```
In[175]:= solution=( smallOsc[T,V,{theta1[t],theta2[t]},{0,0},t]
                 //ExpandAll
                 //Simplify
                 //PowerExpand
            )

Out[175]=   g len m1 + g len m2 - g len Sqrt[m2] Sqrt[m1 + m2]
          {{--------------------------------------------------,
                                 2
                              len  m1

             g len m1 + g len m2 + g len Sqrt[m2] Sqrt[m1 + m2]
             --------------------------------------------------},
                                 2
                              len  m1

                       Sqrt[m1 + m2]                 Sqrt[m2]
            {{a[2] -> ----------------, a[1] -> ----------------},
                      Sqrt[m1 + 2 m2]           Sqrt[m1 + 2 m2]

                       Sqrt[m1 + m2]                   Sqrt[m2]
             {a[2] -> ----------------, a[1] -> -(----------------)}}}
                      Sqrt[m1 + 2 m2]             Sqrt[m1 + 2 m2]
```

Part (b) To construct the general solution, define the **eigenmatrix**:

```
In[176]:= eigenmatrix= Array[a,2] //.solution[[2]]

Out[176]=      Sqrt[m2]         Sqrt[m1 + m2]
          {{----------------, ----------------},
            Sqrt[m1 + 2 m2]   Sqrt[m1 + 2 m2]

                 Sqrt[m2]           Sqrt[m1 + m2]
            {-(----------------), ----------------}}
               Sqrt[m1 + 2 m2]    Sqrt[m1 + 2 m2]
```

The general solution will be linear combinations of normal modes with relative amplitudes $\{c1, c2\}$:

```
In[177]:= thetaRule= ( {theta1[t],theta2[t]} ->
            Transpose[eigenmatrix].{c1 Cos[freq1 t + phase1]
                                   ,c2 Cos[freq2 t + phase2]}
                //Thread
                //Simplify )

Out[177]= {theta1[t] ->

            Sqrt[m2] (c1 Cos[phase1 + freq1 t] - c2 Cos[phase2 + freq2 t])
            -------------------------------------------------------------- ,
                              Sqrt[m1 + 2 m2]

          theta2[t] ->

            Sqrt[m1 + m2] (c1 Cos[phase1 + freq1 t] + c2 Cos[phase2 + freq2 t])
            -------------------------------------------------------------------
                                 Sqrt[m1 + 2 m2]

          }
```

where **freq1** and **freq2** are the eigenfrequencies and **phase1** and **phase2** are the phases of the normal modes. The frequencies are given by the first list of **solution** returned from **smallOsc**:

```
In[178]:= freqrule=Thread[ {freq1,freq2}->Sqrt[ solution[[1]] ] ]

Out[178]=
                        g len m1 + g len m2 - g len Sqrt[m2] Sqrt[m1 + m2]
          {freq1 -> Sqrt[--------------------------------------------------],
                                            2
                                         len  m1

                        g len m1 + g len m2 + g len Sqrt[m2] Sqrt[m1 + m2]
           freq2 -> Sqrt[--------------------------------------------------]}
                                            2
                                         len  m1
```

Plot the motion of the two mass points for the values

```
In[179]:= values={m1->3, m2->1, len->1, g->1, phase1->0, phase2->0};
```

Collect all relevant rules into **allRules**:

```
In[180]:= allRules=Join[thetaRule,values,freqrule];
```

Then create a graphics object that represents the two rods of the pendulum. The line comes from connecting the origin $\{0, 0\}$ with the coordinates of the first rod $\{x1[t], y1[t]\}$, and the second rod $\{x2[t], y2[t]\}$. Make this graphics object a function of $\{c1, c2\}$ so that you can vary the relative weight of the normal modes.

```
In[181]:= line[c1_,c2_]=
            Line[{{0,0},{x1[t],-y1[t]},{x2[t],-y2[t]}}] //.allRules;
```

The graph for the normal mode described by $\{c1 = 1,\ c2 = 0\}$ is

```
In[182]:= Show[ Graphics[ Table[line[1,0] //N,{t,0,2,0.1}] ] ];
```

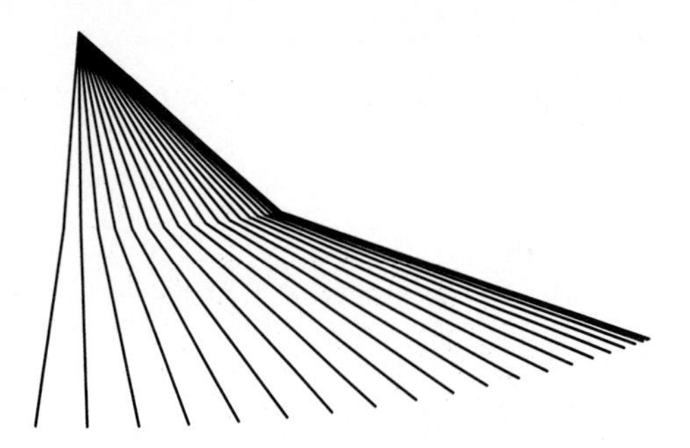

Note that this is a symmetric mode in the sense that the two rods are moving in the same direction.

The graph for the second normal mode described by $\{c1 = 0, c2 = 1\}$ is

```
In[183]:= Show[ Graphics[ Table[line[0,1] //N,{t,0,2.5,0.2}] ] ];
```

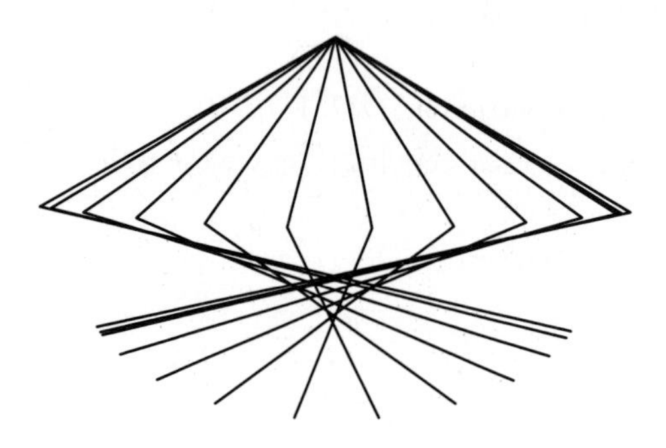

Note that this is an antisymmetric mode in the sense that the two rods are moving in opposite directions.

The graph for a mixture of normal mode states $\{c1 = 1/2, c2 = 1/2\}$ is

```
In[184]:= Show[ Graphics[ Table[line[0.5,0.5] //N,{t,0,6,0.2}] ] ];
```

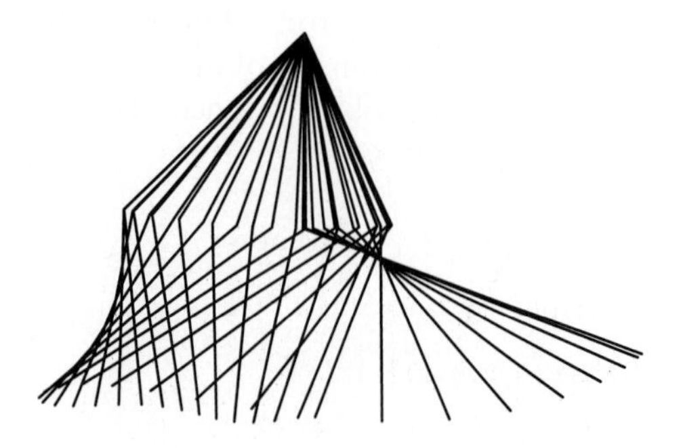

This motion appears complex; however, this solution can be understood more com

pletely by animating this mode.

Part (c) The animation of this last case follows from enclosing the **Show** command inside a **Do** loop:

```
In[185]:= frames=1;
          Do[
            Show[ Graphics[ line[0.5,0.5] //N ,PlotRange->{{-2,2},{0,-2}}] ]
          ,{t,1,25,25/frames}];
```

Note the characteristic behavior. The top rod leads, and then the bottom rod follows. It appears as though only one rod is moving significantly at any time.

3.4 Unsolved Problems

Exercise 3.1: Small oscillations of one-dimensional systems

Assume the motion of a particle can be described by a potential V.

Part (a) Expand the solution around the points of equilibrium and reduce the equation of motion to simple harmonic equations.

Part (b) Write a procedure that, given the potential, will find the frequency of oscillation.

Part (c) Using the procedure found in Part (b), find the frequencies for the following potentials:

$$V = \frac{a}{x} + bx$$
$$V = \frac{a}{x^3} - \frac{b}{x}$$
$$V = \frac{a}{x} + bx^3$$
$$V = \frac{ax^2}{(x^2 + b^2)}$$

Exercise 3.2: Perturbation solution

Consider the satellite equation

$$u''[t] + a^2 u[t] = k u[t]^2.$$

Solve the equation by assuming a perturbative solution in k. Keep terms to order k^2.

Exercise 3.3: Damped harmonic oscillator and driving forces

Consider the motion of a forced, damped, harmonic oscillator described by

$$q''[t] + \omega_0^2 q[t] + \gamma q'[t] = Q_0 \mathrm{Cos}[\omega_d t].$$

Solve for $q[t]$ using Laplace transforms.

Exercise 3.4: Phase plane

The motion of a magnet suspended by a spring above a large iron plate is described by the equation

$$x''[t] + a x[t] = \frac{a}{(b - x[t])},$$

where a and b are positive constants. Discuss the phase plane trajectories of this equation.

Exercise 3.5: Period of nonlinear pendulum

Consider the pendulum equation

$$\theta''[t] + \frac{g}{L}\mathrm{Sin}[\theta[t]] = 0.$$

Find the exact expression for the period of the pendulum.

Exercise 3.6: Elliptical equation

Consider the equation

$$y''[t] = A + By[t] + Cy^2[t] + Dy^3[t].$$

Plot the potential and phase diagrams and discuss the solution of the equation. Program *Mathematica* to recognize the solution of this equation.

Exercise 3.7: Linear motion of three masses

Three particles of mass m are coupled in such a way that the potential energy is given by

$$V = k_1(x_1^2 + x_3^2) + k_2 x_2^2 + k_3(x_1 x_2 + x_2 x_3).$$

Find the eigenfrequencies and eigenvalues.

Exercise 3.8: Three masses constrained to move on a circle

Three equal mass points have equilibrium positions at the vertices of an equilateral triangle. They are connected by equal springs that lie along the arcs of the circle circumscribing the triangle. Mass points and springs are constrained to move only on the circle so that the potential energy of a spring is determined by the arc length considered. Determine the eigenfrequencies and normal modes for small oscillations in the plane of the circle.

Exercise 3.9: Two coupled harmonic oscillators

Consider the vibrations of two particles of mass m connected to each other by a spring of stiffness k_1. Each particle is also connected to the wall by two more springs with equal spring stiffness k_2. Let x_1 and x_2 be the displacements from the equilibrium configuration for the two particles, respectively. Consider only vibrations along the line of the particles.

Part (a) Find the eigenfrequencies and eigenvectors for small vibrations using the command **`Eigensystem`** and the commands **`Eigenvalues`** and **`Eigenvectors`**.

Part (b) Rewrite the user-defined function **`smallOsc`** using the commands **`Eigenvalues`** and **`Eigenvectors`**.

Exercise 3.10: Massive bar with two springs

A rigid uniform bar of mass m and length L is supported in equilibrium in a horizontal position by two identical springs attached one to each end. Assume the motion is constrained to the vertical plane.

Part (a) Find the eigenvectors and eigenfrequencies for small oscillations.

Part (b) Plot the motion if initially one end of the bar is displaced, with the other remaining in its equilibrium position, and the system is released from rest.

Exercise 3.11: A triple pendulum

Consider a triple pendulum with equal masses and string lengths. Assuming small oscillations in a plane, determine the eigenfrequencies and eigenvalues. Plot the normal mode solutions. Write a procedure that will animate an arbitrary solution.

CHAPTER FOUR

Lagrangians and Hamiltonians

4.1 Introduction

Classical mechanics serves as the starting point for several branches of physics. This chapter illustrates the application of *Mathematica* to problems in advanced mechanics. We consider problems solved with Lagrangians, Hamiltonians, and Hamilton-Jacobi equations. This chapter is divided into three sections:

1. Introduction
2. Problems
3. Unsolved Problems

The problems section is subdivided into those on Lagrangians, orbiting bodies, and Hamiltonians and Hamilton-Jacobi methods. Our approach to mechanics closely parallels the discussions found in *Classical Mechanics*, Herbert Goldstein, Second Edition, Addison-Wesley Publishing Company 1980 and *Classical Mechanics*, Edward A. Desloge, John Wiley & Sons 1982.

4.1.1 Lagrange's Equations

Generalized Coordinates and Constraints

A system of n particles is described by at most $3n$ coordinates $\{x_1, \ldots, x_{3n}\}$. Not all these coordinates are independent; constraints may restrict them. These constraints are easiest to handle when they are of the form $f[x_1, x_2, \ldots, t] = 0$, in which case, they are called **holonomic**. Constraints not expressible in this fashion are called **nonholonomic**. If there are k holonomic constraints, then k of the $3n$ coordinates can be eliminated, leaving $3n - k$ independent variables $\{q_1, \ldots, q_{3n-k}\}$ known as the generalized coordinates for the motion. The transformation from the $3n$ coordinates x_i to the $3n - k$ generalized coordinates q_i can be represented by the transformation equations $x_i = x_i[q_1, \ldots, q_{3n-k}, t]$. The generalized coordinates can be distances, angles, or other quantities relating to the description of the motion. The number of generalized coordinates is the number of independent degrees of freedom.

Lagrangian

In terms of generalized coordinates q_i, the equations of motion follow from the $3n - k$ equations

$$\frac{\partial}{\partial t}\left(\frac{\partial T}{\partial q_i'[t]}\right) - \left(\frac{\partial T}{\partial q_i[t]}\right) = Q_i,$$

where T is the kinetic energy and Q_i are the generalized forces. If F_j is the j-component of the force, then the generalized force Q_i is defined by

$$Q_i = \sum_{j,1,3n} F_j \frac{\partial x_j[q_1, \ldots, q_{3n-k}, t]}{\partial q_i}.$$

If some of the forces are conservative, they can be expressed in terms of a potential V, where $F = -\nabla V$. Expressing the conservative forces by a potential V and nonconservative forces by the generalized forces Q_i, the equations of motion follow from Lagrange's equations:

$$\frac{\partial}{\partial t}\left(\frac{\partial L}{\partial q_i'[t]}\right) - \left(\frac{\partial L}{\partial q_i[t]}\right) = Q_i.$$

The Lagrangian $L = T - V$ describes the conservative forces.

Nonholonomic Constraints and Lagrangian Multipliers

For nonholonomic systems, the coordinates q_i are not independent of each other and it is not possible to reduce them by means of constraint equations. However, if there are k constraints of the form

$$\sum_{i,1,3n} A_{ji} dq_i = 0,$$

where $j = \{1, \ldots, k\}$, then Lagrange multipliers can be used to describe the constraints. The equations of motion that follow from these constraints are

$$\frac{\partial}{\partial t}\left(\frac{\partial L}{\partial q_i'[t]}\right) - \left(\frac{\partial L}{\partial q_i[t]}\right) = Q_i,$$

where

$$Q_i = \sum_{j,1,k} y_j A_{ji}.$$

The k parameters y_j are called Lagrange multipliers. Lagrangian's equations give a coupled set of second-order differential equations.

4.1.2 Hamilton's and Hamilton-Jacobi Equations

Hamilton's Equations

Consider those systems described by n generalized coordinates q_i and the Lagrangian $L[q', q, t]$. The Hamiltonian is defined in terms of the Lagrangian $L[q', q, t]$ by

$$H = \sum_{i,1,n} p_i[t] q_i'[t] - L[q', q, t].$$

The p_i are the generalized momentum and are related to the generalized coordinates by

$$p_i = \frac{\partial L[q', q, t]}{\partial q_i'[t]}.$$

By solving these n equations for the q_i' in terms of the p and q, you can write q_i' as a function of q and p, $q' = q'[p, q]$. Eliminate the q_i' variables in H and express H in terms of the canonical coordinates. The equations of motion follow from

$$p_k'[t] = \frac{-\partial H[q, p]}{\partial q_k[t]}$$

$$q_k'[t] = \frac{+\partial H[q, p]}{\partial p_k[t]}.$$

These first-order differential equations are called **Hamilton's canonical equations.**

Hamilton-Jacobi Technique

The Hamilton-Jacobi technique seeks a transformation to a new set of canonical variables $\{Q, P\}$, where the new Hamiltonian is zero. These new canonical variables $\{Q, P\}$ will be constants, and the problem is reduced to solving algebraic equations. If $\{q, p\}$ are the old canonical coordinates and $\{Q, P\}$ are the new coordinates, the transformation between the coordinates is given by $Q = Q[q, p, t]$ and $P = P[q, p, t]$. The transformation is said to be canonical if there exists a new Hamiltonian $K[P, Q]$, where the new coordinates obey Hamilton's equations:

$$P' = \frac{-\partial K}{\partial Q}$$

$$Q' = \frac{+\partial K}{\partial P}.$$

Canonical transformations can be obtained from generating functions. In particular for the Hamilton-Jacobi technique, consider generating functions of the form $S[q, P, t]$. The transformation between the variables $\{q, p\}$ and $\{Q, P\}$ follows from derivatives of the generating function:

$$p = \frac{\partial S[q, P, t]}{\partial q}$$

$$Q = \frac{\partial S[q, P, t]}{\partial P}.$$

Solve the first set of equations to get $P[q, p, t]$. The remaining equations give the expression for $Q[q, p, t]$. Next, search for a canonical transformation whereby the new Hamiltonian K is zero; this will render the equations of motion for $\{P, Q\}$ trivial. The new and old Hamiltonians are related by

$$K = H[q, p, t] + \frac{\partial S[q, P, t]}{\partial t}.$$

If K is zero, then it follows from Hamilton's equations,

$$\frac{+\partial K}{\partial P} = Q' = 0$$

$$\frac{-\partial K}{\partial Q} = P' = 0,$$

that the new variables Q and P are constants. The constraint $K = 0$ implies the generating function S satisfies

$$H[q, p, t] + \frac{\partial S[q, P, t]}{\partial t} = 0.$$

Replacing p in this equation with $\partial S[q, P, t]/\partial q$, you get the Hamilton-Jacobi equation

$$H\left[q, \frac{\partial S[q, P, t]}{\partial q}, t\right] + \frac{\partial S[q, P, t]}{\partial t} = 0.$$

The generating function for the Hamilton-Jacobi variables is called **Hamilton's principal function.**

The Hamilton-Jacobi equation is a partial differential equation for the generating function $S[q, P, t]$. It contains $n + 1$ variables: one time variable and n q_i variables. A complete solution of the partial differential equation involves $n + 1$ independent constants of integration. One of the $n + 1$ constants is just a trivial additive constant and can be set equal to zero. The remaining n constants of integration $\{a_1, \ldots, a_n\}$ are taken to be the new canonical momentum P. The problem of finding Hamilton's principal function, $S[q, P, t]$, is simplified if the Hamilton-Jacobi equation is separable. That is,

$$S[q, P, t] = W_1[q[1]] + W_2[q[2]] \ldots + W_n[q[n]] - at.$$

The W's are called Hamilton's characteristic functions. The Hamilton-Jacobi equation is reduced to a set of ordinary differential equations, one for each W. The solution for W can then be integrated. In the separable case, the problem is reduced to integrals. The transformations between the old coordinates $\{q, p\}$ and the new ones $\{Q, P\}$ follow from

$$p = \frac{\partial S[q, a, t]}{\partial q}$$

$$Q = \frac{\partial S[q, P, t]}{\partial P}.$$

The coordinate q is solved by inverting the second equation to give $q = q[Q, a, t]$.

4.1.3 *Mathematica* Commands

Packages

First, turn off the spell checker:

```
In[1]:=  Off[General::spell ];
         Off[General::spell1];
```

This chapter needs the following packages. The packages should be activated now to avoid shadowing. The packages are:

```
In[2]:=  Needs["Algebra`SymbolicSum`"]
         Needs["Algebra`Trigonometry`"]
         Needs["Calculus`VectorAnalysis`"]
         Needs["Graphics`Graphics`"]
         Needs["Graphics`Graphics3D`"]
```

User-defined Rules

HyperbolicToComplex and ComplexToHyperbolic

To convert hyperbolic functions to exponential functions and vice versa, it is useful to define **HyperbolicToComplex** and **ComplexToHyperbolic**. These functions are just like their counterparts, **TrigToComplex** and **ComplexToTrig**.

```
In[3]:=  HyperbolicToComplexRule={Cosh[x___]:>(Exp[x]+Exp[-x])/2,
                                 Sinh[x___]:>(Exp[x]-Exp[-x])/2};

         ComplexToHyperbolicRule={Exp[x___]:>Cosh[x]+Sinh[x]};

In[4]:=  HyperbolicToComplex[expression_]:=
                 expression //.HyperbolicToComplexRule
         ComplexToHyperbolic[expression_]:=
                 expression //.ComplexToHyperbolicRule
```

Example: HyperbolicToComplex and ComplexToHyperbolic

Transform $a_1\text{Cosh}[x] + a_2\text{Sinh}[x]$:

```
In[5]:=  temp= ( a1 Cosh[x] + a2 Sinh[x]
                    //HyperbolicToComplex
                    //Simplify )

Out[5]=
                        2 x        2 x
         a1 - a2 + a1 E     + a2 E
         ---------------------------
                      x
                   2 E
```

Next, invert this transform as a cross check of the rules:

```
In[6]:=  temp //ComplexToHyperbolic //Simplify

Out[6]=  a1 Cosh[x] + a2 Sinh[x]
```

User-defined Procedures

It is useful to create user-defined procedures for those calculations that are used repeatedly. To understand the procedures, it is best to redefine the function without any local variables. (This is described in the next section.) Then you can examine the intermediate steps to understand the procedure. This is an essential technique for debugging errors.

As an exercise, you might try to make the procedures more time efficient and add default conditions and options. Some of these procedures will fail in particular cases; you are encouraged to write conditional statements to handle these cases.

Finding Lagrange's equations

Throughout this chapter, you will need to compute the Lagrange equations of motion. The **Lag** function will compute this in one step in the case of one or more variables.

The Lagrangian equations are given by

$$\frac{\partial}{\partial t}\left(\frac{\partial L}{\partial q_i'[t]}\right)-\left(\frac{\partial L}{\partial q_i[t]}\right)=Q_i.$$

```
In[7]:=  Lag::usage=" Lag[qi,L,Qi:0,t:t] computes Lagrange's
         equations of motion given the canonical coordinate
         qi,the Lagrangian L, and the generalized force Qi.
         The generalized coordinate qi can be entered as a
         list or as a single coordinate.";

In[8]:=  Lag[qi_,L_,Qi_:0,t_:t] :=
           (D[ D[L,(D[qi[t],t])],t]-D[L,qi[t]]==Qi);

In[9]:=  Lag[qi_List,L_,Qi_:0,t_:t] :=
           (D[ D[L,(D[#[t],t])],t]-D[L,#[t]]==Qi)& /@ qi;
```

The definition of **Lag** is straightforward enough so that you will not need to decompose this function. Rather, simply give an example in one and two dimensions.

Example: Lagrange's equation for a particle in a potential V[x]

Consider the one-dimensional Lagrangian

```
In[10]:= L=  (m/2) x'[t]^2 - V[x[t]];
```

The equations of motion are

```
In[11]:= Lag[x,L]

Out[11]= V'[x[t]] + m x''[t] == 0
```

For the two-dimensional case, **Lag** also works:

```
In[12]:= L=  (m/2) (x'[t]^2 + y'[t]^2) - V[x[t],y[t]];
         Lag[{x,y},L]

Out[12]=                   (1,0)
         {m x''[t] + V          [x[t], y[t]] == 0,

                            (0,1)
           m y''[t] + V          [x[t], y[t]] == 0}
```

Finding the canonical momentum, Hamiltonian, and equations of motion

You will also need to compute Hamilton's equations of motion. The **Hamilton** function will compute this in one step for the case of one or multiple variables.

```
In[13]:= Clear[Hamilton];
         Hamilton::usage=" Hamilton[L,{xList},{pList},t:t]
          returns, \n
              i.   the expression for the canonical momentum; \n
              ii.  the hamiltonian; and   \n
              iii. Hamilton's equation of motion.";
```

```
In[14]:= Hamilton[L_,xList_List,pList_List,t_:t]:=
          Module[{xx,vv,pp,sol,ham,eqp,eqx,eqs},
            xx = Map[ #[t]&,xList];
            vv = Map[#'[t]&,xList];
            pp = Map[ #[t]&,pList];
            sol= Solve[ (D[L,#]& /@ vv) == pp,vv] //Flatten;
            ham= pp.vv-L //.sol //Simplify //Expand;
            eqp= D[pp,t]==-Map[D[ham,#]&,xx] //Thread;
            eqx= D[xx,t]==+Map[D[ham,#]&,pp] //Thread;
            eqs= Join[eqp,eqx];
            Return[{sol,ham,eqs}]
           ]
```

Example 1: Hamilton's equations of motion in one dimension

Given the simplest one-dimensional Lagrangian, you can easily compute the Hamilton's equations of motion:

```
In[15]:= L=  (m/2) x'[t]^2 - V[x[t]];
         {pRule, hamiltonian, eqMotion} = Hamilton[L,{x},{px}];
         pRule
         hamiltonian
         eqMotion

Out[15]=            px[t]
         {x'[t] -> -----}
                      m

Out[15]=       2
         px[t]
         ------ + V[x[t]]
          2 m

Out[15]=                                   px[t]
         {px'[t] == -V'[x[t]], x'[t] == -----}
                                             m
```

Note how the three separate expressions returned by **Hamilton** are extracted.

Example 2: Hamilton's equations of motion in two dimensions

Next, consider a two-dimensional Lagrangian:

```
In[16]:= L=  (m/2) (x'[t]^2 + y'[t]^2) - V[x[t],y[t]];
         Hamilton[L,{x,y},{px,py}]

Out[16]=                                              2         2
                   px[t]               py[t]    px[t]     py[t]
         {{x'[t] -> -----, y'[t] -> -----},  ------ + ------ + V[x[t], y[t]],
                     m                   m      2 m       2 m

                         (1,0)                                (0,1)
           {px'[t] == -V      [x[t], y[t]], py'[t] == -V      [x[t], y[t]],

                      px[t]               py[t]
             x'[t] == -----, y'[t] == -----}}
                        m                  m
```

Example 3: How **Hamilton** works

To understand how **Hamilton** works, create a function **xHamilton** that is identical to **Hamilton** except it has no local variables (compare the second line: **Module[{},...]** with the original definition):

```
In[17]:= xHamilton[L_,xList_List,pList_List,t_:t]:=
           Module[{},
             xx = Map[ #[t]&,xList];
             vv = Map[#'[t]&,xList];
             pp = Map[ #[t]&,pList];
             sol= Solve[ (D[L,#]& /@ vv) == pp,vv] //Flatten;
             ham= pp.vv-L //.sol //Simplify //Expand;
             eqp= D[pp,t]==-Map[D[ham,#]&,xx] //Thread;
             eqx= D[xx,t]==+Map[D[ham,#]&,pp] //Thread;
             eqs= Join[eqp,eqx];
             Return[{sol,ham,eqs}]
            ]
```

Next, execute **xHamilton** as in Example 1 so that you can examine the internal variables:

```
In[18]:= L=  (m/2) x'[t]^2 - V[x[t]];
         xHamilton[L,{x},{px}];
```

First, see how the lists of position, velocity, and momentum are represented:

```
In[19]:= {xx,vv,pp}

Out[19]= {{x[t]}, {x'[t]}, {px[t]}}
```

The equation that relates velocity and momentum is:

```
In[20]:= (D[L,#]& /@ vv) == pp

Out[20]= {m x'[t]} == {px[t]}
```

The solution **sol** gives a way of converting from velocity **x'[t]** to momentum **px[t]**:

```
In[21]:= sol

Out[21]=
                  px[t]
         {x'[t] -> -----}
                    m
```

The Hamiltonian is then defined by $H = pv - L$:

```
In[22]:= ham

Out[22]=
              2
         px[t]
         ------ + V[x[t]]
          2 m
```

and the equations of motion are given by $\partial p/\partial t = -\partial H/\partial x$ and $\partial x/\partial t = +\partial H/\partial p$:

```
In[23]:= {eqp,eqx}

Out[23]=
                                                 px[t]
         {{px'[t] == -V'[x[t]]}, {x'[t] == -----}}
                                                   m
```

Finding the canonical momentum, Hamilton's principal function, and Hamilton-Jacobi equations

You will also need to compute the Hamilton-Jacobi equations of motion. Define the **HamiltonJacobi** function to compute this systematically:

```
In[24]:= Clear[HamiltonJacobi];
         HamiltonJacobi::usage=
          " HamiltonJacobi[L,{xList},{pList},{wList[...]},a,t:t]
          returns,  \n\n
             i. the expression for the canonical momentum; \n
            ii. the Hamilton's Principal Function; and  \n
           iii. Hamilton-Jacobi Equation.  \n\n
         The procedure assumes that the Hamilton Principal
         Function can be expressed in the form W[q]-a t.";
```

```
In[25]:=  HamiltonJacobi[L_,xList_List,pList__List,wList_List,a_,t_:t]:=
           Module[{xx,vv,pp,sol,ham,ww,hamPrincFunc,pNewRule,HJeq},
             xx = Map[ #[t]&,xList];
             vv = Map[#'[t]&,xList];
             pp = Map[ #[t]&,pList];
             sol= Solve[ (D[L,#]& /@ vv) == pp,vv] //Flatten;
             ham= (pp.vv-L) //.sol //Simplify //Expand;

             ww = Map[ #[t]&,wList,{2}];
             hamPrincFunc = (Plus @@ ww) - a t;
             pNewRule =
               {pp -> (D[hamPrincFunc,#]& /@ xx) //Thread }//Flatten;
             HJeq = (ham//.pNewRule ) - a ==0;

             Return[{sol,hamPrincFunc,HJeq}]
            ]
```

Example 1: One-dimensional particle in a potential V[x]

Consider a system described by the Lagrangian:

```
In[26]:= L=  (m/2) x'[t]^2 - V[x[t]];
         {pRule,hamPrincFunc,eqHamJac}=
            HamiltonJacobi[L, {x}, {p}, {W[x]}, a];
         pRule
         hamPrincFunc
         eqHamJac

Out[26]=
                    p[t]
         {x'[t] -> ----}
                     m

Out[26]= -(a t) + W[x[t]]

Out[26]=
                                         2
                             W'[x[t]]
         -a + V[x[t]] + ------------ == 0
                              2 m
```

Again note how the separate parts returned by **HamiltonJacobi** are extracted.

Example 2: Two-dimensional particle in a potential V[x]

Consider a system described by the Lagrangian:

```
In[27]:= L=  (m/2) (x'[t]^2 + y'[t]^2) - V[x[t],y[t]];
         HamiltonJacobi[L, {x,y}, {px,py}, {W1[x],W2[y]}, a]

Out[27]=              px[t]            py[t]
         {{x'[t] -> -----, y'[t] -> -----}, -(a t) + W1[x[t]] + W2[y[t]],
                        m                m

                                              2              2
                                    W1'[x[t]]      W2'[y[t]]
            -a + V[x[t], y[t]] + ---------- + ---------- == 0}
                                      2 m            2 m
```

Example 3: How **HamiltonJacobi** works

As with Hamilton, define a function identical to **HamiltonJacobi**, **xHamiltonJacobi**, which has no local variables:

```
In[28]:= xHamiltonJacobi[L_,xList_List,pList__List,wList_List,a_,t_:t]:=
           Module[{},
             xx = Map[ #[t]&,xList];
             vv = Map[#'[t]&,xList];
             pp = Map[ #[t]&,pList];
             sol= Solve[ (D[L,#]& /@ vv) == pp,vv] //Flatten;
             ham= (pp.vv-L) //.sol //Simplify //Expand;

             ww = Map[ #[t]&,wList,{2}];
             hamPrincFunc = (Plus @@ ww) - a t;
             pNewRule =
               {pp -> (D[hamPrincFunc,#]& /@ xx) //Thread }//Flatten;
             HJeq = (ham//.pNewRule ) - a ==0;

             Return[{sol,hamPrincFunc,HJeq}]
           ]
```

Next, execute **xHamiltonJacobi** so that you can examine the internal variables:

```
In[29]:= L=  (m/2) x'[t]^2 - V[x[t]];
         xHamiltonJacobi[L, {x}, {p}, {W[x]}, a];
```

The first few lines, which generate an expression for the Hamiltonian **ham**, are identical to the **Hamilton** function. To see how it is internally stored, start with the **wList**.

```
In[30]:= ww

Out[30]= {W[x[t]]}
```

Then it is simple to compute Hamilton's principle function **hamPrincFunc**:

```
In[31]:= hamPrincFunc

Out[31]= -(a t) + W[x[t]]
```

Next, compute a rule to eliminate the momentum in terms of **W'[x[t]]**:

```
In[32]:= pNewRule

Out[32]= {p[t] -> W'[x[t]]}
```

The Hamilton-Jacobi equation is then given by $H - a$ with this momentum substitution:

```
In[33]:= HJeq

Out[33]=
                          2
                 W'[x[t]]
-a + V[x[t]] + ————————— == 0
                   2 m
```

Series expansion solution for second-order equation

This function is duplicated from Chapter 3. Refer to that chapter for details.

```
In[34]:= diffSeriesOne::usage="
         diffSeriesOne[eq,z,t,order,{initialList}]
         eq is an equation, z is the function, t is the independent
         variable, order is the order of the Series solution, and
         {initialList} is a list of initial conditions such as
         {z[0]->z0}.
         This procedure assumes the solution can be written as
         a power series.";

In[35]:= Clear[diffSeriesOne];
         diffSeriesOne[eqin_,z_,t_,order_,initList__List]:=
          Module[{eq,zSer,eSer,eqs,vars,sol,zSol},
            eq=eqin[[1]]-eqin[[2]]==0;
            zSer= Series[z[t],{t,0,order}];
            eSer= eq/.(z[t]->zSer)//ExpandAll;
            eqs = Thread[CoefficientList[eSer[[1]],t]==0];
            vars=(Table[D[z[t],{t,j}],{j,2,order+2}]/.t->0);
            sol = Solve[eqs,vars][[1]];
            zSol= zSer/.sol/.initList;
            Return[ zSol]
          ]
```

Series expansion solution for two first-order equations

This function is closely related to **diffSeriesOne**, except it takes two coupled, first-order equations instead of one, second-order equation:

```
In[36]:= Clear[firstDiffSeries];
         firstDiffSeries::usage="
         firstDiffSeries[eqin,{z,p},initRule_List,order,t_:t]:=
          returns and expansion for the solution of two first order
          equations in powers of t.";

In[37]:= firstDiffSeries[eqin_,{z_,p_},initRule_List,order_,t_:t]:=
          Module[{eq,zSer,pSer,eSer,vars,sol,zpsol},
            eq=Table[eqin[[i,1]]-eqin[[i,2]]==0,{i,2}];
            zSer= Series[z[t],{t,0,order}];
            pSer= Series[p[t],{t,0,order}];
            eSer= eq//.({z[t]->zSer,p[t]->pSer})//ExpandAll;
            vars=(Table[{D[z[t],{t,j}],
                         D[p[t],{t,j}]},{j,1,order+1}]/.t->0)//Flatten;
            sol = Solve[eSer,vars] //Flatten;
            zpSol= {zSer,pSer} //.sol //.initRule;
            Return[ zpSol]
          ]
```

Example 1: Expansion of harmonic oscillator

Consider the harmonic oscillator described by the two coupled, first-order equations:

```
In[38]:= eq1= {p'[t] + w^2 x[t] == 0, p[t] == x'[t]};
         firstDiffSeries[eq1,{x,p},{x[0]->x0,p[0]->p0},4,t]

Out[38]=
                      2   2        2 3     4    4
                     w  x0 t    p0 w  t   w  x0 t        5
         {x0 + p0 t - -------- - ------- + -------- + O[t] ,
                        2          6         24

                           2 2     4    3       4 4
                2       p0 w  t   w  x0 t   p0 w  t       5
          p0 - w  x0 t - ------- + ------- + ------- + O[t] }
                           2         6         24
```

firstDiffSeries returns a perturbative series expansion for **x** and **p**.

Example 2: How firstDiffSeries works

In the usual manner, define a function identical to **firstDiffSeries** but without any local variables:

```
In[39]:= xfirstDiffSeries[eqin_,{z_,p_},initRule_List,order_,t_:t]:=
          Module[{},
            eq=Table[eqin[[i,1]]-eqin[[i,2]]==0,{i,2}];
            zSer= Series[z[t],{t,0,order}];
            pSer= Series[p[t],{t,0,order}];
            eSer= eq//.({z[t]->zSer,p[t]->pSer})//ExpandAll;
            vars=(Table[{D[z[t],{t,j}],
                         D[p[t],{t,j}]},{j,1,order+1}]/.t->0)//Flatten;
            sol = Solve[eSer,vars] //Flatten;
            zpSol= {zSer,pSer} //.sol //.initRule;
            Return[ zpSol]
          ]
```

Then execute the function and examine the internal variables:

```
In[40]:= eq1= {p'[t] + w^2 x[t] == 0, p[t] == x'[t]};
         xfirstDiffSeries[eq1,{x,p},{x[0]->x0,p[0]->p0},2,t];
```

First, you want to rearrange the equations so that all the nonzero terms are on the left-hand side of the equation:

```
In[41]:= eq

Out[41]=   2
         {w  x[t] + p'[t] == 0, p[t] - x'[t] == 0}
```

Next, make a perturbative expansion for the position:

```
In[42]:= zSer

Out[42]=
                                    2
                           x''[0] t        3
         x[0] + x'[0] t + ---------- + O[t]
                               2
```

There is a similar expression for the momentum **pSer**. Substitute these series expressions into the equations of motion to obtain

```
In[43]:= eSer //Short

Out[43]=   2                                                3
         {(w  x[0] + p'[0]) + <<2>> + <<1>> + O[t]  == 0, <<1>>}
```

The unknown terms in the expansion that must be solved for are

```
In[44]:= vars

Out[44]=                                      (3)       (3)
         {x'[0], p'[0], x''[0], p''[0], x   [0], p   [0]}
```

To obtain these unknowns, solve the series equations of motion:

```
In[45]:= sol

Out[45]=  (3)          2          (3)         4
         {x   [0] -> -(w  p[0]), p   [0] -> w  x[0], x'[0] -> p[0],

                     2                     2                      2
           p'[0] -> -(w  x[0]), x''[0] -> -(w  x[0]), p''[0] -> -(w  p[0])}
```

All that remains to be done is to substitute this solution into the series expressions for the position **zSer** and momentum **pSer**, along with the initial conditions. This is what is returned:

```
In[46]:= zpSol

Out[46]=                2   2                                      2 2
                       w  x0 t         3       2             p0 w  t        3
         {x0 + p0 t - ---------- + O[t] , p0 - w  x0 t - ----------- + O[t] }
                          2                                    2
```

First-order perturbation solution

This function returns the first-order perturbative correction. It is closely related to **diffSeriesOne** and **firstDiffSeries** but takes a slightly different approach:

```
In[47]:= Clear[firstOrderPert];
         firstOrderPert::usage=
             "firstOrderPert[eq,x,{x0,v0},eps:eps,t:t]
          returns the first order correction to the term with coefficient eps.";

In[48]:= firstOrderPert[eq1_,x_,{x0_,v0_},eps_:eps,t_:t]:=
           Module[{xRule,eq2,eqEps,initCon0,initCon1,dsol1,dsol2},
              xRule={x->(x[0][#] + eps x[1][#]&)};
              eq2=eq1[[1]]-eq1[[2]] /.xRule;
              eqEps= CoefficientList[eq2,eps]==0 //Thread;
              initCon0={x[0][0]==x0, (x[0])'[0]==v0};
              initCon1={x[1][0]==0 , (x[1])'[0]==0 };
              dsol1=( DSolve[Join[{eqEps[[1]]},initCon0],{x[0][t]},t]
                       //Flatten );
              dsol2=( DSolve[Join[{eqEps[[2]]},initCon1] //.dsol1,{x[1][t]},t]
                       //Flatten );
              Return[ x[t] /.xRule //.Join[dsol1,dsol2] ]
           ]
```

Example 1: Perturbed harmonic oscillator

Applying **firstorderpert** to **eq1**,

```
In[49]:= eq1= x''[t]== - k^2 x[t]+eps x[t]^2;
```

we get

```
In[50]:= ( firstOrderPert[eq1,x,{x0,0},eps,t]
              //ExpandAll //ComplexToTrig //ExpandAll )

Out[50]=          2                            2                     2
         eps x0                         eps x0  Cos[k t]    eps x0  Cos[2 k t]
         ------- + x0 Cos[k t] -  ------------------ - -------------------- +
             2                              2                    2
          4 k                            3 k                  6 k

              2            2          2             2
       eps x0  Cos[2 k t]       eps x0  Sin[2 k t]
       -------------------- + ---------------------
                2                       2
             4 k                     4 k
```

Example 2: Details of firstOrderPert

In the usual manner, define a function identical to **firstOrderPert**, but without any local variables:

```
In[51]:= xfirstOrderPert[eq1_,x_,{x0_,v0_},eps_:eps,t_:t]:=
           Module[{},
              xRule={x->(x[0][#] + eps x[1][#]&)};
              eq2=eq1[[1]]-eq1[[2]] /.xRule;
              eqEps= CoefficientList[eq2,eps]==0 //Thread;
              initCon0={x[0][0]==x0, (x[0])'[0]==v0};
              initCon1={x[1][0]==0 , (x[1])'[0]==0 };
              dsol1=( DSolve[Join[{eqEps[[1]]},initCon0],{x[0][t]},t]
                      //Flatten );
              dsol2=( DSolve[Join[{eqEps[[2]]},initCon1] //.dsol1,{x[1][t]},t]
                      //Flatten );
              Return[ x[t] /.xRule //.Join[dsol1,dsol2] ]
           ]
```

Then execute the function and examine the internal variables:

```
In[52]:= eq1= x''[t]== - k^2 x[t]+eps x[t]^2;
         xfirstOrderPert[eq1,x,{x0,0},eps,t];
```

First, expand the position variable:

```
In[53]:= x[t] /.xRule

Out[53]= x[0][t] + eps x[1][t]
```

and substitute this into the equations of motion. Then, use **CoefficientList** to equate powers of **eps**. (Note that **eq2** is a rearrangement of the input equation so that **CoefficientList** will work properly.) The perturbative equations that you then have to solve are

```
In[54]:= eqEps
Out[54]=  2
         {k  x[0][t] + (x[0])''[t] == 0,

                   2    2
           -x[0][t]  + k  x[1][t] + (x[1])''[t] == 0, -2 x[0][t] x[1][t] == 0,

                   2
           -x[1][t]  == 0}
```

These equations are easily solved with **DSolve**. The unperturbed solution for the position is

```
In[55]:= dsol1
Out[55]=
                    -I k t          I k t
                   E       x0      E      x0
         {x[0][t] -> ------------ + ------------}
                        2               2
```

and the perturbed solution is

```
In[56]:= dsol2
Out[56]=
                      -I k t   2     I k t   2
                   -(E       x0 )   E      x0
         {x[1][t] -> --------------- - ------------ +
                           2              2
                        6 k            6 k

             2
           x0  (Cos[2 k t] - I Sin[2 k t])
           ------------------------------- +
                         2
                     24 k

             2
           x0  (2 - Cos[2 k t] + I Sin[2 k t])
           ----------------------------------- +
                          2
                       8 k

             2
           x0  (Cos[2 k t] + I Sin[2 k t])
           ------------------------------- +
                         2
                     24 k

             2
           x0  (-1 + 2 (Cos[2 k t] - I Sin[2 k t]))
             (Cos[2 k t] + I Sin[2 k t])
           ----------------------------------------------
                                  2
                               8 k

         }
```

All that remains to be done is to substitute these solutions into the perturbative expression for the position. This is what is returned:

```
In[57]:= x[t] /.xRule //.Join[dsol1,dsol2] //ExpandAll //ComplexToTrig //ExpandAll
Out[57]=
               2                      2                  2
         eps x0                 eps x0  Cos[k t]   eps x0  Cos[2 k t]
         ------- + x0 Cos[k t] - ---------------- - ------------------ +
             2                          2                   2
          4 k                        3 k                 6 k

               2            2          2             2
         eps x0  Cos[2 k t]      eps x0  Sin[2 k t]
         -------------------- + --------------------
                   2                       2
                4 k                     4 k
```

4.1.4 Protect User-defined Procedures

```
In[58]:= Protect[
         ,ComplexToHyperbolic ,ComplexToHyperbolicRule
         ,HyperbolicToComplex ,HyperbolicToComplexRule
         ,Lag,Hamilton,HamiltonJacobi
         ,diffSeriesOne,firstDiffSeries,firstOrderPert
         ];
```

4.2 Problems

4.2.1 Lagrangian Problems

Problem 1: Atwood Machine

An Atwood machine consists of two weights of mass m_1 and m_2 connected by a light, inextensible cord of length Len, which passes over a frictionless pulley. Let $x[1]$ be the coordinate of mass m_1 and let $x[2]$ be the coordinate of mass m_2. The coordinates $x[1]$ and $x[2]$ are the vertical distances from the pulley to the masses. Assume the masses move only in a vertical direction.

Part (a) Express the Lagrangian in terms of only $x[1]$. Find the equations of motion and solve.

Part (b) Express the Lagrangian in terms of $x[1]$ and $x[2]$. Find the Lagrange multipliers from the constraint that the cord does not stretch. Find the equations of motion for $x[1]$ and $x[2]$ using Lagrange multipliers. Solve for the motion.

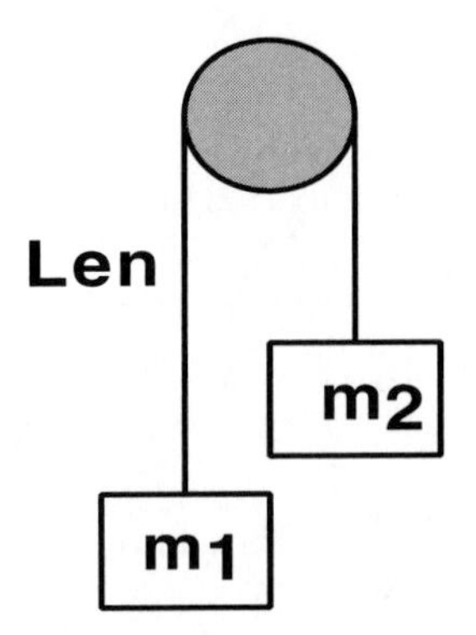

Remarks and outline: Because $Len = x[1] + x[2]$, there is only one independent that can be taken to be $x[1]$. If you express the Lagrangian in terms $x[1]$, the equations of motion follow from the user-defined function **Lag**. The Atwood problem can also be solved with Lagrange multipliers. The variables $x[1]$ and $x[2]$ can be treated as independent variables and the constraint $Len = x[1] + x[2]$ can be used to get a Lagrange multiplier. The user-defined function **Lag** gives two equations. A third equation follows from $Len = x[1]+x[2]$.

Required packages

```
In[1]:=  Needs["Algebra`Trigonometry`"]
         Needs["Calculus`VectorAnalysis`"]
```

Solution

```
In[2]:=  Clear["Global`*"];
```

Part (a) To describe the relation between **x[1]**, **x[2]**, and **Len**, define the rule:

```
In[3]:=  lenRule=x[2]->((Len-x[1][#])&)

Out[3]=  x[2] -> (Len - x[1][#1] & )
```

The substitution was expressed in terms of pure functions so that the rule will replace derivatives of **x[2]**. The potential and kinetic energies are

```
In[4]:=  T= (1/2) Sum[  m[i] D[x[i][t],t]^2 ,{i,1,2}];
         V= (- g) Sum[  m[i]    x[i][t]       ,{i,1,2}];
```

The Lagrangian follows from

```
In[5]:=  L=T-V /.lenRule //Simplify

Out[5]=
                                                                               2
                                                          (m[1] + m[2]) (x[1])'[t]
         g (Len m[2] + m[1] x[1][t] - m[2] x[1][t]) + ------------------------
                                                                    2
```

where the **x[2]** variable with **lenRule** has been eliminated. The equation of motion follows from the user-defined command **Lag**:

```
In[6]:=  eq1=  Lag[x[1] ,L] //Simplify

Out[6]=  g (-m[1] + m[2]) + (m[1] + m[2]) (x[1])''[t] == 0
```

For the initial conditions

```
In[7]:=  initial= {x[1][0]==x0, x[1]'[0]==v0};
```

the solution of **eq1** follows from **DSolve**:

```
In[8]:=  sol= DSolve[Join[{eq1},initial], x[1][t],t][[1]]

Out[8]=
                                     2
                                    t  (-(g m[1]) + g m[2])
         {x[1][t] -> t v0 + x0 - -----------------------}
                                      2 (m[1] + m[2])
```

The last two terms in **sol** can be factored to give

```
In[9]:=  (( sol //Map[Collect[#,{t}]&,#,{2}]& )
                //Map[Simplify       ,#,{2}]& )

Out[9]=
                                      2
                                  g t  (m[1] - m[2])
         {x[1][t] -> t v0 + x0 + ------------------}
                                  2 (m[1] + m[2])
```

The operator **Map** maps the commands **Collect** and **Simplify** on certain parts of **sol** to group the last two terms. Experiment with **Map** and **MapAt** on other equations to learn how they behave.

Part (b) Next, now solve the Atwood problem using two variables and a Lagrange multiplier. The Lagrangian in terms of **x[1]** and **x[2]** is

```
In[10]:= L= T-V
Out[10]=
                                                             2                   2
                                           m[1] (x[1])'[t]  + m[2] (x[2])'[t]
         g (m[1] x[1][t] + m[2] x[2][t]) + -----------------------------------------
                                                            2
```

x[2] has not been eliminated but is treated as an independent variable. The constraint equation follows from the length of the string. The constraint equation **f[1]** is

```
In[11]:= f[1]=x[1][t]+x[2][t]-Len;
```

The differential of **f[1]** gives the constraint forces **{Q[1],Q[2]}** in terms of the Lagrangian multiplier **lambda[1]**:

```
In[12]:= Q[i_]:=Sum[ D[f[j],x[i][t]] lambda[j] ,{j,1,1}];
         Array[Q,2]

Out[12]= {lambda[1], lambda[1]}
```

Two equations of motion follow from the user-defined command **Lag**:

```
In[13]:= eq2= Table[ Lag[x[i], L, Q[i] ], {i,1,2}]

Out[13]= {-(g m[1]) + m[1] (x[1])''[t] == lambda[1],

          -(g m[2]) + m[2] (x[2])''[t] == lambda[1]}
```

eq2 gives two equations for the three unknowns **x[1]**, **x[2]**, and **lambda[1]**. The third equation follows from the length constraint:

```
In[14]:= f[1]==0

Out[14]= -Len + x[1][t] + x[2][t] == 0
```

Taking the second derivative of length, you get an equation that is more pertinent to the calculation:

```
In[15]:= eq3=D[f[1]==0, {t,2}]

Out[15]= (x[1])''[t] + (x[2])''[t] == 0
```

Applying **Solve** to **eq2** and **eq3**, you get three equations:

```
In[16]:= eq4= ( Solve[Join[eq2,{eq3}],
              {lambda[1], (x[2])''[t],(x[1])''[t] },t ]
          //Simplify
          //Flatten
         ) /.{Rule->Equal}

Out[16]=
                      -2 g m[1] m[2]                        g (-m[1] + m[2])
         {lambda[1] == --------------, (x[2])''[t] == ------------------,
                       m[1] + m[2]                          m[1] + m[2]

                         g (m[1] - m[2])
           (x[1])''[t] == ---------------}
                           m[1] + m[2]
```

These three equations are uncoupled, and it is a trivial task to solve the differential equations. Solve for the **x[1]** variable. Assuming the initial conditions

```
In[17]:= initial={ x[1][0]==x0, (x[1])'[0]==v0 };
```

the solution for **x[1]** follows from part three of **eq4**:

```
In[18]:= sol1=( ( DSolve[ eq4[[{3}]]~Join~initial     ,
                    {(x[1])[t]  },t]
                    //Flatten
                    //Map[Collect[#,{t}]&,#,{2}]&
                  ) //Map[Simplify,#,{2}]&
                 )

Out[18]=
                                       2
                                    g t  (m[1] - m[2])
         {x[1][t] -> t v0 + x0 + ------------------}
                                    2 (m[1] + m[2])
```

The solution for **x[2]** follows from the length constraint:

```
In[19]:= sol2={ x[2][t]->   Len-x[1][t] /.sol1 }

Out[19]=
                                             2
                                          g t  (m[1] - m[2])
         {x[2][t] -> Len - t v0 - x0 - ------------------}
                                          2 (m[1] + m[2])
```

Problem 2: Bead Sliding on a Rotating Wire

The upper end of a straight frictionless wire is fixed at point A. The wire makes an angle θ with the downward vertical drawn from A. The wire rotates around the vertical axis with constant angular velocity ω. A bead of mass m is constrained to move on the wire.

Part (a) Solve for the motion of the bead assuming the initial conditions $r[0] = r_0$ and $r'[0] = v_0$. Express the results in terms of hyperbolic functions.

Part (b) Let r_0 and v_0 be zero at $t = 0$ and find the time it takes for the bead to move a distance Len. Evaluate the time for the special case $\theta = 0$.

Part (c) Let $v_0 = 0$ and $r_0 = 0$ at $t = 0$ and graph the motion of the bead as a function of time.

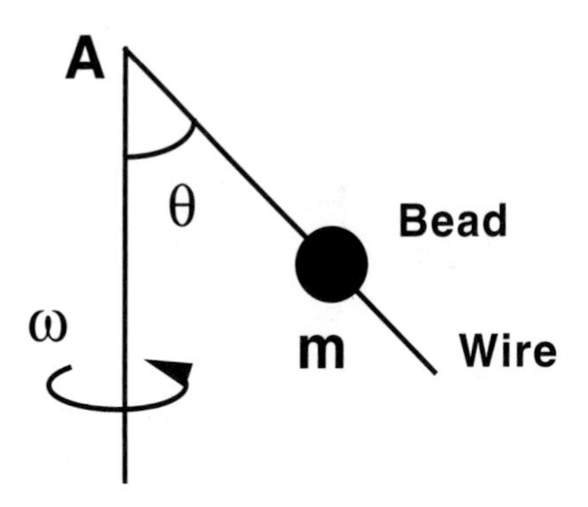

Remarks and outline Use spherical coordinates $\{r, \theta, \phi\}$ to express the position of the bead. The r variable is the distance along the wire, θ is the constant angle the wire makes with the vertical, and $\phi = \omega t$. There is only one generalized coordinate, $r[t]$. The equation for $r[t]$ follows from the user-defined function **Lag**. The equation of motion is solved with **DSolve**.

Required packages

```
In[20]:= Needs["Algebra`Trigonometry`"]
         Needs["Calculus`VectorAnalysis`"]
```

Solution

```
In[21]:= Clear["Global`*"];
```

Part (a) Assume a spherical coordinate system with the origin at the vertex of rotation **A**. Let **r** be the distance of the bead along the wire from point A. Let **theta** be the angular variable, where **theta** is the angle the wire makes with respect to the downward vertical. The **phi** coordinate is given by **w t**. The Cartesian coordinates **{x,y,z}** for the bead can be expressed in terms of **{r,theta}**,

```
In[22]:= x2rRule= {x,y,z}->{ ( r[#] Sin[theta]Cos[w #])&,
                             ( r[#] Sin[theta]Sin[w #])&,
                             (-r[#] Cos[theta]         )& }//Thread;
```

The coordinates have been expressed in their pure form to enable substitutions in differential expressions. **theta** is assumed to be fixed. The only independent variable is **r[t]**. The kinetic and potential energies and the Lagrangian are

```
In[23]:=  T= (m/2) (x'[t]^2 + y'[t]^2 + z'[t]^2);
          V=  m g z[t];
          L= T-V  //.x2rRule  //Simplify

Out[23]=
                                 2      2    2                          2
(m (4 g Cos[theta] r[t] + w  r[t]  - w  Cos[2 theta] r[t]  +

              2
    2 r'[t] )) / 4
```

The equation for **r[t]** follows from the user-defined operator **Lag**:

```
In[24]:= eq1=Lag[r,L]

Out[24]=
                                2               2
         -(m (4 g Cos[theta] + 2 w  r[t] - 2 w  Cos[2 theta] r[t]))
         ---------------------------------------------------------- +
                                   4

           m r''[t] == 0
```

Solve **eq1** for **r''[t]** and use **TrigReduce** to simplify the results:

```
In[25]:= rSol= ((Solve[ Lag[r,L],r''[t]][[1,1]]
                  //TrigReduce
                ) //.{Cos[x_]^2-> 1-Sin[x]^2}
                  //Simplify[#,Trig->False]&
                ) //.Rule->Equal

Out[25]=
                                   2                 2
         r''[t] == g Cos[theta] + w  r[t] Sin[theta]
```

Assume the initial conditions **r[0]=0** and **r'[0]=v0** and apply **DSolve** to **rSol**. Solving for **r[t]**, you get (this takes a while)

```
In[26]:= initialConditions={r[0]==0,r'[0]==v0};

         rRule=(DSolve[ Join[{rSol},initialConditions], r[t],t]
                  //Flatten
                  //PowerExpand
                  //ExpandAll)

Out[26]=
                                         t w Sin[theta]
                  -(v0 Csc[theta])      E               v0 Csc[theta]
         {r[t] -> ------------------ + ------------------------------ -
                   t w Sin[theta]                   2 w
                2 E                w

            g Cot[theta] Csc[theta]    g Cot[theta] Csc[theta]
            ----------------------- + ------------------------ +
                        2               t w Sin[theta]  2
                       w              2 E              w

             t w Sin[theta]
            E               g Cot[theta] Csc[theta]
            ---------------------------------------}
                                 2
                              2 w
```

The results can be expressed in terms of hyperbolic functions if you use the user-defined rule **ComplexToHyperbolic**:

```
In[27]:= rRule= rRule //ComplexToHyperbolic //ExpandAll

Out[27]=           g Cot[theta] Csc[theta]
         {r[t] -> -(-----------------------) +
                               2
                              w

            g Cosh[t w Sin[theta]] Cot[theta] Csc[theta]
            -------------------------------------------- +
                                  2
                                 w
```

```
Out[27]=        v0 Csc[theta] Sinh[t w Sin[theta]]
                ----------------------------------}
                                w
```

Part (b) The time it takes for the bead to go a distance **Len** follows from solving the equation **Len=r[t]** for **t**, where **r[t]** is given by **rRule**:

```
In[28]:= tRule= Solve[Len == r[t] /.rRule /.{v0->0},t][[1]] //Simplify

Out[28]= {t -> (ArcCosh[

                     2
            (Len w  + g Cot[theta] Csc[theta]) Sin[theta] Tan[theta]
            ---------------------------------------------------------]
                                   g

           Csc[theta]) / w}
```

You will get a warning about inverse functions, but that is not a problem. To find the time for the special case in which **theta=0**, you apply the command **Limit** to **t**:

```
In[29]:= Limit[t //.tRule, theta->0] //PowerExpand

Out[29]= Sqrt[2] Sqrt[Len]
         -----------------
              Sqrt[g]
```

This free-fall result, $t = \sqrt{2Len/g}$, is the expected result.

Part (c) Plot the three-dimensional motion of the bead for the parameters:

```
In[30]:= values={theta->Pi/13, g->1, w->2, v0->0};
```

The position of the bead at time **t** is

```
In[31]:= position= ( {x[t],y[t],z[t]} //.x2rRule
                                      //.rRule
                                      //.values
                                      //N
                   )

Out[31]= {0.239316 Cos[2. t] (-4.23829 + 4.23829 Cosh[0.478631 t]),

           0.239316 (-4.23829 + 4.23829 Cosh[0.478631 t]) Sin[2. t],

           -0.970942 (-4.23829 + 4.23829 Cosh[0.478631 t])}
```

Plot this in three dimensions (output suppressed):

```
In[32]:= p1=ParametricPlot3D[ position//Evaluate
                ,{t,0,6 }
                ,Boxed->True
                ,Axes ->False
                ,BoxRatios->{1,1,1}];
```

To embellish this plot, you can plot the positions at discrete time points using

```
In[33]:= discretePoints=Table[position,{t,0,6,0.5}];

         discreteGraphics=
              Graphics3D[{PointSize[.03]
                         ,Point/@discretePoints}];
```

Overlaying **discreteGraphics** with the original plot **p1** results in

```
In[34]:= Show[discreteGraphics,p1,BoxRatios->{1,1,1}];
```

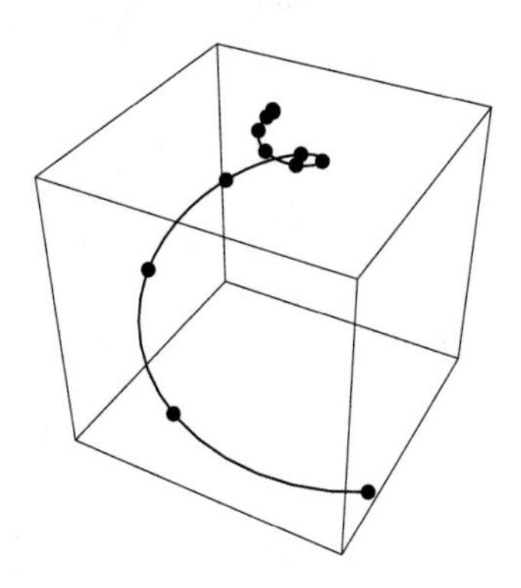

Problem 3: Bead on a Rotating Hoop

A bead of mass m is constrained to move on a hoop of radius r. The hoop rotates with constant angular velocity ω around a vertical axis that coincides with one diameter of the hoop. The only external force is gravity, where g is the acceleration due to gravity.

Part (a) Construct the Lagrangian and derive the equations of motion.

Part (b) Construct the conserved energy and verify that it is conserved.

Part (c) Define a potential for the system and find the equilibrium points. Show the characteristics of the potential change at $\omega = \omega_c$, where $\omega_c^2 = g/r$. Graph the potential for the two characteristic types of potentials.

Part (d) Numerically solve for $\theta[t]$ and $\theta'[t]$ and plot the data.

Part (e) Animate the numerical results.

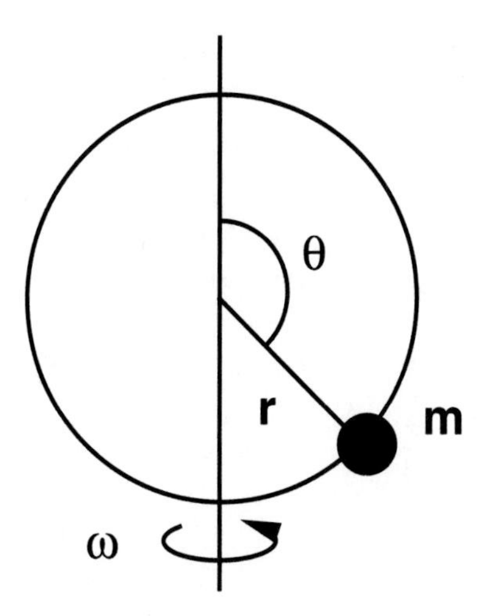

Remarks and outline Choose spherical coordinates $\{r, \theta, \phi\}$ with the origin at the center of the hoop. Let r be the radius of the hoop and ϕ be given by ωt. The position is then described by one independent variable $\theta[t]$. The equations of motion follow from the user-defined function **Lag**. The expression for the conserved energy follows from the expression for the Hamiltonian. An effective potential follows from the energy expression.

Required packages

```
In[35]:= Needs["Algebra`Trigonometry`"]
         Needs["Calculus`VectorAnalysis`"]
```

Solution

```
In[36]:= Clear["Global`*"];
```

Part (**a**) Let the origin of the coordinates be at the center of the hoop and **r** be the radius of the hoop. The coordinates of the bead are given by

```
In[37]:= x2rRule= {x,y,z}->{ ( r Sin[theta[#]]Cos[w #])&,
                             ( r Sin[theta[#]]Sin[w #])&,
                             ( r Cos[theta[#]]         )&}//Thread;
```

The kinetic and potential energies are

```
In[38]:=  T= (m/2)  (x'[t]^2 + y'[t]^2 + z'[t]^2);
          V=  m g z[t];
```

The Lagrangian is

```
In[39]:= L= (( (T-V) //.x2rRule
                     //Simplify
                     //TrigReduce
                   ) //.{Cos[x_]^2->1-Sin[x]^2}
                     //ExpandAll
                 )

Out[39]=
                                    2  2                2      2          2
                                 m r  w  Sin[theta[t]]     m r  theta'[t]
         -(g m r Cos[theta[t]]) + ---------------------- + ---------------
                                           2                      2
```

The equation of motion follows from applying the user-defined function **Lag** to **L**:

```
In[40]:= Lag[theta,L]

Out[40]=
                                    2  2
         -(g m r Sin[theta[t]]) - m r  w  Cos[theta[t]] Sin[theta[t]] +

              2
           m r  theta''[t] == 0
```

Solve this for **theta''[t]**:

```
In[41]:= thetaSol= ( Solve[ Lag[theta,L], theta''[t]]
                          //Flatten
                          //.{g->wc^2 r}
                          //ExpandAll
                     )
```

```
Out[41]=
                   g Sin[theta[t]]      2
{theta''[t] -> ----------------- + w  Cos[theta[t]] Sin[theta[t]]}
                         r
```

where you have set **wc^2=g/r**. The energy of the system follows from the expression for the Hamiltonian, where $H = \theta'[t](\partial L/\partial\theta'[t]) - L$:

```
In[42]:= energy= ( (theta'[t] D[L,theta'[t]]-L)
                   //.{g->wc^2 r}
                   //.{Cos[x_]^2->1-Sin[x]^2}
                   //ExpandAll
                 )
```

```
Out[42]=
                                     2 2            2       2          2
  2   2                          m r  w  Sin[theta[t]]    m r  theta'[t]
m r  wc  Cos[theta[t]] - ------------------------- + ----------------
                                       2                      2
```

Here, verify that the Hamiltonian computed in **energy** is identical to that returned by the user-defined function **Hamilton**:

```
In[43]:= {pRule, hamiltonian, eqMotion}=
                Hamilton[L,{theta},{ptheta}];
```

```
In[44]:= energy==hamiltonian //.{g->wc^2 r} //.pRule
```

```
Out[44]= True
```

To verify that the **energy** is conserved, take the derivative of **energy** and use **thetaSol** to show that the time derivative of the energy is zero:

```
In[45]:= D[energy,t] //.thetaSol //.{g->wc^2 r} //Expand
```

```
Out[45]= 0
```

Part (b) In the expression for the **energy**, the terms independent of velocity give the effective potential, **Veff**:

```
In[46]:= Veff = energy //.{theta'[t]->0}
```

```
Out[46]=
                                    2 2            2
  2   2                         m r  w  Sin[theta[t]]
m r  wc  Cos[theta[t]] - ------------------------
                                      2
```

The equilibrium positions will be where **dVeff** vanishes.

```
In[47]:= dVeff= D[Veff,theta[t]]   //Factor
```

```
Out[47]=
      2    2    2
-(m r  (wc  + w  Cos[theta[t]]) Sin[theta[t]])
```

Two equilibrium positions follow from **Sin[theta]=0**: (1) **theta=0** (top of hoop) and (2) **theta=Pi** (bottom of hoop). An additional equilibrium solution follows from

```
In[48]:= Solve[dVeff/ Sin[theta[t]]==0,theta[t]][[1]]
```

```
Out[48]=
                               2
                             wc
{theta[t] -> ArcCos[-(---)]}
                               2
                             w
```

If **(wc/w)^2** is less than or equal to one, then you will have solutions for real values of **theta**. A more careful analysis of **Veff** reveals an additional solution: **ArcCos[wc^2/w^2]+Pi**. The four relevant equilibrium positions are

```
In[49]:= eqRule=(
             theta[t]->{0,Pi,ArcCos[-wc^2/ w^2]
                          ,ArcCos[+wc^2/ w^2]+Pi}
                             );
```

The last two solutions require **w^2** to be greater than or equal to **wc^2**. Substitute **eqRule** into **dVeff** to verify they are solutions:

```
In[50]:= dVeff==0  /.eqRule //Thread

Out[50]= {True, True, True, True}
```

Part (c) Plot the potential as a function of **theta** and **w**. Without the loss of generality, you set

```
In[51]:= values1={r->1, wc->1, m->1
               ,theta[t]->theta};
```

(note that **theta[t]** is replaced by **theta** to simplify the plotting commands.) Let **theta** range from **-Pi** to **Pi** and **w** range from **wc** to **3 wc**. The plot of the potential is

```
In[52]:= Plot3D[Veff //.values1 //Evaluate
               ,{theta,-Pi,Pi}
               ,{w,1,3}
               ,AxesLabel->{"theta","w","Potential"}];
```

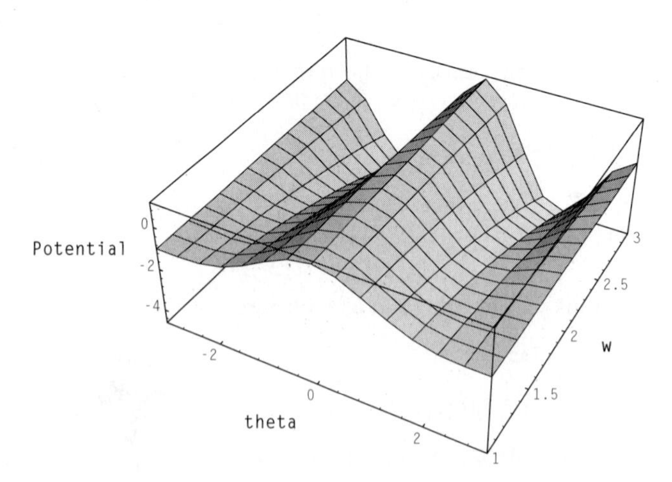

The potential has an unstable equilibrium position at the top of the hoop (**theta=0**) and another equilibrium position at the bottom of the hoop (**theta=Pi** or **theta=-Pi**). The other two equilibrium positions are stable but exist only if **w^2** is greater than **wc^2**. Plot the potential diagram for the two characteristic kinds of potentials. Consider the two values for **w**: **w=0.2** (**w<wc**) and **w=2** (**w>wc**) (output suppressed):

```
In[53]:= p1= Plot[ Veff //.values1 /.{w->0.2} //Evaluate
                   ,{theta,-Pi,Pi}];

In[54]:= p2= Plot[ Veff //.values1 /.{w->2.0} //Evaluate
                   ,{theta,-Pi,Pi}];
```

Overlaying plots **p1** and **p2**, you get

```
In[55]:= Show[p1,p2];
```

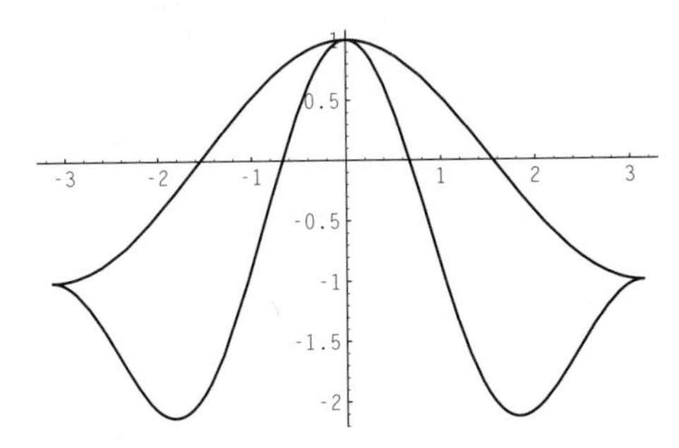

Part (d) Next, numerically solve for **theta**. The equation for **theta** follows from **thetaSol**:

```
In[56]:= eq1= thetaSol //.{g->wc^2 r} /.Rule->Equal

Out[56]=
                        2                         2
{theta''[t] == wc  Sin[theta[t]] + w  Cos[theta[t]] Sin[theta[t]]}
```

Instead of solving this second-order equation, solve two equivalent first-order equations:

```
In[57]:= eq2= {  eq1 /.{theta''[t]->dtheta'[t]}
               ,theta'[t]==dtheta[t]
             } //Flatten

Out[57]=
                        2                         2
{dtheta'[t] == wc  Sin[theta[t]] + w  Cos[theta[t]] Sin[theta[t]],

  theta'[t] == dtheta[t]}
```

Choose the values of the parameters in **eq2** to be

```
In[58]:=  values2=  {w->2, wc->1};
```

Also choose the initial conditions:

```
In[59]:= initial={theta[0]==Pi, dtheta[0]==0.01};
```

Applying **NDSolve** to **eq2**, you get

```
In[60]:= ndsol1= NDSolve[{eq2//.values2,initial} //Flatten
                   ,{theta[t],dtheta[t]},{t,0,30}][[1]]

Out[60]= {theta[t] -> InterpolatingFunction[{0., 30.}, <>][t],

  dtheta[t] -> InterpolatingFunction[{0., 30.}, <>][t]}
```

You will find it most convenient to define the functions:

In[61]:=
```
 theta1[t_]= theta[t] /.ndsol1[[1]];
dtheta1[t_]=dtheta[t] /.ndsol1[[2]];
```

You can then plot the solution for **theta**:

In[62]:=
```
Plot[ theta1[t] ,{t,0,30}];
```

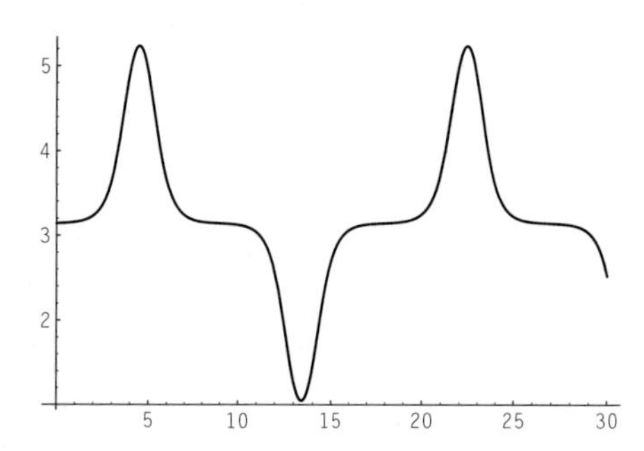

and **dtheta**

In[63]:=
```
Plot[dtheta1[t] ,{t,0,30}];
```

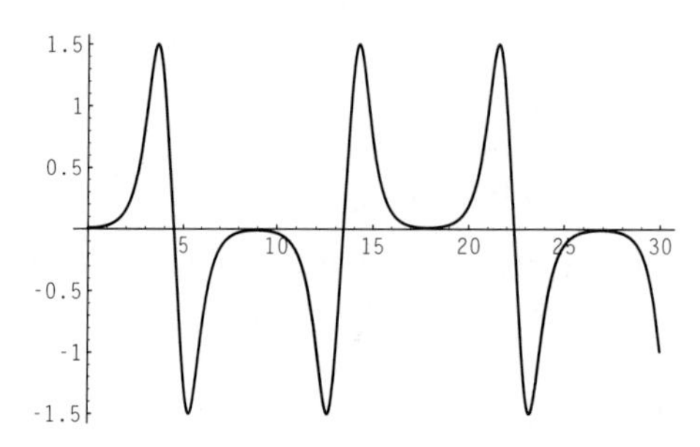

Part (e) To animate the solution, construct the **Graphics** for the bead on the hoop at an instant of time **t**:

In[64]:=
```
bead[t_]:=
  Graphics[
    { PointSize[0.03],
      Point[{Sin[theta1[t]],Cos[theta1[t]]} ]
    }];
```

Generate a set of graphics with a Do loop. This can be animated to see the motion (all but one frame suppressed):

In[65]:=
```
frames=2;
Do[
   Show[
     bead[t]
    ,Graphics[{Hue[.1],Circle[{0,0},1]}
             ,AspectRatio->Automatic]
             ,PlotRange ->{{-2,2},{-2,2}}
             ]
,{t,0,20,20/(frames-1)}]
```

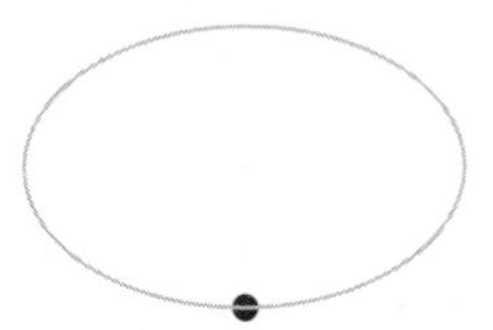

The parameter **frames** determines the number of frames displayed. (**Note:** **frames** must be greater than 1.)

Problem 4: Hoop Rolling on an Incline

Consider the motion of a hoop of radius r rolling without slipping down a fixed incline. Let ϕ be the inclination angle of the incline. Specify the position of the hoop by the distance x traveled by the center of mass along the incline and the angle θ of rotation of the hoop.

Part (**a**) Assume that x and θ are the generalized coordinates and construct the Lagrangian. Write the slippage constraint and use Lagrangian multipliers to obtain the equations of motion for the variables x and θ.

Part (**b**) Solve the equations for $x[t]$ and $\theta[t]$.

Part (**c**) Plot the results.

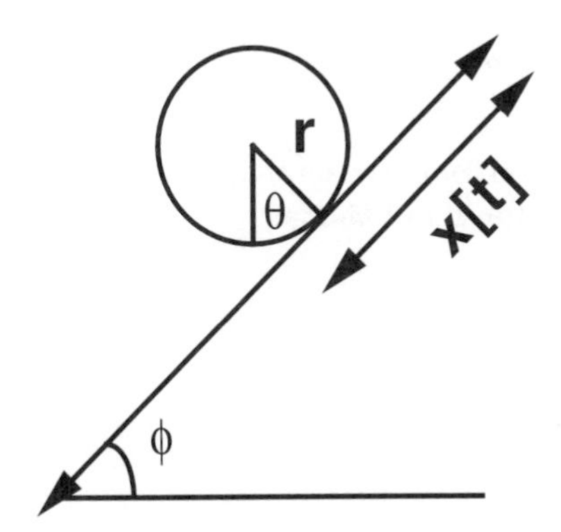

Remarks and outline The problem simplifies if the generalized coordinates are chosen to be the center of mass $x[t]$, and the motion about the center of mass is described by the angular variable $\theta[t]$. The Lagrangian multiplier follows from the condition that there is no slippage, $r\,d\theta - dx = 0$. The equations of motion follow from **Lag** and are solved with **DSolve**. To illustrate the motion, plot a fixed point on the hoop as the hoop rolls down the incline.

Required packages

```
In[66]:= Needs["Algebra`Trigonometry`"]
         Needs["Calculus`VectorAnalysis`"]
```

Solution

```
In[67]:= Clear["Global`*"];
```

Part (a) The kinetic energy can be resolved into the kinetic energy of the motion of the center of mass, plus the kinetic energy of motion about the center of mass. Let **x[t]** be the position vector of the center of mass and **theta[t]** be the angular variable around the center of mass. The kinetic and potential energies are

```
In[68]:= T =(m/2) (x'[t]^2  + r^2 theta'[t]^2);
         V = m g (Len-x[t]) Sin[phi];
         L = T - V

Out[68]=
                                              2          2        2
                                          m (r  theta'[t]  + x'[t] )
         -(g m Sin[phi] (Len - x[t])) + ----------------------------
                                                     2
```

where **Len** is the length of the incline and **phi** is the angle of slant. The constraint that there is no slippage is

```
In[69]:= eq1= r theta[t]-x[t]==0;
```

The derivatives of **eq1** gives the constraint forces **{Q[1],Q[2]}** in terms of the Lagrangian multiplier **lambda**:

```
In[70]:= constraintForce=
              {Q[1]->lambda D[eq1[[1]],x[t]]
              ,Q[2]->lambda D[eq1[[1]],theta[t]]}

Out[70]= {Q[1] -> -lambda, Q[2] -> lambda r}
```

The **constraintForce** gives the expression for the generalized slippage force **{Q[1],Q[2]}**. Applying **Lag**, obtain equations of motion for **x** and **theta**:

```
In[71]:= eqX= Lag[x,L,Q[1]] //.constraintForce

Out[71]= -(g m Sin[phi]) + m x''[t] == -lambda

In[72]:= eqTheta= Lag[theta,L,Q[2]] //.constraintForce

Out[72]=    2
         m r  theta''[t] == lambda r
```

You now have two equations of motion and one constraint equation. These three equations can be solved for the three unknowns **x**, **theta**, and **lambda**.

Part (b) Solve the equations of motion given the following initial conditions:

```
In[73]:= initial={x[0]==x0, x'[0]==v0
                ,theta[0]==x0/r, theta'[0]==v0/r};
```

Next, use **DSolve** to solve for **x[t]** and **theta[t]**:

```
In[74]:= eq2=Join[{eqX,eqTheta},initial];

         xtSol= DSolve[eq2,{x,theta},t][[1]]
```

```
Out[74]=
                                    2                        2
                     -(lambda t )              g t  Sin[phi]
{x -> Function[t, ------------- + t v0 + x0 + ---------------],
                        2 m                          2

                            2
                   lambda t      t v0    x0
  theta -> Function[t, ---------- + ---- + --]}
                     2 m r         r      r
```

The solutions are expressed in pure form for convenience in handling later substitutions. The expression for **lambda** follows from the slippage constraint in **eq1**. Use **xtSol** to eliminate **x'[t]** and **theta'[t]** in **eq1**:

```
In[75]:= lamSol= Solve[eq1 //.xtSol,{lambda}][[1]]

Out[75]=
              g m Sin[phi]
{lambda -> ------------}
                   2
```

Use **lamSol** to eliminate **lambda** in **xtSol**. The complete expressions for **x[t]** and **theta[t]** become

```
In[76]:= xtSol2= xtSol //.lamSol //Simplify

Out[76]=
                                         2
                                    g t  Sin[phi]
{x -> Function[t, t v0 + x0 + ---------------],
                                        4

                                                 2
                        4 t v0 + 4 x0 + g t  Sin[phi]
  theta -> Function[t, -------------------------------]}
                                    4 r
```

Part (c) To illustrate the motion, plot a fixed point on the hoop as the hoop rolls down the incline. Consider the values:

```
In[77]:= values={x0->0, v0->0, g->-0.1, phi->Pi/4, r->2.0};
```

The **x** and **y** coordinates for a fixed point on the hoop are

```
In[78]:= posx = ( x[t] Sin[phi]+ r Sin[theta[t]]);
         posy = ( x[t] Cos[phi]+ r Cos[theta[t]]);
```

and the **{x,y}** point as a function of **t** is

```
In[79]:= point={posx,posy} /.xtSol2 //.values //N

Out[79]=
               2                          2
{-0.0125 t  - 2. Sin[0.00883883 t ],

             2                          2
  -0.0125 t  + 2. Cos[0.00883883 t ]}
```

This is easily displayed with **ParametricPlot**:

```
In[80]:= ParametricPlot[point //Evaluate
                ,{t,0,70}
                ,AspectRatio->1];
```

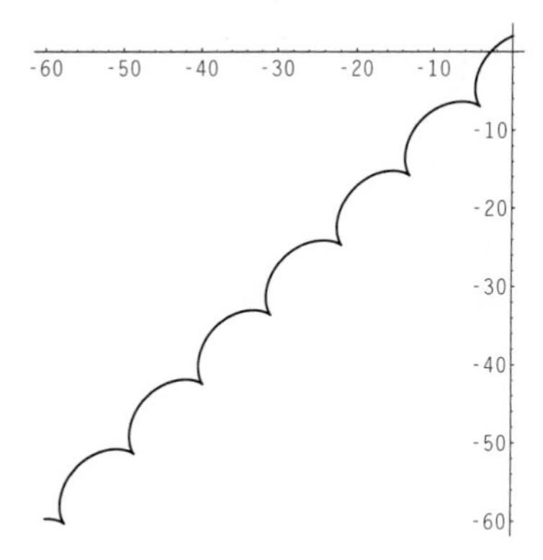

Problem 5: Sphere Rolling on a Fixed Sphere

A sphere of radius a and mass m rests on top of a fixed sphere of radius b. The first sphere is slightly displaced so that it rolls (without slipping) down the second sphere. Let the z-x plane be chosen to pass through the centers of the two spheres and assume the motion is in the z-x plane. Choose a spherical coordinate system at the center of the fixed sphere. Let $r[t]$ be the radial variable, measured from the origin of the coordinates to the center of the moving sphere. Let $\theta_1[t]$ be the angular variable measured from the origin of the fixed sphere. The motion of the moving sphere's center of mass can be described by the variables $\{r[t], \theta_1[t]\}$. The motion of the moving sphere about its center mass can be described by the variables $\theta_1[t]$ and $\theta_2[t]$, where $\theta_2[t]$ is the angular measured relative to spherical coordinates centered at the moving sphere's center of mass.

Part (**a**) Assume the generalized coordinates are $\{r[t], \theta_1[t], \theta_2[t]\}$ and construct the Lagrangian. Write the constraint that there is no slippage and the constraint that the spheres are in contact. Use Lagrangian multipliers to obtain the equations of motion.

Part (**b**) Show that the equation for $\theta_1[t]$ can be reduced to a first integral.

Part (**c**) Find the angle at which the sphere falls off.

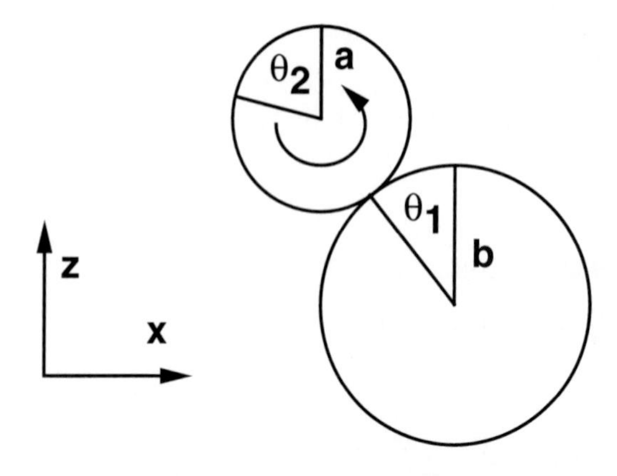

Remarks and outline The total kinetic energy is the kinetic energy of the center of mass plus the kinetic energy of the motion about the center of mass. There are two constraints: (1) the constraint of contact, $r[t] = a + b$, and (2) the constraint of no slippage, $b\theta_1[t] - a\theta_2[t] = 0$. These two constraints give the expression for the generalized constraint forces. The equations of motion follow from **Lag**. The point at which the sphere falls off follows from setting to zero the Lagrangian multiplier associated with the contact constraint.

Required packages

```
In[81]:= Needs["Algebra`Trigonometry`"]
         Needs["Calculus`VectorAnalysis`"]
```

Solution

```
In[82]:= Clear["Global`*"];
```

Part (a) The kinetic energy of the rolling sphere is the energy associated with the motion of its center of mass plus the rotational energy about its center of mass. The kinetic energy of the center of mass of the moving sphere is $T_{cm} = (1/2)m(r'^2(t) + r^2(t)\,\theta_1^2(t))$:

```
In[83]:= Tcm= (m/2)(r'[t]^2+ r[t] ^2 theta1'[t]^2);
```

The rotational energy associated with the motion about the sphere's center of mass is $T_{rot} = (1/2)I_1(\theta_1'(t) + \theta_2'(t))^2$:

```
In[84]:= Trot= (I1/2)( theta1'[t]+theta2'[t] )^2;
```

where **I1** is the moment of inertia of the sphere. The rotational energy is a function of the sum **(theta1 +theta2)**, since the first variable is the contribution from the curved surface of the fixed sphere and the second term is from the motion about the center of the moving sphere. The total kinetic energy is

```
In[85]:= T=Tcm+Trot;
```

The potential energy (to within a constant) is

```
In[86]:= V= m g r[t] Cos[theta1[t]];
```

The Lagrangian is

```
In[87]:= L=T-V
Out[87]=
                                            2        2            2
                                   m (r'[t]  + r[t]  theta1'[t] )
-(g m Cos[theta1[t]] r[t]) + ------------------------------------ +
                                                2

                                 2
  I1 (theta1'[t] + theta2'[t])
  ------------------------------
                2
```

The two constraint equations are

```
In[88]:= constraints= {b theta1[t]- a theta2[t]==0
                      ,r[t]-a-b==0 }

Out[88]= {b theta1[t] - a theta2[t] == 0, -a - b + r[t] == 0}
```

The first equation is the condition that there is no slippage. The second equation is the contact condition and states that the distance between the centers of the two spheres is the sum of their radii.

The algebra simplifies if you express these equations in terms of pure rules (we use **DSolve** to obtain this form):

```
In[89]:= constraintRule=
           { DSolve[constraints[[1]],{theta2},t]
            ,DSolve[constraints[[2]],{r},t]
            } //Flatten

Out[89]=
                                     b theta1[t]
         {theta2 -> Function[t, ------------], r -> Function[t, a + b]}
                                         a
```

The expressions for **Q[i]** follow from taking the differential of the constraints. First, obtain a list of the coordinates as a function of **t**:

```
In[90]:= coordinates={r,theta1,theta2};
         coordinatesT=
           coordinates //Map[Composition[#][t]&,#]&

Out[90]= {r[t], theta1[t], theta2[t]}
```

Then define **Q[i]** via $Q_i = \sum_j \lambda_i (df_j/dx_i)$, where f_j is the j-th constraint equation:

```
In[91]:= Clear[Q];
         Q[i_]:=
           Sum[ lam[j] * D[constraints[[j,1]]
                          ,coordinatesT[[i]] ]
           ,{j,1,2}]

         Array[Q,3]

Out[91]= {lam[2], b lam[1], -(a lam[1])}
```

The Lagrangian is described by three generalized coordinates: **r[t]**, **theta1[t]**, and **theta2[t]**. The equations of motion are

```
In[92]:= eqMotion=
           Table[ Lag[coordinates[[i]], L, Q[i]] ,{i,3}]

Out[92]=
                                                 2
         {g m Cos[theta1[t]] - m r[t] theta1'[t]  + m r''[t] == lam[2],

           -(g m r[t] Sin[theta1[t]]) + 2 m r[t] r'[t] theta1'[t] +

                   2
             m r[t]  theta1''[t] + I1 (theta1''[t] + theta2''[t]) == b lam[1],

           I1 (theta1''[t] + theta2''[t]) == -(a lam[1])}
```

(Note: Use **Table** to deal with the case of multiple **Q[i]**. As an exercise, you may want to modify **Lag** to handle this case automatically.)

These three equations plus the constraint equations give five equations for the five unknowns: **theta1**, **theta2**, **r**, **lam[1]**, and **lam[2]**.

Part (b) Use **constraintRule** to eliminate **theta2** and **r** in **eqMotion** and solve the three equations for **theta1**, **lam[1]**, and **lam[2]**.

```
In[93]:= sol1= ( Solve[eqMotion //.constraintRule
                ,{theta1''[t],lam[1],lam[2]}]
            //Flatten
            //Simplify
          )

Out[93]=
                                                  2                2
         {lam[2] -> m (g Cos[theta1[t]] - a theta1'[t]  - b theta1'[t] ),

                          g I1 m Sin[theta1[t]]
           lam[1] -> -(-------------------------),
                                    2
                              I1 + a  m

                           2
                          a  g m Sin[theta1[t]]
           theta1''[t] -> ------------------------}
                                          2
                           (a + b) (I1 + a  m)
```

The last equation in **sol1** is the equation for the variable **theta1[t]** and the first two are Lagrangian multipliers for the constraints. Convert the rule in **sol1** to an equation for **theta1''[t]** and then transform this equation to a first integral by multiplying by **theta1'[t]** and integrating both sides of the equation:

```
In[94]:= eqTheta=(theta1''[t] //.sol1) - theta1''[t] ==0

Out[94]=  2
         a  g m Sin[theta1[t]]
         ---------------------- - theta1''[t] == 0
                         2
          (a + b) (I1 + a  m)
```

First, set some options in **Limit**. (The best way to see what this does is first to try **intTheta** without this option set.)

```
In[95]:= SetOptions[Limit,Analytic->True];

In[96]:= intTheta= (
           Integrate[eqTheta[[1]] theta1'[t] ,{t,0,t}]
              //.{theta1[0]->0,theta1'[0]->0}
              //Simplify )

Out[96]=     2            2                                         2
         (2 a  g m - 2 a  g m Cos[theta1[t]] - a I1 theta1'[t]  -

                              2    3              2   2               2
             b I1 theta1'[t]  - a  m theta1'[t]  - a  b m theta1'[t] ) /

                              2
           (2 (a + b) (I1 + a  m))
```

(Note that initial conditions are assumed for **theta1**.) Next, solve for **theta1'[t]^2**:

```
In[97]:= solTheta=Solve[intTheta==0,{theta1'[t]}][[1]];

         theta1'[t]^2 //.solTheta //Simplify

Out[97]=    2           theta1[t] 2
         4 a  g m Sin[---------]
                           2
         ------------------------
                           2
           (a + b) (I1 + a  m)
```

Part (c) The sphere loses contact when **lam[2]** goes to zero. **lam[2]** is given by

```
In[98]:= eq1= 0 ==lam[2] //.sol1 //.solTheta //Simplify

Out[98]=                 2                              2
              g m (-2 a  m + I1 Cos[theta1[t]] + 3 a  m Cos[theta1[t]])
         0 == ----------------------------------------------------------
                                          2
                                    I1 + a  m
```

Solving for **Cos[theta1[t]]**, you get

```
In[99]:= sol1=Solve[eq1,Cos[theta1[t]]] //Flatten //Simplify

Out[99]=                                2
                                   2 a  m
         {Cos[theta1[t]] -> ------------}
                                     2
                             I1 + 3 a  m
```

The sphere falls off when **theta1** has the value

```
In[100]:= sol2= ( Solve[ sol1 //.{Rule->Equal} ,theta1[t]]
                  //Flatten
                  //Simplify
                )

Out[100]=                                2
                                    2 a  m
          {theta1[t] -> ArcCos[------------]}
                                      2
                              I1 + 3 a  m
```

Finally, substitute the moment of inertia for a sphere and convert the angle from radians to degrees:

```
In[101]:= (theta1[t]/Degree) //.sol2 //.{I1 -> 2/5 m a^2} //N

Out[101]= 53.9681
```

Problem 6: Mass Falling Through a Hole in a Table

Two mass points m_1 and m_2 are connected by a string of length len passing through a hole in a smooth table. The mass m_1 rests on the table's surface, and m_2 hangs suspended and moves only in a vertical line. Place the origin of the coordinate system at the hole in the table and use polar coordinates $\{r, \phi\}$ to describe the motion of mass m_1. The distance from the hole to m_1 is r, and ϕ is its position angle around the hole. Describe the hanging particle by the vertical distance from the hole, $z = len - r$.

Part (**a**) Find the equations of motion for $r[t]$ and $\phi[t]$.

Part (**b**) Find the two constants of motion that correspond to the conservation of angular momentum and the conservation of energy. Express the constants in terms of the initial conditions $r[0] = r_0$, $r'[0] = 0$, $\phi[0] = 0$, and $\phi'[0] = d\phi_0$.

Part (**c**) Solve the limiting case in which $m1$ describes a circle, $r[t] = r0$.

Part (**d**) If the circular orbit is displaced by $dr[t]$, find the angular frequency ω with which $dr[t]$ oscillates about the circular orbit.

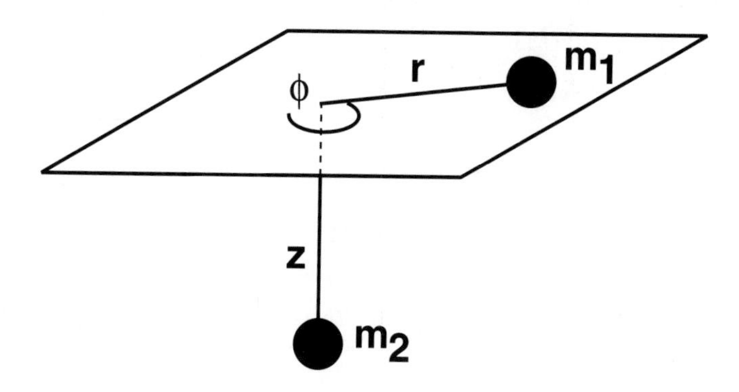

Remarks and outline Because the Lagrangian is independent of ϕ, the momentum conjugate to ϕ will be conserved. The equation for the displacement dr follows from the radial equation of motion. Keeping the lowest-order terms in dr and using the conditions for a circular orbit, **DSolve** returns a solution for dr.

Required packages

```
In[102]:= Needs["Algebra`Trigonometry`"]
          Needs["Calculus`VectorAnalysis`"]
```

Solution

```
In[103]:= Clear["Global`*"];
```

Part (**a**) The kinetic and potential energies for the particle on the table are

```
In[104]:= T1= m1/2 (r'[t]^2 + r[t]^2 phi'[t]^2);
          V1= 0;
```

The hanging particle is described by the coordinate **z[t]=len-r[t]**. The kinetic and potential energies for the hanging particle are

```
In[105]:= T2= m2/2 (r'[t])^2;
          V2= - m2 g(len- r[t]);
```

The Lagrangian for the system is

```
In[106]:= L=(T1+T2)-(V1+V2)
Out[106]=
                                        2          2         2       2
                                m2 r'[t]    m1 (r[t]  phi'[t]  + r'[t] )
          g m2 (len - r[t]) + --------- + ------------------------------
                                   2                     2
```

The equations of motion for the coordinates **r[t]** and **phi[t]** follow from **Lag**:

```
In[107]:= Lag[r,L]
Out[107]=
                                 2
          g m2 - m1 r[t] phi'[t]  + m1 r''[t] + m2 r''[t] == 0
```

The equation for **r''[t]** is

```
In[108]:= eqR= ((  Solve[ Lag[r,L],r''[t]]
                   //Flatten
              )    //.{Rule->Equal}
             )
Out[108]=
                                               2
                        g m2 - m1 r[t] phi'[t]
          {r''[t] == -(-----------------------)}
                               m1 + m2
```

The operator **Solve** has been used to put the equation in a more readable form. Likewise, the equation for **phi''[t]** is

```
In[109]:= eqPhi= (( Solve[ Lag[phi,L],phi''[t]]
                   //Flatten
                ) //.{Rule->Equal}
             )
Out[109]=
                         -2 phi'[t] r'[t]
          {phi''[t] == ----------------}
                              r[t]
```

Part (b) Because the Lagrangian **L** is independent of **phi**, the momentum conjugate to **phi** is a constant:

```
In[110]:= eq1= D[L,phi'[t]]==m1 r0^2 dphi0
Out[110]=
                2                          2
          m1 r[t]  phi'[t] == dphi0 m1 r0
```

where the right-hand side is a constant. The derivative of **eq1** is just **eqR**; verify this by taking the derivative of **eq1** and solving for **theta''[t]**:

```
In[111]:= eqPhi //. Solve[D[eq1,t],phi''[t]]
Out[111]= {{True}}
```

Expressing the angular momentum **phi'[t]** in pure form, you get

```
In[112]:=  solPhi= DSolve[eq1,phi',t] //Flatten
Out[112]=
                                             2
                                     dphi0 r0
          {phi' -> Function[t, ---------]}
                                         2
                                     r[t]
```

An equation for **r'[t]** follows from the expression for the energy:

```
In[113]:= energy= (T1+T2)+(V1+V2) //.solPhi //Simplify

Out[113]=
                           2    4                       2          2
                      dphi0  m1 r0                m1 r'[t]    m2 r'[t]
-(g len m2) + ------------------ + g m2 r[t] + ----------- + -----------
                           2                       2           2
                    2 r[t]
```

To show that **energy** is a constant, take the time derivative of **energy** and use **solPhi** and the equations of motion **eqMotion** to show that this time derivative vanishes:

```
In[114]:= eqMotion={eqR,eqPhi} /.{Equal->Rule} //Flatten

Out[114]=
                                            2
                 g m2 - m1 r[t] phi'[t]                     -2 phi'[t] r'[t]
{r''[t] -> -(------------------------), phi''[t] -> ----------------}
                      m1 + m2                             r[t]

In[115]:= D[energy,t] //.Join[solPhi,eqMotion] //Simplify

Out[115]= 0
```

Write the conserved energy in terms of the initial conditions:

```
In[116]:= initial={ r'[0]->0,   r[0]->r0
                   ,phi[0]->0, phi'[0]->dphi0};
```

Substituting **initial** into **energy**, obtain an expression for the constant energy (**energy0**) in terms of the initial parameters:

```
In[117]:= energy0 = energy //.{t->0} //.initial

Out[117]=
                                  2     2
                             dphi0  m1 r0
-(g len m2) + g m2 r0 + --------------
                                2
```

The expression for **r'[t]** follows from equating **energy** and **energy0** and solving:

```
In[118]:= rSol= Solve[energy==energy0,r'[t]][[1]] //Simplify

Out[118]= {r'[t] -> -(

                                                    2    4
                               2      2       dphi0  m1 r0
     Sqrt[2 g m2 r0 + dphi0  m1 r0  - -------------- - 2 g m2 r[t]]
                                                  2
                                              r[t]
     ------------------------------------------------------------------)}
                              Sqrt[m1 + m2]
```

Part (c) Consider the case in which **m1** describes a circular orbit.

The solution for circular orbits follows from setting **r''[t]->0** and **r[t]->r0** in **eqR**:

```
In[119]:= sol1=Solve[eqR //.{r''[t]->0, r[t]->r0} ,phi'[t]][[2]]

Out[119]=
                  Sqrt[g] Sqrt[m2]
{phi'[t] -> ---------------------}
                  Sqrt[m1] Sqrt[r0]
```

Solving for **phi[t]**, obtain

```
In[120]:= dsol1=DSolve[(sol1 /.{Rule->Equal}) ,phi[t],t]

Out[120]=
                  Sqrt[g] Sqrt[m2] t
          {{phi[t] -> ------------------ + C[1]}}
                  Sqrt[m1] Sqrt[r0]
```

Part (**d**) To solve for the motion resulting from a small displacement about the circular orbit, define

```
In[121]:= radiusRule={r-> (r0 +  dr[#]&)};
```

where **r0** is the radius of a circular orbit and **dr** is a small deviation about **r0.** An equation for **dr** follows from **eqR**. Eliminate **phi'[t]** with **solPhi** and replace **r** with **radiusRule**:

```
In[122]:= eq2= eqR //.solPhi //.radiusRule

Out[122]=
                              2      4
                         dphi0  m1 r0
                  g m2 - ------------
                                   3
                         (r0 + dr[t])
          {dr''[t] == -(--------------------)}
                            m1 + m2
```

You will want to solve **eq2** to lowest order in **dr**. Expanding **dr[t]** about **0** and keeping the lowest-order term, you get

```
In[123]:= rhs= (( Series[eq2[[1,2]],{dr[t],0,1}]
                  //Normal
               )  //.{dphi0->phi'[t]}
                  //.sol1
               )

Out[123]= -3 g m2 dr[t]
          -------------
          (m1 + m2) r0
```

dphi0 has been eliminated using **sol1**. The solution for **dr** follows from applying **DSolve** to **rhs**:

```
In[124]:= displacement= dr''[t]==rhs

Out[124]=
                     -3 g m2 dr[t]
          dr''[t] == -------------
                     (m1 + m2) r0

In[125]:= dsol9=
          ( DSolve[{displacement,dr[0]==dr0,dr'[0]==0},{dr[t]},t]
              //Flatten
              //ExpandAll
              //ComplexToTrig
              //ExpandAll
           )

Out[125]=
                                 Sqrt[3] Sqrt[g] Sqrt[m2] t
          {dr[t] -> dr0 Cos[----------------------------]}
                                 Sqrt[m1 + m2] Sqrt[r0]
```

where the initial conditions **dr[0]=dr0** and **dr'[0]=0** have been assumed. It follows that **dr[t]** oscillates with an angular frequency of

```
In[126]:= ( Solve[(dr[t] /.dsol9) == dr0 Cos[w t],w][[1]]
            //.{ArcCos[Cos[x_]]:>x}
            //Simplify
          )

Out[126]=        Sqrt[3] Sqrt[g] Sqrt[m2]
          {w -> ------------------------------}
                 Sqrt[m1 + m2] Sqrt[r0]
```

4.2.2 Orbiting Bodies

Problem 1: Equivalent One-body Problem

Consider two mass points m_1 and m_2, where the interaction is given by a central force expressible by the potential V. V is only a function of the relative separation $r_2 - r_1$. Define the center of mass vector $R[t]$ and the relative vector $r[t]$ by $R[t] = (m_1 r_1[t] + m_2 r_2[t])/(m_1 + m_2)$ and $r[t] = r_1[t] - r_2[t]$.

Part (a) Express the relative position vector $r[t]$ in terms of spherical coordinates $\{r_0, \theta, \phi\}$ and derive the equations of motion.

Part (b) Place the orbit in the $\theta = \pi/2$ plane and derive expressions for the conservation of angular momentum and energy.

Part (c) Let $u = 1/r$ and derive an equation for $u[\phi]$.

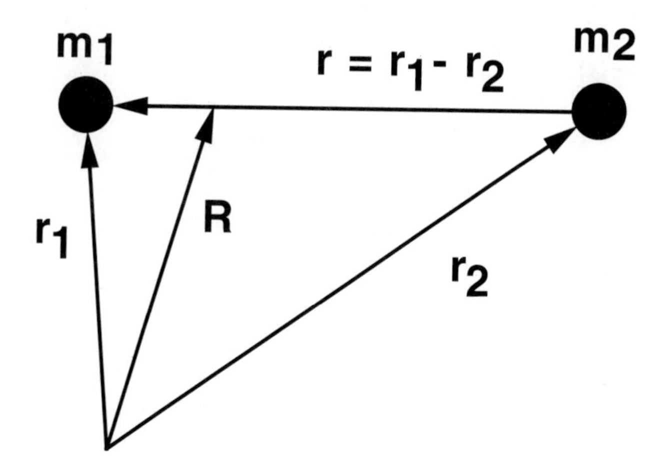

Remarks and outline Write the Lagrangian in terms of r and R and then express vector r in terms of spherical coordinates. The equations of motion follow from applying **Lag** to the Lagrangian. In the $\theta = \pi/2$ plane, the equations simplify and the $\phi''[t]$ equation can be integrated once to get the conserved angular momentum. Likewise, the $r''[t]$ equation can be integrated to get the energy expression.

Required packages

```
In[127]:= Needs["Algebra`Trigonometry`"]
          Needs["Calculus`VectorAnalysis`"]
```

Solution

```
In[128]:= Clear["Global`*"];
```

Part (a) It is convenient to express the equations in terms of the total mass **mt=(m1+m2)** and the reduced mass $\mu = m_1 m_2/(m_1 + m_2)$:

```
In[129]:= eq1= {mt==m1+m2, mu==m1 m2/mt};
```

An expression for **m1** and **m2** in terms of **mt** and **mu** follows from **eq1**:

```
In[130]:= mSol= Solve[eq1,{m1,m2}][[1]]

Out[130]=
                          2                            2
          mt - Sqrt[mt   - 4 mt mu]      mt + Sqrt[mt   - 4 mt mu]
{m1 -> --------------------------, m2 -> --------------------------}
                   2                               2
```

The center of mass and relative displacement vectors are defined by

```
In[131]:= eq2= { R[t] == (m1 r1[t] + m2 r2[t])/mt,
                 r[t] == r1[t] - r2[t] };
```

where **R**, **r1**, **r2**, and **r** are vectors. The inverse relations are

```
In[132]:= sol1= (( Solve[eq2,{r1[t],r2[t]}][[1]]
                //.Rule->Equal
                //.{mt->m1+m2}
              )  //MapAll[Apart,#]&
             )

Out[132]=
            m2 r[t]                          m1 r[t]
{r1[t] == ------- + R[t], r2[t] == -(-------) + R[t]}
            m1 + m2                          m1 + m2
```

It is more convenient to express these equations as rules and in their pure forms:

```
In[133]:= rSol={r1-> (( ((m2 r[#])/(m1 + m2)) + R[#])&),
                r2-> ((-((m1 r[#])/(m1 + m2)) + R[#])&)};
```

The kinetic energy is

```
In[134]:= T= (m1/2) r1'[t]^2  + (m2/2) r2'[t]^2;
```

and the Lagrangian is

```
In[135]:= L= (T - V[r0[t]]) //.Join[rSol,mSol] //Simplify

Out[135]=
                               2           2
          -2 V[r0[t]] + mu r'[t]  + mt R'[t]
          -----------------------------------
                          2
```

where **rSol** and **mSol** have been used to express the results in terms of **r**, **R**, **mt**, and **mu**.

Express the relative position vector in terms of spherical coordinates with **rRule**:

```
In[136]:= rRule=
          r->((CoordinatesToCartesian[
                       {r0[#],theta[#],phi[#]}
                       ,Spherical]
                   //Evaluate
                 )&)

Out[136]= r -> ({Cos[phi[#1]] r0[#1] Sin[theta[#1]],

             r0[#1] Sin[phi[#1]] Sin[theta[#1]], Cos[theta[#1]] r0[#1]} & )
```

Next, replace the notation for the square of a vector with the dot product. First, identify **r** and **R** as vectors:

```
In[137]:= r /: VectorQ[r]=True;
          R /: VectorQ[R]=True;
```

and then introduce a rule to transform vectors. The **/: VectorQ[x]** notations ensure that **vectorRule** is applied only to the vector object, hence avoiding nonsense like having **2^2** replaced by **2.2**.

```
In[138]:= vectorRule= { x_ [t_]^2 :> x [t].x [t] /; VectorQ[x],
                        x_'[t_]^2 :> x'[t].x'[t] /; VectorQ[x]
                      };
```

Illustrate how **vectorRule** transforms the Lagrangian:

```
In[139]:= L //.vectorRule

Out[139]= mu r'[t] . r'[t] + mt R'[t] . R'[t] - 2 V[r0[t]]
          -------------------------------------------------
                                 2
```

The Lagrangian in spherical coordinates becomes (with some extra simplifications):

```
In[140]:= L= ((L //.vectorRule
               //.rRule
               //ExpandAll
               //Simplify
               //TrigReduce
             ) //.{Cos[x_]^2->1-Sin[x]^2}
               //ExpandAll
             )

Out[140]=
                                                  2                 2          2
          mt R'[t] . R'[t]                mu r0[t]  Sin[theta[t]]  phi'[t]
          ---------------- - V[r0[t]] + ------------------------------------ +
                 2                                       2

                   2           2          2
            mu r0'[t]    mu r0[t]  theta'[t]
            ---------- + --------------------
                2                 2
```

The equations for vector **R** are trivial. Replace the vector **R** with its components **{Rx,Ry,Rz}**:

```
In[141]:= rRule2= {R->({Rx[#],Ry[#],Rz[#]}&)};
          L=L //.rRule2
```

```
Out[141]=
                                2                   2        2
                   mu r0[t]  Sin[theta[t]]  phi'[t]
     -V[r0[t]] + ---------------------------------------- +
                                   2

                   2           2            2               2                  2             2
     mt (Rx'[t]  + Ry'[t]  + Rz'[t] )     mu r0'[t]     mu r0[t]  theta'[t]
     ---------------------------------- + ----------- + ---------------------------
                   2                          2                   2
```

and apply **Lag** to obtain the equations of motion:

```
In[142]:= Lag[{Rx,Ry,Rz},L]

Out[142]= {mt Rx''[t] == 0, mt Ry''[t] == 0, mt Rz''[t] == 0}
```

The equations of motion for the center of mass are those of a free particle. Without the loss of generality, set **R=0**. (Note how the components of **R** have been replaced so that **R** *and* its derivatives vanish.)

```
In[143]:= L=L /.{Rx->(0&), Ry->(0&), Rz->(0&)}

Out[143]=
                                2                  2          2             2
                   mu r0[t]  Sin[theta[t]]  phi'[t]      mu r0'[t]
     -V[r0[t]] + ---------------------------------------- + ------------ +
                                   2                          2

                   2             2
      mu r0[t]  theta'[t]
      ---------------------
               2
```

The three equations of motion for the relative displacement components follow from **Lag**:

```
In[144]:= eqMotion=Lag[{r0,theta,phi},L] //Simplify

Out[144]=
                                     2          2                              2
     {-(mu r0[t] Sin[theta[t]]  phi'[t] ) - mu r0[t] theta'[t]  +

        V'[r0[t]] + mu r0''[t] == 0,

                                                            2
       (mu r0[t] (-(r0[t] Sin[2 theta[t]] phi'[t] ) + 4 r0'[t] theta'[t] +

            2 r0[t] theta''[t])) / 2 == 0,

                                        2
       2 mu r0[t] Sin[theta[t]]  phi'[t] r0'[t] +

                  2
        mu r0[t]  Sin[2 theta[t]] phi'[t] theta'[t] +

                  2                  2
        mu r0[t]  Sin[theta[t]]  phi''[t] == 0}
```

Part (b) Without loss of generality, the orbit can be oriented in the **theta=Pi/2** plane. Define a rule to reduce the equations to motion in the plane:

```
In[145]:= thetaRule= theta->( (Pi/2)& );
```

Eliminating the **theta** variable in the equations of motion, you get

```
In[146]:= eqMotion= eqMotion //.thetaRule
Out[146]=
                                  2
          {-(mu r0[t] phi'[t] ) + V'[r0[t]] + mu r0''[t] == 0, True,

                                                    2
            2 mu r0[t] phi'[t] r0'[t] + mu r0[t]  phi''[t] == 0}
```

The **theta** equation is automatically satisfied, as verified by **True** in the second part of **eqMotion**. The first and third equations in **eqMotion** can be reduced to first integrals. The conservation of angular momentum follows from the third equation in **eqMotion**:

```
In[147]:= phiEq= eqMotion[[3]]
Out[147]=
                                                  2
          2 mu r0[t] phi'[t] r0'[t] + mu r0[t]  phi''[t] == 0
```

Applying **DSolve** to the variable **r0**, you get

```
In[148]:= dSol= DSolve[{phiEq,r0[0]==Sqrt[el/(mu phi'[0])]}
                       ,{r0[t]},t][[2]]
Out[148]=
                               Sqrt[el]
          {r0[t] -> ----------------------------}
                     Sqrt[mu] Sqrt[phi'[t]]
```

The angular momentum **el** is a constant, and at **t=0**, **el=mu r[0]^2 phi'[0]**. Use this initial condition to eliminate the integration constant in **DSolve**.

Use the above relation (**dSol**) to find a differential equation for **phi'[t]**:

```
In[149]:= phiRule= Solve[dSol /.{Rule->Equal} ,phi'[t]][[1]]
Out[149]=
                          el
          {phi'[t] -> ----------}
                              2
                       mu r0[t]
```

Substituting this relation into the first equation of motion given by **eqMotion**, you get

```
In[150]:= rEq=eqMotion[[1]]/.phiRule
Out[150]=
                2
              el
          -(----------) + V'[r0[t]] + mu r0''[t] == 0
                    3
            mu r0[t]
```

Notice that the **r** and **phi** dependence of the equations of motion has been successfully uncoupled.

The conserved energy is obtained by integrating **r0'[t]** times the left-hand side of **rEq**:

```
In[151]:= energy= Integrate[rEq[[1]] r0'[t],t]
Out[151]=
               2                                   2
             el                          mu r0'[t]
          ------------ + V[r0[t]] + -------------
                     2                     2
          2 mu r0[t]
```

Part (c) An equation for **u=1/r** can be obtained with the rule

```
In[152]:= uRule={r0->(1/u[phi[#]]&)};
```

Apply this rule to **rEq** and replace **V'[r]** by the central force **V'[r]=-f[r]**:

```
In[153]:=  eq3=  ( rEq
                   //.{V'[x_]->-f[x]}
                   //.uRule
                   //Simplify
                 )

Out[153]=
                                    2            3
                1             el  u[phi[t]]
         -f[-----------] - ------------------- +
            u[phi[t]]               mu

                           2               2
               2 phi'[t]  u'[phi[t]]
          mu (------------------------ -
                              3
                     u[phi[t]]

                                                 2
             u'[phi[t]] phi''[t] + phi'[t]  u''[phi[t]]
             --------------------------------------------) == 0
                                      2
                             u[phi[t]]
```

To eliminate **phi'** and **phi''**, define

```
In[154]:= phiRule2={phiRule, D[phiRule,t]} //.uRule //Flatten

Out[154]=
                                        2
                      el u[phi[t]]
        {phi'[t] -> -----------------, phi''[t] ->
                           mu

         2 el u[phi[t]] phi'[t] u'[phi[t]]
         ---------------------------------}
                        mu
```

Applying **phiRule2** to **eq3** and solving for **u''[phi]**, you get

```
In[155]:= Solve[eq3 //.phiRule2 ,u''[phi[t]] ][[1]] //Simplify

Out[155]=
                                            1
                                 mu f[-----------]
                                       u[phi[t]]
        {u''[phi[t]] -> -(-----------------------) - u[phi[t]]}
                                  2            2
                                el  u[phi[t]]
```

Problem 2: Kepler Problem

The orbit equation for a particle moving in a central potential can be written as

$$u''[\phi] = -u[\phi] - \frac{m\ f\,(1/u[\phi])}{\ell^2\ u[\phi]^2},$$

where $u = 1/r$. The orbit is assumed to be in the equatorial plane ($\theta = \pi/2$), and u is considered a function of the angular variable ϕ.

Part (a) Solve for $u[\phi]$ when the gravitational potential is $V = -k/r$. Assume the initial condition $u'[0] = 0$.

Part (b) Plot the elliptical and hyperbolic orbits.

Part (c) Use the user-defined function **diffSeriesOne** to get a power series solution for the $u[\phi]$ equation. Sum the series to get the exact solution.

Remarks and outline Find the expression for the force term from the potential and substitute it into the u-equation. Using **DSolve**, find the solution. The command **PolarPlot** is used to graph the solutions.

Required packages

```
In[156]:= Needs["Algebra`SymbolicSum`"]
          Needs["Algebra`Trigonometry`"]
          Needs["Calculus`VectorAnalysis`"]
          Needs["Graphics`Graphics`"]
```

Solution

```
In[157]:= Clear["Global`*"];
```

Part (a) Consider the orbit equation

```
In[158]:= eq1= u''[phi] == -u[phi] - m f[1/u[phi]]/(el^2 u[phi]^2);
```

The force is related to the potential by **f[r]=-V'[r]**. Construct a rule that expresses **f** in terms of **u** for the gravitational potential **V=-k/r**:

```
In[159]:= forceRule={ V -> (-k /#&)
                    ,f[ 1/u[phi] ] -> -V'[r]
                    ,r -> 1/u[phi]
                   };
```

Substituting the gravitational potential into **eq1**, you get

```
In[160]:= eq2= eq1 //.forceRule

Out[160]=
                    k m
          u''[phi] == ---- - u[phi]
                      2
                    el
```

Note that to complete the substitution, you must use **//.** and not just **/.** command. Without the loss of generality, choose the initial condition **u'[0]=0**. Applying **DSolve** to **eq2**, you get

```
In[161]:= eq3= DSolve[{eq2,u'[0]==0,u[0]==u0(1+e)},{u[phi]},phi][[1]]

Out[161]=
                     k m        k m
          {u[phi] -> ---- + (-(----) + u0 + e u0) Cos[phi]}
                      2          2
                    el         el
```

The general form for the orbit can be written $u[\phi] = u_0(1 + e\,\mathrm{Cos}[\phi])$. When **phi=0**, you can solve for **u0**:

```
In[162]:= u0Rule= Solve[(u[phi] /.eq3 /.{Cos[phi]->0} ) == u0 ,u0][[1]]

Out[162]=         k m
          {u0 -> ----}
                   2
                 el
```

Using this in **eq3**, find the familiar equation for **eq3**:

```
In[163]:= eq4= eq3 //.u0Rule //Simplify

Out[163]=               k m (1 + e Cos[phi])
          {u[phi] -> ---------------------}
                               2
                             el
```

The value of **e** determines the three kinds of conic sections: (1) **e>1**, hyperbolic; (2) **e<1**, elliptic; and (3) **e=1**, parabolic. It is common to introduce the parameter **a**, implicitly defined by **el^2/(k m)=a(1-e^2)**, so that the canonical form of **u[phi]** becomes

```
In[164]:= eq5= eq4 //.{el -> Sqrt[a k m (1 - e^2)]}

Out[164]=               1 + e Cos[phi]
          {u[phi] -> ---------------}
                               2
                         a (1 - e )
```

Part (b) Use **PolarPlot** to graph the elliptical and hyperbolic orbits. Define a general plot function:

```
In[165]:= Clear[plotOrbit];
          plotOrbit[aa_,ee_]:=
            PolarPlot[( 1/u[phi] //.eq5
                                 //.{a->aa, e->ee}
                                 //Evaluate )
                       ,{phi,0,4 Pi}];
```

(*Note*: We must introduce the rule **a->aa** and **e->ee** so that these values are properly substituted.) Now you can view the plots of the orbits for a range of **e** values (intermediate output suppressed):

```
In[166]:= Show[
            Table[plotOrbit[1,eIn] ,{eIn,0.1,0.9,0.2}]
               //Evaluate
             ];
```

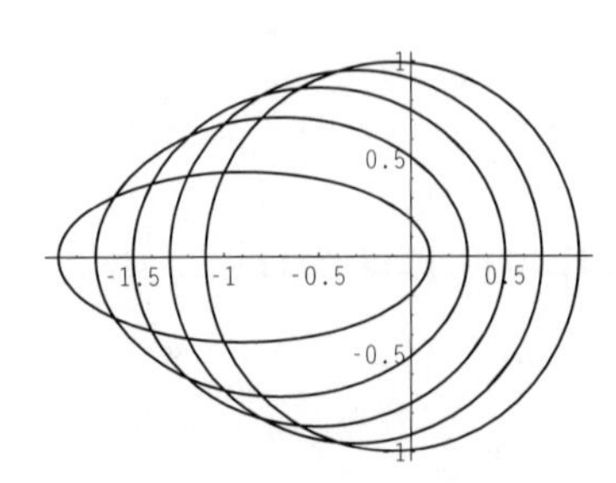

To plot the hyperbolic curves, you can in fact use the same command:

```
In[167]:= plotOrbit[1,1.1];
```

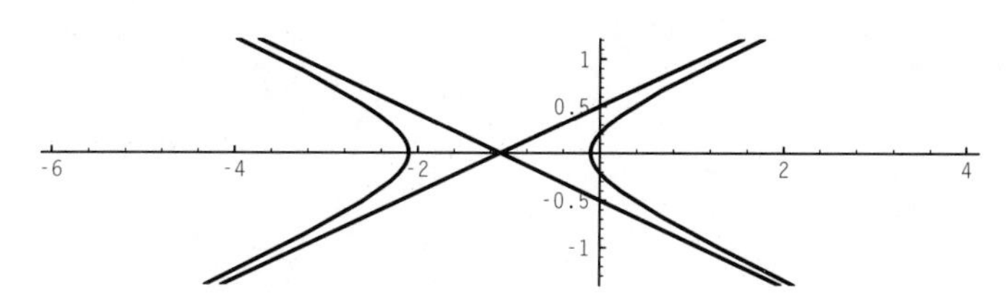


```
In[168]:= plotOrbit[1,1.7];
```

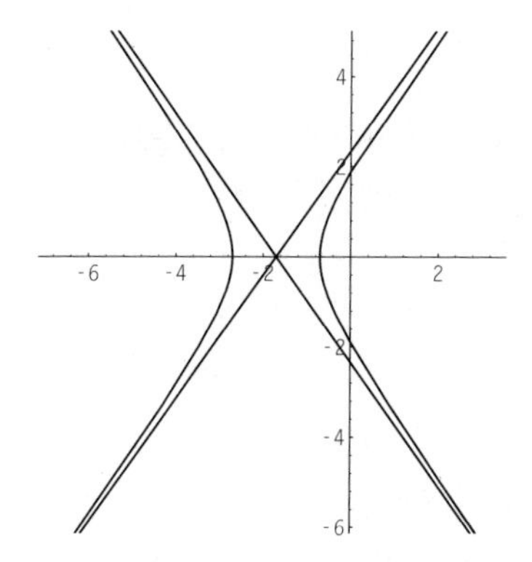

Part (c) Use the user-defined function **diffSeriesOne** to derive a power series solution for **eq2** and compare it with the exact solution found in Part (a).

To agree with the initial conditions in Part (a), use these initial conditions:

```
In[169]:= initial={u'[0]->0,u[0]->k m /el^2(1+e)};
```

Applying **diffSeriesOne** to **eq2**, you get

```
In[170]:= pertSol= ( diffSeriesOne[eq2,u,phi,6,initial]
                      //Normal
                      //ExpandAll
              )

Out[170]=
                                 2              4               6
          k m     e k m    e k m phi     e k m phi      e k m phi
          ---  +  -----  - ---------  +  ---------  -  ----------
            2       2          2              2              2
          el      el       2 el          24 el          720 el
```

Notice that the general expression for the nth term is

```
In[171]:= cterm[n_]= (e k m /el^2) (-1)^n phi^(2 n)/(2 n)!;
```

Verify this general expansion:

```
In[172]:= pertSol==(k m/el^2 + Sum[cterm[i],{i,0,3}])
```

```
Out[172]= True
```

Summing the series solution (with the help of **Algebra`SymbolicSum`**), you get the exact answer:

```
In[173]:= ( (k m/el^2+ Sum[cterm[n],{n,0,Infinity}])
              //PowerExpand
              //Simplify
          )

Out[173]= k m (1 + e Cos[phi])
          --------------------
                    2
                  el
```

This result agrees with the answer given in Part (a).

Problem 3: Precessing Ellipse and Generalized Kepler Problem

The orbit equation for a particle moving in a central potential can be written as

$$u''[\phi] = -u[\phi] - \frac{mf\,(1/u[\phi])}{\ell^2 u[\phi]^2},$$

where $u = 1/r$. The orbit is assumed to be in the equatorial plane ($\theta = \pi/2$), and u is considered a function of the angular variable ϕ.

Part (a) Solve for $u[\phi]$ when the potential is given by $V = -k/r - b/r^2$. Express the solution in the form $u_0(1 + e\,\mathrm{Cos}[\phi(1-\delta)])$ where $u_0 = km/(\ell^2 - 2bm)$.

Part (b) Plot the results using the command **PolarPlot**.

Part (c) Use the user-defined function **firstOrderPert** to find the first-order perturbative solution in the parameter **b**. Show that the result agrees with the expansion of the solution derived in Part (a).

Remarks and outline Find the expression for the force in terms of u. Substituting the force relation into the equations of motion, **DSolve** yields the solution. The command **PolarPlot** is used to graph the solutions. The perturbative solution follows from **firstOrderPert**.

Required packages

```
In[174]:= Needs["Algebra`Trigonometry`"]
          Needs["Calculus`VectorAnalysis`"]
          Needs["Graphics`Graphics`"]
```

Solution

```
In[175]:= Clear["Global`*"];
```

Part (a) Consider the orbit equation

```
In[176]:= eq1=u''[phi] == -u[phi]-m f[1/u[phi]]/(el^2 u[phi]^2);
```

The force is related to the potential by **f[r]=-V'[r]**. The rule that will find **f** for the potential is

```
In[177]:= forceRule= {  V -> ( -(k/#) - (b/#^2) &)
                      ,f[1/u[phi]] -> -V'[r]
                      ,r -> 1/u[phi]
                    };
```

Substituting the **forceRule** into **eq1**, you get

```
In[178]:=  eq2= eq1 //.forceRule //ExpandAll

Out[178]=
                     k m                2 b m u[phi]
          u''[phi] == --- - u[phi] + ------------
                       2                    2
                     el                   el
```

Use these initial conditions:

```
In[179]:= initial={u'[0]==0,u[0]==u0(1+e)};
```

Applying **DSolve** to **eq2**, you get

```
In[180]:= uRule=((( DSolve[ Join[{eq2},initial] ,u[phi],phi]
                    //Flatten
                  ) /.{Sqrt[x_]-> I Sqrt[-x]}
                 ) //ComplexToTrig
                    //Simplify
              )

Out[180]=
                                                          2 b m
          {u[phi] -> (-(k m) + k m Cos[Sqrt[1 - -----] phi] -
                                                             2
                                                           el

               2                      2 b m
             el  u0 Cos[Sqrt[1 - -----] phi] -
                                      2
                                    el

                 2                     2 b m
             e el  u0 Cos[Sqrt[1 - -----] phi] +
                                        2
                                      el

                                       2 b m
             2 b m u0 Cos[Sqrt[1 - -----] phi] +
                                        2
                                      el

                                         2 b m                 2
             2 b e m u0 Cos[Sqrt[1 - -----] phi]) / (-el  + 2 b m)}
                                          2
                                        el
```

with some simplifications added. Although this expression is rather cumbersome, we will show that it can be written in the form $u_0(1+e\text{Cos}[\phi(1-\delta_{shift})]$ where $u_0 = km/(\ell^2 - 2bm)$. This will be recognized as the usual solution to the Kepler problem (c.f., previous problem) with a shift term (δ_{shift}) in the argument of the Cos.

Identify the shift:

```
In[181]:= shiftRule =  {Sqrt[1-(2 b m)/el^2]->1-shift};
```

Substituting this into **u[phi]** gives you the right-hand side **rhs** of the necessary equation:

```
In[182]:= rhs= u[phi] //.uRule //.shiftRule
Out[182]=
                                                  2
(-(k m) + k m Cos[phi (1 - shift)] - el  u0 Cos[phi (1 - shift)] -

        2
    e el  u0 Cos[phi (1 - shift)] + 2 b m u0 Cos[phi (1 - shift)] +

                                               2
    2 b e m u0 Cos[phi (1 - shift)]) / (-el  + 2 b m)
```

Clearly, the **shift** has been identified correctly. We must now find a relation for **u0**. For the left-hand side **lhs** of the equation, take the canonical form:

```
In[183]:= u1Rule={u[phi] -> u0 (1 + e Cos[phi (1 - shift)])};
          lhs= u[phi] //. u1Rule;
```

Equating the **rhs** and **lhs**, we determine the value of u0:

```
In[184]:= u0Rule= Solve[lhs==rhs,u0][[1]]
Out[184]=
                   k m
{u0 -> -(-------------)}
              2
           -el  + 2 b m
```

Note that in the limit **b->0**, **u0** reduces the unperturbed solution obtained in Problem 2.

Part (b) Use **PolarPlot** to graph the orbits and define a general plot function, **plotOrbit**:

```
In[185]:= Clear[plotOrbit];
          plotOrbit[start_,finish_,values_List]:=
            PolarPlot[ u[phi] //.u1Rule //.values //Evaluate
               ,{phi,start, finish}]
```

Here, three revolutions of an orbit are plotted with a small value of the shift parameter:

```
In[186]:= plotOrbit[0,6 Pi,{u0->1, e->0.5, shift->0.1}];
```

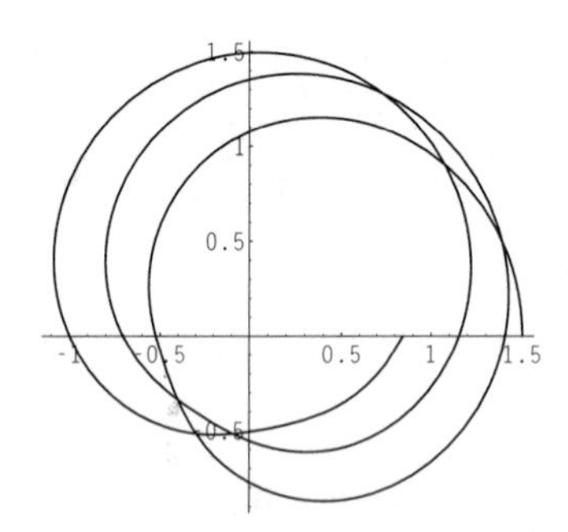

Note that the resulting curve is a precessing ellipse (**shift**$\neq$0).

Part (c) Use the user-defined function **firstOrderPert** to solve **eq2** to first order in the parameter **b** (this takes a while):

```
In[187]:= pert=( firstOrderPert[eq2,u,{u0(1+e),0},b,phi]
                  //.u0Rule
                  //Simplify )
```

```
Out[187]=       4      2  2        4                 2  2
         (k m (el   - 4 b   m  + e el  Cos[phi] + 4 b   m  Cos[phi] +

                   2                   2  2
            b e el   m phi Sin[phi] + 2 b   m  phi Sin[phi])) /

             6          4
           (el   - 2 b el  m)
```

Expanding the result to lowest order in **b**, you get

```
In[188]:= pertSer= Series[pert,{b,0,1}]//Normal//ExpandAll
Out[188]=            2                               2
         k m   2 b k m    e k m Cos[phi]   2 b e k m  Cos[phi]
         --- + ------- + -------------- + ------------------- +
           2      4             2                   4
         el     el            el                  el

                    2
           b e k m   phi Sin[phi]
           ----------------------
                      4
                    el
```

Verify that this solution agrees with the exact solution by expanding the latter in a power series to lowest order in **b**:

```
In[189]:= exactSer= Series[u[phi] //.uRule //.u0Rule ,{b,0,1}]//Normal //Expand
Out[189]=            2                               2
         k m   2 b k m    e k m Cos[phi]   2 b e k m  Cos[phi]
         --- + ------- + -------------- + ------------------- +
           2      4             2                   4
         el     el            el                  el

                    2
           b e k m   phi Sin[phi]
           ----------------------
                      4
                    el
```

and demonstrating that these series match:

```
In[190]:= pertSer==exactSer //ExpandAll

Out[190]= True
```

Problem 4: Numerical Solution for Orbits with Central Forces

The orbit equation for a particle moving in a central potential can be written as

$$u''[\phi] = -u[\phi] - \frac{m\, f\,(1/u[\phi])}{\ell^2\, u[\phi]^2},$$

where $u = 1/r$. The orbit is assumed to be in the equatorial plane ($\theta = \pi/2$), and u is considered a function of the angular variable ϕ.

Part (a) Assume the Keplerian potential $V = -k/r$. Numerically solve the equation $u[\phi]$ with boundary conditions that correspond (1) to an ellipse and (2) a hyperbola. Plot both solutions.

Part (**b**) Make a user-defined procedure that will return a **PolarPlot** of r for an arbitrary potential. Compare the results from this user-defined procedure with the graphs in Part (a).

Part (**c**) Use the user-defined procedure to compare the orbits with the potential $V = -k/r^{1.1}$ and $V = -k/r^{0.9}$. Also compare the orbits with the potential $V = -k/r - b/r^3$ and $V = -k/r + b/r^3$.

Remarks and outline First, express the force in terms of u. The numerical solution follows from **NDSolve**. The choice of the initial conditions determines whether the orbit is an ellipse or hyperbola. Use the command **PolarPlot** to graph the results. The steps for plotting the numerical solution with a generalized potential can be summarized in a user-defined command.

Required packages

```
In[191]:= Needs["Algebra`Trigonometry`"]
          Needs["Graphics`Graphics`"]
          Needs["Calculus`VectorAnalysis`"]
```

Solution

```
In[192]:= Clear["Global`*"];
```

Part (**a**) Consider the orbit equation

```
In[193]:= eq1= u''[phi]==-u[phi]- m f[1/u[phi]]/(el^2 u[phi]^2);
```

and find the numerical solution of **eq1** for the Kepler potential **V=-k/r**. The force **f** expressed in terms of **u** follows from the rule,

```
In[194]:= forceRule= {  V -> (-k/#&)
                       ,f[1/u[phi]] -> -V'[r]
                       ,r->1/u[phi]
                     };
```

Substituting the **forceRule** in **eq1**, you get

```
In[195]:= eq2= eq1 //. forceRule
Out[195]=
                     k m
          u''[phi] == ---- - u[phi]
                       2
                     el
```

Choose the parameters:

```
In[196]:= values ={k->1,el->1,m->1 };
```

you obtain an ellipse if you assume the initial conditions to be

```
In[197]:= initial1={u[0]==1.6,u'[0]==0};
```

The numerical solution of **eq2** is

```
In[198]:= ndsol1=
          NDSolve[{eq2/.values,initial1}//Flatten
                  ,u[phi],{phi,0,2 Pi}]//Flatten;
```

For convenience, define the function **u1[phi]** to be the solution for the ellipse case:

```
In[199]:= u1[phi_]=u[phi] //.ndsol1;
          u1[1.1]

Out[199]= 1.27216
```

Evaluate **u1** at a sample point to verify the solution was properly obtained. The plot of **r** as a function of **phi** follows from applying **PolarPlot** to **1/u[phi]**:

```
In[200]:= PolarPlot[1/u1[phi] //Evaluate,{phi,0,2 Pi}];
```

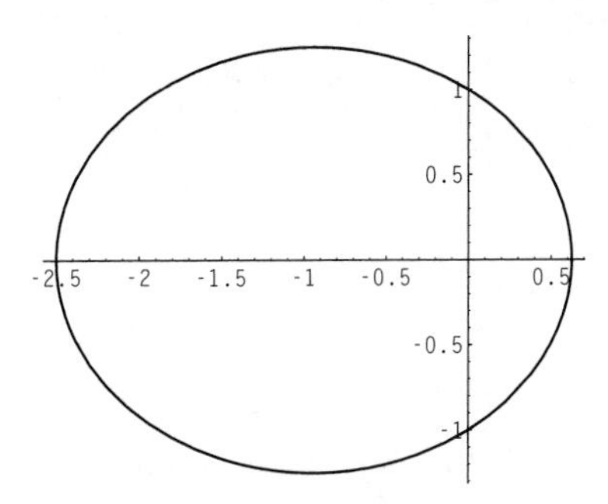

Hyperbolic orbits follow from the initial conditions:

```
In[201]:= initial2= {u[0]==3,u'[0]==0};
```

Again applying **NDSolve** to **eq2**, you get

```
In[202]:= ndsol2=
          NDSolve[{eq2/.values,initial2}//Flatten
                  ,u[phi],{phi,0,2 Pi}]//Flatten;
```

Define the function **u2[phi]** to be the solution for the hyperbolic case:

```
In[203]:= u2[phi_]=u[phi] //.ndsol2;
          u2[1.1]

Out[203]= 1.90719
```

The plot of **r** is

```
In[204]:= PolarPlot[1/u2[phi] //Evaluate,{phi,0,2 Pi}];
```

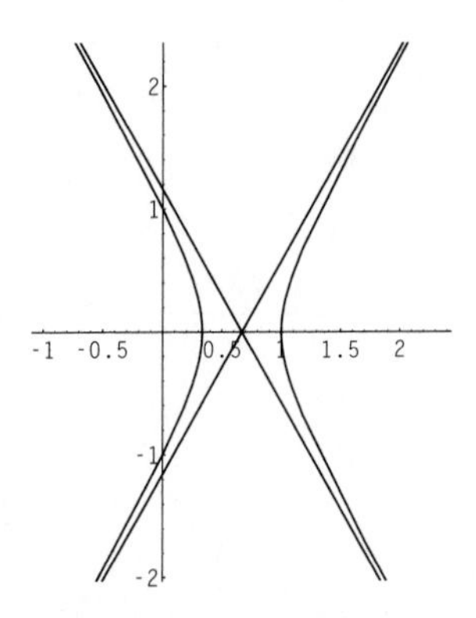

Part (b) The routine for plotting the numerical solution can be summarized in four steps:

1. Find the force in terms of **u** from the potential.
2. Make a list of the orbit equation with the initial conditions.
3. Numerically solve the orbit equation.
4. Use **PolarPlot** to generate the graphics.

These four steps are combined in the user-defined command:

```
In[205]:= Clear[numOrbit];
          numOrbit[V_,r_,r0_:1,dr0_:0,el_:1,m_:1,angle_:(6 Pi)]:=
           Module[{u,phi,uforce,eq1,initial,eqs,ndsol},
             uforce=-D[V,r]/.r->1/u[phi];
             eq1= (u''[phi]==
                     -u[phi]-m uforce/(el^2 u[phi]^2)
                   //ExpandAll);
             initial={u[0]==1/r0,u'[0]==-dr0/r0^2};
             eqs = Join[{eq1},initial];
             ndsol= NDSolve[eqs ,u[phi],{phi,0,angle}]//Flatten;
             PolarPlot[1/u[phi] /.ndsol //Evaluate
              ,{phi,0,angle}]
          ]
```

uforce represents the force as a function of **V'[1/u]**; **eq1** is the equation of motion for **u**; **initial** contains the initial conditions; **eqs** is the collection of equations for **NDSolve**, whose solution is **ndsol**. Then return the **PolarPlot** of **r=1/u**. Note that default values for **r0**, **dr0**, **el**, **m**, and **angle** are incorporated.

Applying **numorbit** to the Kepler potential, you get

```
In[206]:= numOrbit[-1/r, r, 0.6];
```

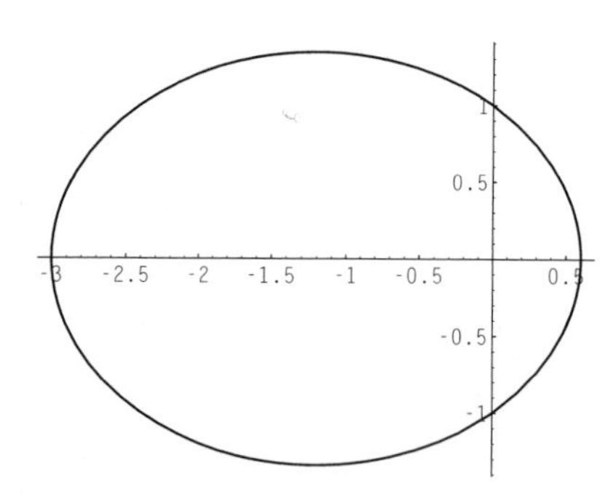

Part (c) Compare the orbits of $V = -k/r^{1.1}$ and $V = -k/r^{0.9}$:

```
In[207]:= numOrbit[-1/r^1.1,r,0.6];
```

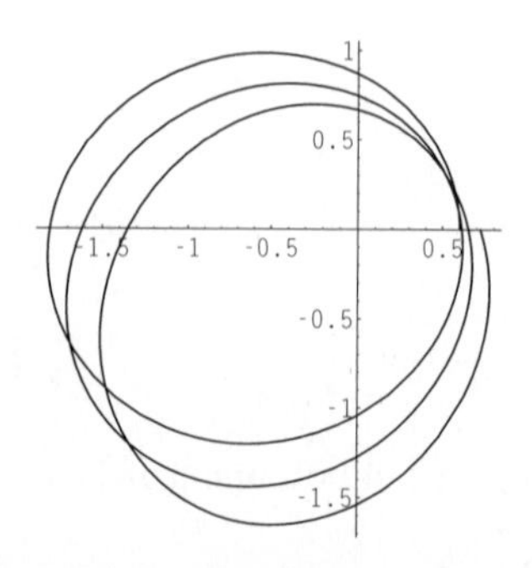

```
In[208]:= numOrbit[-1/r^0.9,r,0.6];
```

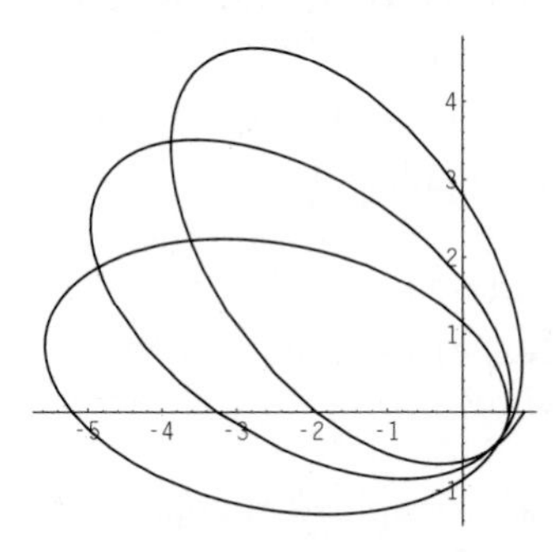

Similarly, the orbits with $V = -1/r - b/r^3$ and $V = -1/r + b/r^3$ with $b = 0.02$ are

```
In[209]:= numOrbit[-1/r - 0.02/r^3,r,0.6];
```

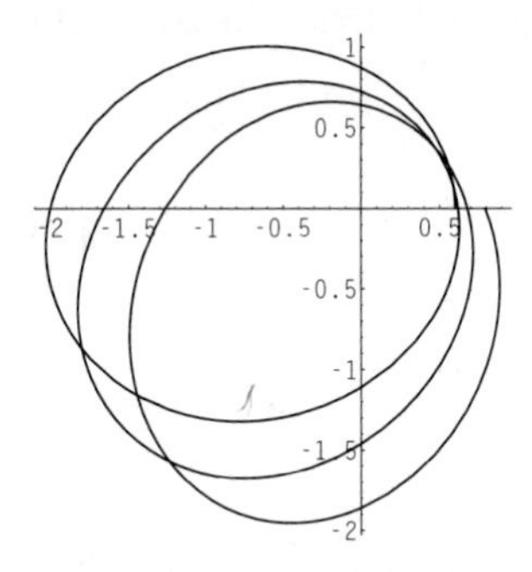


```
In[210]:= numOrbit[-1/r + 0.02/r^3,r,0.6];
```

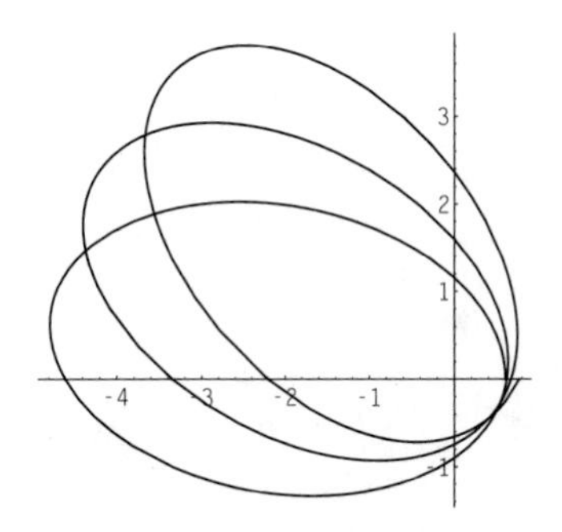

Problem 5: Quadripole Potential and Perturbative Solutions

Consider the orbit for a satellite in the equatorial plane of an oblate solid of revolution. The gravitational potential is

$$V = \frac{-GM}{r}\left(1 + \frac{J_2}{2r^2}(1 - 3\mathrm{Cos}[\theta]^2)\right).$$

Part (a) Assume the satellite is in the equatorial plane $\theta = \pi/2$ and write the equations of motion. Eliminate the ϕ variable and get a radial equation for the orbit.

Part (b) Let $u = 1/r$ and derive an equation for $u = 1/r$.

Part (c) Consider the elliptical solution that follows from the equation with $J_2 = 0$. Find the first-order perturbation correction due to the quadripole term J_2 by expanding the equations of motion about J_2 and solving. Keep only the lowest-order terms in e. Show that the results can be expressed as a precessing ellipse.

Part (d) Use the user-defined function **firstOrderPert** to find the perturbative solution and show that it agrees with the solution found in Part (c).

Remarks and outline Express the orbit in terms of spherical coordinates. **Lag** gives the orbit equations. An equation for **u** follows from letting $r = 1/u[\phi]$. To find the first-order solution, assume the solution is of the form $u = u_0 + J_2 u_1$, where u_0 is the $J_2 = 0$ elliptical solution. Substituting u into the equations of motion and setting $J_2 = 0$ and $e^2 = 0$, you get an equation for u_1. Solve the equation for u_1 by using the command **DSolve**. This perturbative solution follows immediately from the user-defined function **firstOrderPert**.

Required packages

```
In[211]:= Needs["Algebra`Trigonometry`"]
          Needs["Calculus`VectorAnalysis`"]
```

Solution

```
In[212]:= Clear["Global`*"];
```

Part (a) First, express the Lagrangian in spherical coordinates and then apply **Lag** to get the equations of motion. The position vector for the satellite in spherical coordinates is

```
In[213]:= rRule=
            r->(Evaluate[
                   CoordinatesToCartesian[
                      {r0[#],theta[#],phi[#]}
                      ,Spherical
                   ]
                ]&);
```

The gravitational potential is assumed to be

```
In[214]:= V= -G m/r0[t](1+J2/(2 r0[t]^2)(1-3 Cos[theta[t]]^2));
```

The Lagrangian is

```
In[215]:= T= 1/2  r'[t].r'[t];
          L=T-V

Out[215]=
                                                       2
                                 J2 (1 - 3 Cos[theta[t]] )
                        G m (1 + --------------------------)
                                                2
r'[t] . r'[t]                            2 r0[t]
------------- + ------------------------------------------
      2                           r0[t]
```

Expanding the **r** vector in components and simplifying the Lagrangian, you get

```
In[216]:= L=(( L //.rRule
               //ExpandAll
               //Simplify
               //TrigReduce
             ) //.{Cos[x_]^2->1-Sin[x]^2}
               //ExpandAll
            )

Out[216]=
                                                    2
      G J2 m      G m     3 G J2 m Sin[theta[t]]
    -(------) + ----- + ------------------------- +
          3      r0[t]                3
      r0[t]                    2 r0[t]

          2                 2         2          2           2             2
      r0[t]  Sin[theta[t]]  phi'[t]       r0'[t]     r0[t]  theta'[t]
      -------------------------------- + -------- + --------------------
                     2                       2               2
```

We are interested in the equations only when the motion is in the equatorial plane, so define

```
In[217]:= thetaRule= { theta->((Pi/2)&) };
```

The equations of motion are

```
In[218]:= eqMotion= Lag[{r0,theta,phi},L] //.thetaRule

Out[218]=  3 G J2 m      G m                      2
          {-------- + ------- - r0[t] phi'[t]  + r0''[t] == 0, True,
                  4         2
          2 r0[t]     r0[t]

                                            2
           2 r0[t] phi'[t] r0'[t] + r0[t]  phi''[t] == 0}
```

The second equation (the theta equation) is automatically satisfied in the equatorial plane. The two nontrivial equations are the **r** and **phi** equations. The **phi** equation can be integrated to get a constant of motion. This result can be expressed by the rule

```
In[219]:= phiRule={phi -> (Integrate[dphi[#],#]&)};
```

Next, substitute this into the **phi** equation of motion (**eqMotion[[3]]**):

```
In[220]:= phiEq= eqMotion[[3]] /.phiRule

Out[220]=     2
          r0[t]  dphi'[t] + 2 dphi[t] r0[t] r0'[t] == 0
```

and then solve for **dphi** (**=phi'[t]**). (This is simply a consequence of angular momentum conservation.)

```
In[221]:= dphiSol= DSolve[{phiEq,dphi[0]==el/(m r0[0]^2)}
                           ,dphi,t][[1]]

Out[221]=                              el
          {dphi -> Function[t, ----------]}
                                       2
                                m r0[t]
```

You can then use **dphiSol** to eliminate **phi** in favor of **r**:

```
In[222]:= phiRule2={phi'[t]-> dphi[t],  phi''[t]-> dphi'[t]} //.dphiSol

Out[222]=
                    el                       -2 el r0'[t]
{phi'[t] -> ----------, phi''[t] -> -------------}
                      2                            3
              m r0[t]                      m r0[t]
```

Verify the validity of the **phiRule2** by applying it to the equations of motion:

```
In[223]:= eqMotion2=eqMotion //.phiRule2

Out[223]=
                          2
  3 G J2 m          el            G m
{----------  -  -----------  +  ------  + r0''[t] == 0, True, True}
          4       2       3          2
  2 r0[t]        m  r0[t]       r0[t]
```

whereby you are left with the radial equation as the only remaining nontrivial equation.

Part (b) You get an equation for **u=1/r** as a function **phi** if you define the rule

```
In[224]:= uRule= {r0 -> ( 1/u[phi[#]]& )};
```

The rule has been expressed in pure form so that the substitution of derivatives will work.

It follows from the radial equation that

```
In[225]:=  eq1=  ( eqMotion2[[1]]
                      //.Join[phiRule2,uRule]
                      //ExpandAll
                 )

Out[225]=
                               2             3                          4
                   2     el  u[phi[t]]       3 G J2 m u[phi[t]]
G m u[phi[t]]  -  ------------------  +  ----------------------  -
                            2                       2
                           m

     2             2
  el  u[phi[t]]  u''[phi[t]]
  -------------------------- == 0
               2
              m
```

Use the **phiRule2** to eliminate the **phi'** and **phi''** terms. Solving **eq1** and solving for **u''**, you get

```
In[226]:= eq2= ((Solve[eq1,u''[phi[t]]][[1]]
                  //.{Rule->Equal}
                  //ExpandAll
               ) //.{phi[t]->phi}
              )

Out[226]=
                       3                         3        2
                  G m                   3 G J2 m  u[phi]
{u''[phi] == ------- - u[phi] + ---------------------}
                    2                         2
                  el                      2 el
```

Part (c) You want to find a perturbative solution of **eq2** about **J2**. At the zeroth order (**J2=0**), the solution corresponds to an ellipse. Then find the first-order correction due to **J2** and choose these initial conditions:

```
In[227]:= initial= {u'[0]==0, u[0]== u0(1+e)};
```

Next, solve for **u[phi]**:

```
In[228]:= dsol= ( DSolve[{(eq2/.{J2->0}),initial}//Flatten
                    ,{u[phi]},phi][[1]]
          )

Out[228]=
                      3          3
                   G m        G m
         {u[phi] -> ---- + (-(----) + u0 + e u0) Cos[phi]}
                      2          2
                    el         el
```

When the **Cos[phi]** term vanishes, you have **u[phi]=u0**. Therefore you can determine **u0**:

```
In[229]:= uSol= Solve[(u[phi] /.dsol /.{Cos[phi]->0})== u0,u0][[1]]
Out[229]=
                  3
               G m
         {u0 -> ----}
                  2
                el
```

To find the first-order correction due to **J2**, assume the perturbative solution is of the form

```
In[230]:= pertRule={u->(( u0 ( 1+e Cos[#1])  + J2 u1[#]) &)}
Out[230]= {u -> (u0 (1 + e Cos[#1]) + J2 u1[#1] & )}
```

where **J2 u1** is the correction to zeroth-order solution. (Also keep only first-order terms in **e**.) Substituting **pertRule** into **eq2**, you get

```
In[231]:= eq3=eq2 /.pertRule
Out[231]= {-(e u0 Cos[phi]) + J2 u1''[phi] ==

              3
           G m
           ---- - u0 (1 + e Cos[phi]) - J2 u1[phi] +
             2
           el

                   3                                      2
          3 G J2 m  (u0 (1 + e Cos[phi]) + J2 u1[phi])
          --------------------------------------------}
                                 2
                             2 el
```

which can be easily solved for **u''[phi]**:

```
In[232]:= eq4= (( Solve[eq3,u1''[phi]][[1]]
                 //.uSol
                 //ExpandAll
                ) //.{J2->0, e^2->0}
                 //Simplify
               ) //.{Rule->Equal}
```

```
Out[232]=
                        3  9         3  9
                     3 G  m     3 e G  m  Cos[phi]
         {u1''[phi] == ------ + ------------------- - u1[phi]}
                         6              6
                      2 el            el
```

(Notice that higher-order **J2** and **e^2** terms have been dropped.) With initial conditions of

```
In[233]:= initial={u1[0]==0, u1'[0]==0};
```

for the perturbative function **u1[phi]**, you can then find

```
In[234]:= dsol4= ( DSolve[Join[eq4,initial],u1[phi],phi][[1]]
                    //ExpandAll
                    //Simplify
                 ) //Map[Expand[#,Trig->True]&,#,{2}]&

Out[234]=
                     3  9        3  9                    3  9
                  3 G  m      3 G  m  Cos[phi]     3 e G  m  phi Sin[phi]
         {u1[phi] -> ------ - ----------------- + ----------------------}
                       6             6                      6
                   2 el           2 el                   2 el
```

Adding the zeroth- and first-order terms, you get

```
In[235]:= seq5 =    u[phi]/.pertRule/.dsol4

Out[235]=
                                           3  9       3  9
                                        3 G  m     3 G  m  Cos[phi]
         u0 (1 + e Cos[phi]) + J2 (------ - ----------------- +
                                          6              6
                                      2 el           2 el

               3  9
           3 e G  m  phi Sin[phi]
           ----------------------)
                      6
                  2 el
```

seq5 is a precessing ellipse. To show this, assume a solution of the form

```
In[236]:= eq5= u01(1+e1 Cos[(1- J2 alpha)phi] )

Out[236]= u01 (1 + e1 Cos[(1 - alpha J2) phi])
```

and also expand this to terms linear in **J2**:

```
In[237]:= seq6=Series[eq5,{J2,0,1}] //Normal

Out[237]= u01 (1 + e1 Cos[phi]) + alpha e1 J2 phi u01 Sin[phi]
```

By equating **seq5** and **seq6**, you can systematically solve for the unknown parameters, **u01**, **e1**, and **alpha**. Take advantage of the orthogonality of Sin and Cos to extract three pieces of information from the single equation **seq5==seq6**:

```
In[238]:= u01Rule= (
            Solve[seq5==seq6 //.{Cos[__]->0, Sin[__]->0},u01]
                 //Flatten
                 //ExpandAll
               )
```

```
Out[238]=            3    9
                 3 G  J2 m
        {u01 -> ----------- + u0}
                      6
                  2 el

In[239]:= e1Rule= (
            Solve[seq5==seq6 //.{Sin[__]->0} //.u01Rule,e1]
                 //Flatten
                 //Simplify
               )

Out[239]=            3    9          6
                 3 G  J2 m  - 2 e el  u0
        {e1 -> --------------------------}
                   3    9        6
                -3 G  J2 m  - 2 el  u0

In[240]:= alphaRule= (
            Solve[seq5==seq6 //.{Cos[__]->0}
                  //.u01Rule//.e1Rule,alpha]
                 //Flatten
                 //Simplify
               ) //.uSol //Simplify

Out[240]=                   2  6
                      3 e G  m
        {alpha -> --------------------}
                         4      2     6
                   2 e el  - 3 G  J2 m
```

As a cross check, note in the limit `J2->0` that `u01`, `e1`, and `J2 alpha` reduce to the appropriate limit to yield the unperturbed Kepler solution:

```
In[241]:= ( {u01,e1,J2 alpha}
                 //.Join[u01Rule,e1Rule,alphaRule]
                 //.{J2->0}
               )

Out[241]= {u0, e, 0}
```

Part (d) The first-order solution also follows from the user-defined function `firstOrderPert`. Applying the command to `eq2`, you get (this takes a while)

```
In[242]:= test1=( firstOrderPert[eq2[[1]],u,{u0(1+e),0},J2,phi]
                          //.uSol
                          //.{J2^2->0, e^2->0}
                          //Simplify
                        ) //Map[Expand[#,Trig->True]&,#,{1}]&

Out[242]=  3       3    9          3                3    9
        G m     3 G  J2 m     e G m  Cos[phi]   3 G  J2 m  Cos[phi]
        ---- + ----------- + --------------- - -------------------- +
          2          6               2                   6
        el       2 el              el                2 el

            3    9
         3 e G  J2 m  phi Sin[phi]
         ------------------------
                       6
                   2 el
```

To show that **test1** is the same as the solution **seq5** derived in Part (c), we eliminate **u0** with the rule **uSol**:

```
In[243]:= test2= (Series[seq5,{J2,0,1}] //Normal) //.uSol //ExpandAll;
```

and verify that these solutions are equal:

```
In[244]:= test1==test2 //ExpandAll

Out[244]= True
```

4.2.3 Hamilton and Hamilton-Jacobi Problems

Problem 1: Harmonic Oscillator and Hamilton's Equations

Consider a particle of mass m moving in a harmonic potential, $V = (1/2)kx^2$.

Part (a) Find the Hamiltonian for the system and solve Hamilton's equation.

Part (b) Use the user-defined **Hamilton** procedure to obtain the Hamiltonian and the equations of motion.

Part (c) Consider the nonlinear oscillator described by $V = (1/2)kx^2 + \epsilon x^5$. Write Hamilton's equations. Assume the solution can be written as a power-series in t and use the user-defined procedure **firstDiffSeries** to find the power-series solution. Find the first-order perturbation solution in the parameter ϵ using the user-defined procedure **firstOrderPert**.

Remarks and outline Express the Lagrangian in terms of $\{x[t], x'[t]\}$. The canonical momentum p follows from taking the derivative of the Lagrangian relative to the variable $x'[t]$. The Hamiltonian is $H = px' - L[x, x']$. The equation of motion follows from Hamilton's equations.

Required packages

```
In[245]:= Needs["Algebra`Trigonometry`"]
          Needs["Calculus`VectorAnalysis`"]
```

Solution

```
In[246]:= Clear["Global`*"];
```

Part (a) The kinetic and potential energies are

```
In[247]:= T=(m/2) x'[t]^2;
          V=(k/2)  x[t]^2;
```

and the Lagrangian is

```
In[248]:= L= T - V

Out[248]=           2          2
          -(k x[t] )   m x'[t]
          ---------- + --------
              2           2
```

The momentum conjugate to **x[t]** is

```
In[249]:= eq1= p[t]==D[L,x'[t]]

Out[249]= p[t] == m x'[t]
```

Solving **eq1** for **x'[t]**, you get

```
In[250]:= xpRule= Solve[eq1,x'[t]][[1]]

Out[250]=
                 p[t]
          {x'[t] -> ----}
                  m
```

Expressing the Hamiltonian in terms of the canonical variables **{x,p}**, you get

```
In[251]:= H= (p[t] x'[t]- L) //.xpRule

Out[251]=
              2          2
          p[t]     k x[t]
          ----- + --------
           2 m       2
```

Hamilton's equations of motion are

```
In[252]:= eq2= { p'[t]== -D[H,x[t]]
               , x'[t]== +D[H,p[t]] }

Out[252]=
                                              p[t]
          {p'[t] == -(k x[t]), x'[t] == ----}
                                               m
```

For the initial conditions

```
In[253]:= initial ={p[0]==0,x[0]==x0};
```

the solution follows from applying **DSolve** to **eq2**:

```
In[254]:= dsol=((DSolve[ eq2~Join~initial,{p[t],x[t]},t]
                    //Flatten
                    //ComplexToTrig
                  ) //PowerExpand
                    //Simplify
               )

Out[254]=
                                                  Sqrt[k] t
          {p[t] -> -(Sqrt[k] Sqrt[m] x0 Sin[---------]),
                                                   Sqrt[m]

                              Sqrt[k] t
            x[t] -> x0 Cos[---------]}
                              Sqrt[m]
```

Part (b) Use the user-defined **Hamilton** function and apply this to the harmonic oscillator Lagrangian:

```
In[255]:= Hamilton[L,{x},{p}]

Out[255]=
                                  2          2
                       p[t]   p[t]     k x[t]
          {{x'[t] -> ----}, ----- + --------,
                        m      2 m       2

                                               p[t]
            {p'[t] == -(k x[t]), x'[t] == ----}}
                                                m
```

This answer is in agreement with Part (a).

Part (c) As a second example of Hamilton's equation, consider the nonlinear oscillator potential and Lagrangian:

```
In[256]:= V = 1/2 k x[t]^2;
          V1=   eps x[t]^5;
          L1= T - (V + V1)

Out[256]=
                 2                         2
          -(k x[t] )           5   m x'[t]
          ---------- - eps x[t]  + --------
              2                       2
```

Hamilton's equations follow from the user-defined function **Hamilton**:

```
In[257]:= {xprule,hamiltonian,eqMotion}= Hamilton[L1,{x},{p}]

Out[257]=
                                 2          2
                      p[t]   p[t]    k x[t]            5
          {{x'[t] -> ----},  ----- + ------- + eps x[t] ,
                       m      2 m       2

                                             4           p[t]
           {p'[t] == -(k x[t]) - 5 eps x[t] , x'[t] == ----}}
                                                          m
```

(Note how the necessary parts of **Hamilton** are extracted.)

Use the user-defined command **firstDiffSeries** to solve the equations of motion in a power series:

```
In[258]:= sol1=
           firstDiffSeries[eqMotion,{x,p},{x[0]->x0,p[0]->0},3]

Out[258]=
                            3  2
                x0 (k + 5 eps x0 ) t          4
          {x0 - -------------------- + O[t] ,
                        2 m

                            3
           -(x0 (k + 5 eps x0 ) t) +

               2                     3          2      6   3
           x0 (k  m + 25 eps k m x0  + 100 eps  m x0 ) t          4
           ------------------------------------------------ + O[t] }
                                    2
                                 6 m
```

Convert the output to a rule for **x[t]** and **p[t]**

```
In[259]:= sol2={x[t],p[t]}->sol1 //Thread;
```

and for **x'[t]** and **p'[t]**

```
In[260]:= sol3= sol2 //Map[(D[#,t]&),#,{2}]&
```

```
Out[260]=
                                 3
                   x0 (k + 5 eps x0 ) t           3
         {x'[t] -> -(--------------------) + O[t] ,
                             m

                                    3
          p'[t] -> -(x0 (k + 5 eps x0 )) +

                2                      3              2    6  2
            x0 (k  m + 25 eps k m x0   + 100 eps  m x0 ) t           3
            ----------------------------------------------------- + O[t] }
                                      2
                                   2 m
```

then verify that this solution satisfies the equations of motion to leading order in **eps** and **t**:

```
In[261]:= ( (eqMotion //.Join[sol2,sol3]
                      //Normal
                      //ExpandAll
            ) //.{eps^2->0, t^3->0}
          )

Out[261]= {True, True}
```

Find the perturbative solution that follows from the user-defined function **firstOrderPert**. First, extract the **First** equation of motion and rewrite it as a second-order differential equation in **x[t]**:

```
In[262]:= eq3= (eqMotion//First)/.{p'[t]-> m x''[t]}
Out[262]=
                                                4
          m x''[t] == -(k x[t]) - 5 eps x[t]
```

Apply **firstOrderPert** to **eq3**:

```
In[263]:= pert3= (( firstOrderPert[eq3 ,x,{x0,0},eps]
                    //ComplexToTrig
                    //Simplify
                 )  //PowerExpand
                    //Simplify
                 )

Out[263]=
                        3                 Sqrt[k] t                3       Sqrt[k] t
          (x0 (-45 eps x0  + 24 k Cos[-----------] + 24 eps x0  Cos[-----------] +
                                           Sqrt[m]                          Sqrt[m]

                        3      2 Sqrt[k] t                 3      4 Sqrt[k] t
              20 eps x0  Cos[-------------] + eps x0  Cos[-------------])) /
                                Sqrt[m]                        Sqrt[m]

            (24 k)
```

and then expand **pert3** in **t**:

```
In[264]:= sPert3=  Series[pert3,{t,0,2}]//Normal//ExpandAll
Out[264]=
                  2              2   4
               k t  x0      5 eps t  x0
         x0 - --------- - -------------
                 2 m           2 m
```

so that we can compare with the previous solution

```
In[265]:= test1= x[t] //. sol2 //Normal //ExpandAll;
```

The solutions are identical to this order:

```
In[266]:= test1==sPert3 //ExpandAll

Out[266]= True
```

Problem 2: Hamilton's Equations in Cylindrical and Spherical Coordinates

Consider the orbit of a particle of mass m moving in a time-independent potential V.

Part (a) Consider cylindrical coordinates $\{r, \theta, z\}$ and a potential of the form $V[r, \theta, z]$. Find the expressions for the canonical momentum, the Hamiltonian, and Hamilton's equations.

Part (b) Consider the gravitational potential from a spherical source $V = -k/\sqrt{r^2 + z^2}$. Assume the orbit can be written in the $z = 0$ plane and write Hamilton's equations. Find the condition necessary for circular orbits.

Part (c) Derive the t power-series solution of the equations in Part (b). Verify the validity of the series expansion for circular orbits.

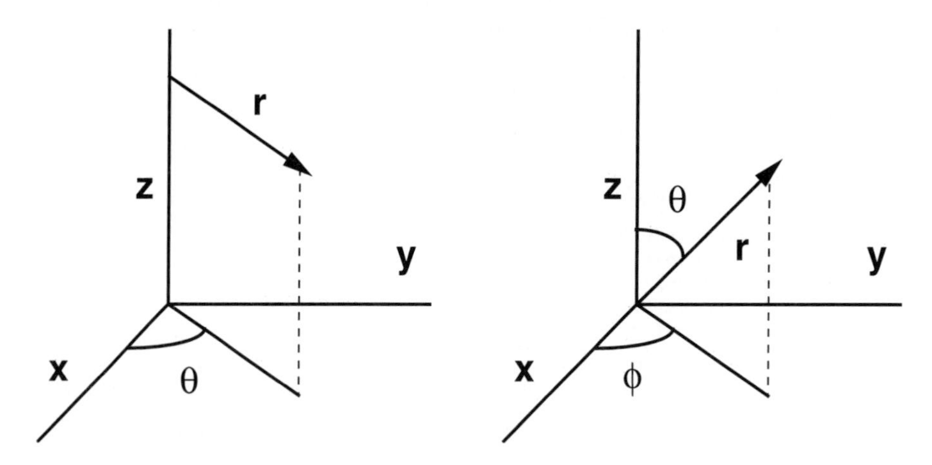

Remarks and outline Express the position vector and the kinetic and potential energies in terms of cylindrical coordinates. The momentum relations, the Hamiltonian, and Hamilton's equation follow from the user-defined function **Hamilton**. The condition for a circular orbit follows from setting **r[t]** equal to a constant. The power-series solution of Hamilton's equations follows from the user-defined function **firstDiffSeries**.

Required packages

```
In[267]:= Needs["Algebra`Trigonometry`"]
          Needs["Calculus`VectorAnalysis`"]
```

Solution

```
In[268]:= Clear["Global`*"];
```

Part (a) The position vector in cylindrical coordinates is described by the rule

```
In[269]:= rRule=
            rVector->( Evaluate[
                          CoordinatesToCartesian[
                               {r[#],theta[#],z[#]}
                                          ,Cylindrical]]& );
```

The kinetic and potential energies are

```
In[270]:= T=(   (1/2 m rVector'[t].rVector'[t])
                     //.rRule
                     //ExpandAll
                     //Simplify
          );

          L= T- V[r[t],theta[t],z[t]]

Out[270]=
                                                  2         2           2         2
                                         m (r'[t]  + r[t]  theta'[t]  + z'[t] )
          -V[r[t], theta[t], z[t]] + -----------------------------------------------
                                                              2
```

The Hamiltonian and Hamilton's equations follow from

```
In[271]:= {pRule,hamiltonian,eqMotion}=
              Hamilton[L,{r,theta,z},{pr,ptheta,pz}]

Out[271]=
                      pr[t]                   ptheta[t]              pz[t]
          {{r'[t] -> -----, theta'[t] -> ----------, z'[t] -> -----},
                        m                          2               m
                                           m r[t]

                 2         2              2
           pr[t]     pz[t]     ptheta[t]
           ------ + ------ + ----------- + V[r[t], theta[t], z[t]],
            2 m       2 m              2
                               2 m r[t]

                                 2
                      ptheta[t]       (1,0,0)
           {pr'[t] == ---------- - V        [r[t], theta[t], z[t]],
                              3
                        m r[t]

                                  (0,1,0)
            ptheta'[t] == -V        [r[t], theta[t], z[t]],

                             (0,0,1)                                       pr[t]
            pz'[t] == -V        [r[t], theta[t], z[t]], r'[t] == -----,
                                                                             m

                         ptheta[t]              pz[t]
            theta'[t] == ----------, z'[t] == -----}}
                                 2                m
                           m r[t]
```

The first part of the list is the expression for the canonical momentum; the second is the Hamiltonian; the last are the six Hamilton first-order equations.

Part (b) Express gravitational potential in pure form with the rule

```
In[272]:= vRule={V -> ( (-k/Sqrt[#1^2+#3^2])& )};

          V[r[t],theta[t],z[t]] //.vRule
Out[272]=           k
          -(——————————————————————)
                     2        2
            Sqrt[r[t]  + z[t] ]
```

Restrict the orbit to the **z=0** plane with

```
In[273]:= zRule= {z->(0&), pz->(0&)};
```

(Use this form to eliminate derivatives of **z** and **pz** also.) Applying **vRule** and **zRule** to Hamilton's equations of motion, you get

```
In[274]:= eq1= (eqMotion) //.vRule //.zRule //PowerExpand
Out[274]=
                           2
                 ptheta[t]        k                                              pr[t]
          {pr'[t] == ————————— - ————, ptheta'[t] == 0, True, r'[t] == —————,
                            3        2                                             m
                    m r[t]      r[t]

                          ptheta[t]
            theta'[t] == —————————, True}
                                2
                          m r[t]
```

The **z** equations are automatically satisfied. Note that **ptheta[t]** must be a constant. Setting **ptheta[t]=ptheta0**, Hamilton's equations reduce to

```
In[275]:= eq2= eq1 //. ptheta->(ptheta0&)
Out[275]=
                           2
                 ptheta0         k                            pr[t]
          {pr'[t] == ——————— - ————, True, True, r'[t] == —————,
                          3        2                            m
                  m r[t]      r[t]

                         ptheta0
            theta'[t] == ———————, True}
                              2
                        m r[t]
```

The condition for circular orbits are described by the rule

```
In[276]:= constantR= {pr->(0&), r->(r0&)};
```

Applying these rules and solving for **ptheta0**, you get the conditions for circular orbits:

```
In[277]:= circleRule=
            Solve[(eq2//First) //.constantR ,ptheta0][[2]]
Out[277]= {ptheta0 -> Sqrt[k] Sqrt[m] Sqrt[r0]}
```

(Be sure to take the positive root here.)

Part (c) The user-defined function **firstDiffSeries** returns a power-series solution for two first-order equations. Apply **firstDiffSeries** to the **r** equations of motion:

```
In[278]:= rEq=eq2[[{1,4}]]
Out[278]=
                      2
             ptheta0         k                pr[t]
{pr'[t] == ---------- - -------, r'[t] == -------}
                 3           2                m
             m r[t]      r[t]
```

with the initial conditions :

```
In[279]:= initial= {r[0]->r0,pr[0]->0};
```

The results are

```
In[280]:= pert= ( firstDiffSeries[rEq,{r,pr},initial,4,t]
                  //Simplify
          )
Out[280]=
                  2             2
         (ptheta0  - k m r0) t
{r0 + ------------------------- +
                  2   3
              2 m  r0

           4                    2          2  2   2  4
  (-3 ptheta0  + 5 k m ptheta0  r0 - 2 k  m  r0 ) t        5
  -------------------------------------------------- + O[t] ,
                          4   7
                      24 m  r0

            2
  (ptheta0  - k m r0) t
  --------------------- +
              3
          m r0

            4                    2          2  2   2  3
   (-3 ptheta0  + 5 k m ptheta0  r0 - 2 k  m  r0 ) t        5
   -------------------------------------------------- + O[t] }
                           3  7
                        6 m  r0
```

To verify the validity of this solution for circular orbits, apply **circleRule**:

```
In[281]:=     pert //.circleRule
Out[281]=
                   5       5
          {r0 + O[t] , O[t] }
```

Clearly this yields a constant radius to the proper order of **t**.

Problem 3: Spherical Pendulum and Hamilton's Equations

Consider a pendulum of mass m suspended by a rigid weightless rod. The motion is not restricted to a plane. Use the spherical angular variables $\{\theta, \phi\}$ to describe the motion and the conjugate momentum $\{p_\theta, p_\phi\}$.

Part (a) Find Hamilton's equations for the variables $\{\theta, \phi\}$. Eliminate the ϕ variable and reduce Hamilton's equations to a nonlinear equation for θ. Restrict the pendulum motion to a plane and show that the solution reduces to the standard pendulum equation.

Part (b) Assume the pendulum equation can be expanded in a power series and solve for the first few terms. Assume $g = 1$, $m = 1$, $Len = 1$, and initial conditions $\{\theta[0] = \pi/4, \theta'[0] = 0\}$. Find the numerical solutions for θ when $p_\phi[0] = 0$ (planar pendulum) and $p_\phi[0] = 0.2$.

Part (c) On the same graph, plot the following solutions:

1. The power-series solution for the parameters given in Part (b) with $p_\phi[0] = 0.2$
2. The numerical solution found in Part (b) with $p_\phi[0] = 0$
3. The numerical solution found in Part (b) with $p_\phi[0] = 0.2$
4. the linear-pendulum solution given by the function $\theta_0 \mathrm{Cos}[t]$

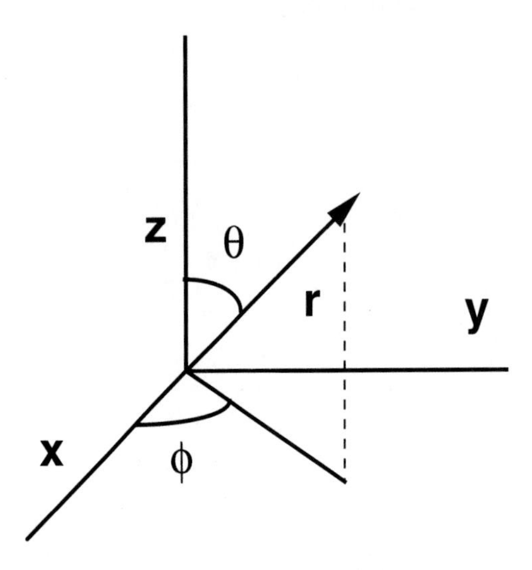

Remarks and outline Write the kinetic and potential energies and then apply the user-defined function **Hamilton**. An equation for the θ variable follows from noticing that the momentum conjugate to ϕ is a constant. The time-series solution follows from the user-defined function **diffSeriesOne**.

Required packages

```
In[282]:= Needs["Algebra`Trigonometry`"]
          Needs["Calculus`VectorAnalysis`"]
```

Solution

```
In[283]:= Clear["Global`*"];
```

Part (a) The kinetic and potential energies and the Lagrangian for a spherical pendulum are

```
In[284]:= T= m/2 Len^2 ( theta'[t]^2
                  + Sin[theta[t]]^2 phi'[t]^2);
          V= m g Len (1 - Cos[theta[t]]);
          L= T - V;
```

Applying **Hamilton** to **L**, you get the equations of motion:

```
In[285]:= {prule,ham,eqMotion}=
            Hamilton[L,{theta,phi},{ptheta,pphi},t];

In[286]:= eqMotion

Out[286]=
                                                   2         2
                   Cot[theta[t]] Csc[theta[t]]  pphi[t]
    {ptheta'[t] == -------------------------------------- -
                                       2
                                    Len  m

                       theta[t]       theta[t]
       2 g Len m Cos[--------] Sin[--------], pphi'[t] == 0,
                          2              2

                                                        2
                      ptheta[t]              Csc[theta[t]]  pphi[t]
     theta'[t] == ---------, phi'[t] == -----------------------}
                         2                          2
                      Len  m                     Len  m
```

Notice that the solution of the **pphi** equation is simply **pphi0** is a constant. Eliminating **pphi[t]** from Hamilton's equations, you get

```
In[287]:= eqMotion=eqMotion //.{pphi->(pphi0&)}

Out[287]=
                        2                             2
                   pphi0  Cot[theta[t]] Csc[theta[t]]
    {ptheta'[t] == ---------------------------------- -
                                     2
                                  Len  m

                       theta[t]       theta[t]
       2 g Len m Cos[--------] Sin[--------], True,
                          2              2

                                                              2
                      ptheta[t]              pphi0 Csc[theta[t]]
     theta'[t] == ---------, phi'[t] == -------------------}
                         2                          2
                      Len  m                     Len  m
```

Extract the two equations in this list that are a set of coupled equations between the variables **ptheta[t]** and **theta[t]**:

```
In[288]:= eqMotion[[{1,3}]]
```

```
Out[288]=
                        2                                2
                   pphi0  Cot[theta[t]] Csc[theta[t]]
     {ptheta'[t] == ------------------------------------ -
                                    2
                                 Len  m

                          theta[t]        theta[t]                   ptheta[t]
      2 g Len m Cos[--------] Sin[--------], theta'[t] == ---------}
                            2               2                         2
                                                                   Len  m
```

Using the first equation, you can make a rule to eliminate **ptheta'[t]**:

```
In[289]:= pthetaRule= eqMotion[[1]] //ToRules
Out[289]= {ptheta'[t] ->

              2                                2
         pphi0  Cot[theta[t]] Csc[theta[t]]
         ------------------------------------ -
                          2
                       Len  m

                          theta[t]        theta[t]
          2 g Len m Cos[--------] Sin[--------]}
                             2               2
```

Obtain an equation for **theta''[t]** by taking the derivative of the third equation of motion and eliminating **ptheta'[t]**:

```
In[290]:= eq1= ((eqMotion[[3]]  //Map[D[#,t]&,#]&)
                                 //.pthetaRule
                                 //ExpandAll
                                 //Simplify
                                )

Out[290]=
                        2                                2
                   pphi0  Cot[theta[t]] Csc[theta[t]]     g Sin[theta[t]]
     theta''[t] == ------------------------------------ - ---------------
                                   4  2                          Len
                                Len  m
```

This is the nonlinear equation for a spherical pendulum. In the limit of planar motion (**phi=0**) and small **theta** motion, you get

```
In[291]:= (( eq1 // Map[Series[#,{theta[t],0,1}]&,#]&
              ) //.{pphi0->0}
                //Normal
              )

Out[291]=
                            g theta[t]
     theta''[t] == -(----------)
                               Len
```

This is the well-known linear-pendulum equation.

Part (b) To find the time-series solution of **eq1**, use the user-defined command **diffSeriesOne**. Keeping terms to sixth order in **t**:

```
In[292]:= pert= ( diffSeriesOne[eq1,theta,t,6
                      ,{theta[0]->theta0,theta'[0]->0}]
                  //Simplify
                );
```

(This calculation takes a few minutes, so begin with a lower order.) The expression is sufficiently long, so choose to display only terms to third order in **t**:

```
In[293]:= pert + O[t]^3

Out[293]=
                         2                    2
                  pphi0  Cot[theta0] Csc[theta0]                     2
                 (---------------------------- - g Sin[theta0]) t
                              3  2
                           Len  m                                          3
theta0 + ------------------------------------------------------------ + O[t]
                                   2 Len
```

For the numerical solution, choose

```
In[294]:= values={g->1,m->1,Len->1};
```

and the **initial** conditions

```
In[295]:=  initial={theta[0]==Pi/4, theta'[0]==0};
```

The numerical solution for **pphi[0]=0** (motion in a plane) is

```
In[296]:= ndsol1=
          NDSolve[Join[{eq1},initial] //.values //.{pphi0->0}
                                  ,{theta},{t,0,8}][[1]];
          theta1[t_]=theta[t] //.ndsol1;
          theta1[3.4]

Out[296]= -0.779166
```

The function **theta1[t]** has been defined to yield this numerical solution. For **pphi0=0.2**, do likewise

```
In[297]:= ndsol2=
          NDSolve[Join[{eq1},initial] //.values //.{pphi0->0.2}
                                  ,{theta},{t,0,8}][[1]];
          theta2[t_]=theta[t] //.ndsol2;
          theta2[3.4]

Out[297]= 0.780593
```

Part (c) Plot the following four curves:

1. The numerical solution of **eq1** with **pphi0=0**
2. The power-series solution for **pphi0=0.2**
3. The numerical solution of **eq1** with **pphi0=0.2**
4. The linear-pendulum solution given by the function (**theta0 Cos[t]**)

Define **theta3[t]** to be the numerical solution of **eq1** with **pphi0=0.2**:

```
In[298]:= theta3[t_]= ( pert //.values
                             //.{theta0->Pi/4, pphi0->0.2}
                             //N
                             //Normal
                       )

Out[298]=
                        2              4               6
         0.7854 - 0.31355 t  + 0.026838 t  + 0.0028616 t
```

Define **theta4[t]** to be the solution of the linear pendulum:

```
In[299]:= theta4[t_]= theta0 Cos[t] //.{theta0->Pi/4}
Out[299]= Pi Cos[t]
          ----------
              4
```

Combining these curves in a single plot, you get

```
In[300]:= Plot[{theta1[t], theta2[t], theta3[t], theta4[t]} //Evaluate
               ,{t,0,8}
               ,PlotStyle->{
                   {Dashing[{1.00,0.00}],Thickness[0.010]}
                  ,{Dashing[{0.08,0.01}],Thickness[0.008]}
                  ,{Dashing[{0.06,0.02}],Thickness[0.006]}
                  ,{Dashing[{1    ,0    }],Thickness[0.002]} }
               ,Frame->True
               ,GridLines->Automatic
               ,PlotRange->{-1, 1}
          ];
```

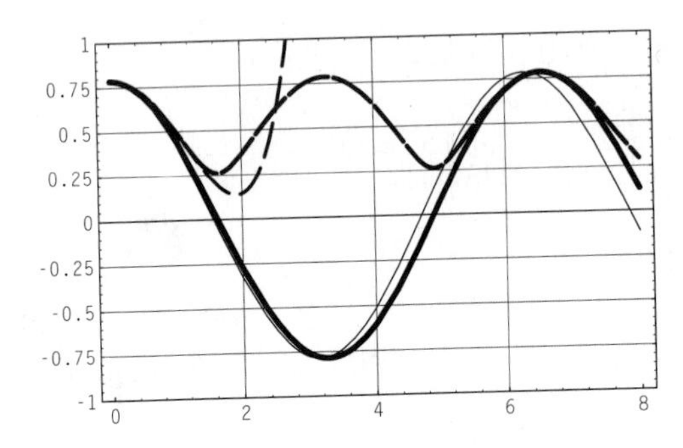

Problem 4: Harmonic Oscillator and Hamilton-Jacobi Equations

Consider a one-dimensional harmonic oscillator.

Part (a) Express the Hamiltonian in canonical variables and derive the Hamilton-Jacobi equation for W using the user-defined function **Hamilton**.

Part (b) Solve the Hamilton-Jacobi equation for W to obtain the oscillator solution for $\{x, p\}$.

Remarks and outline The Hamiltonian $H[q, p]$ follows from the user-defined function **Hamilton**. The Hamilton-Jacobi equation is

$$H\left[q, \frac{\partial S[x,t]}{\partial x}, t\right] + \frac{\partial S[x,t]}{\partial t} = 0.$$

We have $S[x,t] = W[x]+Pt$, where P is the new conjugate momentum and is constant. The transformations between the coordinate systems $\{x, p\}$ and $\{Q, P\}$ are $p = \partial S[x, P, t]/\partial q$ and $Q = \partial S[x, P, t]/\partial P$.

Required packages

```
In[301]:= Needs["Algebra`Trigonometry`"]
          Needs["Calculus`VectorAnalysis`"]
```

Solution

```
In[302]:= Clear["Global`*"];
```

Part (**a**) The kinetic and potential energies for a harmonic oscillator are

```
In[303]:= T= m/2 x'[t]^2;
          V= k/2 x[t]^2;
          L= T - V

Out[303]=        2          2
          -(k x[t] )   m x'[t]
          ---------- + --------
              2           2
```

The Hamiltonian follows from

```
In[304]:= {xpRule,ham,eqMotion}=
                Hamilton[L,{x},{p}] //.{x_[t]->x}

Out[304]=
                      2     2
                p    p     k x                            p
          {{x' -> -}, --- + ----, {p' == -(k x), x' == -}}
                m    2 m    2                             m
```

where **{x,p}** are the canonical coordinates. The momentum **p** and Hamilton's principal function **S[x,t]** are related by

```
In[305]:= pRule= {p-> D[S[x,t],x]}

Out[305]=         (1,0)
          {p -> S      [x, t]}
```

The Hamilton-Jacobi equation follows:

```
In[306]:= eqHJ = (ham + D[S[x,t],t]) ==0  //.pRule

Out[306]=   2                         (1,0)        2
          k x      (0,1)             S      [x, t]
          ---- + S      [x, t] + ---------------- == 0
           2                            2 m
```

Assume the solution for **S** is of the form

```
In[307]:= sRule= {S -> (W[#1]- P #2 &)};
          S[x,t] //.sRule

Out[307]= -(P t) + W[x]
```

where the new canonical momentum **P** is a constant. Substitute **S** into **eqHJ** to get an equation for **W'[x]**:

```
In[308]:= eqW= eqHJ //.sRule

Out[308]=        2        2
               k x    W'[x]
          -P + ---- + ------ == 0
                2      2 m
```

Part (**b**) Solving **eqW** for **W'[x]**, you get

```
In[309]:= wPrimeRule= Solve[eqW,W'[x]][[2]]

Out[309]=                                    2
          {W'[x] -> Sqrt[m] Sqrt[2 P - k x ]}
```

(Be sure to choose the positive root here.) Integrate **wPrimeRule** to get an expression for **W[x]**:

```
In[310]:= wRule={W[x]->(Integrate[W'[x]/.wPrimeRule,x]
                            //Hold)}

Out[310]= {W[x] -> Hold[Integrate[W'[x] /. wPrimeRule, x]]}
```

The **Hold** command keeps the integration from being evaluated. There is a reason not to evaluate this integration at this step of the calculation. In general, we are interested in the derivative of **W** with respect to some parameter, such as **P**. The algebra simplifies if we interchange the order of integration and derivative. We retain **W[x]** in its unevaluated form until the appropriate derivatives are taken.

The following identification interchanges the order of integration and differentiation and then performs the integration:

```
In[311]:= D[Hold[Integrate[a_,x_]],b_] ^:=
                        Integrate[ D[a,b], x]
```

The notation **^:=** assigns the right-hand side to be the delayed value of the left-hand side. The **^** assigns the rule to **Hold** instead of **D**. Hamilton's principal function follows from the **sRule** and **wRule**:

```
In[312]:= S[x,t]= S[x,t] //.sRule //.wRule

Out[312]= -(P t) + Hold[Integrate[W'[x] /. wPrimeRule, x]]
```

The new canonical position coordinate **Q** (constant) follows from taking the derivative of **S** with respect to **P**:

```
In[313]:= eq4= Q== D[S[x,t],P]

Out[313]=
                                Sqrt[k] x
                 Sqrt[m] ArcTan[-----------------]
                                            2
                                Sqrt[2 P - k x ]
          Q == -t + -----------------------------------
                              Sqrt[k]
```

You can solve **eq4** for **x**:

```
In[314]:= xSol= Solve[eq4,x][[2]] //Simplify //PowerExpand //Simplify

Out[314]=
                                 Sqrt[k] (Q + t)
              Sqrt[2] Sqrt[P] Sin[---------------]
                                    Sqrt[m]
          {x -> ---------------------------------------}
                          Sqrt[k]
```

The expression for the canonical momentum **p** follows from the equation of motion:

```
In[315]:= prule2= Solve[eqMotion[[2]],p][[2]]

Out[315]= {{p -> m x'}}[[2]]
```

You can easily solve for **p** given the solution for **x**:

```
In[316]:= pSol= {p->m D[(x/.xSol),t] } //Simplify
```

```
Out[316]=
                                            Sqrt[k] (Q + t)
          {p -> Sqrt[2] Sqrt[m] Sqrt[P] Cos[---------------]}
                                               Sqrt[m]
```

Problem 5: Kepler's Problem and Hamilton-Jacobi Equations

Consider a bound particle moving in the Keplerian potential. Use cylindrical coordinates $\{r, \theta, z\}$ and restrict the motion to the $z = 0$ plane. Call the momentum variables conjugate to $\{r, \theta\}$ and $\{p_r, p_\theta\}$, respectively. Let $S[t, r, \theta]$ be Hamilton's principal function, and $W_r[r]$ and $W_\theta[\theta]$ be the Hamilton characteristic functions.

Part (a) Derive an expression for Hamilton's characteristic functions, $W_r[r]$ and $W_\theta[\theta]$.

Part (b) Solve for the elliptical orbits using Hamilton's principal function $S[t, r, \theta]$. Show that the expression for r can be written as $r = a\,(1 - e^2)/(1 + e\,\mathrm{Cos}[\phi])$.

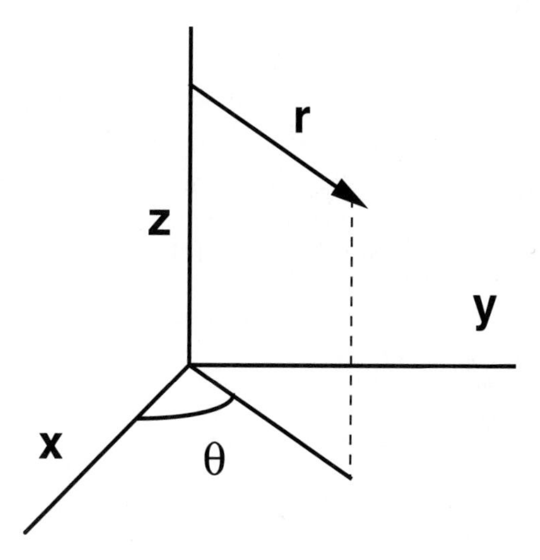

Remarks and outline Express the position vector in cylindrical coordinates using the command **CoordinatesToCartesian** found in the package **Calculus`VectorAnalysis`**. Find the expression for the kinetic and potential energies. The Hamilton-Jacobi equation follows from the user-defined function **HamiltonJacobi**. Because of the lack of θ dependence in the Hamilton-Jacobi equation, it follows that $W_\theta'[\theta]$ can be set equal to a constant. The solution for the orbit follows from solving the equation $Q = \partial S[r, P, t]/\partial P$ for r.

Required packages

```
In[317]:= Needs["Algebra`Trigonometry`"]
          Needs["Calculus`VectorAnalysis`"]
```

Solution

```
In[318]:= Clear["Global`*"];
```

Part (a) Use cylindrical coordinates and restrict the motion to the **z=0** plane. The position vector in cylindrical coordinates, called **rvector**, is

```
In[319]:= rRule=  rvector->(Evaluate[CoordinatesToCartesian[
                  {r[#],theta[#],z[#]} ,Cylindrical]]&);
```

The kinetic and potential energies for the particle moving in the **z=0** plane are

```
In[320]:= T = ( (m/2 rvector'[t].rvector'[t])
                //.rRule
                //.{z'[t]->0}
                //ExpandAll
                //Simplify
              )

Out[320]=
                2         2         2
          m (r'[t]  + r[t]  theta'[t] )
          -----------------------------
                       2

In[321]:= V= -k/r[t];
          L= T-V

Out[321]=
                         2        2          2
           k      m (r'[t]  + r[t]  theta'[t] )
          ---- + -----------------------------
          r[t]                2
```

The Hamilton-Jacobi equation follows from the user-defined function **HamiltonJacobi**. The momentum variables conjugate to **{r,theta}** are called **{pr,ptheta}**. The Hamilton characteristic functions are called **Wr[r]** and **Wth[theta]**. Applying **HamiltonJacobi** to **T** and **PE**, you get

```
In[322]:= {momRule,hamPF,eqMotion}=
             HamiltonJacobi[L,{r,theta},{pr,ptheta}
                         ,{Wr[r],Wth[theta]},En]

Out[322]=
                    pr[t]                  ptheta[t]
          {{r'[t] -> -----, theta'[t] -> ---------},
                      m                         2
                                          m r[t]

            -(En t) + Wr[r[t]] + Wth[theta[t]],

                                   2                   2
                  k      Wr'[r[t]]     Wth'[theta[t]]
            -En - ---- + ---------- + ---------------- == 0}
                  r[t]      2 m                 2
                                         2 m r[t]
```

The first set of brackets contain the momentum relationships, the second term is Hamilton's principal function, and the third term is the Hamilton-Jacobi equation. Because of the lack of **theta** dependence in the Hamilton-Jacobi equation, it follows that **Wth'[theta]** can be set equal to a constant. Set **Wth'[theta]=el** and solve for **Wth[theta]**:

```
In[323]:= wthrule=
            DSolve[{Wth'[theta]==+el
            ,Wth[0]==0},{Wth},theta][[1]]

Out[323]= {Wth -> Function[theta, el theta]}
```

The equation for **Wr'[r]** becomes

```
In[324]:= eq1= Solve[ eqMotion //.wthrule, Wr'[r[t]] ][[2]]
Out[324]=
                                    2
                                  el      2 k m
          {Wr'[r[t]] -> Sqrt[2 En m - ----- + -----]}
                                    2     r[t]
                                  r[t]
```

(Be sure to take the positive root here.) Integrate **Wr'[r]** to get an expression for **Wr[r]**:

```
In[325]:= wrRule= Wr[r[t]]-> Hold[ Integrate[Wr'[r[t]]//.eq1,r[t]] ]
Out[325]= Wr[r[t]] -> Hold[Integrate[Wr'[r[t]] //. eq1, r[t]]]
```

The algebra simplifies if you do not evaluate this integration at this step of the calculation. In general, you are interested in the derivative of **W** with respect to some parameter. The algebra simplifies if you interchange the order of integration and derivative, so we define the following rule:

```
In[326]:= D[Hold[Integrate[a_,x_]],b_] ^:=
                          Integrate[D[a,b],x]
```

Part (b) Hamilton's principal function now follows from applying **wrRule** and **wthrule** to **S**, where **S** is given by **hamPF[[2]]**:

```
In[327]:= S= hamPF //.wrRule //.wthrule
Out[327]= -(En t) + Hold[Integrate[Wr'[r[t]] //. eq1, r[t]]] + el theta[t]
```

To get a solution for **r**, first consider the expression for the **Q** coordinate, which follows from taking the derivative of **S** relative to **el**:

```
In[328]:= eq4=  Q==D[S,el] //PowerExpand
Out[328]=
                                   2               2
                             -(el  r[t]) + k m r[t]
          Q == -ArcTan[------------------------------------------------] +
                                   2                               2
                   el r[t] Sqrt[-el  + 2 k m r[t] + 2 En m r[t] ]

           theta[t]
```

You will find it convenient to work with the variable $u = 1/r$. Solving for **u**, you get

```
In[329]:= sol4= Solve[(eq4 /.{r[t]->1/u} //Simplify), u][[1]]
Out[329]=
                  2                                   2
          {u -> (-(el  k m (-2 - 2 Tan[Q - theta[t]] )) -

                    4  2  2                          2 2
               Sqrt[el  k  m  (-2 - 2 Tan[Q - theta[t]] )  -

                  4                            2
                4 el  m (1 + Tan[Q - theta[t]] )

                   2        2                      2
                 (k  m - 2 el  En Tan[Q - theta[t]] )]) /

                  4                         2
              (2 el  (1 + Tan[Q - theta[t]] ))}
```

The expression simplifies when you replace **Tan[Q - theta[t]]->tan** and then **tan->sin/Sqrt[1-sin^2]**:

```
In[330]:= sol5= (((sol4 //.{Tan[Q - theta[t]] -> tan}
                      //.{tan->sin/Sqrt[1-sin^2]}
                   ) //Simplify
                     //PowerExpand
                   ) //Simplify
                 ) /.{sin->Sin[Q - theta[t]]}

Out[330]=
                                        2          2
           Sqrt[m] (k Sqrt[m] + Sqrt[2 el  En + k  m] Sin[Q - theta[t]])
     {u -> -------------------------------------------------------------}
                                         2
                                       el
```

The solution for **u** can be expressed in the canonical form $(1 + e\text{Cos}[\theta[t]])/(a(1-e^2))$. Without loss of generality, set **Q=Pi/2** and equate **u** given by **sol5** to its canonical form:

```
In[331]:= eq2= (u /.sol5 /.{Q->Pi/2}) == (1+e Cos[theta[t]])/( a(1-e^2))

Out[331]=
                                    2          2
     Sqrt[m] (k Sqrt[m] + Sqrt[2 el  En + k  m] Cos[theta[t]])
     --------------------------------------------------------- ==
                                   2
                                 el

       1 + e Cos[theta[t]]
       -------------------
                  2
          a (1 - e )
```

You can obtain two independent equations from **eq2** by choosing two values for **Cos[theta[t]]**:

```
In[332]:= eq3a= eq2 /.{Cos[__]->0}

Out[332]= k m           1
          --- == -----------
            2             2
          el      a (1 - e )

In[333]:= eq3b= eq2 /.{Cos[__]->1}

Out[333]=
                                      2          2
          Sqrt[m] (k Sqrt[m] + Sqrt[2 el  En + k  m])         1 + e
          ------------------------------------------- ==  ----------
                              2                                    2
                            el                             a (1 - e )
```

Then solve for **e** and **a**:

```
In[334]:= Solve[{eq3a,eq3b},{a,e}]

Out[334]=
                        2         2
                Sqrt[2 el  En + k  m]          -k
          {{e -> ---------------------, a -> ----}}
                     k Sqrt[m]                 2 En
```

4.3 Unsolved Problems

Exercise 4.1: Double Atwood machine

A string of length L_1 passes over a fixed light pulley supporting a mass m_1 on one end and a pulley of mass m_2 on the other. Over this second pulley passes a string of length L_2 that supports a mass m_3 on one end and m_4 on the other.

Part (**a**) Write the Lagrangian and derive the equations of motion.

Part (**b**) Solve for the motion.

Part (**c**) Animate the solution.

Exercise 4.2: Particle sliding on a movable inclined plane

Consider a particle sliding on a smooth inclined plane that itself is free to slide on a smooth horizontal surface.

Part (**a**) Write the Lagrangian and derive the equations of motion.

Part (**b**) Solve the equations of motion.

Part (**c**) Animate the solution.

Exercise 4.3 : Spring pendulum

A spring pendulum has a mass m suspended by an elastic spring of stiffness k and force-free length L. Assume the motion is confined to a plane.

Part (**a**) Write the Lagrangian and obtain the equations of motion.

Part (**b**) Assume the oscillations are small and analytically solve for the motion.

Part (**c**) Numerically solve the nonlinear equations and graph the results.

Part (**d**) Animate the results.

Exercise 4.4: Lagrangian plot procedure

Consider a system that has at most two degrees of freedom. Given the kinetic energy, potential, and initial conditions, write a procedure that will numerically solve Lagrangian's equations and return a plot of the coordinates. Apply this to the case of a spring pendulum.

Exercise 4.5: Series expansion procedure

Write a procedure that changes the function and variable in a differential equation and returns a series solution for the new function in terms of the new variable. Call this user-defined procedure **`diffChange`**. Apply **`diffChange`** to the radial equation for a particle in a potential $-k/r$. Assume the Keplerian orbit is of the form $r[t] = p/(1 - \mathrm{Cos}[f[t]])$ and find the series solution for $f[t]$.

Exercise 4.6: Central force problems

A particle is acted on by a central force for which the potential is $V = -ae^{-br}/r$, where a and b are positive constants.

Part (a) Use potential and phase diagrams to discuss the nature of the motion.

Part (b) Find the solution for circular orbits.

Part (c) Find the period of small radial oscillations about the circular motion.

Part (d) Solve Parts (a), (b), and (c) for the potential $V = -a/r^n$, where a is a positive constant and $n > 2$.

Exercise 4.7: Central force procedure

Consider the problem of a particle acted on by a central force with potential $V[r]$. Write a procedure that does the following:

Part (a) Plots the potential

Part (b) Finds the extremum points

Part (c) Returns the solution for the circular orbits that lie at the stable equilibrium points

Part (d) Returns the correction to the circular orbit for small radial oscillation about the circular motion

Exercise 4.8: Central forces and elliptical solutions

A particle is acted on by a central force for which the potential is $V = -ar^{(n+1)}$, where $n = \{+5, +3, 0, -4, -5, -7\}$.

Part (a) Plot the effective potential and discuss the motion.

Part (b) Show that the solutions can be reduced to elliptical equations.

Part (c) Teach *Mathematica* to recognize the solution of these elliptical equations.

Part (d) Expand the solution in a power series and compare the results with the power series that follows from the elliptical solutions.

Exercise 4.9: Kepler problem with drag

A uniform distribution of dust in the solar system adds to the gravitational attraction of the sun on a planet an additional radial force of the form $-mCr$, where m is the mass of the planet, C is a constant, and r is the radius vector from the sun to the planet. This additional force is small compared to the direct sun/planet gravitational force.

Part (a) Calculate the period for a circular orbit of radius r_0 for a planet in this combined field.

Part (b) Calculate the period of radial oscillations for slight disturbances from a circular orbit.

Part (**c**) Show that nearly circular orbits can be approximated by precessing ellipses and find the precessional frequency.

Exercise 4.10: Hamilton-Jacobi in parabolic coordinates

A particle of mass m is moving in a force whose potential is $V = A/r - Bz$. This produces a force that is a combination of a uniform force in the z-direction and an inverse square central force.

Part (**a**) Derive the Hamilton-Jacobi equation in parabolic coordinates.

Part (**b**) Solve the equations and discuss the solution.

Exercise 4.11: Hamilton-Jacobi procedure in separable coordinates

Consider the problem of a particle acted on by a force with a potential of the form $V[r, \theta, \phi]$. When the user-defined function **`HamiltonJacobi`** is applied to this problem, it will return an expression for the Hamilton-Jacobi partial differential equation.

Part (**a**) Generalize the user-defined function **`HamiltonJacobi`** so that it will return the Hamilton-Jacobi in any coordinate system supported by the package **`Calculus'VectorAnalysis'`**.

Part (**b**) Generalize **`HamiltonJacobi`** so that it not only will write the Hamilton-Jacobi equations in other coordinate systems, but also will check to see if each equation is separable. If it is separable, then have the procedure return the ordinary differential equations for Hamilton's characteristic functions $W_1[q_1]$, $W_2[q_2]$, and $W_3[q_3]$.

CHAPTER FIVE

Electrostatics

5.1 Introduction

Electrostatics is the basis of electrodynamics and is the study of time-independent distributions of charge and fields. The problems in this chapter were chosen to illustrate *Mathematica*'s analytic and graphics abilities when applied to vector and scalar field problems. The level of the problems is appropriate for the typical advanced undergraduate or beginning graduate student. The methods used to solve these problems are not unique. You are encouraged to find procedures that will illuminate the physics or to find other commands that speed up the calculations.

5.1.1 Electric Field and Potential

Electric field

When an infinitesimal electric charge experiences a force, we say an electric field exists. The electric field is a vector equal to the force per unit charge acting on a positive charge placed at that point. If an arbitrary charge q is placed in an electric field $\vec{E}$, it will experience a force $\vec{F}$ given by $\vec{F} = q\vec{E}$. The electric field is described by an irrotational vector field, $\nabla \times E = 0$.

A fundamental property of the electric field is described by Gauss's equation. If the charges lie interior to a closed surface S, then the total charge Q enclosed by S is related to the electric field by $\int \vec{E} \cdot d\vec{S} = 4\pi Q$. The differential form of Gauss's law is $\nabla \cdot \vec{E} = 4\pi\rho$, where ρ is the charge density. The electric field can also be expressed as an integral over the charge density:

$$\vec{E} = \int d\vec{r}\,'\rho(\vec{r}\,')\frac{(\vec{r} - \vec{r}\,')}{|\vec{r} - \vec{r}\,'|^3}.$$

A common problem in electrostatics is determining the electric field due to a given surface distribution of charges. If a surface S has a unit normal $\vec{n}$ directed from side one to side two of the surface, a surface-charge density of σ, and electric field E_1 and E_2 on either side of the surface, then according to Gauss's law the normal component of the electric field is discontinuous. The discontinuity is related to the surface charge by the relation $(E_2 - E_1) \cdot \vec{n} = 4\pi\sigma$.

Electrostatic potential

It is often simpler to deal with scalar fields than with vector fields. The vector electric field is related to a scalar potential by the equation $\vec{E} = -\nabla U$, where ∇ is the gradient operator. In terms of a charge density, the scalar potential is $\Delta U = -4\pi\rho$. This equation is known as *Poisson's equation*. It follows in a region in which the charge density is zero that $\Delta U = 0$. This equation is called *Laplace's equation*. The problem of finding the potential corresponds to that of finding the solution to either Laplace's or Poisson's equation that satisfies the boundary conditions of the problem. The operators **`Grad`**, **`Div`**, and **`Laplacian`** are found in the package **`Calculus`VectorAnalysis``**. The operators can be expressed in many different coordinate systems, the three most common of which are Cartesian, cylindrical, and spherical.

5.1.2 Laplace's Equation

Many electrostatic problems involve boundary surfaces on which either the potential or the surface-charge density is specified. The solution of such problems reduces to the solution of Laplace's equation. If the problem is assumed to be separable, the solution of Laplace's equation can be expressed as a product of three terms, each of which is a function of only one coordinate. The solution is obtained by expanding in orthogonal functions. The coefficients of this expansion are obtained using the boundary conditions and the orthogonality properties. The solution of Laplace's equation in the three coordinate systems are discussed next.

Cartesian coordinates

In Cartesian coordinates, the solution of Laplace's equation can be represented by a product of three functions $X[x]\,Y[y]\,Z[z]$ of the form

$$\begin{aligned} X[x] &= \{\mathrm{Sin}[\alpha x], \mathrm{Cos}[\alpha x]\} \\ Y[y] &= \{\mathrm{Sin}[\beta y], \mathrm{Cos}[\beta y]\}\ , \\ Z[z] &= \{e^{+\gamma z}, e^{-\gamma z}\} \end{aligned}$$

where $\alpha^2 + \beta^2 = \gamma^2$. The constants are determined by the boundary conditions. Note that these equations only appear to be nonsymmetric. Substituting $\gamma \to i\gamma$ makes the equation for $Z[z]$ a trigonometric equation. Likewise, substituting $\alpha \to i\alpha$ makes the equation for $X[x]$ an exponential equation.

Cylindrical coordinates

In cylindrical coordinates, the solution of Laplace's equation can be expressed as a product of $R[r]\,\Phi[\phi]\,Z[z]$ of the form

$$Z[z] = \begin{cases} ae^{+kz} + be^{-kz} & \text{for } k \neq 0 \\ a + bz & \text{for } k = 0 \end{cases}$$

$$R[r] = \begin{cases} a_n J[n, kr] + b_n N[n, kr] & \text{for } k \neq 0 \\ (a_n r^n + b_n r^{-n}) + (a_0 \mathrm{Log}[r] + b_0) & \text{for } k = 0 \end{cases}$$

$$\Phi[\phi] = (c_n \mathrm{Cos}[n\phi] + d_n \mathrm{Sin}[n\phi]) + (c_0\phi + d_0).$$

Spherical coordinates

In spherical coordinates, the solution of Laplace's equation is a product of $R[r]\,\Theta[\theta]\,\Phi[\phi]$. Laplace's equation is satisfied if

$$R[r] = a_n r^n + b_n r^{-(n+1)}$$
$$\Phi[\phi] = (a_m \mathrm{Sin}[m\phi] + b_m \mathrm{Cos}[m\phi]) + (a_0 \phi + b_0)$$
$$\Theta[\theta] = c_n P[n, m, \mathrm{Cos}(\theta)] + d_n Q[n, m, \mathrm{Cos}(\theta)].$$

5.1.3 *Mathematica* Commands

Packages

You may want to turn off the spell checker before starting this chapter:

```
In[1]:=  Off[General::spell ];
         Off[General::spell1];
```

There are several packages that are used in this chapter:

```
In[2]:=  Needs["Algebra`SymbolicSum`"]
         Needs["Algebra`Trigonometry`"]
         Needs["Calculus`VectorAnalysis`"]
         Needs["Graphics`Graphics3D`"]
         Needs["Graphics`ParametricPlot3D`"]
         Needs["Graphics`PlotField`"]
         Needs["Graphics`PlotField3D`"]
```

Care must be taken in loading these packages to avoid shadowing. It is safest to load the packages at the start.

User-defined procedures

It is useful to create user-defined procedures for calculations that are used repeatedly. We illustrate how to analyze these procedures by examining the steps one at a time. Some procedures will fail in particular cases, and you are encouraged to write conditional statements to handle them.

Operator: **TrigToY**

For convenience, you can represent a function $f[\theta, \phi]$ as an expansion in spherical harmonics. **TrigToY** will expand $f[\theta, \phi]$ to order n in spherical harmonics:

```
In[3]:=  TrigToY::usage="The  user defined operator TrigToY[expression,n] will
         expand an expression of theta and phi to order n in spherical harmonics.";
```

```
In[4]:=  TrigToY[expression_,terms_:2]:=
         Module[{el,m,result},
           result=
             Sum[
                Integrate[ (-1)^(m) SphericalHarmonicY[el,-m,theta,phi]
                          * expression  Sin[theta]
                       ,{theta,0,Pi},{phi,0,2 Pi}]
```

```
      * Y[el,m,theta,phi]
    ,{el,0,terms},{m,-el,el}];
  Return[result]
];
```

Note that the orthogonality relation has been used:

$$\int_0^{2\pi} d\phi \int_0^{\pi} d\theta \; \sin(\theta)\, Y^*[\ell', m', \theta, \phi]\; Y[\ell, m, \theta, \phi] = \delta_{\ell,\ell'}\, \delta_{m,m'}$$

and the conjugate of the spherical harmonic explicitly replaced using

$$Y^*[\ell, m, \theta, \phi] = (-1)^m\, Y[\ell, -m, \theta, \phi].$$

The default for the number of terms is set equal to 2.

Specifically, $f[\theta, \phi] = \sum_{\ell,m} a(\ell, m) Y[\ell, m, \theta, \phi]$, where

$$a(\ell, m) = \int_0^{\pi} d\theta\, \mathrm{Sin}[\theta] \int_0^{2\pi} d\phi\, (-1)^m\, Y[\ell, -m, \theta, \phi]\, f[\theta, \phi].$$

Example

Convert a simple trigonometric function to spherical harmonics:

```
In[5]:=  eq1= Cos[theta]^2  //TrigToY

Out[5]=  2 Sqrt[Pi] Y[0, 0, theta, phi]   4 Sqrt[5 Pi] Y[2, 0, theta, phi]
         ------------------------------ + ---------------------------------
                       3                                 15
```

where **Y** denotes the spherical harmonic. You can verify that this conversion is correct by reproducing the original expression:

```
In[6]:=  eq1 //.Y->SphericalHarmonicY //Simplify

Out[6]=             2
         Cos[theta]
```

Operator: TrigToP

In some cases, it is convenient to represent a function $f[\theta]$ as an expansion in Legendre polynomials. **TrigToP** will expand $f[\theta]$ to order n in a Legendre polynomial series:

```
In[7]:=  TrigToP::usage="The  user defined operator TrigToP[expression,n]
         will expand  an expression of theta to order n in Legendre Functions.";

In[8]:=  TrigToP[expression_,terms_:2]:=
         Module[{el,result},
           result=
             Sum[ Integrate[ LegendreP[el,Cos[theta]] expression  Sin[theta]
                      ,{theta,0,Pi}]
                 * (2 el+1)/2 P[el,Cos[theta]]
               ,{el,0,terms}];
           Return[result]
         ];
```

The default value of the number of terms is 2. **P** denotes the Legendre polynomials.

Specifically, $f[\theta] = \sum_\ell a(\ell)P[\ell,\theta]$, where

$$a(\ell) = \frac{(2\ell+1)}{2} \int_0^\pi d\theta \, \mathrm{Sin}[\theta] \; P[\ell, \mathrm{Cos}[\theta]] \; f[\theta].$$

Example

Convert a simple trigonometric function to Legendre polynomials:

```
In[9]:=  eq1= Sin[theta]^2 //TrigToP

Out[9]=  2 P[0, Cos[theta]]   2 P[2, Cos[theta]]
         ------------------ - ------------------
                 3                    3
```

Converting back to trigonometric functions, you get

```
In[10]:=    eq1 //.P->LegendreP //Simplify

Out[10]=             2
         Sin[theta]
```

Monopole

Complex charge distributions can always be represented by a sum of point charge distributions. We use this superposition principle repeatedly, so it is useful to define a function to compute the potential distribution of a charge in a general manner. The **Monopole** function will do this in any dimension:

```
In[11]:= Monopole::usage="Monopole[q,{x0,y0,z0},{x,y,z}]
         computes the potential due to a monopole of charge q
         located at position {x0,y0,z0} in {x,y,z} coordinates.";

In[12]:= Monopole[q_:1,r0_:{0,0,0},r_:{x,y,z}]:=
            q/Sqrt[ Sum[(r0[[i]]-r[[i]])^2,{i,1,Length[r0]}] ];
```

The specific formula is

$$V = \frac{q}{\sqrt{\sum_{i=0}^{D} (r_0(i) - r(i))^2}},$$

where **D** is the dimension of the vector.

Example

Consider a simple two-dimensional example with specified coordinates:

```
In[13]:= Monopole[charge,{1,2},{x1,x2}]

Out[13]=                charge
         ---------------------------------
                       2             2
         Sqrt[(1 - x1)  + (2 - x2) ]
```

A dipole is easily formed from two monopoles:

```
In[14]:= dipole= ( Monopole[+1,{0,0,-a/2}] +
                   Monopole[-1,{0,0,+a/2}] )
```

```
Out[14]=                1                              1
         ------------------------------ - ------------------------------
                 2    2    -a     2              2    2    a     2
         Sqrt[x  + y  + (-- - z) ]       Sqrt[x  + y  + (- - z) ]
                            2                              2
```

PotentialExpansion

One common operation converts a potential distribution to spherical coordinates and expands the distribution in the limit of large **r**:

```
In[15]:= PotentialExpansion::usage=
         "potentialExpansion[potential,norder] takes an
         electrostatic potential, converts from Cartesian
         coordinates {x,y,z} to spherical coordinates
         {r,theta,phi}, and then performs a series expansion in
         1/r to order norder as r->Infinity.";
```

```
In[16]:= PotentialExpansion[potentialX_,order_:3]:=
             Module[{potentialR,series,x2rRule},
               x2rRule= Thread[{x,y,z}->
                       {r Cos[phi] Sin[theta],
                        r Sin[phi] Sin[theta],
                        r          Cos[theta]}];
               potentialR=potentialX //.x2rRule //Simplify;
               series=(Series[potentialR,{r,Infinity,order}]
                       //Normal
                       //ExpandAll
                       //Simplify
                       //Collect[#,r]&
                    );
               Return[series]
           ];
```

x2rRule converts Cartesian coordinates to spherical coordinates. **potentialR** is the potential in spherical coordinates, and **potentialX** is the potential in Cartesian coordinates. **series** is the series expansion.

Example: Asymptotic potential of two-point charges

Consider the potential expansion of a dipole. The asymptotic expression for the potential is

```
In[17]:= dipole= ( Monopole[+1,{0,0,-a/2}] +
                   Monopole[-1,{0,0,+a/2}] );

         dipole //PotentialExpansion

Out[17]=   a Cos[theta]
         -(------------)
                2
               r
```

VEPlot

You will find it useful to view the equipotential lines and the electric field lines on the same plot:

```
In[18]:= VEPlot::usage=
         "VEPlot[potential,xlim:{x,-1,1},ylim:{y,-1,1}]
         plots the electric field and equipotential surfaces
         xlim and ylim.";
```

```
In[19]:= Clear[VEPlot];
         VEPlot[potential_,xlim_:{x,-1,1.1},ylim_:{y,-1,1.1}]:=
         Module[{plot1,plot2},

           plot1=PlotGradientField[-potential
                   ,xlim,ylim
                   ,Graphics`PlotField`ScaleFunction->(1&)
                   ,DisplayFunction->Identity];

           plot2=ContourPlot[potential
                   ,xlim,ylim
                   ,ContourShading->False
                   ,ContourSmoothing->True
                   ,DisplayFunction->Identity];

           Return[  Show[plot1,plot2
                       ,DisplayFunction->$DisplayFunction]
                 ]
         ];
```

VEPlot simply combines the plot of the equipotential lines obtained from **ContourPlot** with the electric field obtained from **PlotGradientField**. Set **DisplayFunction** to **Identity** to suppress plots and to **$DisplayFunction** to show plots.

Caution: Set the default limits of **PlotGradientField** slightly asymmetric to try to avoid hitting any singularities. If the **VEPlot** function fails, **PlotGradientField**probably was unable to compute the necessary numerical gradients. If this happens, try using **PlotGradientField** and **ContourPlot** separately to isolate the problem.

Example: Equipotential surface and electric field of two-point charges

Consider the equipotential surfaces and electric field of a dipole:

```
In[20]:= dipole= ( Monopole[+1,{0,-1/2}] +
                   Monopole[-1,{0,+1/2}] );

         VEPlot[dipole];
```

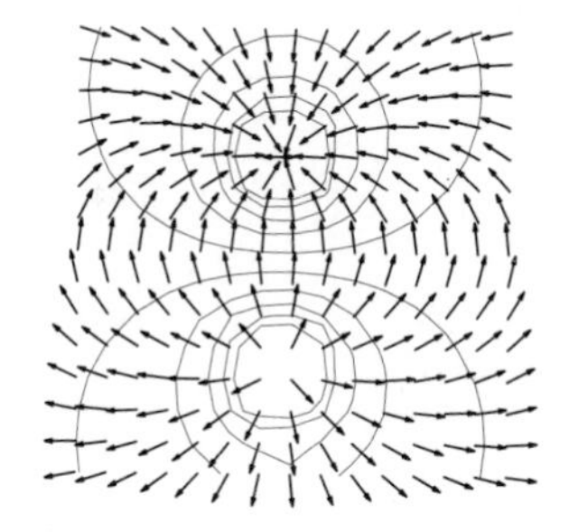

MultipoleSH

Given a continuous charge distribution, you will find it is useful to expand this in spherical harmonics. Note this function shares similar elements with **TrigToY**:

```
In[21]:= MultipoleSH::usage=
"The procedure MultipoleSH[distribution,rlimit,n]
takes a charge  distribution of {r,theta,phi}
and returns a potential expanded
to n terms in spherical harmonics denoted by Y[l,m,theta,phi].
The radial extension of the charge density is given by
rlimit. To convert to trigonometric functions let
Y->SphericalHarmonicY.";
```

```
In[22]:= Clear[MultipoleSH];
MultipoleSH[distribution_,rlimit_,n_]:=
 Module[{q,V},
   q:=Integrate[
          (Integrate[ distribution r^2,{r,0,rlimit}])*
          (-1)^(m) SphericalHarmonicY[el,-m,theta,phi]*
           Sin[theta]
        ,{theta,0,Pi},{phi,0,2 Pi}];
   V=Sum[ 4 Pi /(2 el+1) q/r  (rlimit/r)^(el) *
             Y[el,m,theta,phi]
         ,{el,0,n},{m,-el,el}];
   Return[V]
 ];
```

The specific formula implemented is

$$V = \sum_{\ell=0}^{n} \sum_{m=-\ell}^{+\ell} \frac{4\pi}{(2\ell+1)} \frac{q(\ell,m)}{r} \left(\frac{r_{limit}}{r}\right)^{\ell} Y[\ell,m,\theta,\phi]\, q[r,\theta,\phi],$$

where

$$q(\ell,m) = \int_0^{r_{limit}} r^2 dr \int_0^{\pi} d\theta\, \mathrm{Sin}[\theta] \int_0^{2\pi} (-1)^m\, Y[\ell,-m,\theta,\phi]\, q[r,\theta,\phi].$$

Example: Potential of non-axially symmetric charge density

Consider the charge distribution

```
In[23]:= density = (Q /a^4) (r-a) Sin[phi]^2 Sin[theta]^2;
```

for **r<a** and zero for **r>a**. With the number of terms as 2, the potential expanded in spherical harmonics is (this takes a while)

```
In[24]:= temp1= MultipoleSH[density,a,2] //Simplify //Apart

Out[24]=
        3/2
  -2 Pi     Q Y[0, 0, theta, phi]
  ------------------------------- +
              9 r

      2            3
   2 a  Sqrt[5 Pi ] Q Y[2, 0, theta, phi]
   -------------------------------------- +
                       3
                  225 r
```

```
                3
      2       5 Pi
 2 a  Sqrt[------] Q (Y[2, -2, theta, phi] + Y[2, 2, theta, phi])
               6
 ----------------------------------------------------------------
                                   3
                               75 r
```

This procedure is slow because of the number of integration variables and the number of terms in the sum. Converting the results to trigonometric functions, you get

```
In[25]:= temp2= ((temp1
                   //.Y->SphericalHarmonicY
                   //ComplexToTrig
                   //TrigReduce
                 ) //.Cos[x_]->Sqrt[1-Sin[x]^2]
                   //Simplify[#,Trig->False]&
                 ) //Apart

Out[25]=        2      2    2              2             2
         Pi Q (a  - 5 r )   a  Pi Q Sin[phi]  Sin[theta]
         ---------------- - ----------------------------
                  3                      3
              45 r                   15 r
```

MultipoleP

The number of terms in the **MultipoleSH** increases dramatically with larger values of ℓ-terms, so the calculation time becomes prohibitively large. If the charge distribution is axially symmetric, then one of sum can be dropped, making the calculation time much shorter:

```
In[26]:= MultipoleP::usage=
         "The procedure multipoleP[distribution,rlimit,n]
         takes a charge  distribution that is
         expressed in spherical coordinates {r,theta,0} (symemtric in phi)
         and returns a potential. The result is expanded
         to n terms in Legendre polynomials denoted by P[el,Cos[theta]].
         The radial extension of the charge density is given by
         rlimit. To convert to trigonometric functions let
         P->LegendreP.";

In[27]:= MultipoleP[distribution_,rlimit_,n_]:=
          Module[{q,pot},
            q:=Integrate[ 2 Pi *
                   (Integrate[ distribution r^2,{r,0,rlimit}])*
                   Sqrt[(2 el+1)/(4 Pi)] LegendreP[el,Cos[theta]]
                 * Sin[theta]
                 ,{theta,0,Pi}];
            pot=Sum[ Sqrt[4 Pi/(2 el+1)] q/r (rlimit/r)^(el)
                    * P[el,Cos[theta]]
                 ,{el,0,n}];
            Return[pot]
          ];
```

The specific formula implemented is

$$V = \sum_{\ell=0}^{n} \frac{\sqrt{4\pi}}{\sqrt{2\ell+1}} \frac{q(\ell,m)}{r} \left(\frac{r_{limit}}{r}\right)^{\ell} P[\ell,\theta],$$

where

$$q(\ell) = 2\pi \int_0^{r_{limit}} r^2 dr \int_0^{\pi} d\theta \mathrm{Sin}[\theta]\, Y[\ell, 0, \theta, 0]\, q[r, \theta].$$

Example: Potential of an axially symmetric charge distribution

Consider the charge distribution given by

```
In[28]:= density = (Q /a^4) (r-a Cos[theta]^2);
```

for **r<a** and zero for **r>a**. The potential expanded in Legendre polynomials is

```
In[29]:= test1= MultipoleP[density,a,2]

Out[29]=
                                       2
5 Pi Q P[0, Cos[theta]]       8 a  Pi Q P[2, Cos[theta]]
-----------------------  -  ---------------------------
         9 r                               3
                                       45 r
```

The same calculation using **MultipoleSH** takes longer:

```
In[30]:= test2= MultipoleSH[density,a,2]

Out[30]=
        3/2
10 Pi      Q Y[0, 0, theta, phi]
-------------------------------- -
              9 r

       2             3
  16 a  Sqrt[5 Pi ] Q Y[2, 0, theta, phi]
  ---------------------------------------
                       3
                  225 r
```

They both agree with each other, as can be verified by reducing the expressions to trigonometric functions:

```
In[31]:= test1==test2 //.{P->LegendreP,Y->SphericalHarmonicY}

Out[31]= True
```

5.1.4 Protect User-defined Procedures

```
In[32]:= Protect[TrigToY,TrigToP,Monopole,PotentialExpansion
              ,VEPlot,MultipoleSH,MultipoleP];
```

5.2 Problems

5.2.1 Point Charges, Multipoles, and Image Problems

Problem 1: Superposition of point charges

Consider the following charge distributions: (1) a linear quadrupole, (2) a square quadrupole, and (3) three charges $\{-q, -q, 2q\}$ located on the corners of a triangle.

Part (a) Compute the potential fields for these charge distributions.

Part (b) Graph the equipotential lines and electric field lines in two dimensions.

Part (c) Find the potential to the lowest order in the expansion of $1/r$.

Part (d) Make a three-dimensional plot of the electric field for the linear quadrupole distribution.

Remarks and outline This problem introduces you to *Mathematica*'s various graphics tools for visualizing electric and magnetic fields.

Suggestions Experiment with different charge distributions. Try using different functions for the **ScaleFunction** in the **PlotField** command. Compare the potential field between distributions. Which one falls off faster with increasing radius?

Required packages

```
In[1]:= Needs["Calculus`VectorAnalysis`"]
        Needs["Graphics`PlotField`"]
        Needs["Graphics`PlotField3D`"]
```

Solution

```
In[2]:= Clear["Global`*"];
```

Part (a) The potential at **r** from a single charge located at **r0** is given by **Monopole**. Consider the potential for two-point charges. Assume that one particle has charge **q** and is located at **z=d/2** and the other particle has charge **-q** is located at **z=-d/2**. The potential is the superposition of the individual potentials:

```
In[3]:= quad=( Monopole[ 2q,{0,0,   0} ]+
               Monopole[ -q,{0,0, d/2} ]+
               Monopole[ -q,{0,0,-d/2} ]);
```

```
In[4]:= triangle=( Monopole[+2q, {   0,0,   0} ]+
                   Monopole[ -q, {+d/2,0,-d/2} ]+
                   Monopole[ -q, {-d/2,0,-d/2} ]);
```

```
In[5]:= square=( Monopole[+q,{+d/2,0, d/2} ]+
                 Monopole[-q,{+d/2,0,-d/2} ]+
                 Monopole[+q,{-d/2,0,-d/2} ]+
                 Monopole[-q,{-d/2,0, d/2} ]);
```

Part (b) Plot the electric field and potential of these distribution using **VEPlot**. First, for the linear quadrupole, plot the $\{x, z\}$ plane:

```
In[6]:= VEPlot[quad //.{q->1, d->1, y->0},{x,-1,1.1},{z,-1,1.1}];
```

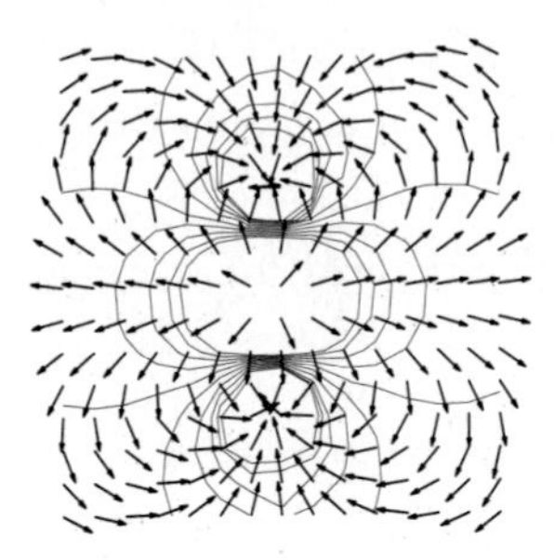

Another way to reduce the problem to the $\{x, z\}$ plane is to first set **y** to zero and then swap **z** for **y**. (*Caution*: Note the use of **/.** instead of **//.**) For the **triangle**, you get

```
In[7]:=  VEPlot[triangle /.{q->1, d->1, y->0, z->y}];
```

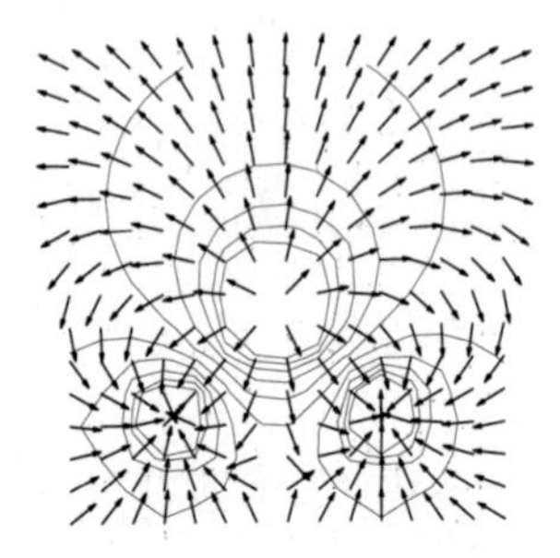

Finally, the square quadrupole:

```
In[8]:=  VEPlot[square /.{q->1, d->1, y->0, z->y}];
```

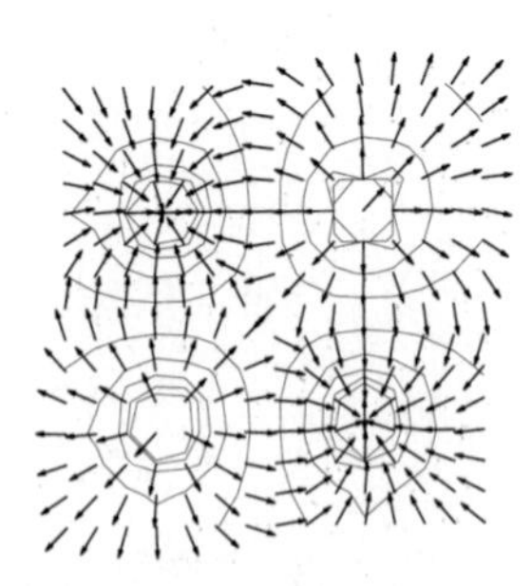

Part (c) Next, expand the potentials in the limit of large **r** using the user-defined function **PotentialExpansion**. For the linear quadrupole (**quad**), you get

```
In[9]:=  seriesQ= quad //PotentialExpansion
```

```
Out[9]=    2
        -(d  q (1 + 3 Cos[2 theta]))
        ----------------------------
                      3
                   8 r
```

Alternatively, you can rewrite this as a Legendre polynomial using the user-defined function **TrigToP**:

```
In[10]:= seriesQ //TrigToP //Simplify

Out[10]=    2
         -(d  q P[2, Cos[theta]])
         ------------------------
                      3
                   2 r
```

Likewise for the **square** configuration:

```
In[11]:= seriesS= square //PotentialExpansion

Out[11]=    2
         3 d  q Cos[phi] Sin[2 theta]
         ----------------------------
                       3
                    2 r

In[12]:= seriesS //TrigToY //Simplify

Out[12]=    2      Pi
         6 d  Sqrt[--] q (Y[2, -1, theta, phi] - Y[2, 1, theta, phi])
                   30
         ------------------------------------------------------------
                                      3
                                     r
```

And finally for the triangle configuration:

```
In[13]:= seriesT= triangle //PotentialExpansion //Simplify

Out[13]= d q Cos[theta]      2
         -------------- + (d  q (-2 - 6 Cos[2 phi] + 3 Cos[2 (phi - theta)] -
                2
               r

                                                            3
             6 Cos[2 theta] + 3 Cos[2 (phi + theta)])) / (32 r )
```

Note the triangle begins at order **1/r^2** because, unlike the two quadrupoles, it contains a dipole component.

Part (d) As a final example, show the three-dimensional vector plot of the linear quadrupole electric field:

```
In[14]:= p1=PlotGradientField3D[
            quad//.{q->1, d->1} //Evaluate
            ,{x,-1,1.1},{y,-1,1.1},{z,-1,1.1}
            ,Graphics`PlotField3D`ScaleFunction -> (1 & )
            ,VectorHeads->True
            ,PlotPoints->6
         ];
```

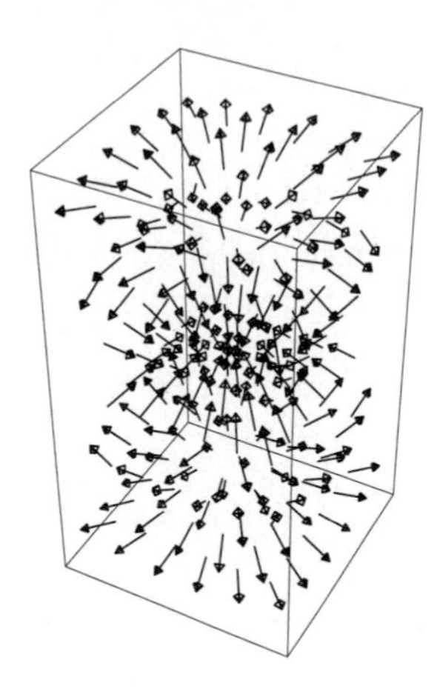

Here it is again from another viewpoint:

```
In[15]:= Show[p1,ViewPoint->{0,-4,0}];
```

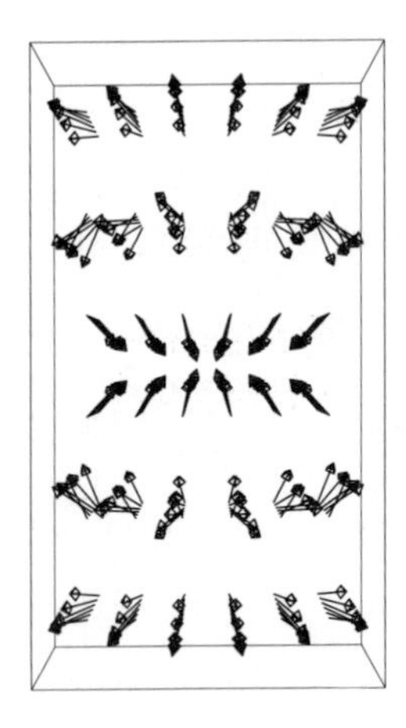

Problem 2: Point charges and grounded plane

Consider a point charge q at a distance d from an infinite conducting plane that is grounded.

Part (a) Find the potential. Expand the potential in powers of **1/r** and find the leading-order term. Find the electric field to this same order.

Part (b) Find the induced charge density on the plane and verify that the total induced charge is $-q$.

Part (c) Plot the induced charge using **ContourPlot**.

Remarks and outline This is a basic boundary value problem. It will serve as a warm-up for the following problems that have more-complex boundary conditions. *Mathematica*'s ability to handle three-vectors as well as to compute gradients in different coordinate systems will enable you to solve this problem in a trivial manner. Using the symbolic integration, you can integrate the answer for the charge density to verify the total charge.

Plot the charge density on the plane using **ContourPlot**. Viewing this plot, you can see that you obtained an answer that makes physical sense. (This is always a powerful check of any calculation.)

To solve the problem, use the method of images to simplify the problem. Remove the conductor and try to find the position and magnitude of an image charge that will make the potential zero on the plane of the conductor. Note, the potential is just the superposition of the potentials from the charge and image charge. The induced charge is proportional to the normal derivative of the potential on the surface of the conductor.

Suggestions Plot the potential field arising from the point charge and its image charge. Do the boundary conditions appear to be satisfied in the region of the conducting plane?

Try the same problem for multiple charges. **ContourPlot** the charge density to see if the answer is plausible.

Required packages

```
In[16]:= Needs["Calculus`VectorAnalysis`"]
         Needs["Graphics`PlotField`"]
```

Solution

```
In[17]:= Clear["Global`*"];
```

Part (a) The potential on the grounded conductor is zero. Using the method of images, remove the plane conductor and find the position and magnitude of an image charge that makes the potential zero at the conductor's position. On the basis of the problem's symmetry, you can observe that the image charge has magnitude **-q** and is located at a distance **d** behind the plane.

The total potential for the charge and image charge is the superposition of their individual potentials:

```
In[18]:= dipole= ( Monopole[+q,{0,0,+d}] +
                   Monopole[-q,{0,0,-d}] )

Out[18]=
                   q                                  q
         -(------------------------------) + -----------------------------
                    2    2          2                2    2         2
          Sqrt[x  + y  + (-d - z) ]        Sqrt[x  + y  + (d - z) ]
```

The plane will have zero potential, since every point on the plane is equidistant from **q** and **-q**. To verify this conclusion, evaluate the potential at **z=0**:

```
In[19]:=  dipole/.z->0

Out[19]= 0
```

The asymptotic expansion of the potential follows from **PotentialExpansion**:

```
In[20]:= dipoleR= dipole //PotentialExpansion
Out[20]= 2 d q Cos[theta]
         ----------------
                 2
                r
```

TrigToP is used to express the answer in Legendre polynomials:

```
In[21]:= dipoleR //TrigToP
Out[21]= 2 d q P[1, Cos[theta]]
         ----------------------
                   2
                  r
```

The electric field follows from applying **Grad** to **dipoleR**:

```
In[22]:= efieldR= -Grad[dipoleR,Spherical[r,theta,phi]]
Out[22]=  4 d q Cos[theta]  2 d q Sin[theta]
         {----------------, ----------------, 0}
                 3                 3
                r                 r
```

Part (b) The induced charge density on the conductor is $-1/(4\pi)$ times the normal derivative of the potential evaluated on the plane:

```
In[23]:= charge= -1/(4 Pi)  D[dipole,z]      //.z->0
Out[23]=            -(d q)
         ---------------------
                2    2    2 3/2
         2 Pi (d  + x  + y )
```

It follows from Gauss's theorem that the total induced charge on the conductor must be **-q**. Integrating the charge density over the plane, you can verify this result:

```
In[24]:= Integrate[charge,{y,-Infinity,+Infinity}
                      ,{x,-Infinity,+Infinity}]//PowerExpand
Out[24]= -q
```

Part (c) A **ContourPlot** of the induced charge on the conductor is as follows:

```
In[25]:= ContourPlot[(charge/.{d->1,q->1}),{y,-1,1},{x,-1,1}];
```

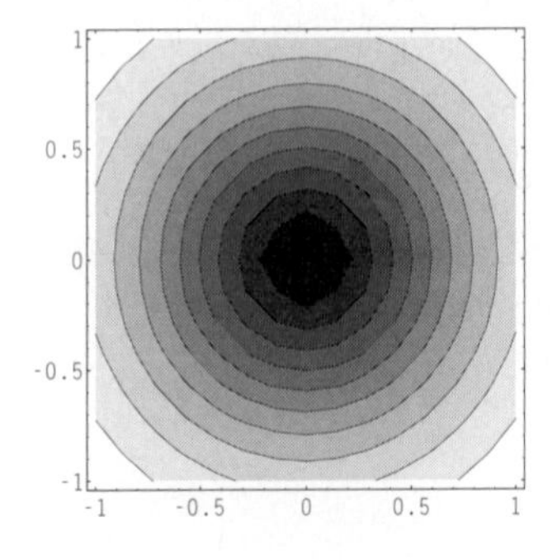

Problem 3: Point charges and grounded sphere

Consider a point charge q located a distance d from the center of a grounded conducting sphere of radius a.

Part (a) Find the potential outside the sphere. Expand the potential in a $1/r$ series and find the electric field to this same order.

Part (b) Derive an expression for the induced charge density. Integrate the induced charge density over the surface of the sphere and show that the total induced charge is equal to the image charge.

Part (c) Plot the induced charge on a sphere's surface.

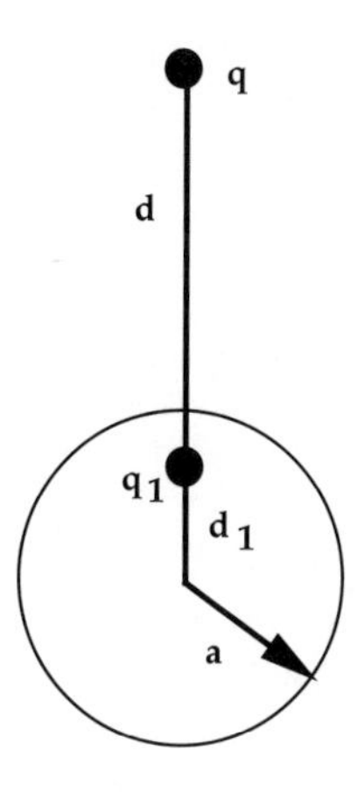

Remarks and outline This problem introduces the next level of complication for boundary value problems. However, the solution is again simple. Plot the electric field to verify that the boundary conditions are satisfied and that the solution is plausible. Using the symbolic integration, integrate your answer for the charge density to verify the total charge.

Also use **ParametricPlot3D** to show the charge density induced on the sphere. From this, you can see that you obtained an answer that makes physical sense. (This is always a powerful check of any calculation.)

To solve this problem, remove the conductor and try to find the position and magnitude of an image charge that will make the potential zero on the spherical surface. By symmetry, the image charge must lie on the line connecting q and the center of the sphere. To get two equations for the unknowns, consider one point to be on the equator of the sphere nearest the charge and the second point on the equator farthest from the charge. The potential must be zero at these two points. Note, the induced charge is proportional to the normal derivative of the potential on the surface of the conductor.

Required packages

```
In[26]:= Needs["Calculus`VectorAnalysis`"]
         Needs["Graphics`PlotField`"]
```

Solution

```
In[27]:= Clear["Global`*"];
```

Part (a) Solve the problem using the method of images. Remove the conductor and find the position **d1** and charge **q1** of an image that will make the potential zero on a sphere of radius **a**. You will need two equations to solve for {**q1**,**d1**}. The two equations follow from picking two points on the sphere: the point nearest the particle and the point farthest from the particle. Requiring the potential to be zero at these two points, you get

```
In[28]:= eq1= {      q/(d+a)    +    q1/(d1+a)==0,
                     q/(d-a)    +    q1/(a-d1)==0}

Out[28]=   q        q1               q          q1
         {───── + ────── == 0, ─────── + ────── == 0}
          a + d   a + d1        -a + d   a - d1
```

Solving these two equations for **q1** and **d1**, it follows that

```
In[29]:= imageRule= Solve[eq1,{d1,q1}] //ExpandAll //Flatten

Out[29]=          2
                 a              a q
         {d1 -> ──, q1 -> -(───)}
                 d               d
```

The image charge has a value of **-a/d q** and is located at a distance **a^2/d** from the origin. The potential is the superposition of the potentials from the charge and image charge. The potential for a single charge is given by **Monopole**:

```
In[30]:= potential=( Monopole[ q,{0,0,d }]
                   + Monopole[q1,{0,0,d1}]
                       //.imageRule
                       //ExpandAll
                       //PowerExpand
                   )

Out[30]=                      q
         ─────────────────────────────────────── -
                2    2    2              2
         Sqrt[d  + x  + y  - 2 d z + z ]

                                 a q
           ──────────────────────────────────────────────
                  4    2  2    2  2      2         2  2
           Sqrt[a  + d  x  + d  y  - 2 a  d z + d  z ]
```

To change to spherical coordinates, define the rule:

```
In[31]:= x2rRule=Thread[{x,y,z}->
                   CoordinatesToCartesian[{r,theta,phi},Spherical]]

Out[31]= {x -> r Cos[phi] Sin[theta], y -> r Sin[phi] Sin[theta],

           z -> r Cos[theta]}
```

The potential in spherical coordinates is

```
In[32]:= potentialR=potential //.x2rRule //Simplify //PowerExpand
```

```
Out[32]=                    q
         ------------------------------------ -
                2    2
         Sqrt[d  + r  - 2 d r Cos[theta]]

                            a q
           ---------------------------------------
                4    2 2      2
           Sqrt[a  + d  r  - 2 a  d r Cos[theta]]
```

Note, the potential on the sphere (**r=a**) is zero, as it should be:

```
In[33]:=  potentialR //.r->a //Simplify //PowerExpand

Out[33]= 0
```

The expansion in powers of **1/r** follow from the user-defined function **PotentialExpansion** found in Section 5.3, "*Mathematica* Commands":

```
In[34]:= ( potentialR //PotentialExpansion
                //PowerExpand
                //Simplify
         )

Out[34]=     a q
         q - ---
              d      (-a^3 + d^3) q Cos[theta]    (-a^5 + d^5) q (1 + 3 Cos[2 theta])
         -------- + -------------------------- + ------------------------------------
            r                 2  2                               3  3
                             d  r                             4 d  r
```

Part (**b**) The plot of the electric field and equipotential lines in the **y=0** plane follow from **VEPlot**:

```
In[35]:= VEPlot[potential /.{a->.2, d->.4, q->1, y->0, z->y}];
```

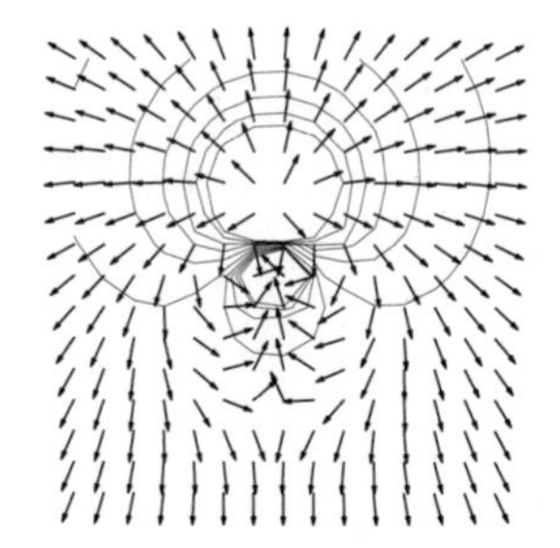

Part (**c**) The induced charge density is $-1/(4\pi)$ times the normal derivative of the potential evaluated at **r=a**. Taking the radial derivative (normal to the surface of the sphere) of the potential, you get

```
In[36]:= chargedensity=(  -1/(4 Pi) D[potentialR,r]
                               /.{r->a}
                               //Simplify
                               //PowerExpand
                               //Together
                               //Simplify
                         )
```

```
Out[36]=             2    2
                 (a  - d ) q
        ----------------------------------------------
                2    2                        3/2
        4 a Pi (a  + d  - 2 a d Cos[theta])
```

To get the total induced charge, integrate the charge density over the surface of the sphere. (*Caution:* Using **PowerExpand** will give incorrect results because of the sign of the square roots.)

```
In[37]:= totalcharge =
         ( ( 2 Pi a^2 Integrate[chargedensity Sin[theta],{theta,0,Pi}]
               //Simplify
           )   /.{1/Sqrt[(a-d)^2]->1/(d-a), 1/Sqrt[(a+d)^2]->1/(d+a)
                ,  Sqrt[(a-d)^2]->  (d-a),    Sqrt[(a+d)^2]->  (d+a)}
               //Simplify
         )

Out[37]=   a q
         -(---)
            d
```

Part (d) To plot the induced charge on a sphere, normalize the charge density so that it ranges from $\{0, 1\}$ as theta ranges over $\{0, 2\pi\}$. Call this function **gray** and define it by

```
In[38]:= thetaMin=+(
         FindMinimum[+chargedensity/.{a->1,q->1,d->2},{theta,0,1}]//First);

         thetaMax=-(
         FindMinimum[-chargedensity/.{a->1,q->1,d->2},{theta,1.0}]//First);

         gray=((chargedensity-thetaMin)/(thetaMax-thetaMin)/.{a->1,q->1,d->2});

         {thetaMin,thetaMax}

Out[38]= {-0.238732, -0.00884194}
```

Note how **FindMinimum** is used to determine the limits of the function so that you can map it onto the interval $\{0, 1\}$. Indicating the induced charge by the **gray** scale, the graph of the induced charge plotted on a spherical surface is

```
In[39]:= radius=1;
         ParametricPlot3D[{Abs[radius] Sin[theta] Cos[phi],
                           Abs[radius] Sin[theta] Sin[phi],
                           Abs[radius] Cos[theta],
                           GrayLevel[gray ]}
                   ,{theta,0,Pi }
                   ,{phi,0,2 Pi }
                   ,Lighting->False
                   ,Axes->False
                   ,Boxed->False];
```

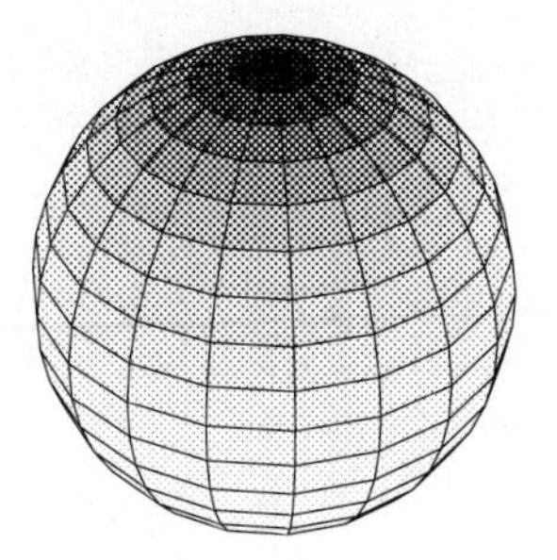

Problem 4: Line charge and grounded plane

The potential from a single charged line parallel to the z-axis and located at $x = x_0$ and $y = y_0$ is

$$V[x, y] = -q \operatorname{Log}\left[(x - x_0)^2 + (y - y_0)^2\right],$$

where q is the charge per unit length. Consider a line with charge q per unit length located along the z-axis, in front of an infinite plane conductor that is located in the $\{x, z\}$ plane and kept at zero potential.

Part (a) Find the potential and plot the equipotential surfaces in the plane of symmetry. Show that the Laplacian of the potential is zero.

Part (b) Find the induced charge. Plot the induced charge density on the plane.

Part (c) Show that integrating over the charge density gives a charge per unit length that equals the image charge per unit length.

Part (d) Find the electric field and use **`PlotVectorField`** to plot the electric field lines:

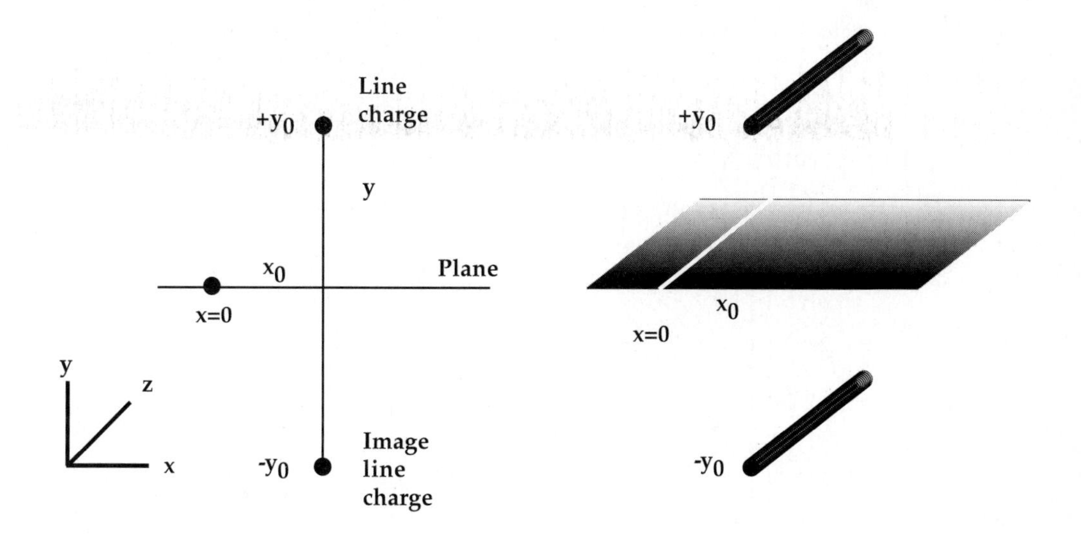

Remarks and outline Use the method of images and reduce the problem to a superposition of two line charges. By symmetry, the image line will be located behind the plane and will have a charge per unit length that is the negative of the line charge. The equipotential surfaces follow from the user-defined function **VEPlot**. The plot of the induced charge follows from **ContourPlot**. The induced charge is proportional to the normal derivative of the potential evaluated on the conductor's surface. The electric field follows from the gradient of the potential.

Required packages

```
In[40]:= Needs["Calculus`VectorAnalysis`"]
         Needs["Graphics`PlotField`"]
```

Solution

```
In[41]:= Clear["Global`*"];
```

Part (a) The solution for the line charge and plane equals the superposition of a line charge at $\{x = 0, y = y0\}$ and a negative image line charge at $\{x = 0, y = -y0\}$. The sum of the two line potentials is

```
In[42]:= potential=(-(+q) Log[x^2+(y-(+y0))^2]
                   -(-q) Log[x^2+(y-(-y0))^2]);
         potential= potential //.{a_. Log[b_] -a_. Log[c_]:> a Log[b/c]}

Out[42]=        2          2
               x  + (y + y0)
         q Log[--------------]
                2          2
               x  + (y - y0)
```

The potential is zero on the plane (**y=0**), as it should be:

```
In[43]:= potential //.{y->0}

Out[43]= 0
```

Use **VEPlot** to graph the equipotential surfaces and the electric field lines in the plane of symmetry:

```
In[44]:= VEPlot[potential //.{q->1, y0->1},{x,-1,1},{y,0,2.1}];
```

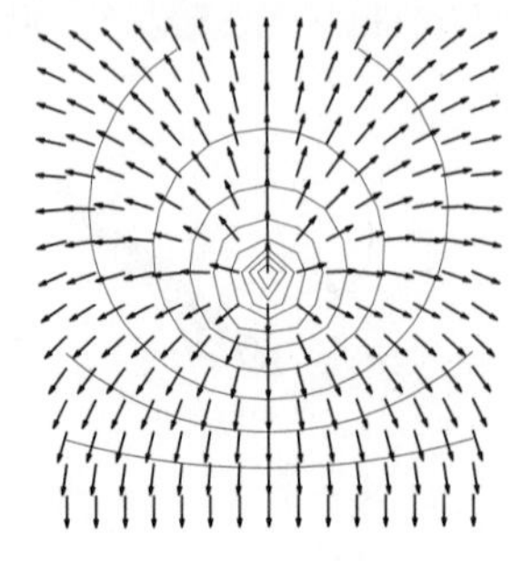

Show that the **Laplacian** of the potential is zero:

```
In[45]:= Laplacian[ potential,Cartesian[x,y,z]]//Together

Out[45]= 0
```

Part (b) The induced charge is proportional to the normal derivative at **z=0**:

```
In[46]:= charge= -1/(4 Pi) D[potential,y] /.{y->0}

Out[46]=         q y0
         -(-------------)
                2     2
           Pi (x  + y0 )
```

Plotting the induced charge on the plane, you get

```
In[47]:= ContourPlot[ charge/.{q->1, y0->1} //Evaluate
                     ,{x,-2,2},{y,-2,2}
                     ,ContourSmoothing -> Automatic ];
```

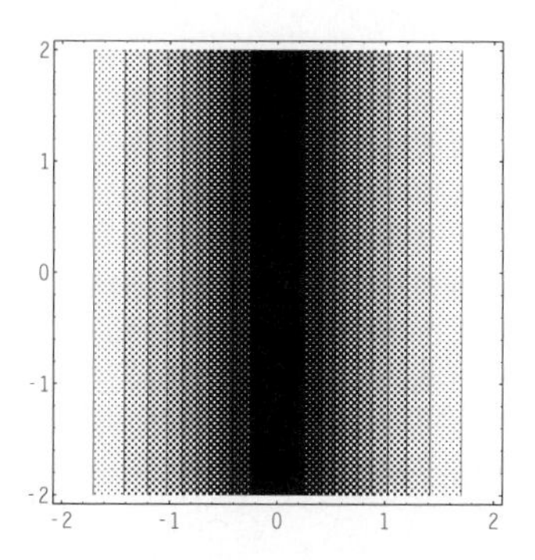

Part (c) Integrate charge along the x-direction to get the induced charge per unit length:

```
In[48]:= Integrate[charge,{x,-Infinity,Infinity}] //PowerExpand

Out[48]= -q
```

Part (d) Take the gradient of the potential to find the electric field, $\vec{E} = -\nabla V$:

```
In[49]:= eField= -Grad[potential,Cartesian[x,y,z]] //Simplify

Out[49]=                       8 q x y y0
         {---------------------------------------------------,
            4      2  2    4      2   2      2   2     4
           x  + 2 x  y  + y  + 2 x  y0  - 2 y  y0  + y0

                              2     2     2
                  4 q y0 (-x  + y  - y0 )
           ---------------------------------------------------, 0}
            4      2  2    4      2   2      2   2     4
           x  + 2 x  y  + y  + 2 x  y0  - 2 y  y0  + y0
```

The electric field lines in the **z=0** plane are

```
In[50]:= PlotVectorField[ eField[[{1,2}]] //.{q->1, y0->1}
              ,{x,-5,5}
              ,{y,0,3}
              ,PlotPoints->16
              ,Graphics`PlotField`ScaleFunction -> (1&)];
```

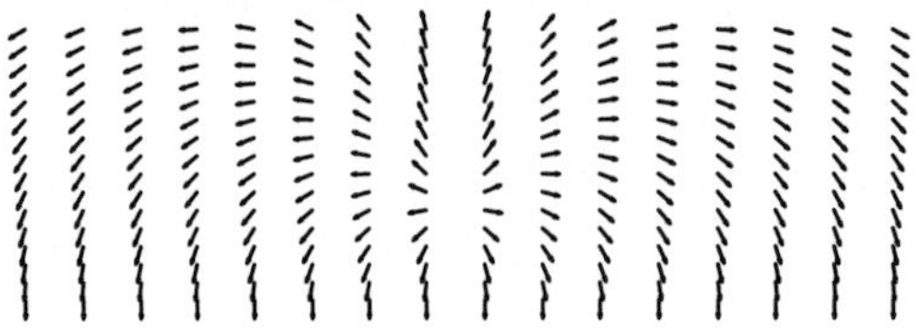

Problem 5: Multipole expansion of a charge distribution

Consider the charge densities

$$\rho_1 = q_0(a-r)/a^4(\mathrm{Sin}[\theta] - \pi/4)$$
$$\rho_2 = q_0(a-r)/a^4(e^{(a-r)/a})\mathrm{Cos}[\theta]$$
$$\rho_3 = q_0(a-r)^2/a^5\mathrm{Sin}[\theta]^2\mathrm{Sin}[2\phi]$$

for $r < a$ and $\rho_i = 0$ for $r \geq a$.

Part (a) Use **ContourPlot** to make a two-dimensional plot of the charge densities.

Part (b) Find the multipole expansion for the potentials.

Part (c) Find the electric fields for the multipole expansions of the potentials.

Part (d) Plot potential contours and electric field lines of the multipole expansions using **VEPlot**.

Remarks and outline

This problem introduces the multipole expansion. Again, because it is simple for *Mathematica* to expand complex charge distributions into multipoles, you can experiment with more-complex and realistic charge distributions than you could analytically. Further, the ability to plot the vector field associated with the charge distribution will enable you to explicitly see why the leading terms in the multipole expansion yield the dominant characteristics of the charge distribution.

Experiment with different charge distributions. Take a complex charge distribution of net charge +1 and compare it with a point charge of +1 as you increase the scale of the plot. Are the potential fields comparable in the limit $r \to \infty$? Repeat the exercise using a single dipole and then a collection of dipoles with random orientation. (*Hint:* Try using different functions for the **ScaleFunction** in the **PlotField** command, including the default function.)

Compare the potential field between a monopole, a dipole, and a quadrupole. Which one falls off faster with increasing radius? Is this consistent with the expansion in powers of r?

Required packages

```
In[51]:= Needs["Algebra`Trigonometry`"]
         Needs["Calculus`VectorAnalysis`"]
         Needs["Graphics`PlotField`"]
```

Solution

```
In[52]:= Clear["Global`*"];
```

Part (**a**) The three charge distributions are

```
In[53]:= chargeDensity1= q0 (a-r)/a^4( Sin[theta]- Pi/4);
         chargeDensity2= q0 (a-r)/a^4( Exp[(a-r)/a ] ) Cos[theta];
         chargeDensity3= q0 (a-r)^2/a^5  Sin[theta]^2 Sin[2 phi];
```

To change from spherical coordinates to Cartesian coordinates (so you can use **ContourPlot**), define

```
In[54]:= r2xRule=Thread[{r,theta,phi}->
                  CoordinatesFromCartesian[{x,y,z},Spherical]]

Out[54]=
                  2    2    2                              z
{r -> Sqrt[x  + y  + z ], theta -> ArcCos[-----------------],
                                                  2    2    2
                                           Sqrt[x  + y  + z ]

  phi -> ArcTan[x, y]}
```

Choose the values

```
In[55]:= values={q0->1, a->1};
```

and plot **chargeDensity1** in the **y->0** plane:

```
In[56]:= ContourPlot[chargeDensity1  /.r2xRule //.values /.{y->0}
           ,{x,-1,1},{z,-1,1}
           ,PlotPoints->20];
```

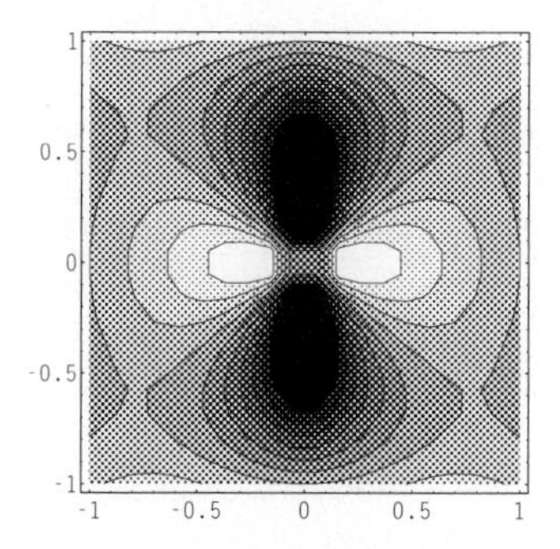

Likewise for **chargeDensity2**:

```
In[57]:= ContourPlot[chargeDensity2  /.r2xRule //.values /.{y->0}
           ,{x,-1,1},{z,-1,1}
           ,PlotPoints->20];
```

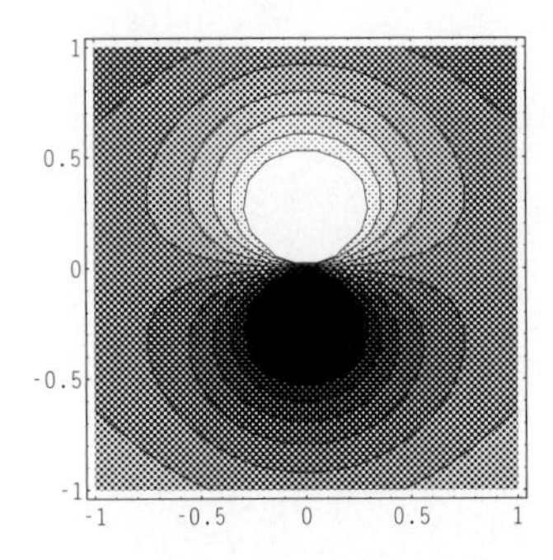

For **chargeDensity3**, choose the **z->0** plane:

```
In[58]:= ContourPlot[chargeDensity3  /.r2xRule //.values /.{z->0}
           ,{x,-1,1},{y,-1,1}
           ,PlotPoints->20];
```

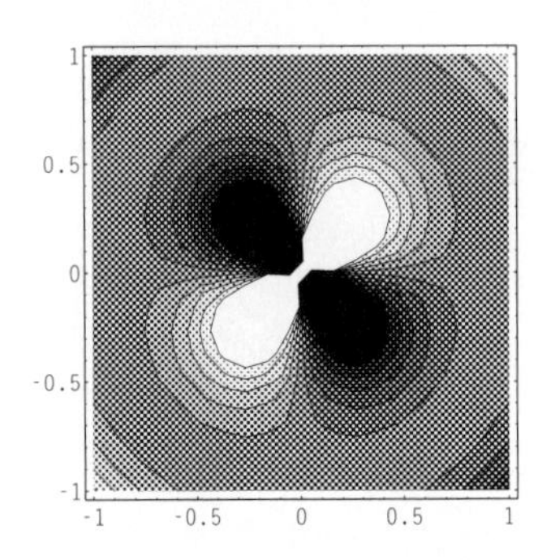

Part (b) Applying **MultipoleP** to **chargeDensity1**, you get the leading-order term of the multipole expansion:

```
In[59]:= multi1= MultipoleP[chargeDensity1,a,2]

Out[59]=     2   2
         -(a  Pi  q0 P[2, Cos[theta]])
         -----------------------------
                         3
                     96 r
```

As a crosscheck, also use **MultipoleSH**:

```
In[60]:= multi1check=MultipoleSH[chargeDensity1,a,2]

Out[60]=     2           5
         -(a  Sqrt[5 Pi ] q0 Y[2, 0, theta, phi])
         ----------------------------------------
                              3
                         240 r
```

and verify that these yield the same results:

```
In[61]:= multi1==multi1check //.{P->LegendreP, Y->SphericalHarmonicY}
```

```
Out[61]= True
```

Applying **MultipoleP** to **chargeDensity2**, you get

```
In[62]:= multi2= MultipoleP[chargeDensity2,a,2]

Out[62]= -4 a (-11 + 4 E) Pi q0 P[1, Cos[theta]]
         ---------------------------------------
                            2
                         3 r
```

For **chargeDensity3**, you must use **MultipoleSH**, since this charge density is not symmetric in ϕ:

```
In[63]:= multi3= MultipoleSH[chargeDensity3,a,2] //Simplify

Out[63]=
                         3
          -8 I  2      5 Pi
          ---- a  Sqrt[----] q0 (-Y[2, -2, theta, phi] + Y[2, 2, theta, phi])
          375           6
          ---------------------------------------------------------------------
                                           3
                                          r
```

Part (c) The electric fields follow from $\vec{E} = -\nabla V$. In spherical coordinates,

```
In[64]:= eField1= -Grad[multi1,Spherical]

Out[64]=    2   2
          -(a  Pi  q0 P[2, Cos[theta]])
         {-----------------------------,
                        4
                    32 r

             2   2                  (0,1)
           -(a  Pi  q0 Sin[theta] P      [2, Cos[theta]])
           ------------------------------------------------, 0}
                                4
                            96 r
```

Or, evaluating the Legendre polynomials explicitly,

```
In[65]:= eField1 //.P->LegendreP //Simplify

Out[65]=    2   2                                   2   2
          -(a  Pi  q0 (1 + 3 Cos[2 theta]))  -(a  Pi  q0 Sin[2 theta])
         {---------------------------------, --------------------------, 0}
                          4                               4
                     128 r                            64 r
```

As a crosscheck, verify that **Grad** properly handled the Legendre polynomials. To do this, first make the subsitution **P->LegendreP** and then take the **Grad**:

```
In[66]:= eField1check= -Grad[multi1//.P->LegendreP,Spherical];
         eField1check==eField1 //.P->LegendreP //Simplify

Out[66]= True
```

The result is the same both ways.

Similarly, you compute the electric field for the second distribution as follows:

```
In[67]:= eField2= -Grad[multi2,Spherical] //.P->LegendreP //Simplify

Out[67]=  8 a (11 - 4 E) Pi q0 Cos[theta]  4 a (11 - 4 E) Pi q0 Sin[theta]
         {-------------------------------, -------------------------------, 0}
                        3                                 3
                     3 r                               3 r
```

And for the third distribution:

```
In[68]:= eField3= -Grad[multi3 //.Y->SphericalHarmonicY ,Spherical];
         eField3 //ComplexToTrig //Simplify

Out[68]=     2                              2
          2 a  Pi q0 Sin[2 phi] Sin[theta]
         {---------------------------------,
                          4
                      25 r

              2
          -2 a  Pi q0 Sin[2 phi] Sin[2 theta]
          -----------------------------------,
                            4
                        75 r

              2
          -4 a  Pi q0 Cos[2 phi] Sin[theta]
          ---------------------------------}
                           4
                       75 r
```

(*Caution:* because of how *Mathematica* handles the derivatives of **SphericalHarmonicY**, you first must make the substitution **Y->SphericalHarmonicY**. This may be a version-dependent feature.)

Part (d) Next, use **VEPlot** to display the potential contours and electric field lines of the multipole expansions. For the first potential, convert to Cartesian coordinates, make **LegendreP** and **SphericalHarmonicY** explicit, and substitute the **values**. Finally, set **y->0** to project to the plane and replace **z** by **y**, since **VEPlot** uses the variables **{x,y}**:

```
In[69]:= pot1= ( multi1  /.r2xRule
                         //.{P->LegendreP, Y->SphericalHarmonicY}
                         //.values
                         /.{y->0,z->y}
                       )

Out[69]=      2    2      2
          -(Pi  (-x  + 2 y ))
          -------------------
                 2    2 5/2
          192 (x  + y )
```

The plot of the potential contours and electric field lines is

```
In[70]:= VEPlot[pot1];
```

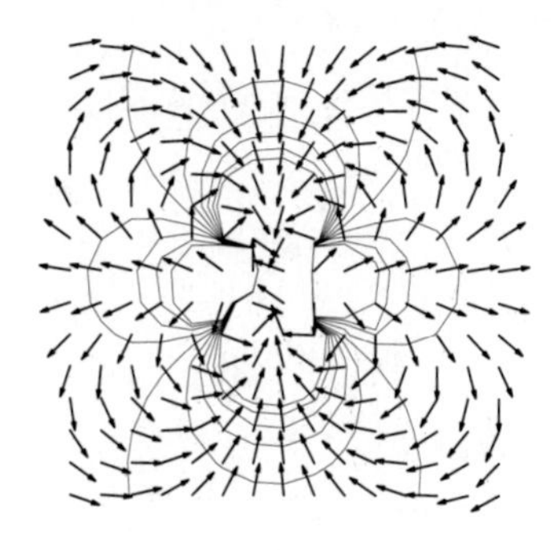

Likewise for the second distribution.

```
In[71]:= pot2= ( multi2  /.r2xRule
                         //.{P->LegendreP, Y->SphericalHarmonicY}
                         //.values
                         /.{y->0,z->y}
                       )

Out[71]= -4 (-11 + 4 E) Pi y
         -------------------
                2    2 3/2
            3 (x  + y )

In[72]:= VEPlot[pot2];
```

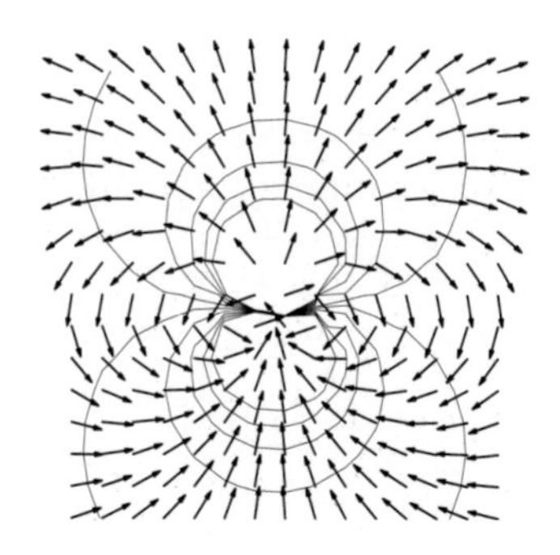

For the third distribution, we plot in the **{x,y}** plane.

```
In[73]:= pot3= ( multi3  /.r2xRule
                         //.{P->LegendreP, Y->SphericalHarmonicY}
                         //.values
                         /.{z->0}
                       ) //ComplexToTrig //ExpandAll //Simplify

Out[73]= 2 Pi Sin[2 ArcTan[x, y]]
         ------------------------
                 2    2 3/2
             75 (x  + y )

In[74]:= VEPlot[pot3];
```

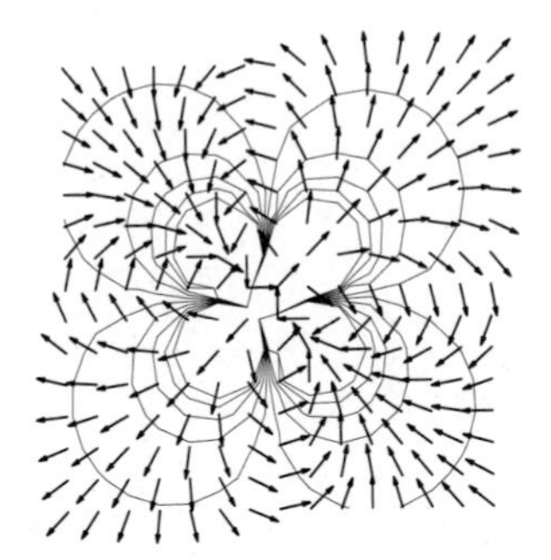

5.2.2 Cartesian and Cylindrical Coordinates

Problem 1 : Separation of variables in Cartesian and cylindrical coordinates

Consider Laplace's equation for the electric potential.

Part (a) For Cartesian coordinates, assume the potential is of the form **vx[x] vy[y] vz[z]** and apply the **Laplacian** to this product. Solve the ordinary differential equation for each function.

Part (b) For cylindrical coordinates, assume the potential is of the form **R[r] P[phi] Z[z]** and apply the **Laplacian** to this product. Solve the ordinary differential equation for each function.

Remarks and outline The Laplacian can be expressed in Cartesian, cylindrical, and many other coordinate systems. The equations can be separated by grouping terms with the same variable and setting the groups equal to a separation constant.

Required packages

```
In[75]:= Needs["Algebra`Trigonometry`"]
         Needs["Calculus`VectorAnalysis`"]
         Needs["Graphics`PlotField`"]
```

Solution

```
In[76]:= Clear["Global`*"];
```

Part (a) Assume the potential is of the form **vx[x] vy[y] vz[z]** and apply the Laplacian to this product. The equation for the potential follows from setting the Laplacian of the potential (**lapEq**) equal to zero:

```
In[77]:= potential= vx[x] vy[y] vz[z];

In[78]:= SetCoordinates[Cartesian[x,y,z]];
         lapEq= Laplacian[potential]

Out[78]= vy[y] vz[z] vx''[x] + vx[x] vz[z] vy''[y] + vx[x] vy[y] vz''[z]
```

If you divide both sides of **lapEq** by **vx[x] vy[y] vz[z]**, this equation separates into terms that depend on only one variable:

```
In[79]:= eq1= lapEq/( vx[x] vy[y] vz[z] ) //Expand
Out[79]= vx''[x]     vy''[y]     vz''[z]
         ------- + ------- + -------
         vx[x]       vy[y]       vz[z]
```

You get three equations by equating terms with the same variable to a separation constant. You will find it convenient to define the **select** (lowercase "s") command:

```
In[80]:= select[eq_,var_]:=Select[ eq,(!FreeQ[#,var])&]
```

Then **select** the terms of **eq1**, which are *not* **Free** of **x**, **y**, and **z**:

```
In[81]:= xEq= select[eq1,x] == -cx^2
Out[81]= vx''[x]        2
         ------- == -cx
         vx[x]

In[82]:= yEq= select[eq1,y] == -cy^2
Out[82]= vy''[y]        2
         ------- == -cy
         vy[y]

In[83]:= zEq= select[eq1,z] == +cz^2
Out[83]= vz''[z]       2
         ------- == cz
         vz[z]
```

To satisfy Laplace's equation, the constants must satisfy **-cx^2-cy^2+cz^2=0**. (Note the different sign for **cz^2**. We discuss this later.) Express the **x** equation in a more convenient form:

```
In[84]:= xEq2= Solve[xEq,{vx[x]}][[1]] //.Rule->Equal
Out[84]=
                        vx''[x]
         {vx[x] == -(--------)}
                           2
                         cx
```

Group this form with the initial conditions:

```
In[85]:= xEquations= {xEq2,vx[0]==vx0, vx'[0]==vx0p} //Flatten
Out[85]=
                       vx''[x]
        {vx[x] == -(--------), vx[0] == vx0, vx'[0] == vx0p}
                          2
                        cx
```

and you can then solve this differential equation:

```
In[86]:= dsolX= ( DSolve[xEquations,{vx[x]},x][[1]]
                  //ComplexToTrig
                  //Simplify
                )
Out[86]=
                                         vx0p Sin[cx x]
        {vx[x] -> vx0 Cos[cx x] + --------------}
                                             cx
```

The solutions are trignometric functions. If you had chosen the opposite sign for **cx^2**, you would have obtained hyperbolic functions. You can illustrate this by substitution: **cx -> I cx**, that is, **cx^2 -> -cx^2**:

```
In[87]:= dsolX /.{cx-> I cx}

Out[87]=                             vx0p Sinh[cx x]
         {vx[x] -> vx0 Cosh[cx x] + ---------------}
                                           cx
```

The form of the **y** and **z** equations is identical, so you need not solve the equations explicitly here.

Part (b) Assume the potential is of the form **R[r] P[phi] Z[z]** and apply the Laplacian to this product. The equation for the potential follows from setting the Laplacian of the potential (**lapEq**) equal to zero:

```
In[88]:= potential= R[r] P[phi] Z[z];

In[89]:= SetCoordinates[Cylindrical[r,phi,z]];
         lapEq= Laplacian[potential]

Out[89]=                          R[r] Z[z] P''[phi]
         (P[phi] Z[z] R'[r] + ------------------ + r P[phi] Z[z] R''[r] +
                                          r

            r P[phi] R[r] Z''[z]) / r
```

This equation for **Z[z]** separates if you divide both sides of **lapEq** by **R[r] P[phi] Z[z]**:

```
In[90]:= eq1= lapEq/( R[r] P[phi] Z[z] ) //Expand

Out[90]= R'[r]     P''[phi]     R''[r]    Z''[z]
         ------- + ---------- + ------- + ------
         r R[r]     2            R[r]      Z[z]
                   r  P[phi]
```

Select the terms that are *not* **Free** of **z** and set them equal to a separation constant:

```
In[91]:= zEq= Select[ eq1,(!FreeQ[#,z])&] == +cz^2

Out[91]= Z''[z]        2
         ------ == cz
          Z[z]
```

The form of this equation is identical to that in Part (a), so you need not solve the equation here. **Select** the terms that *are* **Free** of **z** and account for the missing **Z[z]** terms by adding in the separation constant:

```
In[92]:= eq2= Select[ eq1,(FreeQ[#,z])&] +cz^2

Out[92]=   2    R'[r]    P''[phi]     R''[r]
         cz  + ------ + ---------- + -------
               r R[r]    2            R[r]
                        r  P[phi]
```

(As a crosscheck, you should obtain the original **eq1** by adding **zEq** and **eq2**.) You now can separate the **R[r]** terms by multiplying by **r^2**:

```
In[93]:= eq3= r^2 eq2 //Expand

Out[93]=
                                          2
  2  2     r R'[r]      P''[phi]     r   R''[r]
cz   r  + ---------- + ---------- + -----------
           R[r]        P[phi]          R[r]
```

Select the terms that are *not* **Free** of **phi** and set these to a different separation constant:

```
In[94]:= phiEq= Select[ eq3,(!FreeQ[#,phi])&] == -cphi^2

Out[94]= P''[phi]            2
         --------- == -cphi
          P[phi]
```

This equation is of the form discussed in Part (a), so you need not solve it here. Finally, **Select** the terms that *are* **Free** of **phi** and account for the missing **P[phi]** terms by adding in the separation constant:

```
In[95]:= rEq= Select[ eq3,(FreeQ[#,phi])&] -cphi^2

Out[95]=
                                     2
      2       2  2     r R'[r]      r   R''[r]
-cphi   + cz   r   + ---------- + -----------
                        R[r]          R[r]
```

Expressing **rEq** in a more convenient form, you get

```
In[96]:= rEq= rEq R[r] //Expand

Out[96]=
          2                2  2                               2
-(cphi    R[r]) + cz   r    R[r]  + r R'[r]  + r    R''[r]
```

Combining this equation with boundary conditions,

```
In[97]:= rEquations={ rEq==0, R[0]==r0, R'[0]==0};
```

you get the solution for **R[r]** in terms of Bessel functions:

```
In[98]:= rSol= DSolve[rEquations[[1]],{R[r]},r][[1]] //PowerExpand

Out[98]= {R[r] -> BesselJ[cphi, cz r] C[1] + BesselY[cphi, cz r] C[2]}
```

Problem 2 : Potential and a rectangular groove

Consider a rectangular groove that runs from $z = -\infty$ to $+\infty$ and is open in the positive y-direction. The groove is bounded by three walls: $y = 0$, $x = 0$, and $x = a$. The walls at $x = 0$ and $x = a$ are at zero potential and the end at $y = 0$ is at a specified potential, $V[x, 0]$ (independent of the z-coordinate).

Part (a) Find a series expansion for the potential inside the groove for an arbitrary $V[x, 0]$.

Part (b) Write the first few terms of the potential if $V[x, 0] = \mathrm{Sin}[x]$ and separately, $V[x, 0] = 1$. Plot the potentials and use **VEPlot** to plot the equipotential and electric field lines.

Part (c) Sum the series for $V[x, 0] = V0$ and get a closed form for the solution:

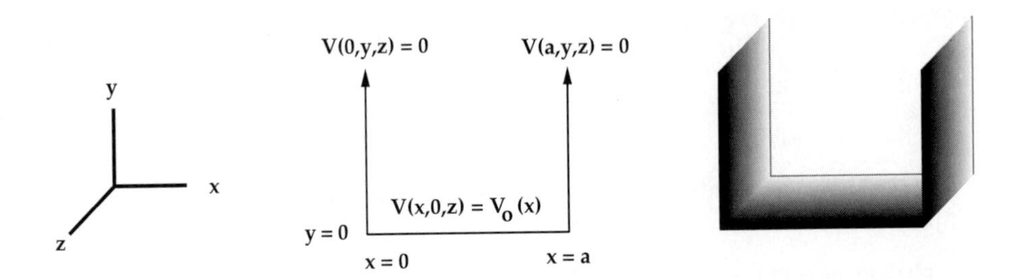

Remarks and outline (1) Use Cartesian coordinates and express the answer as a product **X[x] Y[y]**. (2) The boundary conditions suggest that **X[x]** is proportional to Sin and **Y[y]** is proportional to a decreasing exponential. The orthogonality properties of Sin allow the expansion coefficients to be expressed as integrals over the boundary at $y = 0$. (3) For $V[x, 0] = V0$, the command **Sum** will analytically sum the infinite potential series.

Using *Mathematica*'s ability to work with symbolic integrals, you can actually solve an entire class of problems using only two lines of code. Having solved the two one-dimensional problems, you can reconstruct these into a single two-dimensional problem and then graph the solution to verify its accuracy. Intermediate results are shown using series that have varying numbers of terms.

Compare the accuracy of the solutions with the specified boundary conditions as a function of the number of terms used in the series. Note that it is particularly difficult to reproduce the "sharp corners" using sine and exponential expansions. Experiment with other choices for the specified potential, such as Sin$[nx]$ and Cos$[nx]$.

Required packages

```
In[99]:= Needs["Algebra`SymbolicSum`"]
         Needs["Algebra`Trigonometry`"]
         Needs["Calculus`VectorAnalysis`"]
         Needs["Graphics`PlotField`"]
```

Solution

```
In[100]:= Clear["Global`*"];
```

Part (a) The boundary conditions suggest that you should use Cartesian coordinates and that the solution is independent of **z**. The potential can be expressed as an expansion of orthogonal functions of variables **x** and **y**,

```
In[101]:= pot[f_,n_]:= Sum[A[f,m ] Sin[Pi m x/a] Exp[-Pi m y/b],{m,1,n}];
```

with coefficients **A[f,m]**, which are determined using the orthogonality property of the **Sin** function:

```
In[102]:= A[f_,m_]:= 2/a Integrate[f Sin[Pi m x/a],{x,0,a}];
```

The form of the potential is chosen to ensure that it vanishes at **x=0** and **x=a**.

The function **pot[f,n]** gives the first **n** terms of the potential if the end wall at **y=0** has a potential **V[x,0]**. (The operator **pot** works only for the special case in which the integral in **A** can be done analytically; otherwise, **Integrate** must be replaced by **NIntegrate**.)

Part (b) Consider the case in which the potential on the end of the groove is **V[x,0]=Sin[x]**. The first six terms of the potential are (this takes a while)

```
In[103]:= pot1= pot[Sin[x],6];
          pot1 //Short

Out[103]=
                           Pi x
          2 Pi Sin[a] Sin[----]
                            a
          --------------------------- + <<5>>
           (Pi y)/b     2      2
          E         (-a   + Pi )
```

To see this visually, choose some values and plot the potential in the $\{x, y\}$ plane:

```
In[104]:= values={a->1,b->1,V0->1};

          Plot3D[ pot1 //.values //Evaluate ,{x,0,1},{y,0,0.5}];
```

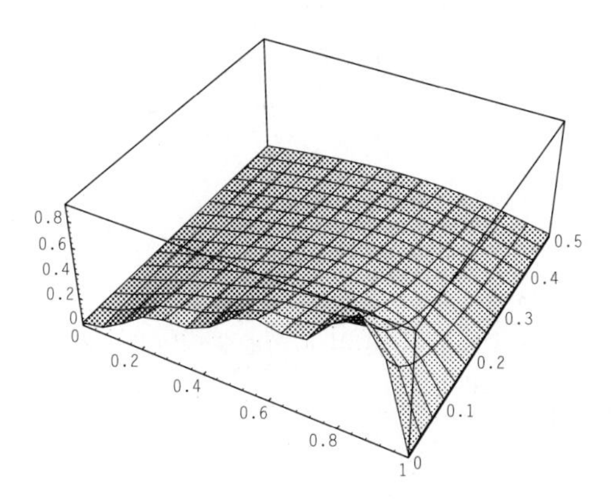

Clearly, the boundary conditions are satisfied on all surfaces except **y=0**. If you increase the number of terms in the potential series, the solution at the **y=0** surface is closer to the prescribed function. Next, plot the equipotential lines and electric field lines with **VEPlot**:

```
In[105]:= VEPlot[ pot1 //.values ,{x,0,1},{y,0,0.5}];
```

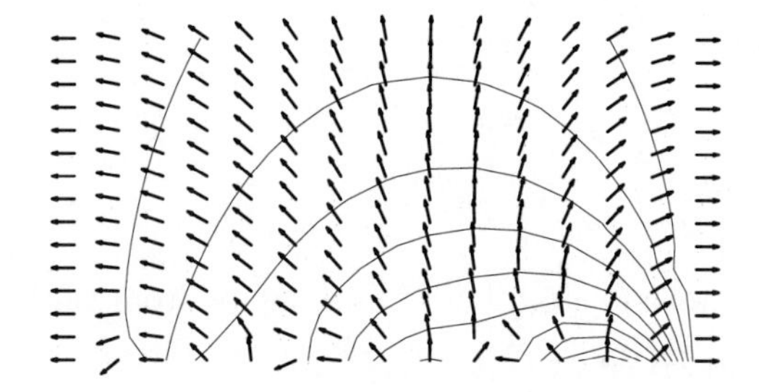

When the potential at **y=0** is **V[x,0]=1**, the first few terms of the potential are seen to be

```
In[106]:= pot2= pot[1, 4] //.values
Out[106]= 4 Sin[Pi x]     4 Sin[3 Pi x]
          ------------ + --------------
            Pi y            3 Pi y
           E     Pi      3 E       Pi
```

To see this visually, plot the potential in the $\{x, z\}$ plane:

```
In[107]:= Plot3D[ pot2,{x,0,1},{y,0,0.5}];
```

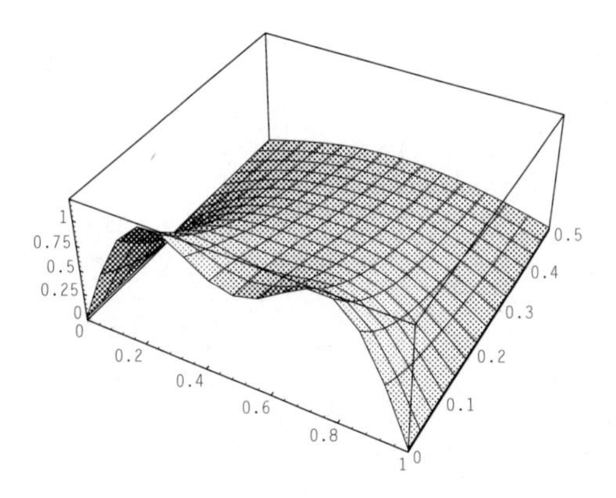

If you increase the number of terms, the potential at the **y=0** surface is closer to the prescribed value:

```
In[108]:= Plot3D[ pot[1,15] //.values //Evaluate ,{x,0,1},{y,0,0.5}];
```

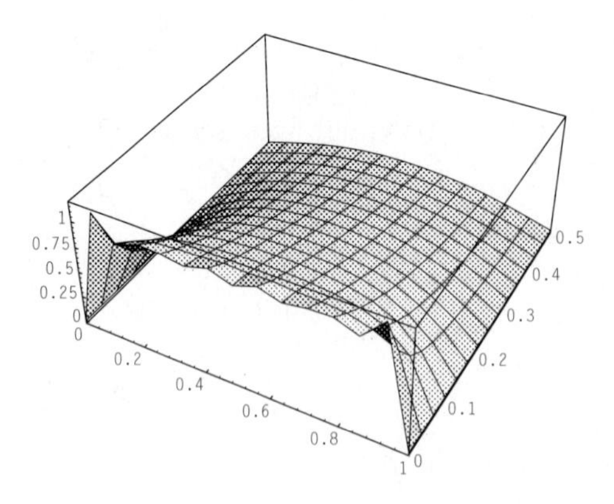

As before, the equipotential lines and electric field lines follow from **VEPlot**:

```
In[109]:= VEPlot[ pot[1,15] //.values //Evaluate ,{x,0,1},{y,0,0.5}];
```

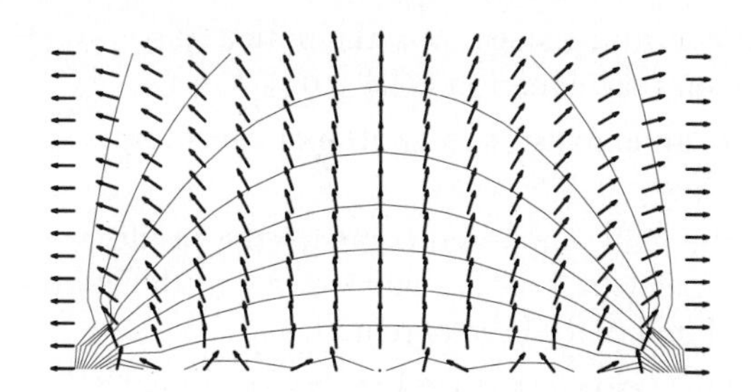

Part (c) The case **V[x,0]=V0** can be summed exactly with the help of the package **Algebra`SymbolicSum`**:

```
In[110]:= pot3= pot[V0,Infinity] //TrigToComplex //Simplify
Out[110]=
                        (-I Pi x)/a - (Pi y)/b
(-I V0 (Log[1 - E                              ] -

                      (-I Pi x)/a - (Pi y)/b
          Log[1 + E                            ] -

                      (I Pi x)/a - (Pi y)/b
          Log[1 - E                           ] +

                      (I Pi x)/a - (Pi y)/b
          Log[1 + E                           ])) / Pi
```

(Note, you must help **Sum** with **TrigToComplex**.) As you can see, the result agrees exactly with **V[x,0]=1** on the **y=0** boundary:

```
In[111]:= Plot3D[ pot3 //.values //Evaluate ,{x,0.001,0.999},{y,0,0.5}];
```

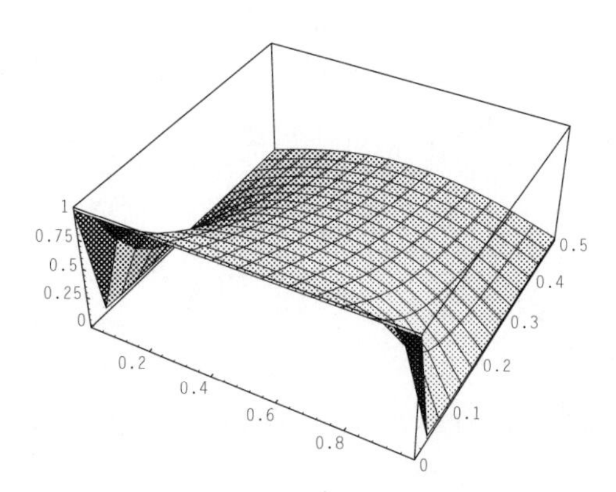

Problem 3: Rectangular conduit

Consider a rectangular conduit that is bounded by walls on four sides located at $x = 0$, $x = a$, $y = 0$, and $y = b$. The conduit has infinite length along the z-axis. The potential is zero on the sides $y = 0$ and $y = b$. On the side $x = a$, the potential is $V[a, y] = V_0$ and on the side $x = 0$, the potential satisfies the equation $V'[x] = 0$.

Part (a) Express the potential as an expansion in Cartesian coordinates. Determine the expansion coefficients.

Part (b) Plot the potential on the wall $x = a$ and compare the results with the boundary condition potential. Experiment by keeping more terms in the expression for the potential. Make a three-dimensional plot of the potential.

Part (c) Plot the equipotential lines and electric field in the $\{x, y\}$ plane.

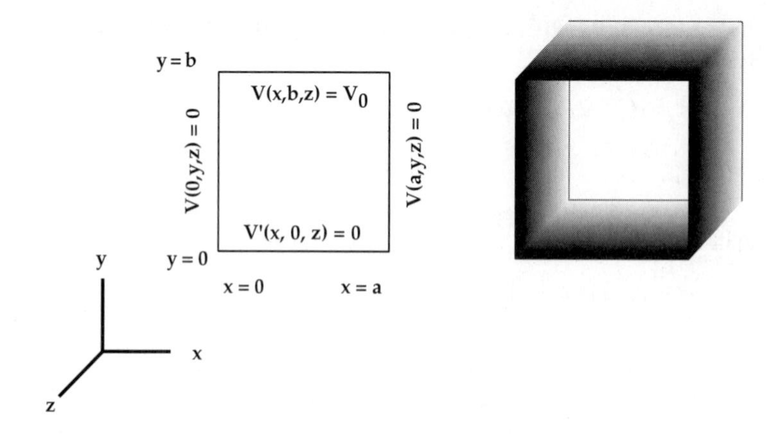

Remarks and outline The symmetry in z implies that the solution is independent of the z-variable. Use Cartesian coordinates and express the answer as a product $X[x]Y[y]$. The boundary conditions on the walls require the product to be of the form $\mathrm{Cosh}[m\pi x/b]\mathrm{Sin}[m\pi y/b]$, where $m = \{1, 2, 3, \ldots\}$. The orthogonality of $\mathrm{Sin}[m\pi y/b]$ enables you to express the expansion coefficients as integrals over the boundary at $x = a$.

Required packages

```
In[112]:= Needs["Calculus`VectorAnalysis`"]
          Needs["Graphics`PlotField`"]
```

Solution

```
In[113]:= Clear["Global`*"];
```

Part (a) Start by defining the potential (**pot**) as a sum over the orthogonal functions:

```
In[114]:= term[m_]= A[m] Cosh[m Pi x/b] Sin[m Pi y/b];
          pot[n_]:= Sum[ term[m] ,{m,1,n}];
          pot[n]

Out[114]=
                         m Pi x       m Pi y
          Sum[A[m] Cosh[------] Sin[------], {m, 1, n}]
                           b            b
```

Verify the boundary condition at $V'[x] = 0$ at **x=0**:

```
In[115]:= D[ pot[n] ,x] /.{x->0}

Out[115]= 0
```

At the boundary **x=a**, the form of the potential is

```
In[116]:= pot[n] /.{x->a}

Out[116]=
                       a m Pi        m Pi y
          Sum[A[m] Cosh[------] Sin[------], {m, 1, n}]
                         b             b
```

You can determine the coefficients **A[m]** using the orthogonality properties of the sine function. At the **x=a** boundary, the potential will be given by a sum of terms of the form

```
In[117]:= eq1=  V0 == term[m] /.{x->a}

Out[117]=
                       a m Pi        m Pi y
          V0 == A[m] Cosh[------] Sin[------]
                         b             b
```

(There is an implied sum on the right-hand side (**rhs**), that is, $V_0 = \sum_m A_m \mathrm{Cosh}(m\pi a/b)\,\mathrm{Sin}(m\pi y/b)$.) You can determine **A[m]** by multiplying both sides of **eq1** by **Sin[n Pi y/b]** and then integrating. Do this first to the **rhs.** Because the sine functions are orthogonal, the only surviving term will be when **n=m**, so you can spare *Mathematica* the effort of integrating all terms except this one. (That is, $\int \mathrm{Sin}(n\pi y/b)\,\mathrm{Sin}(m\pi y/b) = 0$ except for $n = m$.) The **rhs** is then

```
In[118]:= rhs= ((Integrate[ eq1[[2]] Sin[m Pi y/b] ,{y,0,b}]
                  //Simplify
               )  //.{Sin[2 m Pi]->0}
              )

Out[118]=
                         a m Pi
          b A[m] Cosh[------]
                           b
          -------------------
                   2
```

Likewise, for the left-hand side (**lhs**), you have

```
In[119]:= lhs= ((Integrate[ eq1[[1]] Sin[m Pi y/b] ,{y,0,b}]
                )   //.{Cos[m Pi]->(-1)^m}
              ) //Simplify

Out[119]=
                    m
          (1 - (-1) ) b V0
          ----------------
               m Pi
```

Setting the **lhs** equal to the **rhs**, you find **A[m]**

```
In[120]:= aSol= Solve[rhs==lhs,A[m]][[1]] //Simplify;
          A[m_]= A[m] /. aSol

Out[120]=
                       m                a m Pi
          2 (1 - (-1) ) V0 Sech[------]
                                  b
          -------------------------------
                      m Pi
```

and thus, the potential:

```
In[121]:= pot[n]

Out[121]=
                       m              m Pi x          a m Pi          m Pi y
          2 (1 - (-1) ) VO Cosh[——————] Sech[——————] Sin[——————]
                                        b               b               b
     Sum[———————————————————————————————————————————————————————————————, {m, 1, n}]
                                       m Pi
```

Part (b) You can see the effect of using different numbers of terms in the expansion of the potential. Here, make a table of potentials with $\{2, 4, 6, 8\}$ terms at the boundary **x=a** and plot the results:

```
In[122]:= Plot[ Table[pot[n],{n,2,8,2}]
              //.{b->1, VO->1, a->1, x->a} //Evaluate
              ,{y,0,1}
              ,PlotRange->{0,2}];
```

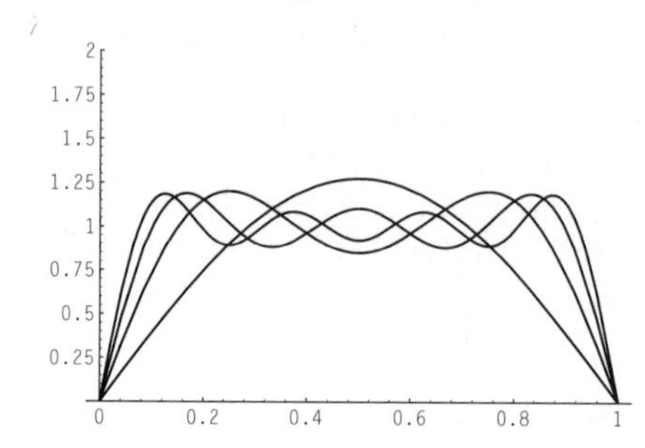

The curves with more inflection points contain more terms in the expansion and more closely represent the correct potential at the boundary. Taking **n=4**, the potential in the $\{x, y\}$ plane is

```
In[123]:= Plot3D[ pot[4]  //.{b->1, VO->1, a->1} //Evaluate
              ,{y,0,1}
              ,{x,0,1}
              ,ViewPoint->{2,2,1}
           ];
```

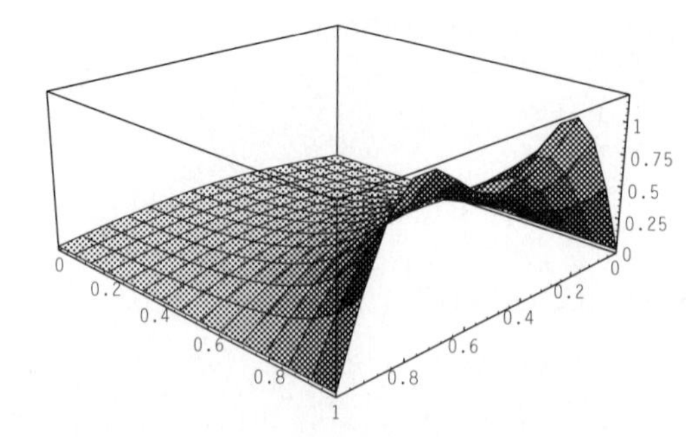

It is clear from this plot that the boundary conditions are satisfied except at the **x=a** boundary.

Part (c) Further, you can plot the electric field and the equipotential lines with **VEPlot**:

```
In[124]:= VEPlot[pot[4] //.{b->1, V0->1, a->1} ,{x,0,1},{y,0,1}];
```

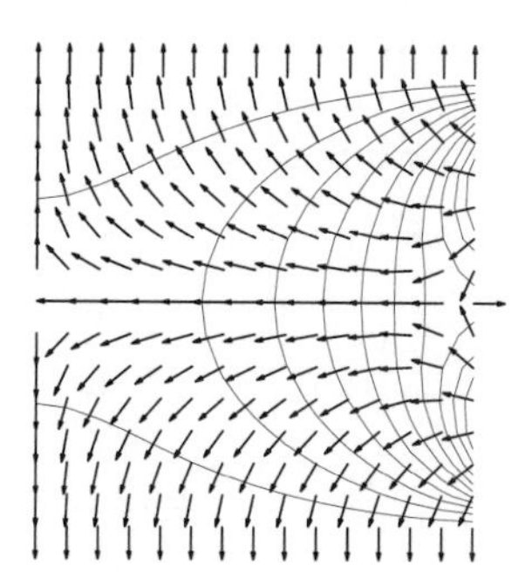

Problem 4: Potential inside a rectangular box with five sides at zero potential

Consider a rectangular enclosure with sides at $x = 0$ and $x = a$, $y = 0$ and $y = b$, and $z = 0$ and $z = c$. The side at $z = c$ has a specified potential $V[x, y, c] = V_0$; the other five sides have zero potential.

Part (a) Express the potential for general $V[x, y, z]$ as an expansion of functions expressed in Cartesian coordinates. Find the expansion coefficients.

Part (b) Find the first few terms of the potential expansion when $V[x, y, c] \equiv V_0 = 1$. Illustrate the accuracy of the solution by comparing the results with the boundary potential on the wall $z = c$.

Part (c) Let $x = 1/2$ and plot the potential in the $\{y, z\}$ plane. Let $x = 1/2$ and use **VEPlot** to plot the equipotential lines and the electric field in the $\{y, z\}$ plane:

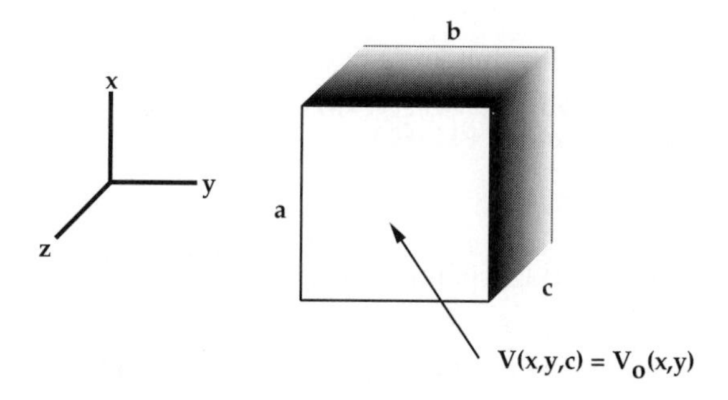

Remarks and outline The boundary conditions are satisfied if the potential is a product of the form $\mathrm{Sin}[\alpha x]\mathrm{Sin}[\beta y]\mathrm{Sinh}[\gamma z]$. The orthogonality properties of these functions allow the expansion coefficients to be expressed in terms of the potential on the $z = c$ side.

Even though you generalized your solution to a three-dimensional problem, *Mathematica* allows you to write the solution in a form that is identical to the previous one- and two-dimensional problems. Again, plot the result to verify its accuracy.

Compare the solution's accuracy with the specified boundary condition as a function of the number of terms used in the series. Note that it is particularly difficult to reproduce the "sharp corners" using sin and exponential expansions.

Experiment with potentials on the $z = c$ wall. For more complicated functions, replace **Integrate** by **NIntegrate** to generate purely numerical solutions.

Required packages

```
In[125]:= Needs["Algebra`Trigonometry`"]
          Needs["Calculus`VectorAnalysis`"]
          Needs["Graphics`PlotField`"]
```

Solution

```
In[126]:= Clear["Global`*"];
```

Part (a) The boundary conditions suggest that you should use Cartesian coordinates. Write the potential as a product of three functions $X[x]Y[y]Z[z]$. The vanishing of the potential at $x = \{0, a\}$ requires $X[x] = \mathrm{Sin}[m_x\, \pi x/a]$. The vanishing of the potential at $y = \{0, b\}$ requires $Y[y] = \mathrm{Sin}[m_y\, \pi y/b]$. For the potential to vanish on the wall $z = 0$, the function $Z[z]$ must be $\mathrm{Sinh}[m_z\, \pi z/c]$:

```
In[127]:= Clear[A,pot,term,mz]

In[128]:= term[mx_,my_]:= ( A[mx,my]     Sin[ Pi mx x/a]
                                       * Sin[ Pi my y/b]
                                       * Sinh[Pi mz z/c] )

In[129]:= pot[nx_,ny_]:= Sum[ term[mx,my] ,{mx,1,nx},{my,1,ny}];
          pot[nx,ny]

Out[129]=
                              mx Pi x        my Pi y          mz Pi z
          Sum[A[mx, my] Sin[---------] Sin[---------] Sinh[---------], {mx, 1, nx},
                                 a              b                c

            {my, 1, ny}]
```

Verify that the potential satisfies the boundary conditions:

```
In[130]:= ( pot[nx,ny] /.{{x->0},{y->0},{z->0},{x->a},{y->b}}
                        //.{Sin[mx Pi]->0, Sin[my Pi]->0}
           )

Out[130]= {0, 0, 0, 0, 0}
```

There also is a relation between the separation constants. Use this information to eliminate **mz**:

```
In[131]:= mzSol= Solve[(Pi mx/a)^2 +(Pi my/b)^2 +(Pi I mz/c)^2==0,mz][[2]]

Out[131]=
                       2     2
                     mx    my
          {mz -> c Sqrt[--- + ---]}
                        2     2
                       a     b
```

You can determine the coefficients **A[mx,my]** using the orthogonality properties of the expansion functions. At the **z=c** boundary, the potential will be given by a sum of terms of the form:

```
In[132]:= eq1= V0 == term[mx,my] //.{z->c}

Out[132]=
                           mx Pi x       my Pi y
          V0 == A[mx, my] Sin[-------] Sin[-------] Sinh[mz Pi]
                              a             b
```

(There is an implied sum on the right-hand side, that is, $V_0 = \sum_m A_{m_x,m_y} \mathrm{Sin}(m_x\,\pi x/a)\,\mathrm{Sin}(m_y\,\pi y/b)\,\mathrm{Sinh}[m_z\,\pi]$.) You can determine **A[mx,my]** by multiplying both sides of **eq1** by **Sin[nx Pi x/a] Sin[ny Pi y/b]** and then integrating. Do this first to the **rhs**. Because the sine functions are orthogonal, the only surviving term will be when **nx=mx** and **ny=my**, so you can spare *Mathematica* the effort of integrating all terms except this one. (That is, $\int \mathrm{Sin}(n_x\,\pi x/a)\,\mathrm{Sin}(m_x\,\pi x/a) = 0$ except for $n_x = m_x$.) The **rhs** is then

```
In[133]:= trigRules= {Sin[2 mx Pi]->0, Cos[mx Pi]-> (-1)^mx
                     ,Sin[2 my Pi]->0, Cos[my Pi]-> (-1)^my
                     };

In[134]:= rhs= ( Integrate[ eq1[[2]] Sin[mx Pi x/a]  Sin[my Pi y/b]
                 ,{x,0,a},{y,0,b}]
              //.trigRules
           )

Out[134]=
                                  2
          (a b A[mx, my] (8 mx my Pi  + Cos[2 (mx - my) Pi] -

                                                                2
              Cos[2 (mx + my) Pi]) Sinh[mz Pi]) / (32 mx my Pi )
```

trigRules has been defined to simplify certain trigonometric functions when **mx** and **my** are intergers. Likewise, for the left-hand side, you have

```
In[135]:= lhs= (( Integrate[ eq1[[1]] Sin[mx Pi x/a]  Sin[my Pi y/b]
                 ,{x,0,a},{y,0,b}]
                //TrigReduce
              ) //.trigRules
                //Simplify
           )

Out[135]=
                    mx              my
          (-1 + (-1)  ) (-1 + (-1)  ) a b V0
          ----------------------------------
                             2
                    mx my Pi
```

Setting the **lhs** equal to the **rhs**, you will find **A[m]**:

```
In[136]:= aSol= Solve[rhs==lhs,A[mx,my]][[1]] //Simplify;
          A[mx_,my_]= A[mx,my] /. aSol

Out[136]=
                          mx              my
               32 (-1 + (-1)  ) (-1 + (-1)  ) V0 Csch[mz Pi]
          -------------------------------------------------------
                          2
          8 mx my Pi  + Cos[2 (mx - my) Pi] - Cos[2 (mx + my) Pi]
```

and thus, the potential:

```
In[137]:= pot[nx,ny]
```

```
Out[137]=
                 mx              my                                 mx Pi x
Sum[(32 (-1 + (-1)  ) (-1 + (-1)  ) V0 Csch[mz Pi] Sin[-------]
                                                                   a

      my Pi y          mz Pi z
  Sin[-------] Sinh[-------]) /
         b               c

                   2
  (8 mx my Pi  + Cos[2 (mx - my) Pi] - Cos[2 (mx + my) Pi]),

 {mx, 1, nx}, {my, 1, ny}]
```

Part (b) You can see the potential by choosing some specific values:

```
In[138]:= values={a->1, b->1, c->1, V0->1};
```

and setting **mz** to the constrained value:

```
In[139]:= mz=mz /.mzSol
Out[139]=
             2     2
           mx    my
c Sqrt[--- + ---]
             2     2
           a     b
```

Display the potential with three terms in both the x- and y-coordinates:

```
In[140]:= pot[3,3] //.values
Out[140]= 16 Csch[Sqrt[2] Pi] Sin[Pi x] Sin[Pi y] Sinh[Sqrt[2] Pi z]
          ------------------------------------------------------------ +
                                       2
                                     Pi

  16 Csch[3 Sqrt[2] Pi] Sin[3 Pi x] Sin[3 Pi y]

    Sinh[3 Sqrt[2] Pi z]
  -------------------------------------------------------------------- \
                                        2
                                    9 Pi

     16 Csch[Sqrt[10] Pi] Sin[3 Pi x] Sin[Pi y] Sinh[Sqrt[10] Pi z]
  + ------------------------------------------------------------------ +
                                        2
                                    3 Pi

  16 Csch[Sqrt[10] Pi] Sin[Pi x] Sin[3 Pi y] Sinh[Sqrt[10] Pi z]
  ---------------------------------------------------------------
                                     2
                                 3 Pi
```

To see how accurately you can represent the specified potential at **z=1** with different numbers of terms in the expansion, make a table with $\{2, 4, 6, 8\}$ terms and set **x=1/2**. Plotting the results, you get

```
In[141]:= Plot[ Table[pot[n,n] //.values //.{x->1/2, z->1}
                ,{n,2,8,2}] //Evaluate
          ,{y,0,1}];
```

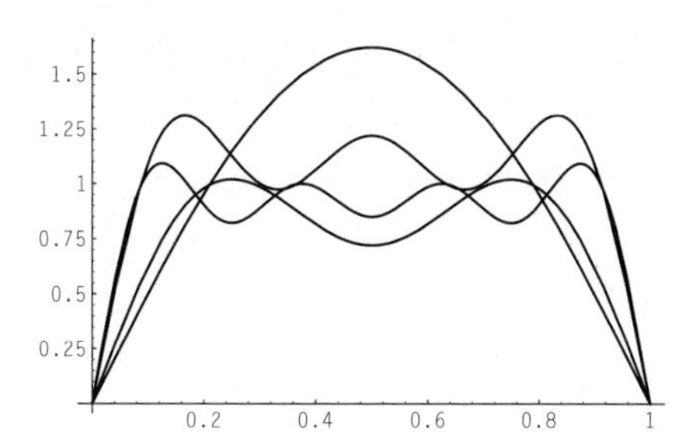

Clearly, as the number of terms increases, the more accurately the desired potential is represented. The above plot examined only the **x=1/2** slice. You can examine the entire $\{x, y\}$ surface at **z=1** using **Plot3D**:

```
In[142]:= Plot3D[pot[8,8] //.values //.{z->1} //Evaluate
                ,{x,0,1},{y,0,1}];
```

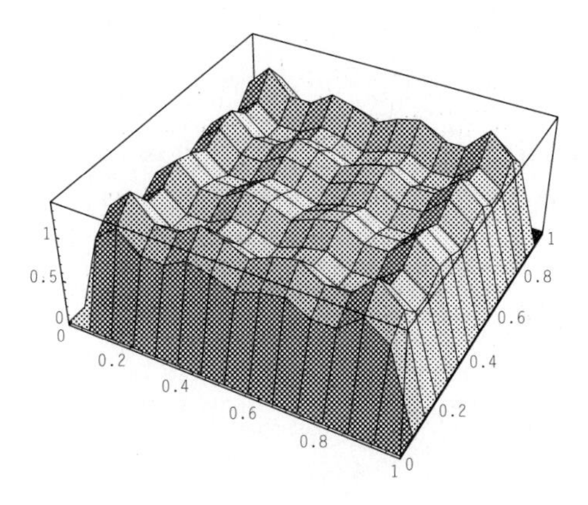

This plot shows how well you approximate the potential $V[x, y, z = 1] = 1$. You can extract similar information from a **ContourPlot**:

```
In[143]:= ContourPlot[pot[8,8] //.values //.{z->1} //Evaluate
                ,{x,0,1},{y,0,1}
                ,ContourSmoothing->True];
```

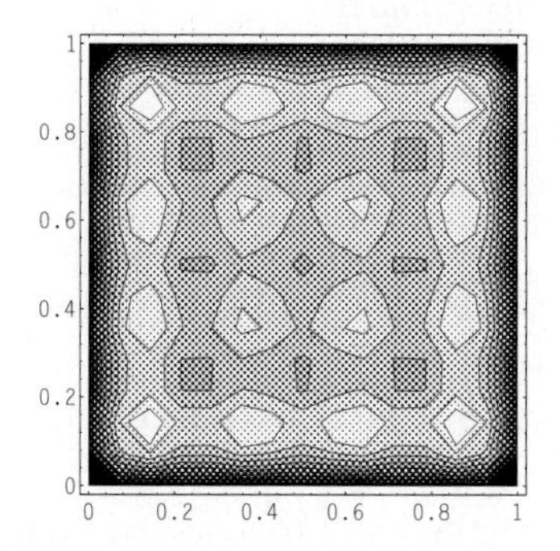

Part (c) You can view the potential in the $\{y, z\}$ plane by potting the first few terms for **x=1/2**:

```
In[144]:= Plot3D[ pot[4,4] //.values //.{x->1/2} //Evaluate ,{y,0,1},{z,0,1}];
```

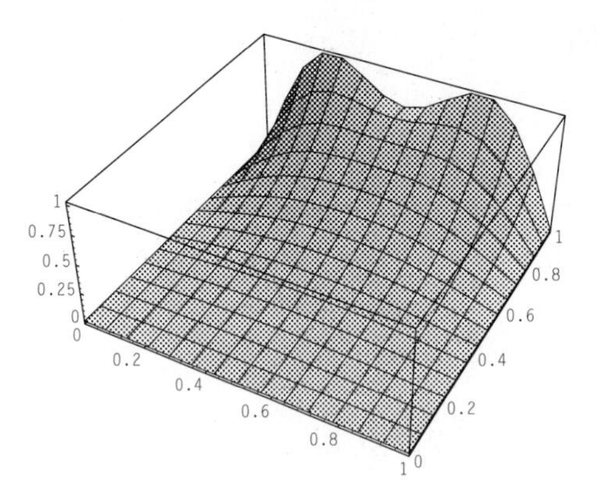

You also can view the equipotential lines and the electric field using **VEPlot**:

```
In[145]:= VEPlot[pot[8,8] //.values //.{x->1/2} //Evaluate,{y,0,1},{z,0,1}];
```

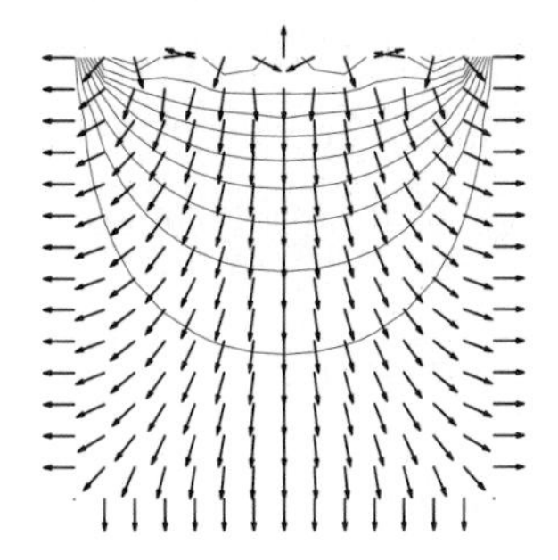

Problem 5: Conducting cylinder with a potential on the surface

Consider a long cylinder of radius $r0$ that is separated by a small lengthwise gap on each side. Use cylindrical coordinates $\{r, \phi, z\}$ and let the length of the cylinder be along the z-axis. One half of the cylinder ($0 < \phi < \pi$) is kept at a potential $V0$; the other half has a potential $-V0$.

Part (a) Express the potential inside and outside the sphere as a power series in r and $1/r$, respectively.

Part (b) Plot the value of the potential as a function of θ for fixed r. Compare the series expansion at $r = r0$.

Part (c) Make a three-dimensional plot of the potential as a function of r and θ.

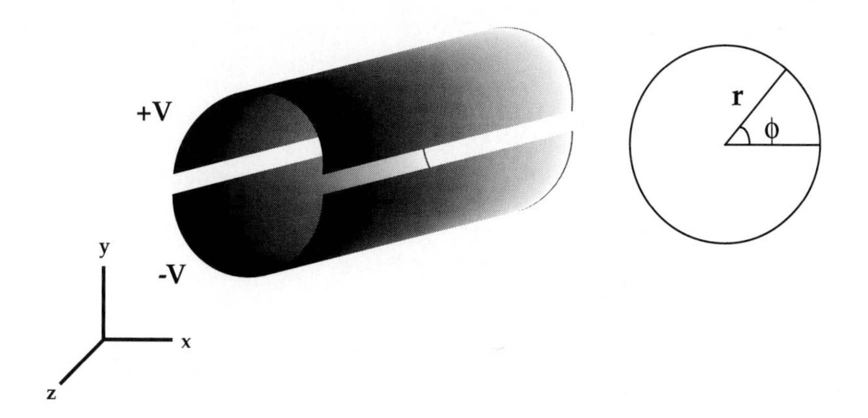

Remarks and outline The solution of Laplace's equation can be represented by a product of two functions $R[r]P[\phi]$. The symmetry of the problem requires $P[\phi]$ to be of the form $\mathrm{Sin}[m\phi]$. The term $R[r]$ is of the form $\{r^m, 1/r^m\}$, depending on whether the field point is inside or outside the cylinder.

Next, take advantage of *Mathematica*'s ability to work in different coordinate systems. Again, plot the result to verify its accuracy as a function of the terms in the series.

Compare the accuracy of the solutions with the specified boundary conditions as a function of the number of terms used in the series. Note that it is particularly difficult to reproduce the "sharp corners" using a finite number of terms in the expansion.

Experiment with other boundary potential functions such as $\mathrm{Sin}[nx]$ and $\mathrm{Cos}[nx]$. For more complicated functions, replace **Integrate** by **NIntegrate** to generate purely numerical solutions.

Try dividing the cylinder into four parts, and then re-compute the expansion coefficients.

Required packages

```
In[146]:= Needs["Calculus`VectorAnalysis`"]
          Needs["Graphics`ParametricPlot3D`"]
          Needs["Graphics`PlotField`"]
```

Solution

```
In[147]:= Clear["Global`*"];
```

Part (a) The symmetry of the problem suggests you should use cylindrical coordinates and that the solution is independent of **z**. The solution of Laplace's equation can be represented by a product of two functions **R[r] P[phi]**, where **R[r]** is of the form **r^m** or **r^-m** and **P[phi]** is of the form **Sin[m phi]**. (The dependence on the **r** variable is determined by requiring the terms in the potential to be finite.) First, construct the potential inside the cylinder. Inside the cylinder, the potential must be finite at **r=0**, so the function **R[r]** is of the form **r^m**. The potential is

```
In[148]:= termIn[m_]= A[m] (r/r0)^m Sin[m phi]
```

```
Out[148]= r  m
         (—)  A[m] Sin[m phi]
          r0

In[149]:= potIn[n_]:= Sum[ termIn[m] ,{m,1,n}]
```

You can determine the coefficients **A[m]** using the orthogonality properties of the sine function. At the **r=r0** boundary, the potential will be given by a sum of terms of the form

```
In[150]:= eq1= V[phi] == termIn[m] /.{r->r0}

Out[150]= V[phi] == A[m] Sin[m phi]
```

(There is an implied sum on the right-hand side, that is, $V_0 = \sum_m A_m \mathrm{Sin}(m\phi)$.) You can determine **A[m]** by multiplying both sides of **eq1** by **Sin[n phi]** and then integrating. Do this first to the **rhs**. Because the sine functions are orthogonal, the only surviving term will be when **n=m**, so you can spare *Mathematica* the effort of integrating all terms except this one. (That is, $\int \mathrm{Sin}(n\phi)\,\mathrm{Sin}(m\phi) = 0$ except for $n = m$.) The **rhs** is then

```
In[151]:= rhs= ( Integrate[ eq1[[2]] Sin[m phi] ,{phi,0,2 Pi}]
                //.{Sin[4 m Pi]->0}
              )

Out[151]= Pi A[m]
```

Likewise, for the **lhs**, you have

```
In[152]:= lhs=(  ( Integrate[ +V0 Sin[m phi] ,{phi, 0,1 Pi}]
                 + Integrate[ -V0 Sin[m phi] ,{phi,Pi,2 Pi}]
               ) //.{Cos[m Pi]-> (-1)^m, Cos[2 m Pi]-> 1}
                 //Simplify
              )

Out[152]=
                 m
          2 (1 - (-1) ) V0
          ----------------
                 m
```

Note that we used the fact that the potential for $0 < \phi < \pi$ is **V0** and the surface potential for $\pi < \phi < 2\pi$ is **-V0**.

Setting the **lhs** equal to the **rhs**, you find **A[m]**:

```
In[153]:= aSol= Solve[rhs==lhs,A[m]][[1]] //Simplify;
          A[m_]= A[m] /. aSol

Out[153]=
                 m
          2 (1 - (-1) ) V0
          ----------------
                m Pi
```

and thus, the potential:

```
In[154]:= potIn[n]

Out[154]=
                      m   r  m
             2 (1 - (-1) ) (—)  V0 Sin[m phi]
                            r0
          Sum[---------------------------------, {m, 1, n}]
                            m Pi
```

Finding the potential outside is done similarly. The **A[m]** coefficients will match, but the expansion will be in powers of $1/r$ such that $V[r \to \infty] = 0$:

```
In[155]:= termOut[m_]= A[m] (r0/r)^m Sin[m phi]

Out[155]=
                 m   r0 m
          2 (1 - (-1) ) (--)  V0 Sin[m phi]
                         r
          ---------------------------------
                       m Pi

In[156]:= potOut[n_]:= Sum[ termOut[m] ,{m,1,n}]
          potOut[n]

Out[156]=
                     m   r0 m
              2 (1 - (-1) ) (--)  V0 Sin[m phi]
                             r
          Sum[---------------------------------, {m, 1, n}]
                           m Pi
```

Part (b) Plot the value of the potential as a function of **r** and **phi** for **r0=1** and **V0=1**,

```
In[157]:= values={r0->1, V0->1};
```

by combining the graphics for the inside and outside region. To do this, create a new function **pot** that has the appropriate conditional properties:

```
In[158]:= Clear[pot];
          pot[n_,x_ /; x<1 ]:=potIn[ n];
          pot[n_,x_ /; x>=1]:=potOut[n];
```

To see the potential as a function of **r**, plot a table of **r** values as a function of **phi**. First, examine values for **r<r0=1**:

```
In[159]:= Plot[Table[ pot[20,r/r0] //.values,{r,0,1,0.2}] //Evaluate
           ,{phi,0,2Pi}];
```

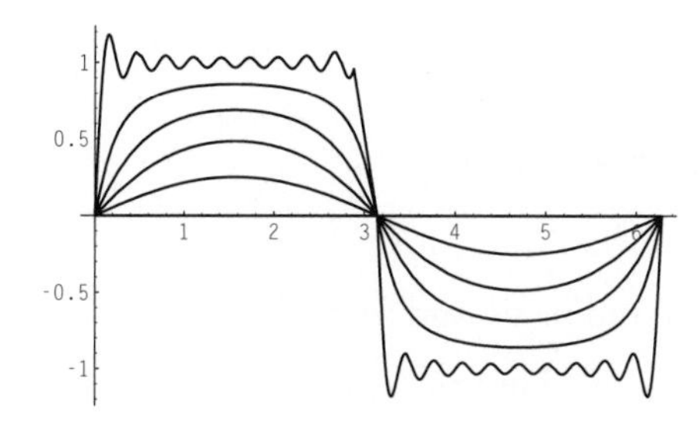

Note, as **r** approaches **r0**, the potential approaches that on the surface. The effect of summing only a finite number of term in the series is obvious as you approach **r=r0**. Next, examine values for **r>r0=1**:

```
In[160]:= Plot[Table[ pot[20,r/r0] //.values,{r,1,2,0.2}] //Evaluate
           ,{phi,0,2Pi}];
```

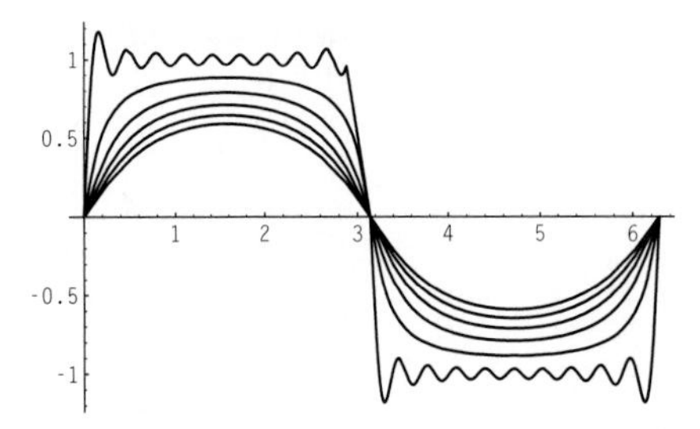

Note, as **r** approaches ∞, the potential approaches zero. The effect of summing only a finite number of term is again obvious.

Part (c) You can view the potential more directly using **CylindricalPlot3D**:

```
In[161]:= CylindricalPlot3D[ pot[20,r/r0] //.values
          ,{r,0,3}
          ,{phi,0,2Pi}
          ,BoxRatios->{1,1,1}
          ,ViewPoint->{4, -2, 0}
          ];
```

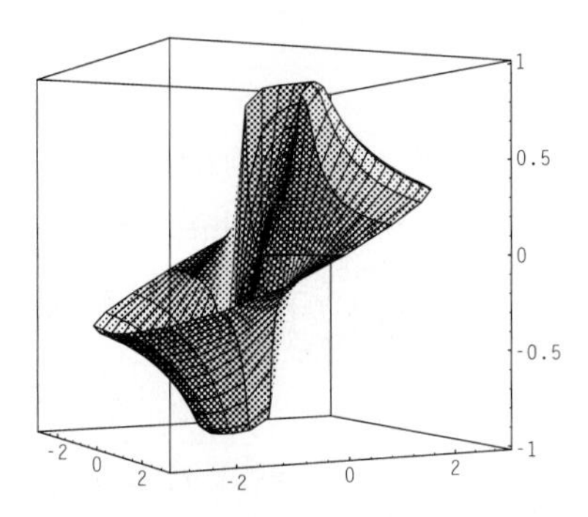

5.2.3 Legendre Polynomials and Spherical Harmonics

Problem 1: A charged ring

Consider a uniformly charged ring with radius **r0** and total charge **q**.

Part (a) Find the potential along the z-axis. Find the expansion of the potential in the limit $r \to 0$ and $r \to \infty$.

Part (b) Find the series expansion for the potential at all points in space.

Part (c) Find the electric field at all points in space.

Part (d) Express the potential in Cartesian coordinates and plot this in two and three dimensions. Plot in the plane of the ring and normal to this plane. Plot the equipotential lines and the electric field lines.

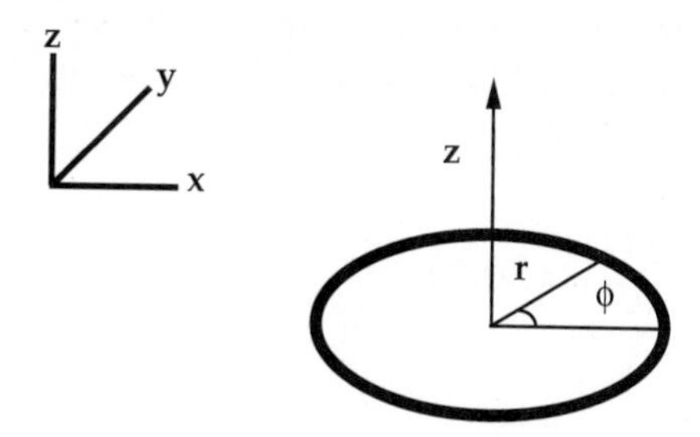

Remarks and outline The potential along the ring's axis is elementary because all points on the ring are equidistant from the axis. Because the solution is axially symmetric, the potential can be written as a sum over Legendre polynomials. The coefficients in the expansion of the potential are determined by comparing them with the exact solution for the potential evaluated on the axis of symmetry.

Required packages

```
In[162]:= Needs["Calculus`VectorAnalysis`"]
          Needs["Graphics`PlotField`"]
```

Solution

```
In[163]:= Clear["Global`*"];
```

Part (a) The potential along the z-axis expressed in spherical coordinates is given by

```
In[164]:= potZ = q/Sqrt[r^2+a^2];
```

Expand the potential in a power series. For **r>a**, expand about **r=Infinity**, and for **r<a**, expand about **r=0**. (Use **PowerExpand** to clean up the results.)

```
In[165]:= potZin[ n_]:= Series[potZ,{r,0,n}]         //PowerExpand //Normal;

          potZout[n_]:= Series[potZ,{r,Infinity,n}] //PowerExpand //Normal;
```

The first nine terms for the potential inside the ring are

```
In[166]:= potZin[9]
```

$$\text{Out[166]=}\quad \frac{q}{a} - \frac{q\,r^2}{2\,a^3} + \frac{3\,q\,r^4}{8\,a^5} - \frac{5\,q\,r^6}{16\,a^7} + \frac{35\,q\,r^8}{128\,a^9}$$

and outside the ring are

```
In[167]:= potZout[9]
```

$$\text{Out[167]=}\quad \frac{35\,a^8\,q}{128\,r^9} - \frac{5\,a^6\,q}{16\,r^7} + \frac{3\,a^4\,q}{8\,r^5} - \frac{a^2\,q}{2\,r^3} + \frac{q}{r}$$

Part (b) The symmetry of the boundary conditions suggests that the general potential depends only on **r** and **theta** and *not* **phi**. The axially symmetric solution of Laplace's equation in **{r,theta}** coordinates is of the form

```
In[168]:= Clear[pot,A,B];
          pot[s_]:= Sum[(B[n] r^n+ A[n](a/r)^(n+1)) *
                          LegendreP[n,Cos[theta]] ,{n,0,s}]
```

The first two terms of the potential are

```
In[169]:= pot[2]

Out[169]=
                                2
          a A[0]               a  A[1]
          ------ + B[0] + (-------- + r B[1]) Cos[theta] +
            r                    2
                                r

             3
            a  A[2]     2                              2
           (-------- + r  B[2]) (-1 + 3 Cos[theta] )
               3
              r
           -------------------------------------------
                              2
```

The series expression for the potential depends on whether **r>a** or **r<a**. For **r<a**, the potential must be finite at the origin, so **A[n]=0**. For **r>a**, the potential must vanish at infinity, so **B[n]=0**. The coefficients in the expansion can be determined by comparing **pot** to the exactly solvable case in which the potential is evaluated on the axis of symmetry (**theta=0**).

For the case **r>a**, you are outside the ring, so evaluate **pot** at **theta->0** and subtract **potZout**. Ignore the **B[i]** terms, since these terms diverge as a power of **r**. You are left with an expansion in powers of **1/r**, so change variables **r->1/w** and use **CoefficientList** to extract powers of **w=1/r**. Doing this will yield **s+1** equations (since you start from **A[0]**), where **s+1** is the number of terms kept in the expansion of the potential. Finally, **Solve** these **s+1** equations for the **s+1** unknown coefficients **A[i]**. All these steps are performed by the function **coeffOut**. (Perform the operations of **coeffOut** in separate steps to better understand its function.)

```
In[170]:= coeffOut[s_]:=
            Solve[0== CoefficientList[
                        ((pot[s]-potZout[s])
                         //.{B[i_]->0, theta->0, r->1/w}),w]
              ,list1[s]] //Flatten
```

Before you can use **coeffOut**, you must specify **list1[s]**, which gives the variables to be solved for:

```
In[171]:= list1[s_]:= A[#]& /@Range[0,s];
          list1[3]

Out[171]= {A[0], A[1], A[2], A[3]}
```

coeffOut produces a solution for **A[i]** when **r>a**:

```
In[172]:= coeffOut[3]

Out[172]=           q                           -q
          {A[0] -> -, A[1] -> 0, A[2] -> ——, A[3] -> 0}
                    a                           2 a
```

You get the potential for **r>a** to order **s** by substituting the expansion coefficients produced by **coeffOut** into **pot**:

```
In[173]:= Clear[potOut];
          potOut[s_]:=pot[s] //.Join[coeffOut[s], {B[x_]->0}]
```

The first three terms of the potential for **r>a** are

```
In[174]:= eq2=  potOut[6]

Out[174]=     2                   2
          q   a  q (-1 + 3 Cos[theta] )
          - - ————————————————————————— +
          r              3
                      4 r

              4                2                  4
           3 a  q (3 - 30 Cos[theta]  + 35 Cos[theta] )
           ————————————————————————————————————————————
                                5
                            64 r
```

To verify the results, compare the **potOut** (the complete potential) with **potZout** (the potential along the **z**-axis) evaluated at **z**:

```
In[175]:= potZout[9]-(potOut[9]/.theta->0)==0

Out[175]= True
```

As expected, they match on the **z**-axis.

Consider the second case **a>r**. The potential follows as above with a few obvious changes:

```
In[176]:= coeffIn[s_]:=
            Solve[0== CoefficientList[(potZin[s]-pot[s])
                                /.{A[x_]->0,theta->0}   ,r]
                ,list2[s]] //Flatten

In[177]:= list2[s_]:= Table[B[n],{n,0,s}];
          list2[3]

Out[177]= {B[0], B[1], B[2], B[3]}
```

The first few **B[i]** coefficients are

```
In[178]:= coeffIn[3]

Out[178]=           q                          -q
          {B[0] -> -, B[1] -> 0, B[2] -> ———, B[3] -> 0}
                    a                          3
                                            2 a
```

and the potential inside the ring is given by

```
In[179]:= Clear[potIn];
          potIn[s_]:=pot[s]//.Join[coeffIn[s], {A[x_]->0}]
```

The first three terms of the potential inside the ring are

```
In[180]:= eq4= potIn[4]

Out[180]=
                2               2
          q   q r  (-1 + 3 Cos[theta] )
          - - ------------------------- +
          a               3
                       4 a

               4                  2                   4
           3 q r  (3 - 30 Cos[theta]  + 35 Cos[theta] )
           ---------------------------------------------
                                 5
                             64 a
```

Part (c) The electric fields for **r>a** and **r<a** are easily given by the relation $\vec{E} = -\nabla V$. **SetCoordinates** to **Spherical**, although **Cylindrical** would also work for this simple example:

```
In[181]:= SetCoordinates[Spherical];
```

The first few terms of the electric field outside the ring are

```
In[182]:= electout[s_]:= -Grad[potOut[s] ];
          electout[3]

Out[182]=
                 2                   2       2
           q    3 a  q (-1 + 3 Cos[theta] )  -3 a  q Cos[theta] Sin[theta]
          {-- - -------------------------, ----------------------------, 0}
            2                4                           4
           r              4 r                         2 r
```

The first few terms of the electric field inside the ring are

```
In[183]:= electin[s_] := -Grad[potIn[s] ];
          electin[3]

Out[183]=
                                 2
           q r (-1 + 3 Cos[theta] )  -3 q r Cos[theta] Sin[theta]
          {------------------------, ----------------------------, 0}
                       3                         3
                    2 a                       2 a
```

Part (d) To plot the results in Cartesian coordinates, define **r2xRule** to convert from spherical coordinates to Cartesian coordinates:

```
In[184]:= r2xRule= ( {r,theta,phi} ->
                     CoordinatesFromCartesian[{x,y,z},Spherical]
                     //Thread
                   )

Out[184]=
                       2    2    2                                   z
          {r -> Sqrt[x  + y  + z ], theta -> ArcCos[-----------------------],
                                                          2    2    2
                                                    Sqrt[x  + y  + z ]

            phi -> ArcTan[x, y]}
```

Choosing some specific values,

```
In[185]:= values={a->1, q->1};
```

evaluate the potential inside and outside the ring in $\{x, y, z\}$ coordinates:

```
In[186]:= potOutXYZ=potOut[4] //.r2xRule //.values //Together;
          potInXYZ =potIn[ 4] //.r2xRule //.values //Together;
```

Define a function **ff** that yields the potential at all values in $\{x, y, z\}$ with the help of the **If** statement. If the radius, given by **Sqrt[x^2+y^2+z^2]**, is greater or equal to **a=1**, then use the result of **potOutXYZ**; otherwise, use **potInXYZ**:

```
In[187]:= Clear[ff];
          ff[x_,y_,z_]=
            If[N[Sqrt[x^2+y^2+z^2]]>=1
            ,(x^2 + 4*x^4 + y^2 + 8*x^2*y^2 + 4*y^4 - 2*z^2 + 8*x^2*z^2 + 8*y^2*z^2 +
              4*z^4)/(4*(x^2 + y^2 + z^2)^(5/2))
            ,(64 + 16*x^2 + 9*x^4 + 16*y^2 + 18*x^2*y^2 + 9*y^4 - 32*z^2 -
              72*x^2*z^2 - 72*y^2*z^2 + 24*z^4)/64
            ];
```

The actual expressions above were not typed in by hand but were generated from **potOutXYZ** and **potInXYZ** directly. (With a Notebook front-end, you can use the *Copy Input From Above* command.) The expressions are entered explicitly in order to make **PlotGradientField** work with **ff**. (This may change with later versions of *Mathematica*.)

Here, the potential along the **x**-axis is shown. Note how the potential inside and outside match at the radius of the ring (**r=a**):

```
In[188]:= Plot[ff[x,0,0],{x,-2,2}];
```

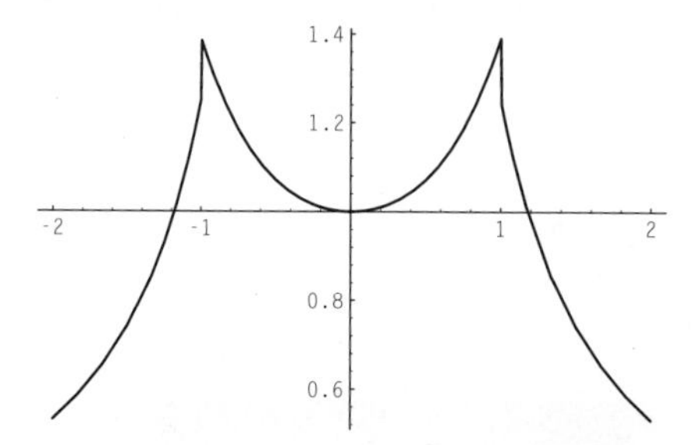

Here is a plot of the potential in the $\{x, z\}$ plane. Note where the ring intersects the plane of the figure (this takes a while; intermediate output suppressed):

```
In[189]:= Show[GraphicsArray[
            {Plot3D[      ff[x,0,z],{x,-2,2},{z,-2,2}
              ,PlotPoints->20]
            ,ContourPlot[ff[x,0,z],{x,-2,2},{z,-2,2}
              ,PlotPoints->20
              ,ContourSmoothing->True]
          }]];
```

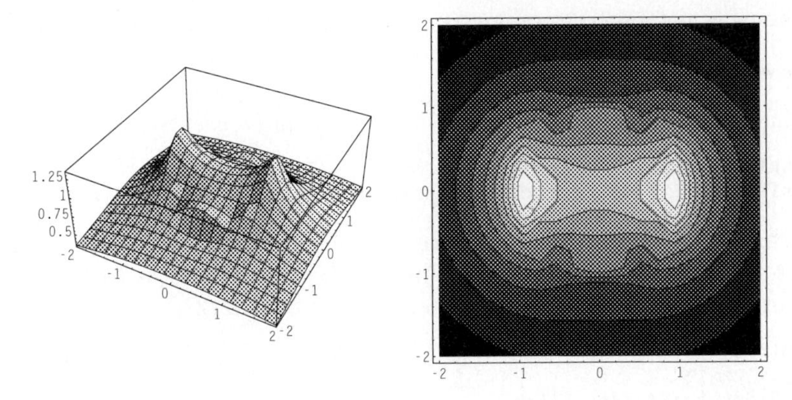

The corresponding **VEPlot** of the equipotential lines and the electric field lines in the plane of the ring (the $\{x, z\}$ plane) is

```
In[190]:= VEPlot[ff[x,0,z],{x,-2,2},{z,-2,2}];
```

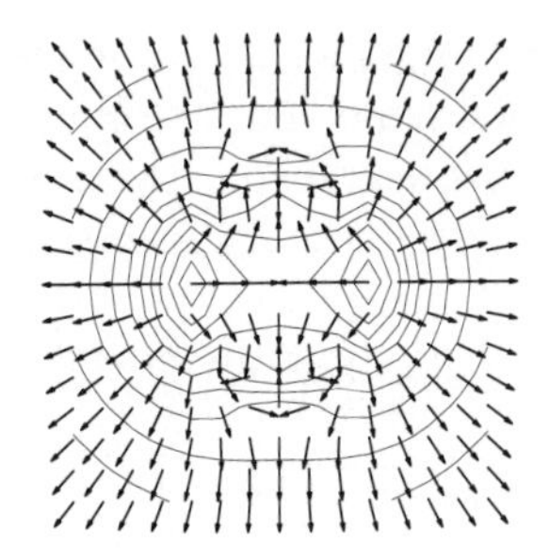

Here is a plot of the potential in the plane of the ring (the $\{x, y\}$ plane) (this takes a while; intermediate output suppressed):

```
In[191]:= Show[GraphicsArray[
            {Plot3D[      ff[x,y,0],{x,-2,2},{y,-2,2}
               ,PlotPoints->20]
            ,ContourPlot[ff[x,y,0],{x,-2,2},{y,-2,2}
               ,PlotPoints->20
               ,ContourSmoothing->True]
          }]];
```

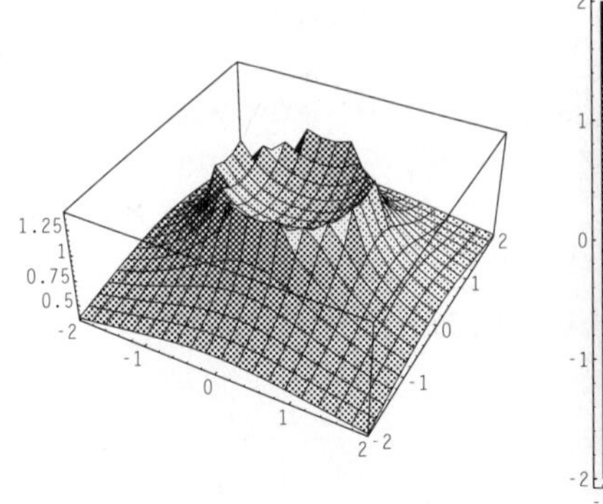

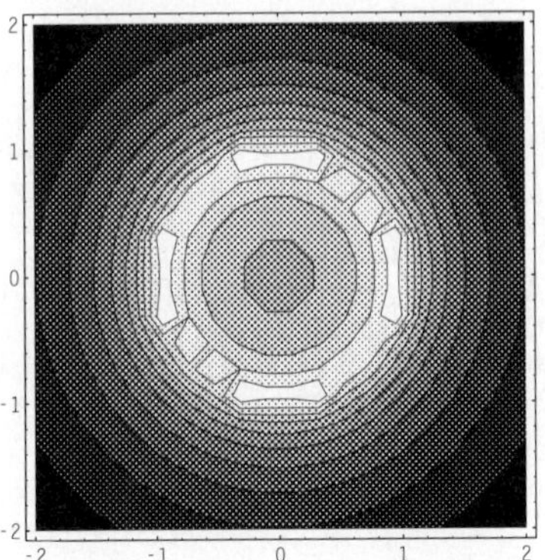

The corresponding **VEPlot** of the equipotential lines and the electric field lines normal to the plane of the ring (the $\{x, y\}$ plane) is

```
In[192]:= VEPlot[ff[x,y,0],{x,-2,2},{y,-2,2}];
```

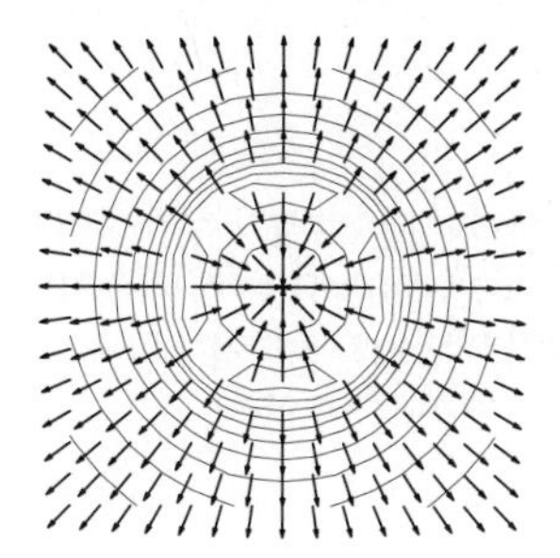

Problem 2: Grounded sphere in an electric field

A conducting sphere of radius a, which is held at zero potential, is placed in a uniform electric field E_0 directed along the z-axis.

Part (a) Find the potential and electric field.

Part (b) Plot the equipotential lines and the electric field.

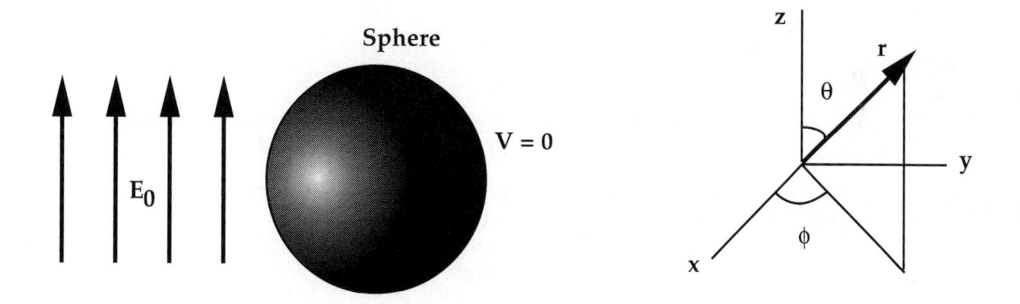

Remarks and outline Expand the potential in Legendre polynomials. The expansion coefficients are determined by requiring the potential to vanish on the surface of the sphere and to reduce to that of the applied electric field as $r \to \infty$. The equations for the coefficients are solved using the command **LogicalExpand**.

Required packages

```
In[193]:= Needs["Calculus`VectorAnalysis`"]
          Needs["Graphics`PlotField`"]
```

Solution

```
In[194]:= Clear["Global`*"];
```

Part (a) Use spherical coordinates and choose the z-axis along the direction of the applied electric field **E0**. The solution is symmetric about the z-axis, so the potential can be written as an expansion in Legendre polynomials:

```
In[195]:= pot[s_]:=( Sum[( A[i] (r/a)^i + B[i] (a/r)^(i+1) )
                        * LegendreP[i,Cos[theta]]
                     ,{i,0,s}]
                     +  O[Cos[theta]]^(s+1)
                   )
```

The command **pot** returns the first **s+1** terms for the potential expanded in Legendre polynomials. The notation **O[Cos[theta]]^(s+1)** generates the series expansion; this is an alternate form of the **Series** command. The unknown coefficients **A** and **B** are to be determined. As an example, consider the first three terms of **pot**:

```
In[196]:= pot[2]  //Expand //Collect[#,{r,1/r}]&

Out[196]=
                                                    2
             a B[0]   r A[1] Cos[theta]     a  B[1] Cos[theta]
     A[0] + ------ + ----------------- + ------------------ +
               r              a                   2
                                                 r

                                           3           3               2
                                         -(a  B[2])   3 a  B[2] Cos[theta]
                                 2       ---------- + --------------------
       2   -A[2]   3 A[2] Cos[theta]         2                 2
      r  (------ + -----------------) + ------------------------------------
              2             2                          3
           2 a           2 a                          r

In[197]:=
```

The coefficients A[i],B[i] are determined from the boundary conditions. The first **s+1** terms of coefficients **A[i]** and **B[i]** are given by **listAB**:

```
In[198]:= listAB[s_]:={Array[A,s+1,0],Array[B,s+1,0]} //Flatten;
          listAB[3]

Out[198]= {A[0], A[1], A[2], A[3], B[0], B[1], B[2], B[3]}
```

At **r=**∞, the potential must go as **-E0 r Cos[theta]**. This condition requires that all **A[i]** vanish except for **A[1]=-E0 a**. This asymptotic boundary condition is described by **boundary1**:

```
In[199]:= boundary1[s_]:= {A[0]==0, A[1]==-E0 a, Table[A[i]==0,{i,2,s}]
                          }//Flatten;
          boundary1[2]

Out[199]= {A[0] == 0, A[1] == -(a E0), A[2] == 0}
```

You will find it convenient to logically join these conditions. **Apply** (**@@**) **And** to the list of equations to obtain. (This operation simply replaces the head **List** with the head **And**.)

```
In[200]:= And @@ boundary1[2]
```

```
Out[200]= A[0] == 0 && A[1] == -(a E0) && A[2] == 0
```

The vanishing of the potential on the sphere requires **pot[s]==0** as **r->a**. Apply **LogicExpand** to extract the equations obtained by setting the coefficients of powers of **r** to zero:

```
In[201]:= boundary2[s_] := LogicalExpand[(pot[s]/.r->a)==0];
          boundary2[2]

Out[201]=
                                                  -A[2] - B[2]
          A[1] + B[1] == 0 && A[0] + B[0] + ------------ == 0 &&
                                                       2

            3 (A[2] + B[2])
            --------------- == 0
                   2
```

You now have sufficient information to solve for **A[i]** and **B[i]**. Apply **Solve** in the complete set of constraint equations to obtain

```
In[202]:= ABrule=
            Solve[(And @@ {boundary2[2],And @@ boundary1[3]})
              ,listAB[2] ]//Flatten

Out[202]= {B[0] -> 0, B[1] -> a E0, A[0] -> 0, A[1] -> -(a E0), B[2] -> 0,

            A[2] -> 0}
```

The complete solution for the potential follows from substituting this solution into the original expansion of the potential (**pot**):

```
In[203]:= potential= ( pot[2] //.ABrule //Normal //ExpandAll)

Out[203]=  3
          a  E0 Cos[theta]
          ---------------- - E0 r Cos[theta]
                 2
                r
```

Observe that the total potential is the superposition of the applied potential (second term) and an induced dipole potential (first term). The electric field follows from $\vec{E} = -\nabla V$:

```
In[204]:= eField= -Grad[potential, Spherical[r,theta,phi]] //ExpandAll

Out[204]=
                                 3
                              2 a  E0 Cos[theta]
          {E0 Cos[theta] + ------------------,
                                    3
                                   r

                                     3
                                    a  E0 Sin[theta]
            -(E0 Sin[theta]) + ----------------, 0}
                                        3
                                       r
```

Part (b) To view the equipotential lines and the electric field, first convert to Cartesian coordinates and then use **VEPlot**. The transformation from spherical coordinates to Cartesian coordinates is accomplished with the rule

```
In[205]:= r2xRule= {r,theta,phi} ->
              CoordinatesFromCartesian[{x,y,z},Spherical]//Thread;
```

Choosing some specific values for the parameters,

```
In[206]:= values={E0->1, a->1};
```

and defining a **sphere** to mask the plot,

```
In[207]:= sphere=Graphics[Disk[{0,0},1]];
```

find the equipotential lines and the electric field (intermediate output suppressed):

```
In[208]:= Show[{
            VEPlot[potential //.r2xRule //.values //.{y->0} //Evaluate
                ,{x,-2,2.1},{z,-2,2.1}]
            ,sphere}];
```

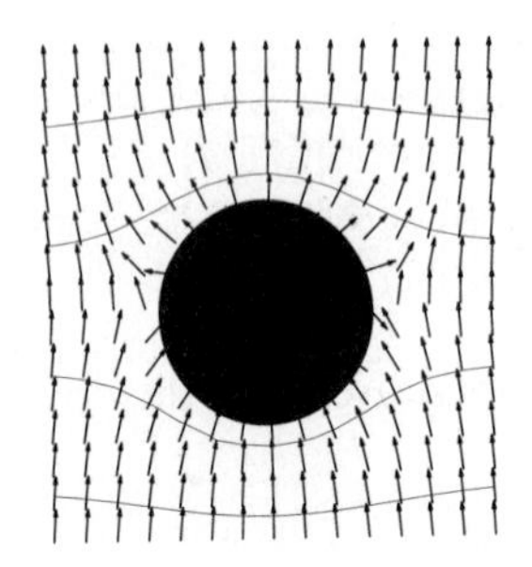

Problem 3: Sphere with an axially symmetric charge distribution

Consider a sphere with an axially symmetric charge distribution. Assume the sphere has a surface-charge density that is expressed in powers of $\mathrm{Cos}^n[\theta]$.

Part (a) Find the potential outside and inside the sphere for an arbitrary charge density.

Part (b) Consider a sphere with a surface-charge density given by $Q/(a^2)(\mathrm{Cos}[\theta]+(1/2)\mathrm{Cos}^2[\theta])$. Find the potential using the results derived in Part (a). Plot the potential in the $\{r, \theta\}$ plane.

Part (c) Plot the equipotential curves and electric field for the potential found in Part (b).

Remarks and outline From Gauss's law, $\nabla \cdot \vec{E} = 4\pi\rho$, the charge density on the sphere is equal to $4\pi(E_n^{outside} - E_n^{inside})$, where E_n is the normal component of the electric field evaluated on the surface of the sphere; E_n is proportional to the radial derivative of the potential, $(V'[r])$. Equate this expression for the charge density to the given charge density and solve for $A[i]$ in terms of the given surface-charge distribution.

Required packages

```
In[209]:= Needs["Calculus`VectorAnalysis`"]
          Needs["Graphics`ParametricPlot3D`"]
          Needs["Graphics`PlotField`"]
```

Solution

```
In[210]:= Clear["Global`*"];
```

Part (a) Use spherical coordinates and assume the solution is independent of the **phi** variable. The potential can be written as an expansion in Legendre polynomials and powers of **r**. Inside the sphere, the potential must be finite at the origin, so the first **s+1** terms of the potential expansion can be expressed as

```
In[211]:= potInA[s_]:=  Sum[( A[i] (r/a)^i  )*
                          LegendreP[i,Cos[theta]],
                              {i,0,s}]
```

The first **s+1** terms for the potential outside the sphere are

```
In[212]:= potOutA[s_]:=  Sum[ (A[i] (a/r)^(i+1) )  *
                          LegendreP[i,Cos[theta]],  {i,0,s}]
```

The coefficients **A[i]** are the same in both expressions, since the potential must be continuous across the sphere. From Gauss's law, $\nabla \cdot \vec{E} = 4\pi\rho$, the charge density on the sphere is equal to $4\pi(E_n^{outside} - E_n^{inside})$, where E_n is the normal component of the electric field evaluated on the surface of the sphere; E_n is proportional to the radial derivative of the potential, $(V'[r])$. The expression for the charge density is therefore

```
In[213]:= chargeDensity[s_]:=
            1/(4Pi)(D[potInA[s],r]-D[potOutA[s],r]) //.{r->a}
```

For example, the first three terms of the charge density are

```
In[214]:= chargeDensity[2]

Out[214]=
                                                           2
          A[0]    3 A[1] Cos[theta]    5 A[2] (-1 + 3 Cos[theta] )
          ---- + ----------------- + ---------------------------
           a             a                       2 a
          ---------------------------------------------------------
                                  4 Pi
```

To obtain relations to solve for the **A[i]** coefficients, equate **chargeDensity** to the given surface charge **Qdensity**. Then expand this to order **s+1** in powers of **Cos[theta]** and use **LogicalExpand** to obtain independent constraints from separate powers of **Cos[theta]**. These steps are all performed by the function **Aequation**:

```
In[215]:= Aequation[s_,Qdensity_]:=
              LogicalExpand[ chargeDensity[s]==Qdensity
                            +O[Cos[theta]]^(s+1)]
```

Qdensity is the charge density to be specified (hence the use of the delayed equals, **:=**). For a surface-charge density of $Q\text{Cos}[\theta]^2$, the first three terms of **Aequation** are

```
In[216]:= Aequation[2,Q Cos[theta]^2]

Out[216]=
                              A[0]   5 A[2]
                              ---- - ------
          3 A[1]               a      2 a                 15 A[2]
          ------ == 0 &&  ----------------- == 0 && -Q + ------- == 0
          4 a Pi                 4 Pi                     8 a Pi
```

Solve these equations to obtain rules for the **A[i]** coefficients:

```
In[217]:= Arule[s_,Qdensity_]:=
            Solve[ Aequation[s,Qdensity], Array[A,s+1,0]]//Flatten
```

For example, the first four **A[i]** coefficients for the above charge density are

```
In[218]:= Arule[4,Q Cos[theta]^2]

Out[218]=
                4 a Pi Q                          8 a Pi Q
          {A[0] -> --------, A[1] -> 0, A[2] -> --------, A[3] -> 0, A[4] -> 0}
                   3                                15
```

In this simple example, all **A[i]** for **i>2** are equal to zero. You can verify this by taking larger values of **s**.

Part (b) Consider the surface-charge density:

```
In[219]:= surfaceQ= Q/a^2 (Cos[theta]+1/2 Cos[theta]^2) //Expand

Out[219]=
                                2
          Q Cos[theta]   Q Cos[theta]
          ------------ + -------------
                2               2
               a             2 a
```

The potential follows from **potInA** and **potOutA** with values of the expansion coefficients given by **Arule**. The potential outside the sphere is

```
In[220]:= potOutQ= potOutA[2] //.Arule[2,surfaceQ]

Out[220]=
                                          2                        2
          2 Pi Q   4 a Pi Q Cos[theta]   2 a  Pi Q (-1 + 3 Cos[theta] )
          ------ + ------------------- + ------------------------------
          3 r                2                         3
                          3 r                      15 r
```

where **s=2**, since higher-order terms will vanish. The potential inside the sphere is

```
In[221]:= potInQ= potInA[2] //.Arule[2,surfaceQ]

Out[221]=
                                                   2                  2
          2 Pi Q   4 Pi Q r Cos[theta]   2 Pi Q r  (-1 + 3 Cos[theta] )
          ------ + ------------------- + ------------------------------
          3 a                2                         3
                          3 a                      15 a
```

Graph the potential in the $\{r, theta\}$ plane by joining the graphs for the potential inside and outside the sphere. Use the values

```
In[222]:= values={Q->1,a->1};
```

The graph for the potential inside the sphere, **0<r<1**, is (output suppressed)

```
In[223]:= p1=CylindricalPlot3D[potInQ //.values //Evaluate
                ,{r,0,1}
                ,{theta,0, 2 Pi}
                ,BoxRatios->{1,1,1}
                ,ViewPoint->{-1.5, -2, 2}];
```

and outside the sphere, **1<r**, is (output suppressed)

```
In[224]:= p2=CylindricalPlot3D[potOutQ //.values //Evaluate
                ,{r,1,3}
                ,{theta,0, 2 Pi}
                ,BoxRatios->{1,1,1}
                ,ViewPoint->{-1.5, -2, 2}];
```

The complete graph for the potential at all values of **r** is

```
In[225]:= Show[p1,p2];
```

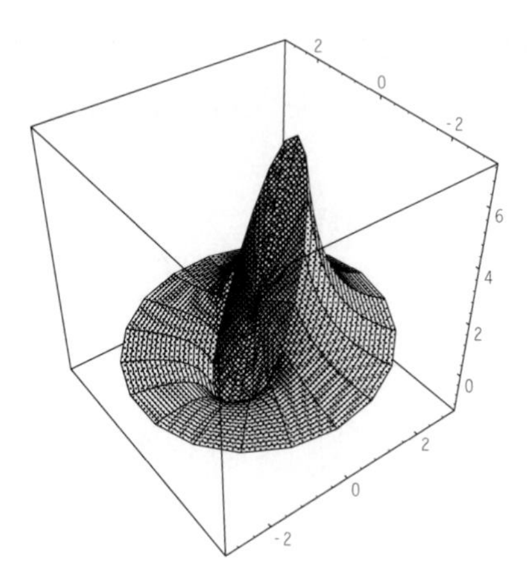

Part (c) Consider the plot for the electric field and equipotential curves in the **x=0** plane. Use **VEPlot** to graph the electric field and equipotential lines. You must express the potential in Cartesian coordinates. The transformation rule from spherical to Cartesian coordinates follows from the rule

```
In[226]:= r2xRule =
           {r,theta,phi}->
                CoordinatesFromCartesian[{x,y,z},Spherical]//Thread;
```

The potentials inside and outside the sphere expressed in Cartesian coordinates and restricted to the **x=0** plane are given by **eq1** and **eq2**, respectively:

```
In[227]:= (* for 0<r<1 *)
          eq1=potInQ/.r2xRule/.{Q->1,a->1}/.x->0//ExpandAll//Together;

          (* for 1<r<3*)
          eq2=potOutQ/.r2xRule/.{Q->1,a->1}/.x->0//ExpandAll ;
```

Use the command **If** to combine **eq1** and **eq2** into one function that is valid for all coordinate values. The condition **y^2+z^2<1** is **True** if you are inside the sphere. The explicit expressions for **eq1** and **eq2** are not entered by hand but are converted from the output of **eq1** and **eq2**. This is so that the **PlotGradientField** operator (inside the **VEPlot** command) will function properly. (This may not be necessary in future *Mathematica* versions.)

```
In[228]:= pot=If[y^2+z^2<1
            ,(2*(5*Pi - Pi*y^2 + 10*Pi*z + 2*Pi*z^2))/15
            ,(2*Pi)/(3*(y^2 + z^2)^(1/2))+
             (4*Pi*z)/(3*y^2*(y^2 + z^2)^(1/2) + 3*z^2*(y^2 + z^2)^(1/2)) -
             (2*Pi)/(15*y^2*(y^2 + z^2)^(1/2) + 15*z^2*(y^2 + z^2)^(1/2)) +
             (6*Pi*z^2)/(15*y^4*(y^2 + z^2)^(1/2) +
              30*y^2*z^2*(y^2 + z^2)^(1/2) + 15*z^4*(y^2 + z^2)^(1/2))];
```

Draw a circle to overlay on the plot (output suppressed):

```
In[229]:= circle= ParametricPlot[{Cos[t],Sin[t]},{t, 0,2Pi}
                    ,Axes->False
                    ,PlotStyle->{Thickness[0.030]}];
```

The graphics for the equipotential lines and electric field follow from (intermediate output suppressed)

```
In[230]:= Show[{VEPlot[pot,{y,-2,2.1},{z,-2,2.1}] ,circle}];
```

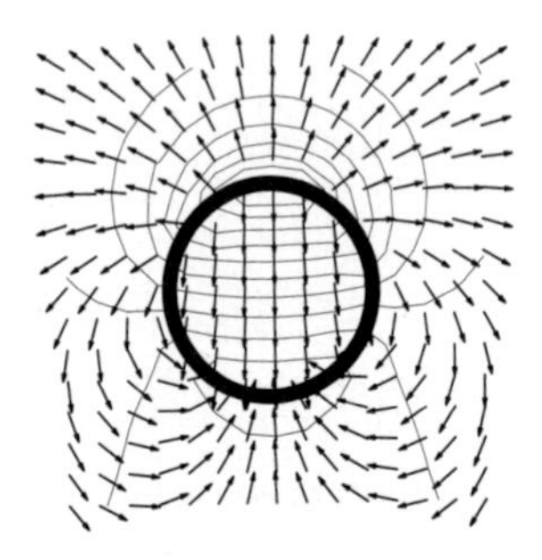

Problem 4: Sphere with a given axially symmetric potential

Consider a sphere whose surface is held at a fixed potential $V[\theta]$.

Part (a) Derive the expression for the potential inside and outside the sphere for an arbitrary surface potential $V[\theta]$. Illustrate with the example $V[\theta] = V_0\mathrm{Cos}[\theta]^2$.

Part (b) Find the potential when the surface potential is $\mathrm{Cos}[\theta]/(1.1 - \mathrm{Cos}[\theta])$. Plot the potential as a function of $\{r, \theta\}$.

Remarks and outline The potential can be written as a sum over Legendre polynomials, powers of **r**, and unknown coefficients **B[i]**. It follows from the orthogonality properties of Legendre polynomials that the coefficients in the expansion can be expressed in terms of an integral over the surface potential.

Required packages

```
In[231]:= Needs["Calculus`VectorAnalysis`"]
          Needs["Graphics`ParametricPlot3D`"]
          Needs["Graphics`PlotField`"]
```

Solution

```
In[232]:= Clear["Global`*"];
```

Part (a) The symmetry of the problem suggests you should use spherical coordinates $\{r, \theta, \phi\}$ and assume the solution to be independent of ϕ. Therefore, expand the potential inside the sphere in Legendre polynomials with positive powers of **r** so that the potential is finite at **r=0**:

```
In[233]:= potIn[V_,s_]:=
            Sum[ B[V,n] (r/r0)^n P[n,Cos[theta]],{n,0,s}]
```

where **r0** is the radius of the sphere and **P** are Legendre polynomials. Outside the sphere, expand the potential in Legendre polynomials with negative powers of **r** so that the potential vanishes at **r=∞**:

```
In[234]:= potOut[V_,s_]:=
            Sum[B[V,n] (r0/r)^(1+n) P[n,Cos[theta]],{n,0,s}]
```

Continuity of the potential on the sphere requires the coefficients **B[i]** to be the same in **potOut** and **potIn**. It follows from the orthogonality of the Legendre polynomials that the coefficients **B[i]** are

```
In[235]:= B[V_,n_]:=
              (2 n+1)/2 Integrate[V LegendreP[n,Cos[theta]]  Sin[theta]
                     ,{theta,0, Pi}]
```

Consider the potential **V0 Cos[theta]^2** as an example. The potential outside the sphere is

```
In[236]:= pIn= potIn[ V0 Cos[theta]^2,2]

Out[236]=
                                         2
          V0 P[0, Cos[theta]]       2 r  V0 P[2, Cos[theta]]
          ------------------- + ----------------------------
                   3                          2
                                          3 r0
```

and the potential inside is

```
In[237]:= pOut= potOut[ V0 Cos[theta]^2,2]

Out[237]=
                                              3
          r0 V0 P[0, Cos[theta]]       2 r0  V0 P[2, Cos[theta]]
          ---------------------- + -----------------------------
                   3 r                           3
                                              3 r
```

To plot the potential, choose the substitutions:

```
In[238]:= values={r0->1, V0->1, P->LegendreP};
```

First, plot the potential inside the sphere (output suppressed):

```
In[239]:= p1=CylindricalPlot3D[pIn /.values //Evaluate
               ,{r,0,1},{theta,0,2Pi}
               ,BoxRatios->{1,1,1}
               ,ViewPoint->{-1.5, -2, 2}];
```

Then plot the potential outside the sphere (output suppressed):

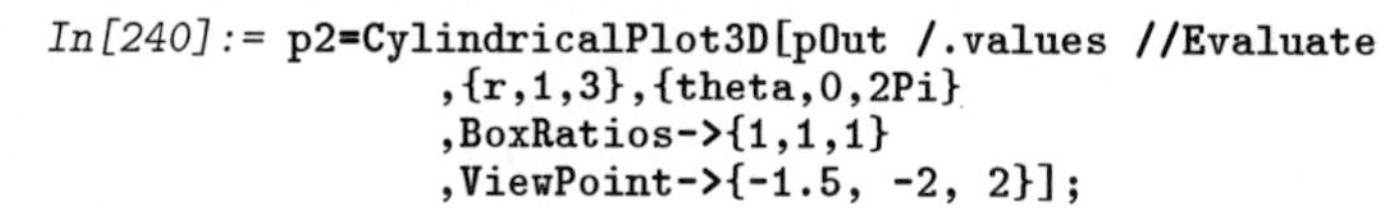

```
In[240]:= p2=CylindricalPlot3D[pOut /.values //Evaluate
                ,{r,1,3},{theta,0,2Pi}
                ,BoxRatios->{1,1,1}
                ,ViewPoint->{-1.5, -2, 2}];
```

```
In[241]:= Show[p1,p2];
```

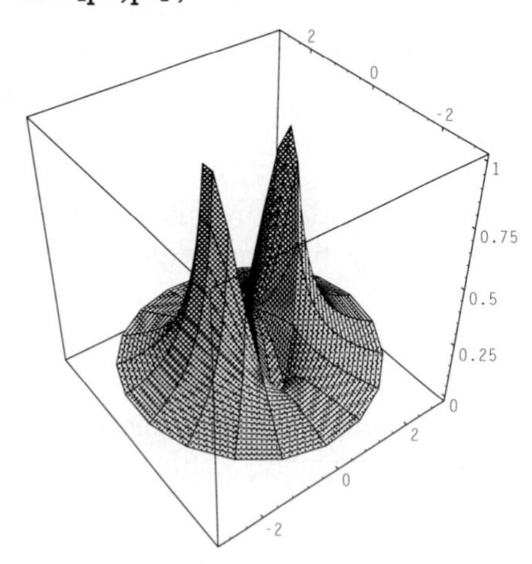

Part (b) Consider the case in which the radius of the sphere is **r0=1** and the surface potential is

```
In[242]:= surfacePot= Cos[theta]/(1.1-Cos[theta]);
```

The first four terms of the potential inside and outside the sphere are

```
In[243]:= {pIn,pOut}= ( {potIn[ surfacePot,4]
                       ,potOut[surfacePot,4]}
                         //. P->LegendreP
                         //.{r0->1}
                       ) //Simplify
```

```
Out[243]= {0.674487 + 2.22581 r Cos[theta] +

                     4                   2                 4
         0.138656 r  (3 - 30 Cos[theta]  + 35 Cos[theta] ) +

                     2
         0.483689 r  (1 + 3 Cos[2 theta]) +

                     3
         0.187938 r  (3 Cos[theta] + 5 Cos[3 theta]),

                                       2            4
        (0.30331 (0.514286 + 1.5947 r  + 2.22376 r  +

                                                 3
            1.85887 r Cos[theta] + 7.3384 r  Cos[theta] +

                                                 2
            1.14286 Cos[2 theta] + 4.78411 r  Cos[2 theta] +

                                                        5
            3.09812 r Cos[3 theta] + 2. Cos[4 theta])) / r }
```

First, plot the potential inside the sphere (output suppressed):

```
In[244]:= p1=CylindricalPlot3D[pIn
            ,{r,0,1},{theta,0,2Pi}
            ,BoxRatios->{1,1,1}
            ,ViewPoint->{-1.5, -2, 2}];
```

and then the potential outside the sphere (output suppressed):

```
In[245]:= p2=CylindricalPlot3D[pOut
            ,{r,1,3},{theta,0,2Pi}
            ,BoxRatios->{1,1,1}
            ,ViewPoint->{-1.5, -2, 2}];
```

Combining plots, you have the potential for all regions of space:

```
In[246]:= Show[p1,p2];
```

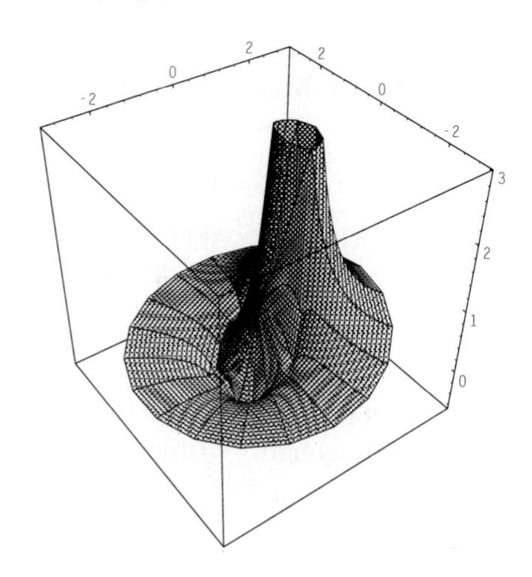

Problem 5: Sphere with upper hemisphere V and lower hemisphere $-V$

Consider a sphere with surface potential that is described by a discontinuous function. The top hemisphere has a potential $+V_0$, and the bottom hemisphere has a potential $-V_0$.

Part (**a**) Find the potential inside and outside the sphere. Plot the potential as a function of $\{r, \theta\}$.

Part (**b**) Plot the equipotential curves in the **x=0** plane.

Remarks and outline The potential can be written as a sum over Legendre polynomials. The expansion coefficients can be expressed in terms of a sum of two integrals: one over the portion of the sphere defined by $0 < \theta < \pi/2$ and the other over the portion $\pi/2 < \theta < \pi$.

Required packages

```
In[247]:= Needs["Calculus`VectorAnalysis`"]
          Needs["Graphics`ParametricPlot3D`"]
          Needs["Graphics`PlotField`"]
```

Solution

```
In[248]:= Clear["Global`*"];
```

Part (a) The symmetry of the problem suggests you should use spherical coordinates $\{r, \theta, \phi\}$ and assume the solution to be independent of ϕ. Therefore, expand the potential inside the sphere in Legendre polynomials with positive powers of **r** so that the potential is finite at **r=0**:

```
In[249]:=  potIn[s_] :=Sum[B[n](r/r0)^n  P[n,Cos[theta]],{n,0,s}]
```

where **r0** is the radius of the sphere and **P** are Legendre polynomials. Outside the sphere, expand the potential in Legendre polynomials with negative powers of **r** so that the potential vanishes at **r=**∞:

```
In[250]:=  potOut[s_] :=Sum[B[ n] (r0/r)^(1+n) P[n,Cos[theta]],{n,0,s}]
```

Continuity of the potential on the sphere requires the coefficients **B[i]** to be the same in **potOut** and **potIn**. It follows from the orthogonality of the Legendre polynomials that the coefficients **B[i]** are

```
In[251]:= B[n_]:=
          (2 n+1)/2(
           +Integrate[ V0 LegendreP[n,Cos[theta]]Sin[theta],{theta,0, Pi/2}]
           +Integrate[-V0 LegendreP[n,Cos[theta]]Sin[theta],{theta,Pi/2,Pi}]
          );
```

Note, we used the information that $V = +V_0$ for $0 < \theta < \pi/2$ and $V = -V_0$ for $\pi/2 < \theta < \pi$. The first few terms for the potential outside the sphere are

```
In[252]:= pOut=potOut[5]

Out[252]=      2                            4
          3 r0  V0 P[1, Cos[theta]]    7 r0  V0 P[3, Cos[theta]]
          -------------------------- - -------------------------- +
                      2                             4
                   2 r                           8 r

                6
           11 r0  V0 P[5, Cos[theta]]
           ---------------------------
                          6
                      16 r
```

and for the potential inside the sphere,

```
In[253]:= pIn=potIn[5]
```

```
Out[253]=
                                         3
          3 r V0 P[1, Cos[theta]]   7 r  V0 P[3, Cos[theta]]
          ----------------------- - ------------------------ +
                   2 r0                        3
                                           8 r0

             5
          11 r  V0 P[5, Cos[theta]]
          -------------------------
                       5
                   16 r0
```

Consider the values of the parameters

```
In[254]:= values={r0->1,V0->1};
```

and plot the potential given by **pIn** and **pOut**. The graphics for the potential when **0<r<1** (**pIn**) for fixed values of **r<1** are

```
In[255]:= Plot[ Table[ (pIn //.values //.P->LegendreP)
                        ,{r,0,1,0.2}]  //Evaluate
            ,{theta,0,2Pi}];
```

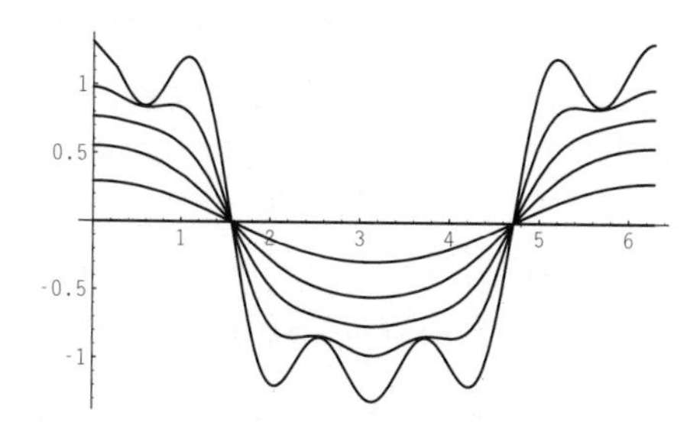

Note that as **r->r0**, the potential approaches the specified potential on the surface. The limits of the finite expansion are obvious as you approach the surface.

You can view the potential for both the inside and outside region by combining the graphics for each region. First, the graph for the potential when **0<r<1** (**pIn**) is (output suppressed)

```
In[256]:= plot1=CylindricalPlot3D[
            (pIn //.values //.P->LegendreP)
            ,{r,0,1}
            ,{theta,0,2Pi}
            ,BoxRatios->{1,1,1}
            ,ViewPoint->{-1.5, -2, 2}];
```

Next, the graph for the potential when **1<r** (**pOut**) is (output suppressed)

```
In[257]:= plot2=CylindricalPlot3D[
            (pOut //.values //.P->LegendreP)
            ,{r,1,3}
            ,{theta,0,2Pi}
            ,BoxRatios->{1,1,1}
            ,ViewPoint->{-1.5, -2, 2}];
```

The complete potential for all regions of space is

```
In[258]:= Show[plot1,plot2];
```

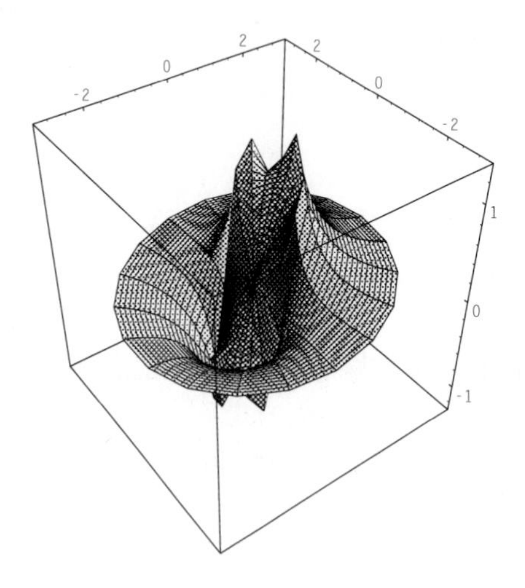

Part (b) To view the equipotential lines and electric field with **VEPlot**, first express the potential in Cartesian coordinates. The transformation from spherical to Cartesian coordinates follows from the rule

```
In[259]:= r2xRule =
           {r,theta,phi}->
                CoordinatesFromCartesian[{x,y,z},Spherical]//Thread;
```

Applying **r2xRule** to **pIn** and **pOut**, you get (with the restriction to the **x=0** plane)

```
In[260]:= allRules=Join[{P->LegendreP,x->0},values,r2xRule];

In[261]:= potTot= If[y^2+z^2<=1, pIn //.allRules ,pOut //.allRules]

Out[261]=      2    2
          If[y  + z  <= 1, pIn //. allRules, pOut //. allRules]
```

Next, draw a circle to overlay on the plot (output suppressed):

```
In[262]:= circle= ParametricPlot[{Cos[t],Sin[t]},{t, 0,2Pi}
                        ,Axes->False
                        ,PlotStyle->{Thickness[0.020]}];
```

Applying **ContourPlot** to **potTot**, you can see the equipotential lines for this potential (intermediate output suppressed):

```
In[263]:= Show[{
            ContourPlot[ potTot //Evaluate ,{y,-2,2},{z,-2,2}
                ,ContourSmoothing->True]
           ,circle
           }];
```

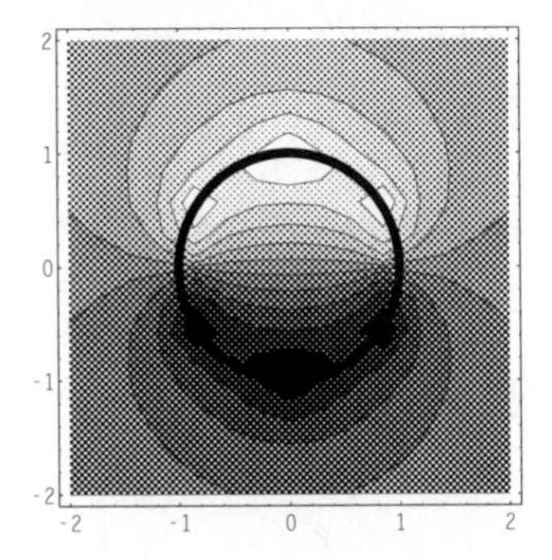

The three "islands" on the top and bottom boundaries of the spherical surface result from the finite number of terms in the potential expansion.

(Note, in principle, the electric field lines can be obtained by applying the **PlotGradientField** command to **potTot**. Unfortunately, this will not work with the present *Mathematica* version, so you must resort to techniques discussed in the previous two problems whereby the evaluated expression is explicitly inserted in the definition of the potential function.)

5.3 Unsolved Problems

Exercise 5.1: Parallel plates and strips

Consider parallel plates located in the $\{x, z\}$ plane. One plate is located at $y = 0$ and the other at $y = L$. The plate at $y = L$ is at zero potential. The plate at $y = L$ is at zero potential except for a strip along the x-axis from $x = \{+d, -d\}$. The strip is at potential V_0).

Part (a) Find the potential and graph the equipotential surfaces and electric field.

Part (b) Generalize the problem to allow for n parallel strips that have the same width, potential, and separation.

Exercise 5.2: Parallel plates and a point charge

Two infinite, grounded conducting planes are at $x = 0$ and $x = L$. A point charge q is placed between the planes.

Part (a) Find the charge induced on the planes and graph the results.

Part (b) Generalize Part (a) to include n charges between the plates. Graph the results for three charges.

Exercise 5.3: Cylinder divided into segments

Consider an infinite, hollow conducting cylinder of radius b. Divide the cylinder into two equal segments that run parallel to the axis of the cylinder. The top segment has constant potential V, and the bottom has potential $-V$.

Part (a) Expand the potential in cylindrical coordinates. Can you sum the series and get an exact expression for the potential?

Part (b) Graph the equipotential surfaces and electric field for the potential.

Part (c) Generalize the problem for a cylinder divided into $2n$ equal segments that have potentials that alternate in sign.

Exercise 5.4: Elliptical cylinder in a uniform electric field

An infinitely long, conducting elliptical cylinder bearing no net charge is placed in an initially uniform electric field E_0. The cylinder is grounded. The direction of E_0 is perpendicular to the cylinder's axis.

Part (a) Find the potential and its electric field.

Part (b) Plot the equipotential surfaces and electric field.

Exercise 5.5: Cylindrical box and n charges

Consider a conducting cylindrical box with one end at $z = 0$ and the other at $Z = L$. The radius of the cylinder is r_0. The walls are kept at zero potential. A positive charge is placed inside the cylindrical box.

Part (a) Find the potential inside the cylindrical box. Plot the equipotential surfaces and electric field.

Part (b) Find the charge induced on the conducting walls and plot the results.

Part (c) Generalize the results to include n charges inside the box.

Exercise 5.6: Distorted sphere with a potential

Consider a distorted sphere whose surface is held at a fixed potential $V[\theta]$. The shape of the surface is described by the radial coordinate $r = r_0(1 + cP[1, \mathrm{Cos}[\theta]])$. The parameter c and the Legendre polynomial P describe the distortion.

Part (a) Derive an expression for the potential inside and outside the sphere.

Part (b) Find the potential when the surface potential is $V[\theta] = \mathrm{Cos}[\theta]$. Plot the potential as a function of $\{r, \theta\}$.

Part (c) Plot the potential found in Part (b) in the two planes defined by $x = 0$ and $z = 0$.

Exercise 5.7: Concentric spheres with different potentials on the two hemispheres

Two concentric spheres have radii a, b ($b > a$) and are each divided into two hemispheres by the same horizontal plane. The upper hemisphere of the inner sphere and the lower hemisphere of the outer sphere are maintained at potential V. The other hemispheres are at zero potential.

Part (a) Determine the potential.

Part (b) Graph the electric field and equipotential surfaces.

Exercise 5.8: Ring charge

Consider a total charge q distributed around a circular ring of radius r_0. The axis of the ring is along the z-axis, and the center of the ring is located at $z = b$. The charge per unit length is a function of ϕ.

Part (a) Find the potential everywhere.

Part (b) For the charge density, $Q/r_0(\text{Cos}[\phi] - \text{Sin}[2\phi])$, find the asymptotic expression for the potential and electric field.

Part (c) Plot the results.

Exercise 5.9: Laplace's equations in other coordinates

Consider the coordinate systems **`EllipticCylindrical`**, **`Bispherical`**, and **`Toroidal`**. For each of these systems solve the following problems.

Part (a) Express Laplace's equation as a product of three functions and derive an ordinary differential equation for each function.

Part (b) Find the solutions for the functions and write the general expression for the potential.

Part (c) Write a procedure that will return the separated Laplace equations when you input a given coordinate system.

CHAPTER SIX

Quantum Mechanics

This chapter contains problems on nonrelativistic quantum mechanics found in standard texts such as *Quantum Physics* by Stephen Gasiorowicz, *Quantum Mechanics* by Leonard Schiff, *Quantum Mechanics* by A. S. Davydov, *Quantum Mechanics* by Philip Stehle, and *Quantum Physics of Atoms, Molecules, Solids, Nuclei, and Particles* by Eisberg and Resnick. We cover the essential points of Schrodinger's theory for both one-dimensional problems and three-dimensional problems for several kinds of coordinate systems. Many kinds of graphics are used to display the results, and you are encouraged to modify the potentials, boundary conditions, and graphics to better understand the solutions.

6.1 Introduction

6.1.1 Foundations of Quantum Mechanics

Historical beginnings

Neils Bohr in 1913 made one of the first attempts to formulate a quantum theory of the atom. Bohr proposed three postulates concerning the structure of an atom:

1. Atomic electrons move in orbits restricted by the requirement that the angular momentum ℓ be an integral multiple (n) of $h/(2\pi) = \hbar$. Furthermore, the electrons in these orbits do not radiate. For an electron of mass m and velocity v moving in a circular orbit of radius r, the angular momentum quantization restricts the electron's velocity according to the relation $\ell = mvr$, where the angular momentum $\ell = n\hbar$.
2. Transitions from one state to another are accompanied by a gain or loss of energy equal to the energy difference between the two states. The energy loss can appear as electromagnetic radiation.
3. If an electron falls from orbit n_i to n_f with a change in energy of $E_{n_i} - E_{n_f}$, the energy difference can appear as radiation with angular frequency $E_{n_i} - E_{n_f} = \hbar\omega$, where $\omega = 2\pi f$ and f is the frequency.

The success of Bohr's provisional theory with hydrogen-like atoms gave impetus for additional research on this quantum approach. The modern theory of quantum mechanics was ushered in with the works of Heisenberg, Born, Jordan, Schrodinger, and Dirac in the 1920s. Modern quantum mechanics uses wave functions to describe the probability that a particle will be found at a space-time point. The wave function is determined from the solution of Schrodinger's partial differential equation (we call this Schrodinger's equation in this book for short) with appropriate boundary conditions that describe the physical problem.

Time-independent quantum mechanics

The solution of Schrodinger's equation gives the wave function for a particle in a potential V. For time-independent potentials, it is possible to express the wave function as a product $Exp[-iE_n t/\hbar]\psi[x,y,z]$, where E_n is the energy of the particle. The time-independent wave function for a single particle of energy E_n in a potential V follows from the solution of Schrodinger's equation:

$$-\hbar^2/(2m)\Delta\psi + (V - E_n)\psi = 0.$$

The probability current or flux is given by the vector equation

$$\vec{J} = \hbar/(2im)(\psi^* \nabla\psi - \nabla\psi^* \psi).$$

The Hamiltonian operator yields the energy eigenvalues and is defined by

$$H = -\hbar^2/(2m)\Delta\psi + V\psi.$$

Schrodinger's equation is simplified if the space coordinates are restricted to one variable and the potential is a constant. The solutions are of the form

$$\psi[x] = c_1 e^{-ikx} + c_2 e^{ikx}.$$

The wave vector k is related to the energy and potential by the relation $k = \sqrt{2m(E_n - V)}/\hbar$, c_1 is the amplitude of a wave traveling to the left, and c_2 is the amplitude of a wave traveling to the right.

6.1.2 *Mathematica* Commands

Packages

You may wish to turn off the spell checker before starting this chapter:

```
In[1]:=  Off[General::spell ];
         Off[General::spell1];
```

Several packages are used in this chapter. You are advised to execute these packages first to avoid variables being shadowed. The packages are

```
In[2]:=  Needs["Algebra`Trigonometry`"]
         Needs["Calculus`VectorAnalysis`"]
         Needs["Graphics`ParametricPlot3D`"]
```

User-defined procedures and rules

It is useful to create user-defined procedures for calculations that are used repeatedly. To completely understand these procedures it may be necessary to decompose them. You are encouraged to make the procedures more time efficient and to add default conditions and options. Many of these procedures will fail in particular cases, and you are encouraged to write conditional statements to handle these cases.

Change of variables procedure

In[3]:=
```
vChange::usage=
 "vChange[dEq,psi,x,z,f] changes the variables in a differential
 equation.  When this operator is applied to a differential
 equation dEq of function psi[x], it will change the variables
 from x to z where x=f[z]";

vChange[dEq_,psi_,x_,z_,f_]:=
 ( dEq /.{ Derivative[n_Integer][psi][x]:>
                 Nest[ (D[#,z]/D[f,z])&, psi[z],n]
           ,psi[x] :> psi[z]
           ,x      :> f
         }
 )
```

Example: Change of variable

Consider the differential equation

In[4]:=
```
eq1= psi''[x] + x^n psi'[x] + psi[x] == 0
```

Out[4]=
```
                n
psi[x] + x  psi'[x] + psi''[x] == 0
```

and make the change of variables **x= alpha z**:

In[5]:=
```
vChange[eq1,psi,x,z,alpha z]
```

Out[5]=
```
                      n
           (alpha z)   psi'[z]    psi''[z]
psi[z] + ------------------- + ---------- == 0
                alpha               2
                                alpha
```

To observe what each step in **vChange** does, notice this step changes the variables in the derivative:

In[6]:=
```
eq2= eq1 /.{ Derivative[n_Integer][psi][x]:>
                 Nest[ (D[#,z]/D[alpha z,z])& ,psi[z],n] }
```

Out[6]=
```
            n
           x  psi'[z]    psi''[z]
psi[x] + ----------- + --------- == 0
            alpha           2
                        alpha
```

The next step changes the function and variable:

In[7]:=
```
eq3= eq2 /.{ psi[x] :> psi[z], x -> alpha z  }
```

Out[7]=
```
                      n
           (alpha z)   psi'[z]    psi''[z]
psi[z] + ------------------- + ---------- == 0
                alpha               2
                                alpha
```

Series expansion solution for second-order equation

This function is copied from Chapter 3. Refer to that chapter for details.

```
In[8]:=  diffSeriesOne::usage="
         diffSeriesOne[eq,z,t,order,{initialList}]
         eq is an equation, z is the function, t is the independent
         variable, order is the order of the Series solution, and
         {initialList} is a list of initial conditions such as
         {z[0]->z0}.
         This procedure assumes the solution can be written as
         a power series.";

In[9]:=  Clear[diffSeriesOne];
         diffSeriesOne[eqin_,z_,t_,order_,initList__List]:=
          Module[{eq,zSer,eSer,eqs,vars,sol,zSol},
            eq=eqin[[1]]-eqin[[2]]==0;
            zSer= Series[z[t],{t,0,order}];
            eSer= eq/.(z[t]->zSer)//ExpandAll;
            eqs = Thread[CoefficientList[eSer[[1]],t]==0];
            vars=(Table[D[z[t],{t,j}],{j,2,order+2}]/.t->0);
            sol = Solve[eqs,vars][[1]];
            zSol= zSer/.sol/.initList;
            Return[ zSol]
          ]
```

`HyperbolicToComplex` and `ComplexToHyperbolic`

This function is copied from Chapter 4. Refer to that chapter for details.

```
In[10]:= HyperbolicToComplexRule={Cosh[x___]:>(Exp[x]+Exp[-x])/2,
                                  Sinh[x___]:>(Exp[x]-Exp[-x])/2};

         ComplexToHyperbolicRule={Exp[x___]:>Cosh[x]+Sinh[x]};

In[11]:= HyperbolicToComplex[expression_]:=
                    expression //.HyperbolicToComplexRule
         ComplexToHyperbolic[expression_]:=
                    expression //.ComplexToHyperbolicRule
```

Complex conjugate rule

We find it convenient to introduce our own definition for **conjugate** rather than use the built-in **Conjugate**:

```
In[12]:= conjugate::usage =
         " A simple method of computing the conjugate of an object
          which is explicitly Complex";

         conjugateRule      = {Complex[re_,im_]:>Complex[re,-im]};
         conjugate[exp__] := exp /. conjugateRule;
```

Note: Using the function ensures we do not apply this in an endless loop.

Example: Complex exponential

When all complex values are explicit, **conjugate** does just what we need:

```
In[13]:= I Exp[I k x] // conjugate

Out[13]=     -I k x
         -I E
```

As expected, **conjugate** is its own inverse:

```
In[14]:= I Exp[I k x] == ( I Exp[I k x] //conjugate //conjugate)

Out[14]= True
```

User-defined solutions of differential equations

It is convenient to define solutions for certain differential equations that are not recognized by *Mathematica*. In this chapter, use the solutions for Hermite and Legendre polynomials.

Solution: Hermite polynomials

```
In[15]:= Unprotect[DSolve];

         DSolve[0 == +2 n_ h_[z_] - 2 z_ h_'[z_] + h_''[z_], {h_}, z_] :=
                  {{h->(HermiteH[n, #]&)}}   /; IntegerQ[n];

         DSolve[h_''[z_] == h_[z_]-(1+2 n_) h_[z_] + 2 z_ h_'[z_], {h_}, z_] :=
                  {{h->(HermiteH[n, #]&)}}   /; IntegerQ[n];

         Protect[DSolve];
```

Example: Hermite solution

Having defined this solution for this differential equation, you will find

```
In[16]:= n /: IntegerQ[n]=True;

         eq1=  0 == 2 n h[z] - 2 z h'[z] + h''[z];

         DSolve[eq1,{h},z][[1]]

Out[16]= {h -> (HermiteH[n, #1] & )}
```

Solution: Legendre polynomials

```
In[17]:= Unprotect[DSolve];

         DSolve[ (n_ h_[z_] + n_^2 h_[z_] + m_^2/(-1+z_^2) h_[z_]
                 - 2 z_ h_'[z_] + (1-z_^2) h_''[z_]) ==0, {h_}, z_]:=
           {{h->((LegendreP[n,m,#])&)}}  /; (IntegerQ[n]&&IntegerQ[m]);

         DSolve[ (n_ h_[z_] + n_^2 h_[z_] + m_^2/(-1+z_^2) h_[z_]
                 - 2 z_ h_'[z_] + (1-z_^2) h_''[z_]) ==0, {h_[z_]}, z_]:=
           {{h->(LegendreP[n,m,z])}}     /;  (IntegerQ[n]&&IntegerQ[m]);

         Protect[DSolve];
```

Example: Legendre solution

Having defined this solution for this differential equation, you will find

```
In[18]:= IntegerQ[el ]^=True;
         IntegerQ[mphi]^=True;

         eq1 = (el Th[z] + el^2 Th[z] + mphi^2/(-1+z^2) Th[z]
                              - 2 z Th'[z] + (1-z^2) Th''[z]==0 );

         DSolve[eq1,{Th},z][[1]]

Out[18]= {Th -> (LegendreP[el, mphi, #1] & )}
```

User-defined one-dimensional wave properties

There are several expressions and operators used extensively in one-dimensional quantum problems. These expressions and operators are defined in this subsection.

One-dimensional wave function with constant potential

```
In[19]:= waveExp::usage=" waveExp[x,k,c1,c2] yields the solution to the
         Schrodinger equation in a constant potential as a function of
         position x and wave number k. c1 and c2 are the coefficients
         of the right and left moving components, respectively.";

         waveExp[x_,k_,c1_,c2_] := c1 Exp[+I k x] + c2 Exp[-I k x]

         waveTrig::usage=" waveTrig[x,k,c1,c2] yields the solution to the
         Schrodinger equation in a constant potential as a function of
         position x and wave number k. c1 and c2 are the coefficients of
         the Sin and Cos components, respectively.";

         waveTrig[x_,k_,c1_,c2_] := c1 Sin[k x] + c2 Cos[k x]
```

6.1.3 User-defined Three-dimensional Quantum Equations

For general quantum problems, it is useful to define Schrodinger, Hamiltonian, and flux operators.

Operator: **schrodinger**

The **schrodinger** operator returns Schrodinger's equation when applied to the wave function:

```
In[20]:= schrodinger::usage = "schrodinger[V,Energy] is an operator to be
         applied to a wave function. The Schrodinger operator uses the
         default coordinates set by SetCoordinates. \n\n Example: \n\n
         SetCoordinates[Cartesian}];\n schrodinger[V[x,y,z],En]@psi[x,y,z]==0";

In[21]:= schrodinger[V_,En_]:= (-hbar^2/(2m) Laplacian[#] + (V-En)#)&
```

Example: Spherical potential and spherical wave function

Consider the wave function $\psi[r]$ and potential $V[r]$. Applying **schrodinger** to this wave function and setting the results to zero, we get Schrodinger's equation for $\psi[r]$:

```
In[22]:= SetCoordinates[Spherical];
         eq1= 0 == schrodinger[V[r],En] @ (psi[r]) //ExpandAll

Out[22]=
                                                   2            2
                                               hbar  psi'[r]   hbar  psi''[r]
         0 == -(En psi[r]) + psi[r] V[r] - ------------- - --------------
                                                 m r              2 m
```

Operator: **hamiltonian**

```
In[23]:= hamiltonian::usage = "hamiltonian[V] is an operator to be applied
         to a wave function. The Hamiltonian operator uses the default
         coordinates set by SetCoordinates. \n\n Example: \n\n
         SetCoordinates[Cartesian}];\n
         hamiltonian[V[x,y,z]]@psi[x,y,z]==En psi[x,y,z]";

In[24]:= hamiltonian[V_]:= (-hbar^2/(2m) Laplacian[#] + V #)&
```

Example: Harmonic oscillator

Consider the wave function $Exp[-a_1x^2]$ with potential $V[x] = (1/2)kx^2$. Applying **hamiltonian**, you get

```
In[25]:= SetCoordinates[Cartesian];
         wave= Exp[- a1 x^2];
         potential=1/2 k x^2;

         hamiltonian[potential][wave] //Simplify //PowerExpand

Out[25]=
                   2        2      2  2          2
         2 a1 hbar  - 4 a1   hbar   x   + k m x
         ----------------------------------------
                                 2
                             a1 x
                         2 E      m
```

Operator: **flux**

Define the flux operator, $(\hbar/(2im))(\psi^* \nabla\psi - \nabla\psi^* \psi)$:

```
In[26]:= flux::usage= "flux[psi] returns the directed flux.
         Coordinates are taken to be the default coordinate.
         All complex quantities are assumed to be explicit.";

          flux[psi_]:=
          ( hbar/(2 I m)
            ( (psi//conjugate) Grad[psi] - Grad[psi//conjugate] (psi) )
          )
```

Example: Plane wave flux

Let the wave function be $Exp[i(k_xx + k_yy + k_zz)]$. Set the coordinates to **Cartesian** and apply **flux**:

```
In[27]:= SetCoordinates[Cartesian];
         flux[Exp[I {kx,ky,kz}.{x,y,z}] ]

Out[27]=  hbar kx  hbar ky  hbar kz
         {-------, -------, -------}
            m        m        m
```

The **flux** operator gives the directed flow of current per unit area. The flux operator assumes all complex quantities are explicit.

6.1.4 Protect User-defined Operators

Next, protect all the user-defined operators. This allows you to use the command **Clear["Global`*"]** to erase all unwanted definitions without erasing the user-defined operators.

```
In[28]:= Protect[{ComplexToHyperbolic ,ComplexToHyperbolicRule
         ,HyperbolicToComplex ,HyperbolicToComplexRule
         ,conjugate ,conjugateRule, schrodinger, hamiltonian
         ,diffSeriesOne, vChange, flux, waveExp, waveTrig}];
```

6.2 Problems

6.2.1 One-dimensional Schrodinger's Equation

Problem 1: Particle bound in an infinite potential well

Suppose a particle is bound by the infinite potential well

$$V(x) = \begin{cases} 0 & \text{if } 0 < x < a; \\ \infty & \text{otherwise.} \end{cases}$$

Part (a) Solve Schrodinger's equation for the general wave function when $V = 0$. Compare this solution with the user-defined wave function **waveTrig**.

Part (b) Use the boundary conditions at the walls to determine the eigenfunctions. Normalize the eigenfunctions. Show that the energy eigenvalues follow from the user-defined operator **hamiltonian**.

Part (c) Explicitly verify that the first four wave functions are orthonormal and graph the wave functions.

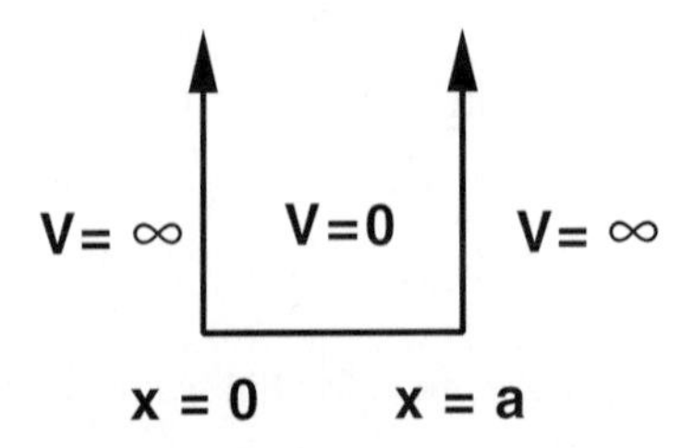

Remarks and outline This simple one-dimensional boundary problem illustrates the eigensolutions of Schrodinger's equation. The boundary conditions require the wave function to vanish at $x = 0$ and a. The energy eigenvalues follow from the user-defined operator **hamiltonian**.

Required packages

```
In[1]:= Needs["Algebra`Trigonometry`"]
        Needs["Calculus`VectorAnalysis`"]
        Needs["Graphics`ParametricPlot3D`"]
```

Solution

```
In[2]:= Clear["Global`*"];
```

Part (a) Schrodinger's equation follows from the user-defined operator **schrodinger**. Set the potential to zero, apply the operator **schrodinger** to the function **wave[x]**, and set the results to zero:

```
In[3]:= SetCoordinates[Cartesian[x,y,z]];

        eq1= 0==schrodinger[0,En] @ wave[x]  // ExpandAll

Out[3]=
                                    2
                                hbar  wave''[x]
        0 == -(En wave[x]) - ----------------
                                   2 m
```

The solution follows from **DSolve**:

```
In[4]:= dsol= ( DSolve[eq1,wave[x],x][[1]]
                //ComplexToTrig
                //ExpandAll
                )

Out[4]=
                                 Sqrt[2] Sqrt[En] Sqrt[m] x
        {wave[x] -> C[1] Cos[-----------------------------] +
                                          hbar

                     Sqrt[2] Sqrt[En] Sqrt[m] x
           C[2] Cos[-----------------------------
                               hbar

                       Sqrt[2] Sqrt[En] Sqrt[m] x
           I C[1] Sin[-----------------------------] +
                                 hbar

                       Sqrt[2] Sqrt[En] Sqrt[m] x
           I C[2] Sin[-----------------------------]}
                                 hbar
```

This solution is in agreement with the user-defined wave function **wavetrig**:

```
In[5]:= test= waveTrig[x,k,I(-C[1]+C[2]),  (C[1]+C[2]) ]//ExpandAll

Out[5]= C[1] Cos[k x] + C[2] Cos[k x] - I C[1] Sin[k x] + I C[2] Sin[k x]
```

where **k** is given by the rule

```
In[6]:= kRule={k-> Sqrt[2 En m]/hbar};
```

To verify this identity, notice that

```
In[7]:=  test == wave[x] /.dsol /.kRule //PowerExpand

Out[7]=  True
```

Part (b) The constants **{C[1],C[2],k}** are determined by the boundary conditions at the infinite barrier and by normalizing the wave function. Requiring the wave function to vanish at **x={0,a}** dictates that only **Sin[k x]** terms be kept and that you set **k=n Pi/a**, where **n** is a positive integer. The allowable eigensolutions are

```
In[8]:=  waveTrig[x,n Pi/a,c1,0] //ExpandAll

Out[8]=           n Pi x
         c1 Sin[------]
                    a
```

The constant **c1** follows from normalizing the wave function. You can get an expression for the normalization constant from

```
In[9]:=  norm[psi_ , x_, a0_,a1_] :=  (
            1/Sqrt[
               Integrate[(psi//conjugate )psi ,{x,a0,a1}]  ]
               //Simplify//PowerExpand
                   )
```

With some simplification, you will find

```
In[10]:= normConst=
         ( norm[ (waveTrig[x,n Pi/a, c1,0]),x,0,a]
              /.{Sin[2 n Pi]->0}
              // PowerExpand
           )

Out[10]=   Sqrt[2]
          ----------
          Sqrt[a] c1
```

The normalized wave function is

```
In[11]:=  psi[x_,n_,a_] = normConst waveTrig[x,n Pi/a, c1,0]

Out[11]=                 n Pi x
          Sqrt[2] Sin[------]
                           a
          -------------------
               Sqrt[a]
```

The energy levels follow from the expression for **k** or from applying the **hamiltonian** operator to the wave function and setting it equal to energy times the wave function:

```
In[12]:= hamEq= energy  psi[ x,n,a] == (hamiltonian[0]@psi[ x ,n,a  ])

Out[12]=                       n Pi x            2  2   2      n Pi x
         Sqrt[2] energy Sin[------]         hbar  n   Pi  Sin[------]
                                a                                a
         -------------------------- == ---------------------------------
                  Sqrt[a]                                 5/2
                                               Sqrt[2] a     m
```

The energy eigenvalues follow by solving for the energy:

```
In[13]:= eSol = Solve[hamEq,energy][[1]]
Out[13]=
                    2  2   2
                hbar  n  Pi
{energy -> ---------------}
                      2
                 2 a  m
```

The energy levels differ by a factor of n^2 and are given by the function

```
In[14]:= Energy[n_]  = energy /.eSol
Out[14]=
    2  2   2
hbar  n  Pi
-----------
        2
   2 a  m
```

Part (c) Integrate over the product of the wave functions to verify that they are orthonormal. Forming a table of the integrations over the products of the first four eigenfunctions, you get

```
In[15]:= table= Table[Integrate[psi[x,i,a] psi[x,j,a],{x,0,a}],{i,4},{j,4}];
         TableForm[table ,TableHeadings->{Table[psi[x,i],{i,4}]
                                          ,Table[psi[x,i],{i,4}]}]
Out[15]=
```

	psi[x, 1]	psi[x, 2]	psi[x, 3]	psi[x, 4]
psi[x, 1]	1	0	0	0
psi[x, 2]	0	1	0	0
psi[x, 3]	0	0	1	0
psi[x, 4]	0	0	0	1

This table verifies that the wave functions are orthonormal.

The graphics for the first four wave functions are

```
In[16]:= waveplot=
           Plot[ Evaluate[ Table[psi[n,x,1],{n,4}] ]
                ,{x,0,1}
                ,PlotStyle->
                  {Dashing[{.01,.07}],Dashing[{.03,.05}],
                   Dashing[{.05,.03}],Dashing[{.07,.01}]},
               PlotLabel-> Evaluate[FontForm[y,{"Symbol",10}]]
               , DisplayFunction->Identity
               ];
```

Define an operator **text** to add labels to the graph:

```
In[17]:= text[name_,position_List] :=
           Text[FontForm[  name ,{"Courier-Bold",10}],position]
```

Combining the wave function graphics and **text**, you get

```
In[18]:= Show[waveplot, Graphics[
         { text["n=1", {.5, 1.5}], text["n=2",{.7,-1.5}],
           text["n=3", {.5,-1.5}], text["n=4",{.3,-1.5}] }]
          , DisplayFunction->$DisplayFunction

          ];
```

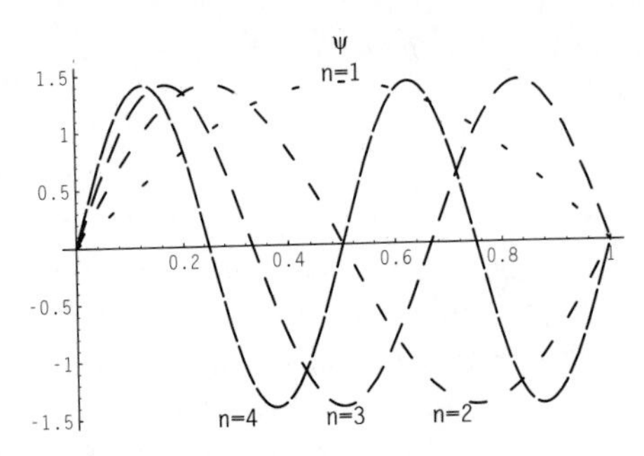

Notice the wave functions with $n = \{1, 3, 5, \ldots\}$ are symmetric in the well and the wave functions with $n = \{2, 4, \ldots\}$ are asymmetric.

Problem 2: Particle bound in a finite potential well

Suppose a particle with negative energy $0 > E_n = -W_n$ is bound by a finite potential well:

$$V(x) = \begin{cases} -V_0 & \text{if } -a < x < a; \\ 0 & \text{otherwise.} \end{cases}$$

We assume $0 < W_n < V_0$.

Part (a) Use the condition that the wave function must vanish as $x \to \infty$ or $x \to -\infty$ and write the general form of the wave in the regions to the left and right and in the center of the well.

Part (b) Require the wave function and its derivative to be continuous across the boundaries of the well and show that there are two classes of solutions: odd and even. Find the relation between the wave function constants and derive an equation for the energy eigenvalues.

Part (c) Express the energy eigenvalues W_n in a form similar to the infinite well by measuring the energy levels from the bottom of the well and parameterizing the values with n, where $W_n = V_0 - n^2\hbar^2\pi^2/(8a^2m)$. Solve for the allowed n values by plotting the two sides of the eigenvalue equation and observe where the curves intersect. Verify that the values of n approach the integer values of the infinite well problem as $V \to \infty$.

Part (d) Solve for the n values a second way by finding the roots of the eigenvalue equation using the command **`FindRoot`**.

Part (e) Plot the various wave functions for a given value of the potential.

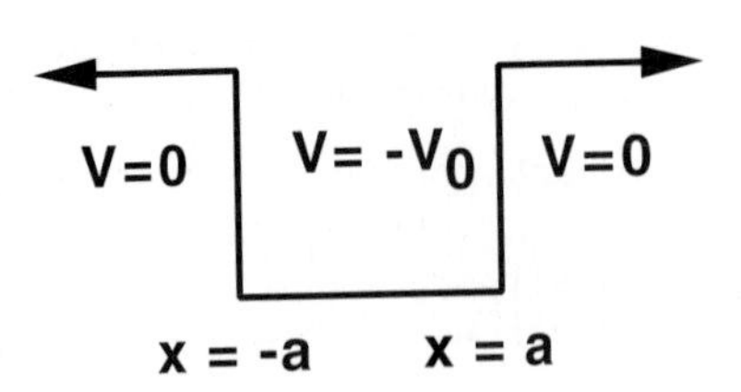

Required packages

```
In[19]:= Needs["Algebra`Trigonometry`"]
         Needs["Calculus`VectorAnalysis`"]
         Needs["Graphics`ParametricPlot3D`"]
```

Remarks and outline This problem is the finite generalization of the infinite well solved in Problem 1. The wave function follows from the solution of the one-dimensional Schrodinger's equation having a constant potential. The solution for the wave function in the three regions $x < -a$, $-a < x < a$, and $a < x$ must be treated separately. The wave function solution for a constant potential is defined in the Introduction. The relevant user-defined functions are **waveTrig** and **waveExp**. Inside the well the wave function oscillates; outside the well, the wave function exponentially decays. The eigenfunctions and energy eigenvalues follow from the boundary condition that the wave function and its derivative must be continuous across the boundaries. The eigenvalue equation is a transcendental equation that can be solved by either graphical or numerical methods.

Solution

```
In[20]:= Clear["Global`*"];
```

Part (a) The solution for the one-dimensional Schrodinger's equation having a constant potential was given in the Introduction. The wave function can be expressed either in trigonometric or exponential form. Take the user-defined function **waveTrig** for the wave in the well and **waveExp** for the wave on either side of the well.

First, consider the solution in the well ($-a < x < a$), where $V < E_n < 0$. The wave function oscillates and is of the form

```
In[21]:= psiW[x_]=  waveTrig[x,ki,A,B]

Out[21]= B Cos[ki x] + A Sin[ki x]
```

where **ki** follows from first solving the equation $\hbar^2 k^2/(2m) = E_n - V$ for k. The solution is trivial:

```
In[22]:= kRule= {k -> Sqrt[2 m(En - V)]/hbar} //PowerExpand;
```

The rule for **ki** follows from

```
In[23]:= kiRule= {ki -> k} /.kRule //.{V->-V0, En->-Wn}

Out[23]=          Sqrt[2] Sqrt[m] Sqrt[V0 - Wn]
         {ki -> -------------------------------}
                             hbar
```

The solution has been expressed in terms of **Wn** and **V0**, where **V0>Wn>0** (recall **En<0**). It will follow that the wave described by **psiW** is symmetric about the center of the well if **A=0** and asymmetric if **B=0**. Outside the well, $x < -a$ and $a < x$, the wave function decays, and the potential is zero. In general, the wave function has both an exponentially decreasing and increasing term:

```
In[24]:= waveExp[x,k,c1,c2]

Out[24]=     -I k x        I k x
         c2 E       + c1 E
```

Choose the values of **c1** and **c2** so that the wave vanishes to the right of the well as **x** goes to $+\infty$ and to the left of the well as **x** goes to $-\infty$. The solution to the left of the well is of the form

```
In[25]:= psiL[x_]= waveExp[x,I k0,0,C]        (* x < -a *)

Out[25]=    k0 x
         C E
```

and to the right of the well,

```
In[26]:= psiR[x_]= waveExp[x,I k0,D,0]        (* x > a *)

Out[26]=  D
         ----
          k0 x
         E
```

where **k0** is

```
In[27]:= k0Rule = ( {k0->-I k} //. kRule
                               //.{V->0, En->-Wn}
                               //PowerExpand
                  )

Out[27]=         Sqrt[2] Sqrt[m] Sqrt[Wn]
         {k0 -> ------------------------}
                          hbar
```

Make **k0** explicitly imaginary so that the wave function is properly damped at infinity.
Part (b) The boundary condition requires the wave function and its derivative to be continuous at the boundaries. Four equations follow from applying these continuity conditions at $x = a$ and $x = -a$:

```
In[28]:= eq2= {  (psiL[ x]-psiW[ x]==0)   /. {x->-a},
                 (psiW[ x]-psiR[ x]==0)   /. {x->+a},
                 (psiL'[x]-psiW'[x]==0)   /. {x->-a},
                 (psiW'[x]-psiR'[x]==0)   /. {x->+a}  };
```

Apply **Reduce** to **eq2** and search for restrictions on **A** (this takes about one min.):

```
In[29]:= eq3= Reduce[eq2,A];
         Short[eq3,4]

Out[29]=            a k0
         ki != 0 && 2 B E     == (C + D) Sec[a ki] && k0 == ki Tan[a ki] &&

                          D Sec[a ki]
           A == I (B - ---------------) && Tan[a ki] == I || <<9>> ||
                              a k0
                             E

           Cos[a ki] != 0 && Sin[a ki] != 0 && A == 0 && B == 0 && ki == 0 &&

           C == 0 && D == 0
```

Display only the short form of the solution because of its length. However, an inspection of the solutions reveal that there are two relevant classes: (1) **A=0** and (2) **B=0**. The **A=0** solutions are the symmetric solutions, and the **B=0** solutions are the asymmetric solutions. Again search for the solutions of **eq2**, but this time impose **{ A==0, C !==0, k0 !==0}**:

```
In[30]:= eq4= Reduce[{eq2, A==0, C !=0, k0 !=0}//Flatten ,B]
Out[30]= ki != 0 && Sin[a ki] != 0 && Cos[a ki] != 0 && C != 0 &&

                                                       C k0 Csc[a ki]
         k0 Cos[a ki] == ki Sin[a ki] && D == C && B == -------------- &&
                                                          a k0
                                                         E     ki

         A == 0
```

The values for the wave function constants follow from solving **eq4** for **{A,B,C}**:

```
In[31]:= symSol= Solve[eq4,{A,B,C}][[1]]
Out[31]=
                         D Sec[a ki]
         {A -> 0, B -> ------------, C -> D}
                            a k0
                           E
```

The wave function constants are expressed in terms **D**, which can be determined by normalizing the solution. An explicit value for **D** is not necessary for this problem, so you do not need to determine its value. The eigenvalue equation for the symmetric solutions is given by the equation **ki Sin[a ki]==k0 Cos[a ki]**. You can use **Select** to extract the proper equation:

```
In[32]:= symRel=
           Select[eq4,((!FreeQ[#,Cos[a ki]])&&(!FreeQ[#,k0]))& ]
Out[32]= k0 Cos[a ki] == ki Sin[a ki]
```

Dividing both sides by **ki Cos[aki]**, you get the eigenvalue equation

```
In[33]:= symEn= Thread[symRel/(ki Cos[a ki]),Equal]
Out[33]= k0
         -- == Tan[a ki]
         ki
```

Similarly, the asymmetric solutions follow from setting **{B==0, C !=0, k0 !=0}** and applying **Reduce** to **eq2**:

```
In[34]:= eq5= Reduce[{eq2, B==0, C !=0, k0 !=0}//Flatten,A]
Out[34]= ki != 0 && Cos[a ki] != 0 && Sin[a ki] != 0 && C != 0 &&

         k0 Sin[a ki] == -(ki Cos[a ki]) && D == -C &&

              C k0 Sec[a ki]
         A == -------------- && B == 0
                 a k0
                E     ki
```

The wave function constants are

```
In[35]:= asymSol= Solve[eq5 ,{A,B,C}]
Out[35]=
                          D Csc[a ki]
         {{B -> 0, A -> ------------, C -> -D}}
                             a k0
                            E
```

and the eigenvalue equation is

```
In[36]:= asymRel= Select[eq5,((!FreeQ[#,Cos[a ki]])&&(!FreeQ[#,k0]))& ]

Out[36]= k0 Sin[a ki] == -(ki Cos[a ki])
```

Dividing both sides by **ki Sin[aki]**, you get the eigenvalue equation

```
In[37]:= asymEn=  Thread[asymRel/(ki Sin[a ki]),Equal]

Out[37]= k0
         -- == -Cot[a ki]
         ki
```

Part (c) The allowed energy values follow from solving **asymEq** and **symEq** for **Wn**, where **ki** and **k0** are given by **kiRule** and **k0Rule**. Combine these two rules into one general rule:

```
In[38]:= kRule2  =  {kiRule ,k0Rule} //Flatten

Out[38]=
                Sqrt[2] Sqrt[m] Sqrt[V0 - Wn]          Sqrt[2] Sqrt[m] Sqrt[Wn]
         {ki -> -----------------------------, k0 -> ------------------------}
                            hbar                               hbar
```

There are two procedures for solving the energy eigenvalue equations. You can plot the left-hand and right-hand sides of the eigenvalue equation as a function of **Wn** and observe where the curves intersect. The intersection of these curves gives the eigenvalues. Or you can directly compute the roots of the eigenvalue equations numerically with **FindRoot**. Both methods are illustrated in this section.

Before solving for the energy eigenvalues, express the energy **Wn** in a form that is similar to the expression for the energy levels in an infinite well. Measure the energy levels from the bottom of the well and parameterize the energy with n, where the energy is similar to that of the Bohr atom, $E_n = n^2\hbar^2\pi^2/(8a^2m)$. The effective n value is related to **Wn** by the rule

```
In[39]:=  nRule= Wn-> V0- n^2 hbar^2 Pi^2/(8 a^2 m) ;
```

First, consider the energy values for the symmetric solutions. The eigenvalue equation for n follows from **symEn**

```
In[40]:= eq6 =symEn/.kRule2/.nRule//PowerExpand

Out[40]=
                                          2  2   2
                                   -(hbar   n   Pi )
         2 Sqrt[2] a Sqrt[m] Sqrt[------------------ + V0]
                                           2
                                        8 a  m                  n Pi
         ------------------------------------------------- == Tan[----]
                              hbar n Pi                          2
```

where **nRule** was used to express the relation in terms of n. The allowed values for n follow from plotting the left-hand and right-hand sides of **eq6** as a function of n and observing where the curves intersect. Plot these curves for the values

```
In[41]:=  values={a->1, m->1, hbar->1, V0->{100,200,500, Infinity} };
```

Four different values have been considered for the potential. Instead of plotting the left-hand and right-hand sides of **eq6**, you can equivalently plot the **ArcTan** of the two sides:

```
In[42]:= plot1=
         Plot[ {ArcTan[eq6[[1]]],ArcTan[eq6[[2]]]}
                 /.values
                 //Flatten
                 //Evaluate
              ,{n,0.01,9}
              ,Frame->True
              ,FrameTicks->{  {{1,"n=1"},{3,"n=3"},{5,"n=5"}
                                  ,{7,"n=7"},{9,"n=9"}}
                            ,{0,0.5,1,1.5}  }
              ,PlotRange->{0,2}
              ,DisplayFunction->Identity
           ];
```

Add the text to the graphics in **plot1**:

```
In[43]:= text ={
          Text[FontForm["V0=Inf",{"Courier-Bold",10}],{2,1.7  }],
          Text[FontForm["V0=500",{"Courier-Bold",10}],{5.7,1.4}],
          Text[FontForm["V0=200",{"Courier-Bold",10}],{7.6,1  }],
          Text[FontForm["V0=100",{"Courier-Bold",10}],{7.7,0.7}]};
```

and display the results:

```
In[44]:= Show[{plot1,Graphics[text]}, DisplayFunction->$DisplayFunction];
```

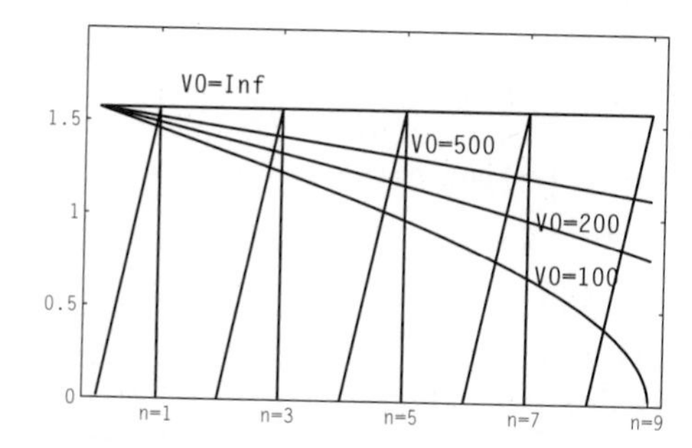

The sloping horizontal curves are one side of the eigenvalue equation for the four different values of the potential. The slanted vertical lines are the other side of the eigenvalue equation. The fully vertical lines correspond to integer n values. As the value of the potential goes to infinity, the intersections approach an odd integer $\{1, 3, 5, 7, \ldots\}$. The asymmetric solutions follow from the equation **asymEn**:

```
In[45]:= eq7 =asymEn /.kRule2 /.nRule //PowerExpand
Out[45]=
                                              2  2   2
                                       -(hbar   n  Pi )
         2 Sqrt[2] a Sqrt[m] Sqrt[---------------- + V0]
                                              2
                                           8 a  m                  n Pi
         ------------------------------------------------- == -Cot[----]
                            hbar n Pi                               2
```

Taking the **ArcTan** of both sides of **eq7** and plotting the results, you get

```
In[46]:= Plot[ {ArcTan[eq7[[1]]],ArcTan[eq7[[2]]]}
              /.values
              //Flatten
              //Evaluate
          ,{n,0.01,9}
          ,Frame->True
          ,FrameTicks->{ {{2,"n=2"},{4,"n=4"}
                          ,{6,"n=6"},{8,"n=8"}}
                        ,{0,.5,1,1.5}  }
          ,PlotRange->{0,2 }
          ,Epilog->Graphics[text][[1]]
     ];
```

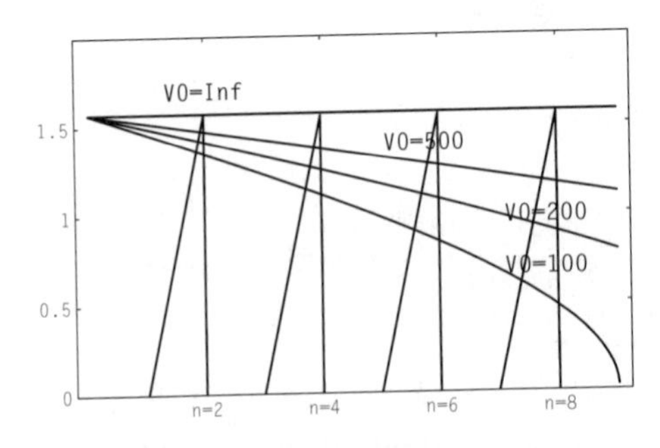

The values of n approach even integers $\{2, 4, 6, \ldots\}$ as **V0** goes to infinity.

Part (d) Next, solve for the symmetric and asymmetric n values by using the command **FindRoot**. The n values follow from the function

```
In[47]:= nValues[eq_, potential_, guess_]:=
         ( n /.FindRoot[eq /.{m->1,a->1,hbar->1,V0->potential }
                           //Evaluate
                       ,{n,guess}][[1]]  )
```

The term **eq** stands for **eq6** in the case of the symmetric solutions and for **eq7** in the case of the asymmetric solutions. To find the roots, an initial **guess** must be given. For the symmetric solutions, consider three different values for the potential, **V0={100, 200, 500}**, and solve for the first five eigenvalues. Make an initial guess for the five values of n:

```
In[48]:= symGuess= {0.97,2.9,4.8,6.8,8.7 };
```

The values of n are

```
In[49]:= symValues= {nValues[eq6,100,#]
                    ,nValues[eq6,200,#]
                    ,nValues[eq6,500,#]}& /@ symGuess;
```

The list for the asymmetric solution follows similarly:

```
In[50]:= asymGuess=    {1.97,3.9,5.8,7.8  };

         asymValues={nValues[eq7,100,#]
                    ,nValues[eq7,200,#]
                    ,nValues[eq7,500,#]}& /@ asymGuess;
```

Combining the solutions in **eq8** and **eq9** and displaying the results in table form, you will find

In[51]:=
```
( Partition[ Sort[{symValues,asymValues} //Flatten
                                          //N[#,3]&  ]
           ,3]
   //TableForm[# ,TableSpacing -> {0,8}
                ,TableHeadings-> {
                    {"n=1","n=2","n=3","n=4","n=5"
                          ,"n=6","n=7","n=8","n=9"}
                    ,{"V0=100","V0=200","V0=500"}  }
              ]&
)
```

Out[51]=

	V0=100	V0=200	V0=500
n=1	0.934	0.952	0.969
n=2	1.87	1.9	1.94
n=3	2.8	2.86	2.91
n=4	3.73	3.81	3.88
n=5	4.65	4.76	4.85
n=6	5.57	5.7	5.81
n=7	6.49	6.65	6.78
n=8	7.39	7.59	7.75
n=9	8.26	8.53	8.71

Part (d) Graph the first nine wave functions for the potential **V0=100**. Choose the parameters to be

In[52]:=
```
symRules= {kRule2,nRule,symSol,
           {a->1,hbar->1,m->1,V0->100,D->100}}//Flatten;

asymRules= {kRule2,nRule,asymSol,
           {a->1,hbar->1,m->1,V0->100,D->100}}//Flatten;
```

Also define a function valid for all values of **x**. The complete wave function to the left and right and in the center of the well becomes

In[53]:=
```
Clear[sPsi,aPsi];

(*Symmetric*)
sPsi[x_ /;     x<-1,n0_]:= (psiL[x]//.symRules//.{n->n0});
sPsi[x_ /; -1<=x<1,n0_]:= (psiW[x]//.symRules//.{n->n0});
sPsi[x_ /;     x>=1,n0_]:= (psiR[x]//.symRules//.{n->n0});

(*Asymmetric*)
aPsi[x_ /;     x<-1,n0_]:= (psiL[x]//.asymRules//.{n->n0});
aPsi[x_ /; -1<=x<1,n0_]:= (psiW[x]//.asymRules//.{n->n0});
aPsi[x_ /;     x>=1,n0_]:= (psiR[x]//.asymRules//.{n->n0});
```

where **n0** are the eigenvalues of the wave function.

The five symmetric energy eigenvalues **Wn** are given by

In[54]:=
```
symEnergy= nValues[eq6,100,#]& /@ symGuess
```

Out[54]=
```
{0.933848, 2.79876, 4.65414, 6.48773, 8.26041}
```

The graphics for the corresponding wave functions are

```
In[55]:= plotsym=

         Plot[ sPsi[x,#] //.symRules //Evaluate
                ,{x,-2,2}
                ,PlotLabel->  N[#,3]
                ,Axes->None
                ,DisplayFunction->Identity ]& /@ symEnergy;
```

Displaying the results with **GraphicsArray**, you get

```
In[56]:= Show[GraphicsArray[{plotsym[[{1,2,3}]],plotsym[[{4,5}]]}]];
```

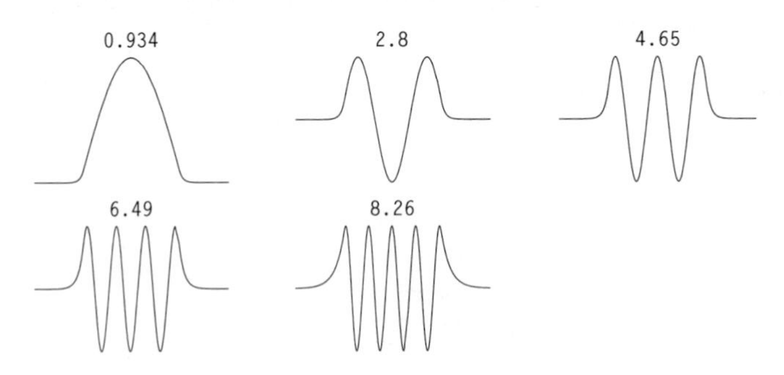

Likewise, the plot for the four asymmetric wave functions follow from

```
In[57]:= asymEnergy= nValues[eq7,100,#]& /@ asymGuess;

         plotasym=
           Plot[ aPsi[x,#] //.asymRules //Evaluate
                 ,{x,-2,2}
                 ,PlotLabel->N[#,3]
                 ,Axes->None
                 ,DisplayFunction->Identity]& /@ asymEnergy;
```

Displaying the results, it follows that

```
In[58]:= Show[GraphicsArray[{plotasym[[{1,2}]],plotasym[[{3,4}]]}]];
```

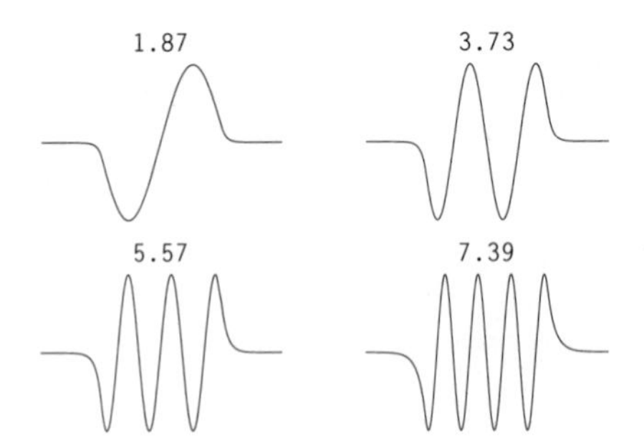

Problem 3: Particle hitting a finite step potential

Consider a particle of energy E approaching a potential barrier

$$V(x) = \begin{cases} 0 & \text{if } x < 0; \\ V_0 & \text{if } 0 < x. \end{cases}$$

The wave function to the left of the barrier ($x < 0$) has an incident wave and a reflected wave. The wave to the right of the barrier ($x > 0$) has only a transmitted wave.

Part (**a**) Write the general form for the wave function in the regions $x < 0$ and $x > 0$. Use the boundary condition at the barrier to evaluate the amplitudes for the reflected and transmitted waves.

Part (**b**) The coefficients of transmission, T, and reflection, R, are defined as the ratio of the transmitted and reflected current densities compared to the incident current density, respectively. Assume that $E_n > V_0$ and find T and R. Plot R and T as a function of E_n.

Part (**c**) Plot the real and imaginary parts of the wave function for energies E_n above and below the barrier.

Part (**d**) Use **Plot3D** to graph the wave function as a function of the two parameters $\{x, E_n\}$.

Part (**e**) Animate the real parts of the incident, reflected, and transmitted waves. Do the animations for energies above and below the barrier.

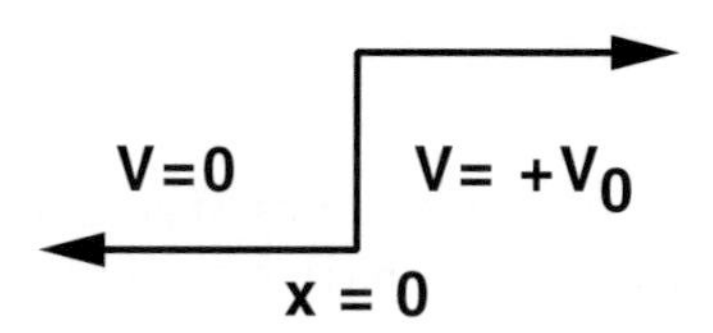

Remarks and outline The general solution of Schrodinger's equation in a constant potential is given by the user-defined function **waveExp** defined in the Introduction. The wave solution in the two regions $x < 0$ and $0 < x$ must be treated separately. The boundary conditions determine the wave function amplitudes in terms of the incident amplitude. The current densities follow from applying the user-defined function **flux** to the appropriate parts of the wave function. The animation is generated by forming a table of wave plots evaluated as a function of time.

Required packages

```
In[59]:= Needs["Algebra`Trigonometry`"]
         Needs["Calculus`VectorAnalysis`"]
         Needs["Graphics`ParametricPlot3D`"]
```

Solution

```
In[60]:= Clear["Global`*"];
```

Part (**a**) A particle approaches from the left ($x < 0$) and is reflected and transmitted by a barrier at $x = 0$. The wave function follows from the solution of Schrodinger's equation having constant potential and is given by the user-defined function **waveExp**:

```
In[61]:=  waveExp[x,k,c1,c2]

Out[61]=     -I k x        I k x
         c2 E       + c1 E
```

where the wave vector is given by the solution of $\hbar^2 k^2/(2m) = (E_n - V)$:

```
In[62]:= kRule= {k -> Sqrt[2 m (En-V)]/hbar}

Out[62]=        Sqrt[2] Sqrt[m (En - V)]
         {k -> ------------------------}
                        hbar
```

Define the value of **k** to the left of the barrier as **k1** and that to the right as **k2**; these are related to the energy and potential by

```
In[63]:= kRule2=  {k1-> k /.kRule /.V->0
                  ,k2-> k /.kRule}

Out[63]=         Sqrt[2] Sqrt[En m]         Sqrt[2] Sqrt[m (En - V)]
         {k1 -> ------------------, k2 -> ------------------------}
                      hbar                          hbar
```

The wave function to the left of the barrier ($x < 0$) has both an incident and reflected wave:

```
In[64]:= psiL[x_]:=waveExp[x,k1,1,R];
         psiL[x]

Out[64]=  I k1 x     -I k1 x
         E       + E        R
```

The term $Exp[+ikx]$ represents the incident wave; without loss of generality, we set its coefficient to one. $Exp[-ikx]$ represents the reflected wave having amplitude **R**. The wave to the right of the barrier (**x>0**) has only a transmitted component traveling to the right and is of the form

```
In[65]:= psiR[x_]:=waveExp[x,k2,T,0];
         psiR[x]

Out[65]=  I k2 x
         E       T
```

The boundary conditions require the wave function and its derivative to be continuous at $x = 0$. The two boundary equations are

```
In[66]:= boundary= {psiL[x] == psiR[x] ,
                    psiL'[x] == psiR'[x]}  //.x->0;
```

Solving for the **T** and **R** amplitudes, you get

```
In[67]:=  RTrule= Solve[boundary ,{R,T}][[1]]  //Simplify

Out[67]=        k1 - k2       2 k1
         {R -> -------, T -> -------}
                k1 + k2       k1 + k2
```

Part (b) Assume **En>V0** and find the transmission and reflection coefficients. The coefficients of transmission and reflection are defined as the ratio of the transmitted and reflected currents to the incident current. The incident current is

```
In[68]:= SetCoordinates[Cartesian[x,y,z]];

         incFlux=flux[ psiL[x][[1]] ]

Out[68]=  hbar k1
         {-------, 0, 0}
            m
```

The reflected and transmitted currents are

```
In[69]:= Rflux=-flux[ (psiL[x][[2]]) ]

Out[69]=
                2
   hbar k1 R
  {-----------, 0, 0}
       m

In[70]:= Tflux= flux[ psiR[x] ]

Out[70]=
                2
   hbar k2 T
  {-----------, 0, 0}
       m
```

The vector currents are in the x-direction, so consider only the first component of the vectors. The transmitted and reflected coefficients follow from dividing the x-components of the reflection and transmitted currents by the x-component of the incident current:

```
In[71]:= {RR,TT}= { Rflux[[1]] /incFlux[[1]],
                     Tflux[[1]] /incFlux[[1]] } /. RTrule

Out[71]=
                2
   (k1 - k2)      4 k1 k2
  {-----------, -----------}
             2            2
   (k1 + k2)    (k1 + k2)
```

To plot the reflected and transmitted coefficients, let

```
In[72]:= Clear[V];
         values={ m->1,hbar->1, V->1 };
```

Plot **RR** and **TT** as a function of **En**:

```
In[73]:= Plot[ {RR,TT}/.RTrule /.kRule2 /.values //Evaluate
             ,{En,1,2}
             ,AxesLabel->{FontForm["En"   ,{"Courier-Bold",10}]
                         ,FontForm["Coef",{"Courier-Bold",10}]}
             ,PlotStyle->{Dashing[{0.1,0.0}],Dashing[{0.01,0.02}]}
             ,Epilog->Graphics[{
                Text[FontForm["TT",{"Courier-Bold",10}],{1.8,.89}],
                Text[FontForm["RR",{"Courier-Bold",10}],{1.8,.15}]}
             ][[1]]];
```

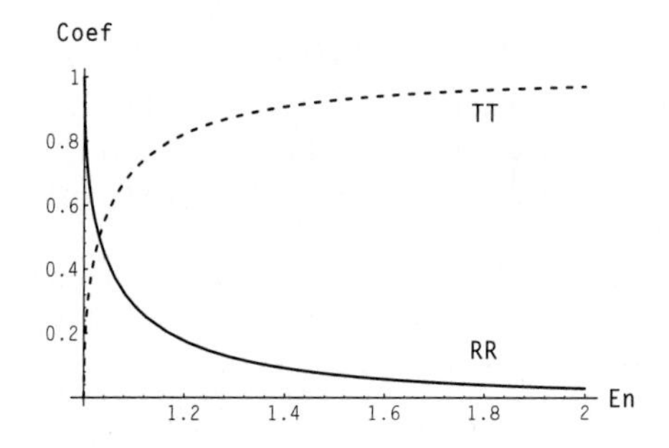

Part (c) Plot the real and imaginary parts of the wave function when the incoming energy is greater than the barrier and for the case in which the energy is below the barrier. Take

```
In[74]:= values={V->1, m->1, hbar->1 };
```

Define a function **psi** that is valid for all values of **x**:

In[75]:=
```
psi[x0_ /; x0<=0,Energy_]:=
   psiL[x] /.RTrule /.kRule2 /.values /.{x->x0,En->Energy};

psi[x0_ /; x0 >0,Energy_]:=
   psiR[x] /.RTrule /.kRule2 /.values /.{x->x0,En->Energy};
```

The particle will be above the barrier if **En>1** and below if **En<1**. To plot the **Re** and **Im** parts of the wave, define the function **waveplot2D**:

In[76]:=
```
waveplot2D[Energy_]:=
Plot[ {Re[psi[x Pi,Energy]],Im[psi[x Pi,Energy]]} //Evaluate
     ,{x,-10,10}
     ,PlotStyle->{Dashing[{0.1,0.0}],Dashing[{0.03,0.02}]}
     ,AxesLabel->{ FontForm["x",{"Courier-Bold",10}]
                  ,FontForm["y",{"Symbol",10}] }
     ,AxesOrigin->{0,0}
     ,DisplayFunction->Identity
   ];
```

The function **waveplot2D** plots the real and imaginary parts of the wave function for given values of **En**. The real part of the wave function is denoted by a solid curve and the imaginary part by a dashed line. We plot the waves for **En=1.2** (above the barrier) and for **En=.9** (below the barrier):

In[77]:=
```
Show@GraphicsArray[{waveplot2D[1.2],waveplot2D[0.9]}];
```

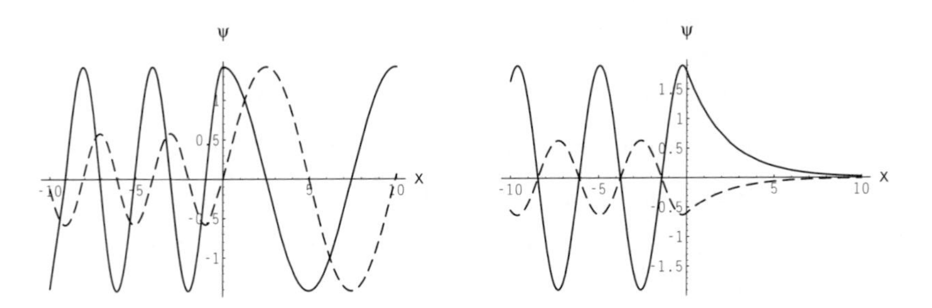

For energies below the barrier, the transmitted wave decays exponentially.

Part (d) Illustrate how the wave function behaves as a function of **En** by using **Plot3D**. Plot the **Re** and **Im** parts of the wave function for values of **En** from 0.5 to 1.5. (Computing these graphics takes several minutes. To speed up the output, let **PlotPoints** be smaller.)

In[78]:=
```
waveplot3D =
Plot3D[
   Evaluate[#[psi[x Pi,Energy]]]
   ,{x,-10,10},{Energy,.5,1.5}
   ,PlotPoints->35
   ,AxesLabel->{"x","Energy","Wave"}
   ,Ticks-> {{-10,-5,{0,"Barrier"},5 ,10}     (* x-Ticks *)
            ,{5,{0.7,"E<V"},{1,"E=V "},{1.2,"E>V"},1.4} (* y-Ticks *)
            ,{-2,-1,0,1,2  }   (* z-Ticks *)  }
   ,DisplayFunction->Identity
         ]& /@ {Re,Im};
```

The function **waveplot3D** contains the graphics for the **Re** and **Im** parts of the wave functions for different values of **{x, En}**. Displaying the results with **GraphicsArray**, you can see

```
In[79]:= Show @ GraphicsArray @ waveplot3D;
```

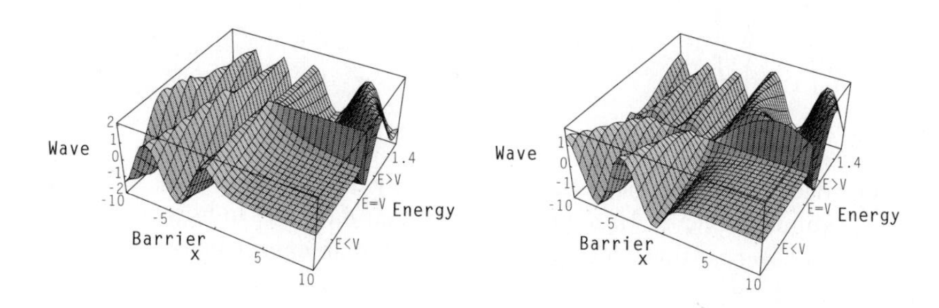

The graph on the left-hand side is the real part of the wave function; that on the right-hand side is the imaginary part.

Part (e) Animate the real part of the incident, reflected, and transmitted waves. To get the time dependence for the wave function, multiply by $Exp[-iE_n t/\hbar]$, where $\hbar$ is set equal to one. The graphics for the time-dependent wave function are generated by the function:

```
In[80]:= timeplot[time_,wave_,Energy_,thickness_,xrange_]:=
           Plot[Evaluate[Re[Exp[-I Energy time]* wave
                  /.RTrule/.kRule2/.values/.En->Energy ] ]
              ,xrange
              ,PlotStyle-> Thickness[thickness]
              ,AxesOrigin->{0,0}
              ,DisplayFunction->Identity
           ];
```

where **Thickness** describes the thickness of the curve and **xrange** are the values of **x**. Apply **timeplot** to **psiL[x][[1]]** (incident wave), **psiL[x][[2]]** (reflected wave), and **psiR[x]** (transmitted wave) for different times. Consider the animation for a wave with an energy above the barrier, **En=1.1**. Forming a time sequence of graphs, you get (all but one plot suppressed)

```
In[81]:= tnumber=6;
         timestep=2 Pi/(tnumber);
         Energy=1.1;
         Do[ Show[{timeplot[time,psiL[x][[1]],Energy,.006,{x,-10,0}]
                 ,timeplot[time,psiL[x][[2]],Energy,.004,{x,-10,0}]
                 ,timeplot[time,psiR[x][[1]],Energy,.001,{x, 0,10}]}
              ,DisplayFunction->$DisplayFunction]
           ,{time,0,2 Pi-timestep,timestep}
           ]
```

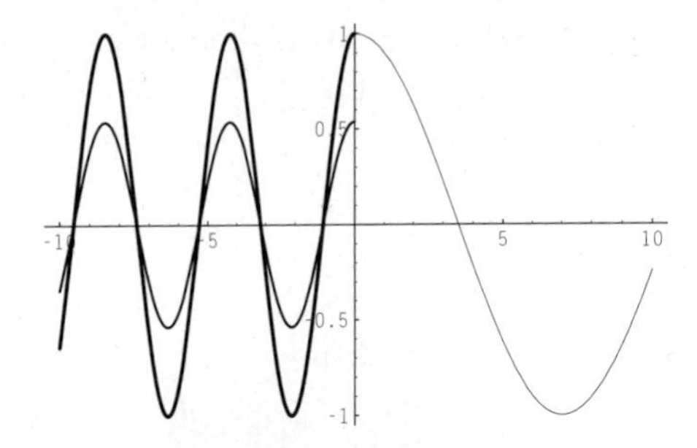

The dark line is the incident wave. Note the transmitted wave is oscillatory. (Refer to your computer-specific *Mathematica* user's manual to learn how to animate the sequence of graphs.) Repeat this exercise for a wave with an energy below the barrier, **En=0.9** (all but one plot suppressed):

In[82]:=
```
tnumber=6;
timestep=2 Pi/(tnumber);
Energy=0.9;
Do[ Show[{timeplot[time,psiL[x][[1]],Energy,.006,{x,-10,0}]
         ,timeplot[time,psiL[x][[2]],Energy,.004,{x,-10,0}]
         ,timeplot[time,psiR[x][[1]],Energy,.001,{x, 0,10}]}
      ,DisplayFunction->$DisplayFunction]
      ,{time,0,2 Pi-timestep,timestep}
  ]
```

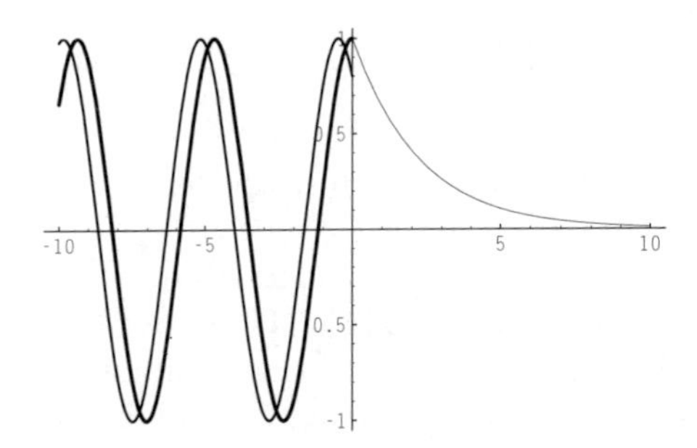

Note that the transmitted wave is exponentially damped.

Problem 4: Particle propagating towards a rectangular potential

Suppose a particle with energy $E_n > 0$ propagates from the left ($x < -a$) towards a rectangular well or barrier described by

$$V(x) = \begin{cases} V_0 & \text{if } -a < x < a; \\ 0 & \text{otherwise.} \end{cases}$$

If $V_0 > 0$, the potential describes a barrier; if $V_0 < 0$, the potential describes a well. The wave to the left ($x < -a$) of the potential has an incoming and reflected component. Set the amplitude of the incoming wave to one and let R be the amplitude of the reflected wave. The wave to the right ($x > a$) of the potential has only a transmitted wave; let its amplitude be T. The wave in the potential can be described by two amplitudes; call them A and B.

Part (a) Write the general form of the wave function to the left and right and in the center of the well.

Part (b) Use the boundary condition to evaluate the wave amplitudes.

Part (c) The coefficients of transmission T and reflection R are defined as the ratio of the transmitted and reflected currents to the incident current. Find the transmission and reflection coefficients. Verify that $T + R = 1$.

Part (d) Plot R and T as a function of E_n for the case of a potential well ($V_0 < 0$) and for the case of a potential barrier ($V_0 > 0$).

Part (e) Use **Plot3D** to plot the reflection and transmission curves as a function of the energy and potential.

Part (f) Plot the real part of the wave for the complete range of x.

Part (g) Animate the time dependence of the real part of the wave as it propagates through a barrier ($V_0 > 0$, $0 < E_n < V_0$).

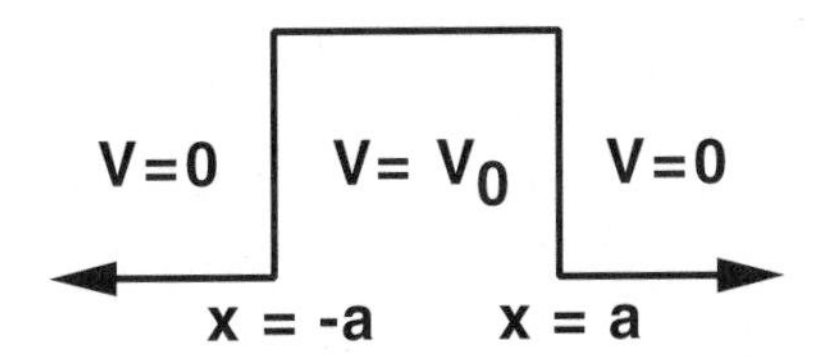

Remarks and outline The general solution of Schrodinger's equation in a constant potential is given by the user-defined function **waveExp**. The wave solutions to the left and right and in the center of the potential must be treated separately. The boundary conditions determine the wave function amplitudes. The coefficients of transmission **TT** and reflection **RR** follow from applying the user-defined function **flux** to the appropriate parts of the wave function. Plots of these coefficients reveal the resonant behavior. You are encouraged to vary the energy, potential, and well size. Define a graphics function to join the wave function in the three distinct potential regions. Finally the animation follows from forming a table of wave plots for different values of the time.

Required packages

```
In[83]:= Needs["Algebra`Trigonometry`"]
         Needs["Calculus`VectorAnalysis`"]
         Needs["Graphics`ParametricPlot3D`"]
```

Solution

```
In[84]:= Clear["Global`*"];
```

Part (a) The general solution for Schrodinger's equation with a constant potential is given by the user-defined function **waveExp**. For $x < -a$, there is an incident wave traveling to the right and a reflected wave traveling to the left. The amplitude for the incident wave is set equal to one without loss of generality, and the amplitude for the reflected wave is called **R**. The form of the wave is

```
In[85]:= psiL[x_]= waveExp[x,k1 ,1,R]

Out[85]=  I k1 x     -I k1 x
         E       + E        R
```

For $x > +a$, there is only a transmitted wave with amplitude **T**.

```
In[86]:= psiR[x_]= waveExp[x, k1 ,T, 0]

Out[86]=  I k1 x
         E       T
```

The wave function in the central region $(-a < x < a)$ is

```
In[87]:= psiW[x_]= waveExp[x, k2, A, B]

Out[87]=    -I k2 x       I k2 x
         B E        + A E
```

The expressions for **k1** and **k2** follow from **kRule**:

```
In[88]:= kRule= { k1->Sqrt[2 m (En  )]/hbar
                ,k2->Sqrt[2 m (En-V)]/hbar
               } //PowerExpand

Out[88]=         Sqrt[2] Sqrt[En] Sqrt[m]          Sqrt[2] Sqrt[m] Sqrt[En - V]
         {k1 -> ------------------------, k2 -> ----------------------------}
                          hbar                              hbar
```

where the wave number is implicitly defined by $k^2\hbar^2/(2m) = (E_n - V)$.

Part (b) The boundary conditions require the wave function and its derivative to be continuous across the boundaries. Applying these conditions at $x = a$ and $x = -a$ yields four equations:

```
In[89]:= eq1= {     (psiL[ x] - psiW[x]==0)    /. {x->-a},
                    (psiW[ x] - psiR[x]==0)    /. {x->+a},
                    (psiL'[x] -psiW'[x]==0)    /. {x->-a},
                    (psiW'[x] -psiR'[x]==0)    /. {x->+a} }

Out[89]=   -I a k1        -I a k2        I a k2    I a k1
         {E        - A E         - B E        + E        R == 0,

            -I a k2        I a k2    I a k1
           B E        + A E       - E       T == 0,

             -I a k1            -I a k2             I a k2
           I E        k1 - I A E         k2 + I B E        k2 -

               I a k1                 -I a k2             I a k2
            I E       k1 R == 0, -I B E         k2 + I A E        k2 -

               I a k1
            I E       k1 T == 0}
```

Solve these equations for the amplitudes **{A,B,R,T}**. (This calculation takes a few minutes. It may be necessary to break up the solution into smaller steps on your particular machine.)

```
In[90]:= ABRTrule=( Solve[eq1,{A,B,R,T}][[1]]
                    //Map[Together,#,{2}]&
                    //ComplexToTrig
                    //Simplify
                  );
```

The results can be simplified if you collect various factors. The simplified results are

```
In[91]:= (( ABRTrule //Map[Collect[#,{Sin[2 a k2]}]&,#,{4}]&
          ) //Map[Simplify,#,{5}]&
         )

Out[91]=
                2     2
            (k1  - k2 ) (Cos[2 a k1] - I Sin[2 a k1]) Sin[2 a k2]
         {R -> --------------------------------------------------,
                                         2     2
               2 I k1 k2 Cos[2 a k2] + (k1  + k2 ) Sin[2 a k2]

               2 k1 k2 Cos[2 a k1] - 2 I k1 k2 Sin[2 a k1]
          T -> ---------------------------------------------,
                                          2     2
              2 k1 k2 Cos[2 a k2] - I (k1  + k2 ) Sin[2 a k2]

              (k1 + k2) (k1 Cos[a (k1 + k2)] - I k1 Sin[a (k1 + k2)])
          A -> -------------------------------------------------------,
                                            2     2
                2 k1 k2 Cos[2 a k2] - I (k1  + k2 ) Sin[2 a k2]

              (k1 - k2) (k1 Cos[a (k1 - k2)] - I k1 Sin[a (k1 - k2)])
          B -> -------------------------------------------------------}
                                            2     2
               -2 k1 k2 Cos[2 a k2] + I (k1  + k2 ) Sin[2 a k2]
```

Part (c) The coefficients of transmission and reflection are defined as the ratio of the transmitted and reflected currents as compared to the incident current. The vector currents are in the x-direction, so consider only the first component. The currents follow from applying the user-defined function **flux** to the wave functions. The x-component of the incident, reflected, and transmitted currents are

```
In[92]:= SetCoordinates[Cartesian[x,y,z]];

         incFlux= flux[  psiL[x][[1]] ]

Out[92]=  hbar k1
         {-------, 0, 0}
            m

In[93]:= refFlux= flux[  psiL[x][[2]] ]

Out[93]=              2
           hbar k1 R
         {-(---------), 0, 0}
               m

In[94]:= tranFlux= flux[  psiR[x] ]

Out[94]=              2
          hbar k1 T
         {----------, 0, 0}
             m
```

You will find the reflection and transmission coefficients to be

```
In[95]:= RR=  refFlux[[1]]/incFlux[[1]] //Abs
         TT= tranFlux[[1]]/incFlux[[1]] //Abs

Out[95]=       2
         Abs[-R ]

Out[95]=      2
         Abs[T ]
```

Note that **{RR,TT}** are defined to be positive definite. You can see that with the normalization chosen for the incident wave, $RR = R R^{*}$ and $TT = T T^{*}$. In terms of the initial variables, you will find that

```
In[96]:= {TT,RR}= (( {T Conjugate[T], R Conjugate[R]}
                  /.ABRTrule
                  /.{Conjugate->conjugate}
                  //Together
                  //Simplify
                ) //Map[Collect[#,{Cos[4 a k2]}]&,#,{3}]&
                  //Simplify
                )

Out[96]=                       2   2
                         8 k1   k2
         {-----------------------------------------------------,
           4       2   2     4       2     2 2
         k1  + 6 k1  k2  + k2  - (k1  - k2 )  Cos[4 a k2]

                     2     2 2             2
               2 (k1  - k2 )  Sin[2 a k2]
         -----------------------------------------------------}
           4       2   2     4       2     2 2
         k1  + 6 k1  k2  + k2  - (k1  - k2 )  Cos[4 a k2]
```

Notice that probability conservation is verified:

```
In[97]:= TT+RR==1 //ExpandAll//Simplify

Out[97]= True
```

Part (d) Plot **RR** and **TT** as a function of **En** for the parameters

```
In[98]:= values={hbar->1,m->1,a->1};
```

Consider the plots for the cases in which **V<0** (potential well) and in which **V>0** (potential barrier). The graphics for **{RR,TT}** follow from the function

```
In[99]:= Clear[plotOperator];
         plotOperator[Potential_, EnRange_] :=
         Plot[
            Evaluate[ {RR,TT} /.kRule/.V->Potential/.values]
           ,EnRange
           ,AxesLabel->{
                     FontForm["En"     ,{"Courier-Bold",10}],
                     FontForm["RR, TT",{"Courier-Bold",10}]}
            ,PlotStyle->
                 {Dashing[{1,0.0}],Dashing[{0.02,0.02}]}
         ];
```

The function **plotOperator** generates the graphics for **{RR,TT}** as a function of energy and potential. The **RR** coefficient is denoted by a solid curve; the **TT** coefficient is denoted by a dashed curve. First consider the case of a potential well with **V=-40** and plot the coefficients for **En** between 0 and 40:

```
In[100]:= plotOperator[-40,{En,1,40}];
```

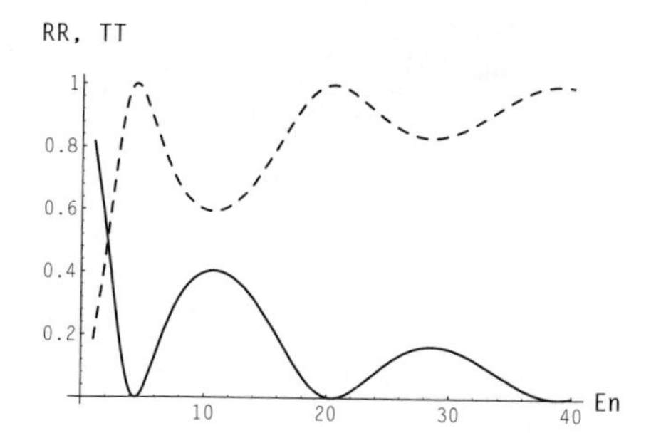

The dashed curve is the **TT**; the solid curve is the **RR**. The striking decrease in the reflection coefficient with increasing energy was noted for gases by Ramsauer in 1920 and independently by Townsend and Bailey later. Consider a potential barrier with **V=1** and let the values of **En** range from 0.01 to 0.9:

```
In[101]:= plotOperator[1,{En,.01,0.99}];
```

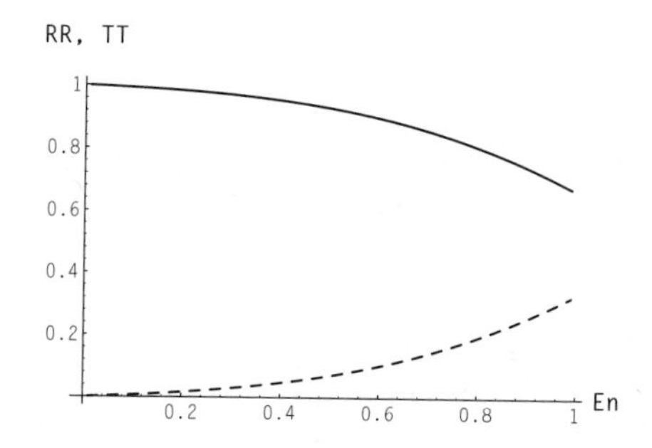

Part (e) Use **Plot3D** to illustrate the reflection and transmission curves for a range of potential and energy values. Applying **Plot3D** to **RR** and **TT** and plotting the curves as a function of energy and potential, you get

```
In[102]:= pt1= Plot3D[
           Evaluate[#/.kRule/.ABRTrule/.values/.a->1]
          ,{En,.01,10}
          ,{V,-20,-.1}
          , DisplayFunction->Identity
              ]& /@ {TT,RR};
```

The function **pt1** contains the graphics for **TT** and **RR** as a function of the two variables **{En,V}**. Let **En** vary between 0.01 and 10 and **V** vary between −20 and −0.1. Displaying the results, you get

```
In[103]:= ShowGraphicsArray[pt1];
```

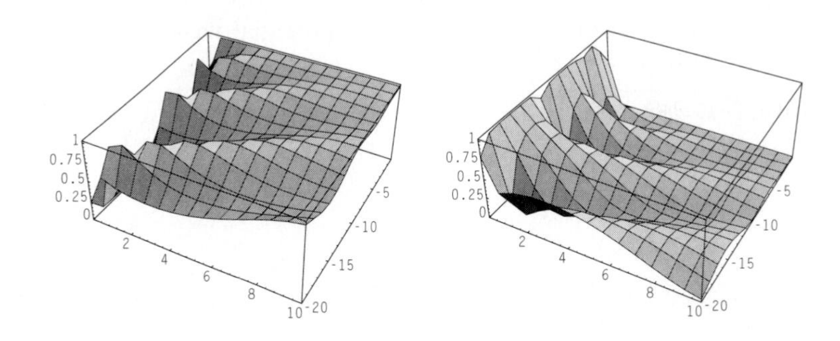

The left-hand graph is for the transmission coefficient; the right-hand graph is for the reflection coefficient. Resonance effects appear as ripples in the diagram.

Part (f) Plot the real part of the wave for the complete range of x. Choose **a=2, m=1**, and **hbar=1** and define **plotOp**:

```
In[104]:= plotOp[wave_,Energy_,V0_,thickness_,xrange_]:=
            Plot[ Re[ wave /.ABRTrule
                            /.kRule
                            //.{m->1,hbar->1,En->Energy,a->2,V->V0} ]
                   //Evaluate
              ,xrange
              ,PlotStyle-> Thickness[thickness]
              ,AxesOrigin->{0,0}
              ,DisplayFunction->Identity
          ];
```

The function **plotOp** generates the graphics for the wave with energy **Energy** and potential **V0**. The variable **thickness** allows you to distinguish the wave to the left and right and in the center of the barrier. Combine graphics with the function

```
In[105]:= wavePlot1[En_,V0_,xmin_,xmax_] :=
            Show[{ plotOp[psiL[x],En,V0,0.01 ,{x,xmin,-2}]
                  ,plotOp[psiR[x],En,V0,0.01 ,{x,2,xmax} ]
                  ,plotOp[psiW[x],En,V0,0.001,{x,-2,2}   ]
                  ,Graphics[
                     {Line[{{-2,Sign[V0]},{2,Sign[V0]}}],
                      Line[{{-2,Sign[V0]},{-2,0}}],
                      Line[{{2, Sign[V0]},{2,0}}] }]
                  }
              ,DisplayFunction-> $DisplayFunction ];
```

The outline of the well is given in the command **Graphics**. Displaying the real part of the wave for **V0=-12** and **En=1**, you get

```
In[106]:= wavePlot1[1,-12,-12,12];
```

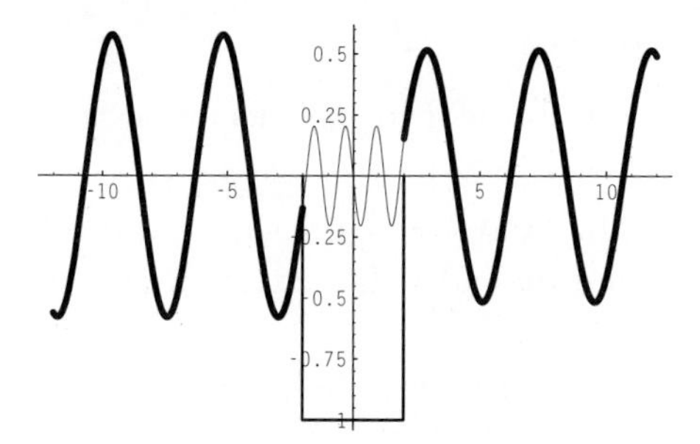

Part (g) Animate the real part of the wave as it propagates through the barrier (**0<En<V0**). To get the time dependence for the wave, multiply the wave by $Exp[-iE_n t/\hbar]$ with **hbar=1**. The time dependence of a wave with energy **Energy** is described by the function

```
In[107]:= Clear[timePlot];
          timePlot[time_,wave_,Energy_,V0_,thickness_,xrange_]:=
            Plot[( Re[Exp[-I Energy time] wave
                       /.ABRTrule
                       /.kRule
                       //.{m->1,hbar->1,En->Energy,a->2,V->V0}
                   ] //Evaluate  )
              ,xrange
              ,Axes->None
              ,PlotStyle->Thickness[thickness]
              ,DisplayFunction->Identity
           ];
```

Consider the case in which **V0=1.1** and **Energy=1**. To animate the time dependence, form a table of graphics (all but one plot suppressed):

```
In[108]:=  tnumber=6;
           timestep=2 Pi/(tnumber);
           V0=1.1;
           Energy=1;
           Do[ Show[ {timePlot[time,psiL[x],Energy,V0,.006,{x,-10,-2}]
                     ,timePlot[time,psiW[x],Energy,V0,.001,{x,-2,  2}]
                     ,timePlot[time,psiR[x],Energy,V0,.006,{x, 2, 10}]}
                ,DisplayFunction->$DisplayFunction    ]
             ,{time,0,2 Pi-timestep,timestep} ]
```

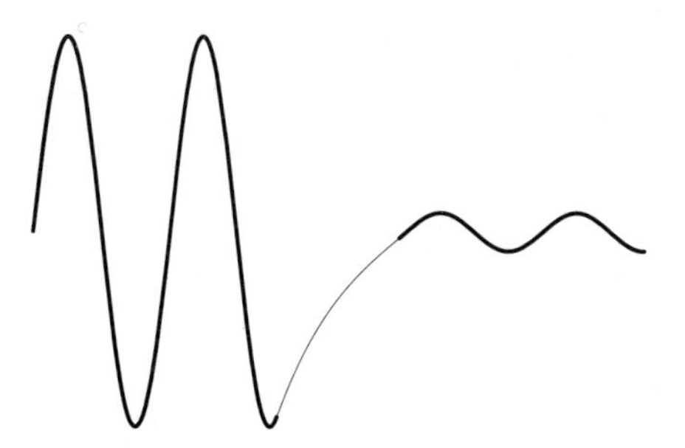

The number of animation frames is determined by **tnumber**. You are encouraged to choose different values for the energy and potential.

Problem 5: The one-dimensional harmonic oscillator

Consider a particle that is bound by a force proportional to its displacement, $F = -kx$; the potential of the restoring force is $V = (1/2)kx^2$.

Part (a) Define $k = \omega^2 m$, $E_n = (1/2)en\hbar\omega$, and $\alpha = \sqrt{\hbar/(\omega m)}$. Express Schrodinger's equation in terms of z, where $x = \alpha z$, and show that this equation reduces to $\psi''[z] == -(en\,\psi[z]) + z^2\psi[z]$.

Part (b) Define a new function $\psi[z] = Exp[-z^2/2]h[z]$ and show that $h[z]$ satisfies the equation $h''[z] == h[z] - en\,h[z] + 2zh'[z]$. Use the user-defined operator **diffSeriesOne** to find a power-series solution for $h[z]$. Show that if the values of en are integers, the power series terminates beyond a certain power of z. Verify that these solutions are Hermite polynomials.

Part (c) Consider the equation $h''[z] == h[z] - enh[z] + 2zh'[z]$. Let $en = 2n + 1$ and declare n to be an integer $\{0, 1, 2, 3, \ldots\}$. Show that the solutions of these equations follow from the user-defined solutions in the Introduction.

Part (d) Normalize the wave function. Plot the normalized wave function and the square of the normalized wave function for the first eight values of n.

Remarks and outline Schrodinger's equation follows from the user-defined operator **schrodinger**. The user-defined operator **vChange** is used to change the variables in the equation. In terms of the dimensionless variable **z** and function **h[z]**, the one-dimensional Schrodinger's equation reduces to an equation for a Hermite polynomial times and exponential. In Section 6.2, "*Mathematica* Commands," you taught *Mathematica* to recognize the solution of Hermite polynomials. Therefore, because *Mathematica* knows the properties of Hermite polynomials, you can verify the correct solution.

Required packages

```
In[109]:= Needs["Algebra`Trigonometry`"]
          Needs["Calculus`VectorAnalysis`"]
          Needs["Graphics`ParametricPlot3D`"]
```

Solution

```
In[110]:= Clear["Global`*"];
```

Part (a) The one-dimensional Schrodinger's equation follows from the user-defined operator **schrodinger**. Applying **schrodinger** to the wave function **psi[x]**, you get

```
In[111]:= SetCoordinates[Cartesian];

          eq1= 0==schrodinger[1/2 k x^2,En] @ psi[x]

Out[111]=
                           2                  2
                        k x               hbar  psi''[x]
          0 == (-En + ----) psi[x] - ---------------
                         2                  2 m
```

It is convenient to define the notation

```
In[112]:= notation={k->w^2 m, En->1/2  en  hbar w};
          alphaRule= {alpha -> Sqrt[hbar/( w  m)]};
```

Use the user-defined function **vChange** to change the variables from **x** to **z**, where **x=alpha z** and **alpha** is given by the **alphaRule**. Apply **vChange** to **eq1**:

```
In[113]:= eq2=vChange[eq1,psi,x,z,alpha z]

Out[113]=
                      2  2                     2
                 alpha  k z                hbar  psi''[z]
          0 == (-En + ----------) psi[z] - --------------
                          2                      2
                                          2 alpha  m
```

and solve for **psi''[z]**:

```
In[114]:= eq3= ( Solve[eq2,psi''[z]][[1,1]]
                //.notation
                //.alphaRule
                //ExpandAll
              ) /.Rule->Equal

Out[114]= psi''[z] == -(en psi[z]) + z  psi[z]
```

Part (b) The form of Schrodinger's equation simplifies if you define a new function **psi[z] = Exp[-z^2/2] h[z]**:

```
In[115]:= psiRule= psi ->( (h[#] Exp[-#^2/2] ) &);
                  psi[x] /.psiRule

Out[115]= h[x]
          ----
            2
           x /2
          E
```

Writing **eq3** in terms of **h[x]**, you get

```
In[116]:= eq4= ( eq3//.psiRule  ) //ExpandAll

Out[116]=
                     2                                                      2
           h[z]     z  h[z]    2 z h'[z]    h''[z]        en h[z]       z  h[z]
          -(----) + ------- - --------- + ------- == -(--------) + -------
             2          2          2           2             2             2
            z /2       z /2       z /2        z /2          z /2          z /2
           E          E          E           E             E             E
```

You can simplify **eq4**:

```
In[117]:= eq5 = Solve[eq4,h''[z]][[1,1]]  /.Rule->Equal

Out[117]= h''[z] == h[z] - en h[z] + 2 z h'[z]
```

Assume the solution of **eq5** can be expanded as a power series in **z** and find the values of **en** for which the series terminates beyond a certain power of **z**. The series solution of **eq5** follows from the user-defined procedure **diffSeriesOne**. Applying **diffSeriesOne** to **eq5** with the initial conditions **{h[0]=h0, h'[0]=hp0}** and keeping the first eight terms, you get

```
In[118]:= diffSol=
          (( diffSeriesOne[eq5,h,z,8,{h[0]->h0, h'[0]->hp0}]
              //Collect[#,{ hp0,h0,z}]&
            ) //Map[Simplify,#,{2}]&
          )
```

```
Out[118]=
                      3                    2   5
          (3 - en) z        (21 - 10 en + en ) z
hp0 (z + ------------- + ----------------------- +
               6                   120

                              2    3  7
       (231 - 131 en + 21 en  - en ) z
       ------------------------------) +
                    5040

                       2                  2   4
            (1 - en) z        (5 - 6 en + en ) z
   h0 (1 + ------------- + --------------------- +
                  2                 24

                            2    3  6
       (45 - 59 en + 15 en  - en ) z
       ---------------------------- +
                   720

                                2        3     4  8
       (585 - 812 en + 254 en  - 28 en  + en ) z
       ----------------------------------------)
                        40320
```

The two independent series are described by the **hp0** and **h0** terms. To determine the values of **en** when the series terminates, let **en=2 n+1**. The **hp0** series terminates for odd **n** $\{n = 1, 2, 3, \ldots\}$ and the **h0** series terminates for even values of **n** $\{n = 0, 2, 4, \ldots\}$, as is verified by

```
In[119]:= serOdd= (Coefficient[diffSol,hp0]
                //.{en ->2 n+1}
                /.n->{1,3,5,7}
                //Factor)
```

```
Out[119]=
                    2                2      4
         z (3 - 2 z )   z (15 - 20 z  + 4 z )
   {z, -------------, ---------------------,
              3                 15

                      2       4      6
        z (105 - 210 z  + 84 z  - 8 z )
        -----------------------------}
                     105
```

and

```
In[120]:= serEven= (Coefficient[diffSol,h0]
                //.{en ->2 n+1}
                /.n->{0,2,4,6}
                //Factor)
```

```
Out[120]=
                          2      4            2       4      6
                2  3 - 12 z  + 4 z    15 - 90 z  + 60 z  - 8 z
   {1, 1 - 2 z , ---------------, --------------------------}
                         3                    15
```

The power of **z** is the value of **n**. To clarify what we mean to say "the series terminates," note (for example) that in **serEven** that for **n**=4, the highest power of **z** is 4, even though we retained terms up to **z^8** in **diffSol**. In this case, the coefficients of **z^6** and **z^8** vanish, as do all higher powers.

These polynomials in **z** are proportional to the Hermite polynomials. The even terms are equal to

```
In[121]:= termEven[n_]= (-1)^(n/2) (n/2)! /(n)! HermiteH[n,z]

Out[121]=      n/2  n
          (-1)     (-)! HermiteH[n, z]
                    2
          ----------------------------
                      n!
```

which can be verified by dividing **serEven** by **termEven**:

```
In[122]:= serEven/ (termEven[#]& /@ {0,2,4,6}) //Factor

Out[122]= {1, 1, 1, 1}
```

Likewise, the odd terms are equal to

```
In[123]:= termOdd[n_]= (-1)^((n-1)/2) ((n-1)/2)! /(n)! /2 HermiteH[n,z]

Out[123]=      (-1 + n)/2  -1 + n
          (-1)            (------)! HermiteH[n, z]
                             2
          ----------------------------------------
                          2 n!
```

which is verified by dividing **serOdd** by **termOdd**

```
In[124]:= serOdd/ (termOdd[#]&/@{1,3,5,7}) //Factor

Out[124]= {1, 1, 1, 1}
```

Part (c) The energy eigenvalues are proportional to integers **n**. Express **eq3** in terms of integer values of **n** and solve. First, declare **n** to be an integer:

```
In[125]:= n /: IntegerQ[n]=True;
```

and express **eq5** in terms of **n**:

```
In[126]:= eq6=  eq5 /.{en->(2 n+1)} //Expand

Out[126]= h''[z] == h[z] - (1 + 2 n) h[z] + 2 z h'[z]
```

where **n**=$\{0, 1, 2, 3, \ldots\}$. In Section 6.2, "*Mathematica* Commands," you taught **DSolve** to recognize this solution:

```
In[127]:= sol= DSolve[eq6, {h},z][[1]]

Out[127]= {h -> (HermiteH[n, #1] & )}
```

Substitute **sol** into **eq6** for eight values of **n** to verify that **sol** is indeed a solution of **eq6**:

```
In[128]:= Table[eq6 /.sol ,{n,0,8}] //ExpandAll

Out[128]= {True, True, True, True, True, True, True, True, True}
```

Part (d) The unrenormalized wave function is

```
In[129]:= psi[z_,n_]= h[z] Exp[-z^2/2] //.sol
```

```
Out[129]= HermiteH[n, z]
          --------------
               2
              z /2
             E
```

To normalize the wave function, define the normalization constant:

```
In[130]:= normConst[n_]:= normConst[n] =
           1/Sqrt[ NIntegrate[ psi[z,n]^2, {z,-Infinity,Infinity}] ]
```

Note that **normConst** is defined such that it stores the values as they are computed. The plot of the first eight normalized wave functions follows from

```
In[131]:= Plot[
            Evaluate[ Table[normConst[n] psi[z,n]+ 2 n,{n,0,7}] ]
            ,{z,-6,6}
            ,Epilog->Plot[1.1 z^2
                          ,{z,-6,6}
                          ,PlotStyle->Thickness[0.01]
                          ,DisplayFunction->Identity
                         ][[1]]
           ];
```

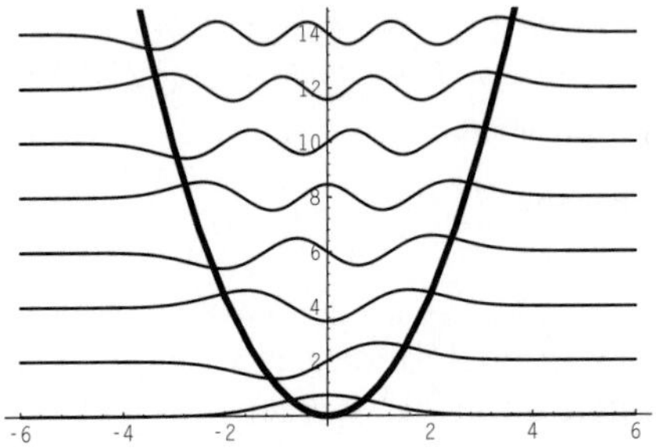

The plots for the squares of the wave functions follow similarly:

```
In[132]:= Plot[
            Evaluate[ Table[(normConst[n] psi[z,n])^2+ 2 n,{n,0,7}] ]
            ,{z,-6,6}
            ,Epilog->Plot[1.1 z^2
                          ,{z,-6,6}
                          ,PlotStyle->Thickness[0.01]
                          ,DisplayFunction->Identity
                         ][[1]]
           ];
```

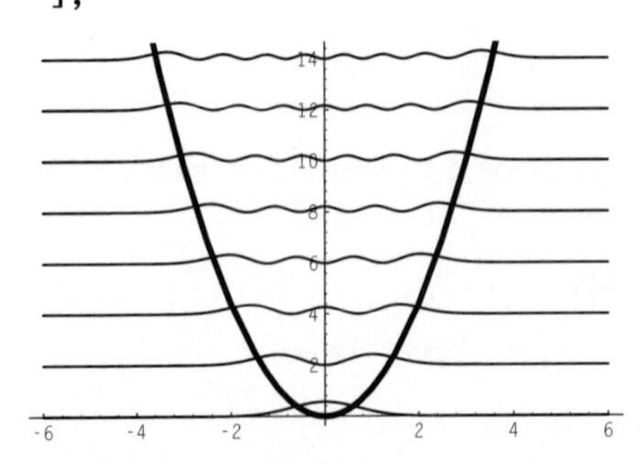

Epilog was used to plot the potential on the same graph.

6.2.2 Three-dimensional Schrodinger's Equation

Problem 1: Three-dimensional harmonic oscillator in Cartesian coordinates

Consider the three-dimensional Schrodinger's equation, where the potential is $V = V_x[x] + V_y[y] + V_z[z]$.

Part (a) Assume the wave function is separable and of the form $X[x]Y[y]Z[z]$ and derive a differential equation for $X[x]$,$Y[y]$, and $Z[z]$.

Part (b) Consider the harmonic oscillator potential $V_x = 1/2k_x x^2$, $V_y = 1/2k_y y^2$, and $V_z = 1/2k_z z^2$. Solve Schrodinger's equation for the wave function using the one-dimensional harmonic oscillator solutions.

Part (c) Plot the two-dimensional wave function $X[x]Y[y]$.

Remarks and outline This is the simplest example of a three-dimensional problem. Schrodinger's equation follows from the user-defined operator **schrodinger**. The three-dimensional Schrodinger's equation can be reduced to three one-dimensional equations by separating the variables and setting the terms with the same variables equal to a constant. The general solution is the product of three, one-dimensional oscillator solutions.

Required packages

```
In[133]:= Needs["Algebra`Trigonometry`"]
          Needs["Calculus`VectorAnalysis`"]
          Needs["Graphics`ParametricPlot3D`"]
```

Solution

```
In[134]:= Clear["Global`*"];
```

Part (a) The potential for the three-dimensional harmonic oscillator is

```
In[135]:= V= 1/2( kx x^2+ ky y^2+ kz z^2)

Out[135]=     2       2       2
          kx x  + ky y  + kz z
          --------------------
                   2
```

Assume the wave function is of the form

```
In[136]:= psi[x,y,z] = X[x] Y[y] Z[z];
```

Applying the user-defined operator **schrodinger** to **psi**, you get Schrodinger's equation:

```
In[137]:= SetCoordinates[Cartesian];

          eq1= schrodinger[V,En] @ psi[x,y,z]
```

```
Out[137]=
                2        2        2
          kx x  + ky y  + kz z
(-En + ---------------------------) X[x] Y[y] Z[z] -
                    2

     2
 hbar  (Y[y] Z[z] X''[x] + X[x] Z[z] Y''[y] + X[x] Y[y] Z''[z])
 ---------------------------------------------------------------
                              2 m
```

Schrodinger's equation follows from setting **eq1** equal to zero.

If you divide by (**X[x] Y[y] Z[z]**), the equation in **eq1** separates into terms that are functions of the same variable:

```
In[138]:= eq2 =  eq1 /(X[x] Y[y] Z[z]) //ExpandAll
Out[138]=
             2        2        2        2              2
         kx x     ky y     kz z     hbar  X''[x]   hbar  Y''[y]
-En + ------ + ------ + ------ - ------------- - ------------- -
           2        2        2       2 m X[x]       2 m Y[y]

     2
 hbar  Z''[z]
 ------------
  2 m Z[z]
```

Three ordinary differential equations follow from **eq2** by grouping terms with the same variable and setting them equal to separation constants **Ex**, **Ey**, and **Ez**. The separation constants are related to **En** by the rule

```
In[139]:= eRule=      En->Ex+Ey+Ez;
```

Expressing **eq2** in terms of **{Ex,Ey,Ez}**, you get

```
In[140]:= eq3= eq2 //. eRule
Out[140]=
                          2        2        2        2              2
                      kx x     ky y     kz z     hbar  X''[x]   hbar  Y''[y]
-Ex - Ey - Ez + ------ + ------ + ------ - ------------- - ------------- -
                     2        2        2       2 m X[x]       2 m Y[y]

     2
 hbar  Z''[z]
 ------------
  2 m Z[z]
```

Define a selection operator that selects terms with the same variable and with the relevant separation constant **Ex**, **Ey**, or **Ez**:

```
In[141]:= select[eq_,En_,var_]:=
           Select[eq, (!FreeQ[#,var] || !FreeQ[#,En ])&]
```

Applying **select** to the three variables and setting the results to zero produces three equations:

```
In[142]:= eq4= (0=={select[eq3,Ex,x],
                    select[eq3,Ey,y],
                    select[eq3,Ez,z]}
               )//Thread
```

```
Out[142]=
                    2          2                              2          2
                kx x      hbar  X''[x]                   ky y       hbar  Y''[y]
 {0 == -Ex + ------- - ---------------, 0 == -Ey + ------- - ---------------,
                  2         2 m X[x]                       2          2 m Y[y]

                    2          2
                kz z      hbar  Z''[z]
   0 == -Ez + ------- - ---------------}
                  2         2 m Z[z]
```

eq4 can be simplified as follows:

```
In[143]:= eq5= {xeq,yeq,zeq} =
          ( Solve[eq4, {X''[x],Y''[y],Z''[z]}][[1]]
                /.Rule->Equal
                //ExpandAll
          )

Out[143]=
                                                2
                -2 Ex m X[x]     kx m x  X[x]
 {X''[x] == ------------- + ---------------,
                      2                  2
                  hbar               hbar

                                                2
                -2 Ey m Y[y]     ky m y  Y[y]
   Y''[y] == ------------- + ---------------,
                      2                  2
                  hbar               hbar

                                                2
                -2 Ez m Z[z]     kz m z  Z[z]
   Z''[z] == ------------- + ---------------}
                      2                  2
                  hbar               hbar
```

The solutions of **eq5** are the one-dimensional harmonic oscillator solutions. For the one-dimensional harmonic oscillator, we defined the notation **k=w^2 m**, **En=1/2 en hbar w**, and **x= Sqrt[hbar/(w m)] z**, where **z** is the dimensionless coordinate. The normalization factor is easily shown to be

```
In[144]:= normConst[n_]= 1/Sqrt[ Sqrt[Pi] 2^n Factorial[n]];
```

The normalized wave function is given by a product of **X[x] Y[y] Z[z]**

```
In[145]:= X[x_,nx_] = normConst[nx] HermiteH[nx,x] Exp[-x^2/2];
          Y[y_,ny_] = normConst[ny] HermiteH[ny,y] Exp[-y^2/2];
          Z[z_,nz_] = normConst[nz] HermiteH[nz,z] Exp[-z^2/2];
```

where **nx**, **ny**, and **nz** are integers. We leave it to you to verify that the wave functions are normalized.

Part (b) Plot a typical two-dimensional wave function that is of the form **X[x] Y[y]**. Consider the wave for **{nx=1, ny=2}**. The graphics for the wave function are generated from

```
In[146]:= wavePlot= Plot3D[ X[x,2] Y[y,1]
                        ,{x,-4,4} ,{y,-4,4}
                        ,PlotPoints->35
                        ,DisplayFunction->Identity ];
```

Add to these graphics the harmonic potential described by the variables **{x,y}**. The potential is proportional to $x^2 + y^2$, and its graphics are

```
In[147]:= potPlot =
            Graphics3D[ Plot3D[ -2 +.15(x^2+y^2)
                          ,{x,-4,4},{y,-4,4}
                          ,PlotPoints->35
                          ,DisplayFunction->Identity]
              ];
```

Combining the wave function and potential, you get

```
In[148]:= Show[{potPlot, wavePlot}
              ,PlotRange-> {-0.5  ,0.7}
              ,Boxed->False
              ,ViewPoint->{1,1,2}
              ,Axes->False
              ,DisplayFunction->$DisplayFunction];
```

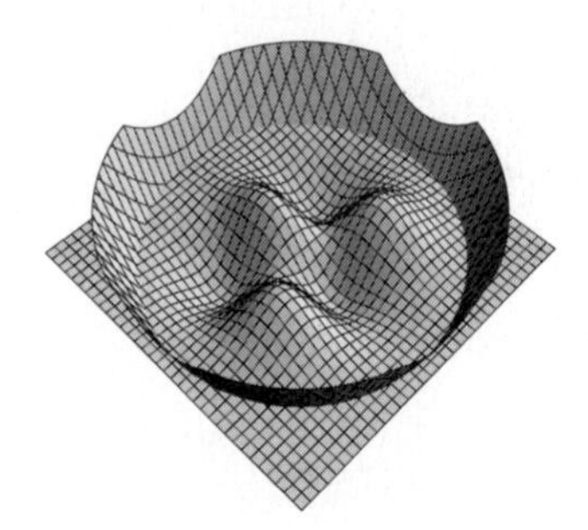

Problem 2: Schrodinger's equation for spherically symmetric potentials

Consider Schrodinger's equation with a spherically symmetric potential $V = V[r]$. Assume the solution separates and is of the form $\psi = R[r]Th[\theta]Phi[\phi]$.

Part (a) Express Schrodinger's equation in spherical coordinates and derive an equation for the angular function $Phi[\phi]$. Solve for $Phi[\phi]$ and show that

$$Phi[\phi] = c_1 Exp[-im_\phi \phi] + c_2 Exp[im_\phi \phi].$$

Part (b) Eliminate the ϕ variable in Schrodinger's equation using the results in Part (a). Separate the remaining r and θ variables and derive equations for $Th[\theta]$ and $R[r]$.

Part (c) Let $z = \mathrm{Cos}[\theta]$ and show that the $Th[\theta]$ equation in Part (b) can be expressed in the canonical form of Legendre's equation:

$$\ell Th[z] + \ell^2 Th[z] + (m_\phi^2 Th[z])/(-1 + z^2) - 2zTh'[z] + (1 - z^2)Th''[z] == 0.$$

Part (d) Declare ℓ and m_ϕ to be integers and solve Legendre's equation. Verify that the solutions satisfy the equation. Show that the product of the angular functions $P[\ell, m, \mathrm{Cos}[\theta]]$, and $Exp[im\phi]$ are related to spherical harmonics by the equation

$$Y[\ell, m, \theta, \phi] == C[\ell, m]P[\ell, m, \mathrm{Cos}[\theta]]Exp[im\phi],$$

where

$$C[\ell, m] = \sqrt{(2\ell + 1)/(4\pi) * (\ell - m)!/(\ell + m)!}$$

and $P[\ell, m, \mathrm{Cos}[\theta]]$ is Legendre's function, and $Y[\ell, m, \theta, \phi]$ is the spherical harmonic function.

Remarks and outline Having solved Schrodinger's equation in Cartesian coordinates, we turn to the next logical set of coordinates—spherical coordinates. Schrodinger's equation expressed in spherical coordinates follows from the user-defined operator **schrodinger**. The problem reduces to a set of ordinary differential equations when the coordinates are separated. If the equation for $Th[\theta]$ is expressed in terms of $z = \mathrm{Cos}[\theta]$, the $Th[\theta]$-equation reduces to the standard form of Legendre's equation. The spherical wave function follows from the product $R[r]Th[\theta]Phi[\phi]$.

Required packages

```
In[149]:= Needs["Algebra`Trigonometry`"]
          Needs["Calculus`VectorAnalysis`"]
          Needs["Graphics`ParametricPlot3D`"]
```

Solution

```
In[150]:= Clear["Global`*"];
```

Part (a) Schrodinger's equation expressed in spherical coordinates follows from the operator **schrodinger**. Assume the wave function is of the form

```
In[151]:= psi[r,theta,phi] =R[r] Th[theta] Phi[phi];
```

Expressing the operator **schrodinger** in spherical coordinates and applying it to **psi**, you get

```
In[152]:= SetCoordinates[Spherical];

          eq1= schrodinger[V[r] ,En] @ psi[r,theta,phi]

Out[152]= Phi[phi] R[r] Th[theta] (-En + V[r]) -

              2
           (hbar  Csc[theta] (2 r Phi[phi] Sin[theta] Th[theta] R'[r] +

                Cos[theta] Phi[phi] R[r] Th'[theta] +

                Csc[theta] R[r] Th[theta] Phi''[phi] +

                 2
                r  Phi[phi] Sin[theta] Th[theta] R''[r] +

                                                                 2
                Phi[phi] R[r] Sin[theta] Th''[theta])) / (2 m r )
```

Schrodinger's equation follows from setting **eq1** to zero. First, separate the **phi** variable and then, the **theta** and **r** variables. To group the **phi** variable, multiply **eq1** by $r^2\mathrm{Sin}[\theta]^2/(\psi[r, \theta, \phi])$:

```
In[153]:= eq2 = ( eq1 (r^2 Sin[theta]^2 / psi[r,theta,phi])
                  //ExpandAll  )

Out[153]=       2             2     2              2
          -(En r  Sin[theta] ) + r  Sin[theta]  V[r] -

              2               2
          hbar  r Sin[theta]  R'[r]
          ------------------------- -
                    m R[r]

              2                                        2
          hbar  Cos[theta] Sin[theta] Th'[theta]   hbar  Phi''[phi]
          -------------------------------------- - ---------------- -
                     2 m Th[theta]                   2 m Phi[phi]

              2  2           2                 2            2
          hbar  r  Sin[theta]  R''[r]   hbar  Sin[theta]  Th''[theta]
          --------------------------- - ------------------------------
                  2 m R[r]                      2 m Th[theta]
```

The **phi** terms are isolated in **eq2**. To **select** those parts of an equation having a certain variable, define the operator:

```
In[154]:= select[eq_, var_]:=  Select[eq,(!FreeQ[#,var])&]
```

Apply **select** to **eq2** and separate the **phi** variable. The results are then set equal to a separation constant $\hbar^2/(2m)m_\phi^2$:

```
In[155]:= phiEq1=  select[eq2,phi ]==hbar^2/(2 m) mphi^2

Out[155]=       2                      2     2
          -(hbar  Phi''[phi])      hbar  mphi
          ------------------- == -----------
             2 m Phi[phi]            2 m
```

Rearranging this, you get

```
In[156]:= phiEq2=  Solve[phiEq1,Phi''[phi] ][[1,1]] /.Rule->Equal

Out[156]= Phi''[phi] == -(mphi  Phi[phi])
```

The form of the separation constant, $\hbar^2/(2m)m_\phi^2$, is defined in anticipation that **mphi** will be an integer. Solving **phiEq** for **Phi**, you get

```
In[157]:= phiRule=DSolve[phiEq2,{Phi},phi][[1]]

Out[157]=
                                   -I mphi phi           I mphi phi
          {Phi -> Function[phi, E            C[1] + E           C[2]]}
```

For the wave function to be periodic, the value of **mphi** must be either zero, a positive integer, or a negative integer.

Part (b) Equations for **Th[theta]** and **R[r]** follow from **eq2.** The **Th[theta]** and **R[r]** functions separate if you divide **eq2** by **Sin[theta]^2** and eliminate the **phi** variable using **phiRule**:

```
In[158]:= eq3= ( (eq2   //.phiRule
                        //Cancel
                  ) /Sin[theta]^2
                        //ExpandAll
               )
```

```
Out[158]=          2     2            2                    2
       2    hbar  mphi   Csc[theta]         2        hbar   r R'[r]
-(En r ) + ---------------------------- + r  V[r] - ---------------- -
                     2 m                                  m R[r]

      2                                        2 2               2
  hbar  Cot[theta] Th'[theta]      hbar  r  R''[r]      hbar  Th''[theta]
  ---------------------------- - ------------------ - ------------------
        2 m Th[theta]                  2 m R[r]           2 m Th[theta]
```

The terms in **eq3** contain either **Th[theta]** or **R[r]** functions; there are no cross terms. The operator **select** will select terms having the same variable. Applying **select** to **eq3** for the **r** variable and setting the results equal to a separation constant $L = -(1/2)\ell(\ell+1)$, the radial equation becomes

```
In[159]:=  rEq1= select[eq3,r]==L hbar^2/m

Out[159]=                      2               2 2              2
       2     2          hbar   r R'[r]    hbar   r  R''[r]    hbar  L
-(En r ) + r  V[r] - --------------- - ----------------- == -------
                          m R[r]            2 m R[r]          m
```

Simplify **rEq1** by solving for **R''[r]** and replacing **L** by **-1/2 el(el+1)**:

```
In[160]:=  rEq= (( Solve[rEq1,R''[r] ][[1,1]]
                  /.Rule->Equal
                  //ExpandAll
                ) /.{L->-1/2 el(el+1)}
                )

Out[160]=
            -2 En m R[r]    el (1 + el) R[r]    2 m R[r] V[r]    2 R'[r]
R''[r] == ------------- + ---------------- + -------------- - -------
                 2                 2                2             r
             hbar                 r             hbar
```

The form of the separation constant **el(1+el)** is chosen in anticipation that the allowed values of **el** will be $\{0, 1, 2, 3, \ldots\}$.

The **theta** equation follows similarly:

```
In[161]:= thetaEq= ( select[eq3,theta]+L hbar^2/m == 0
                  /.{L->-1/2 el(el+1)}
                 )

Out[161]=                  2            2     2            2
 -(el (1 + el) hbar )    hbar  mphi  Csc[theta]
 -------------------- + ------------------------ -
         2 m                     2 m

     2                             2
 hbar  Cot[theta] Th'[theta]   hbar  Th''[theta]
 --------------------------- - ----------------- == 0
       2 m Th[theta]             2 m Th[theta]
```

Part (c) The equation **thetaEq** is Legendre's equation. The canonical form of Legendre's equation is usually written in terms of the variable $z = \mathrm{Cos}[\theta]$. Changing variables using the user-defined function **vChange** and applying **vChange** to **thetaEq**, you get

```
In[162]:= eq4= vChange[thetaEq,Th,theta,z, ArcCos[z]]
```

```
Out[162]=
                        2        2    2           2
 -(el (1 + el) hbar )    hbar  mphi        hbar  z Th'[z]
 --------------------- + ------------- + --------------- +
         2 m                    2           2 m Th[z]
                          2 m (1 - z )

      2              2      z Th'[z]                 2
  hbar  Sqrt[1 - z ] (------------- - Sqrt[1 - z ] Th''[z])
                                2
                        Sqrt[1 - z ]
  --------------------------------------------------------- == 0
                          2 m Th[z]
```

Eliminate the **ArcCos[z]** terms with the rule

```
In[163]:= arcTrigRule={ 1/Sin[ArcCos[z]]-> 1/ Sqrt[1-z^2]
                       ,1/Tan[ArcCos[z]]-> 1/(Sqrt[1-z^2]/z)};
```

Applying **arcTrigRule** to **eq4** and solving for **Th''[z]**, it follows that

```
In[164]:= thEq= ( (-2 m Th[z]/hbar^2 eq4)
                   /.arcTrigRule
                   //Thread[#,Equal ]&
                   //Simplify
                ) //Map[Collect[#,{Th''[z]}]&,#,{1}]&

Out[164]=
                                     2
                  2              mphi  Th[z]                            2
  el Th[z] + el  Th[z] + ------------- - 2 z Th'[z] + (1 - z ) Th''[z] == 0
                                  2
                            -1 + z
```

Part (d) The solutions for **thEq** are Legendre polynomials if **el**={0, 1, 2, 3, ...} and **mphi**={−el, ..., el}. Declare **el** and **mphi** to be integers:

```
In[165]:= IntegerQ[el   ]^=True;
          IntegerQ[mphi]^=True;
```

apply **DSolve** to **thEq**:

```
In[166]:= thRule= DSolve[thEq,{Th},z][[1]]

Out[166]= {Th -> (LegendreP[el, mphi, #1] & )}
```

then verify that **thRule** is the correct solution. Consider **el**=0,1,2,3 with the corresponding **mphi** terms and substitute **thRule** into **thEq**:

```
In[167]:= Table[ thEq //.thRule,{el,0,3},{mphi,0,el}] //Simplify

Out[167]= {{True}, {True, True}, {True, True, True}, {True, True, True, True}}
```

To relate the angular Legendre solutions to spherical harmonics, define the coefficients:

```
In[168]:= Coeff[el_,m_]:=
          Sqrt[(2 el + 1)/(4 Pi)*(el - m)!/(el + m)!]
```

The product of the angular Legendre solutions times **Coeff** coefficients are spherical harmonics. To verify this, consider the equation

```
In[169]:= sphEq :=
           SphericalHarmonicY[el,m,theta,phi] ==
            (Coeff[el,m]  LegendreP[el,m,Cos[theta]]
             * Exp[ I m phi]);
```

It follows for **el**=$\{0, 1, 2, 3\}$ and **m={-el, el}** that

```
In[170]:= Table[sphEq //Simplify
                      //PowerExpand
                      //TrigReduce
                      //N
                ,{el,0,3},{m,0,el}]

Out[170]= {{True}, {True, True}, {True, True, True}, {True, True, True, True}}
```

Problem 3: Particle in an infinite, spherical box

Consider a three-dimensional square well of infinite depth and radius a defined by

$$V(x) = \begin{cases} 0 & \text{if } r < a; \\ \infty & \text{if } a < r. \end{cases}$$

Assume the wave function separates and is of the form $R[r]Th[\theta]Ph[\phi]$. The radial function $R[r]$ satisfies the equation

$$k^2 r^2 R[r] - \ell(\ell+1)R[r] + 2rR'[r] + r^2 R''[r] = 0,$$

where $E_n = ((\hbar k)^2/(2m)) + V$.

Part (a) Solve the radial equation for $R[r]$ and normalize the solution with the criteria that the integral $\int_0^\infty R[r]^2\, r^2\, dr = 1$.

Part (b) Use the boundary condition at the wall to derive an expression for the eigenvalues. Solve for the eigenvalues by plotting the two sides of the eigenvalue equation and observing where the curves intersect. Solve for the eigenvalues a second way by finding the roots of the eigenvalue equation using the command **FindRoot**. Display the results in tabular and graphical forms.

Part (c) Plot $R[r]$ for the S ($\ell = 0$), P ($\ell = 1$), and D ($\ell = 2$) states.

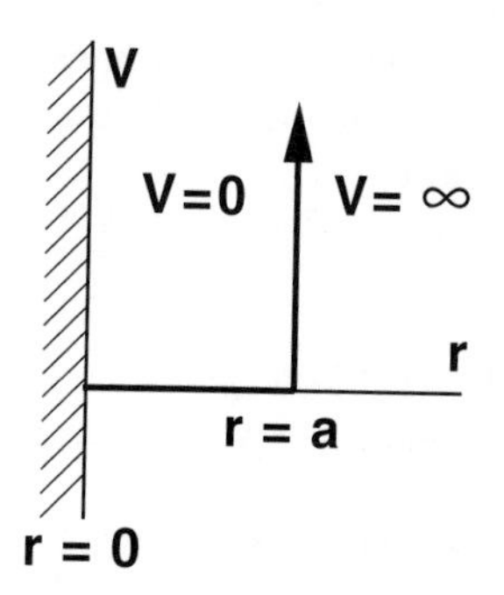

Remarks and outline This is the simplest spherically symmetric problem, in close analogy with the one-dimensional infinite square well. The angular functions are spherical harmonics, so the problem is reduced to solving the radial wave equation. The solutions of the radial equation with $V = 0$ are Bessel functions. Requiring the solutions to be finite at $r = 0$ reduces the wave functions to Bessel functions of the J kind. The energy eigenvalues are a consequence of the boundary condition that require the wave to vanish at $r = a$. Solve the energy eigenvalue equation both graphically and via the **FindRoot** command for the S, P, and D wave functions.

Required packages

```
In[171]:= Needs["Algebra`Trigonometry`"]
          Needs["Calculus`VectorAnalysis`"]
          Needs["Graphics`ParametricPlot3D`"]
```

Solution

```
In[172]:= Clear["Global`*"];
```

Part (a) The radial wave equation for a particle in a well with $V = 0$ is

```
In[173]:= eq1= k^2 r^2 R[r]- el(el+1) R[r]+ 2 r R'[r] + r^2 R''[r]==0 ;
```

where **k** is related to the energy of the particle by

```
In[174]:= kRule= {k -> Sqrt[2 m (En-V)]/hbar}  /.{V->0}

Out[174]=       Sqrt[2] Sqrt[En m]
          {k -> ------------------}
                      hbar
```

The solution of **eq1** follows from **DSolve**:

```
In[175]:= DSolve[eq1, R[r] ,r][[1]] //PowerExpand //Simplify

Out[175]=
                      1                                1
             BesselJ[- + el, k r] C[1] + BesselY[- + el, k r] C[2]
                      2                                2
          {R[r] -> -----------------------------------------------------}
                                       Sqrt[r]
```

The constants **{C[1],C[2]}** are determined by requiring the wave function to be finite at **r=0** and by normalizing the wave function to one. The function **BesselY[el+1/2 ,r]** diverges as **r->0**; therefore, the wave function which satisfies the boundary condition is proportional to only **BesselJ[1/2+el, k r]/Sqrt[r]**.

To normalize the radial functions, define

```
In[176]:= norm[el_,k_,a_]=
            1/Sqrt[
              Integrate[ (r^2 BesselJ[1/2+el, k r]^2/r ) ,{r,0,a}] ];
```

The normalized radial waves follow from

```
In[177]:= rwave[el_,k_,a_,r_]= (norm[el,k,a] BesselJ[1/2+el, k r]/Sqrt[r] );
```

Part (b) The boundary condition requires the wave to vanish at the boundary **r=a**. This condition is equivalent to finding the values of **k** where **BesselJ[el+1/2, k a]** vanishes. There will be a sequence of energy eigenvalues for each choice of **el**. Here, you will find the first four eigenvalues for the S, P, and D states.

There are two procedures for finding where the **BesselJ** function vanishes. You can plot **BesselJ** as a function of **k** and observe where the curves cross the axis. Or you can directly compute the roots of the eigenvalue equation with the command **FindRoot**. Both methods are illustrated next.

First, consider the graphical method. In the following, set **a=1**. Graphing **BesselJ[1/2+el,k]** as a function of **k** for **el**=$\{0, 1, 2\}$, you get

```
In[178]:= Plot[
            Evaluate[
              Table[ BesselJ[1/2+n,k],{n,0,2,1}]
            ],{k,0.01,19}
          ];
```

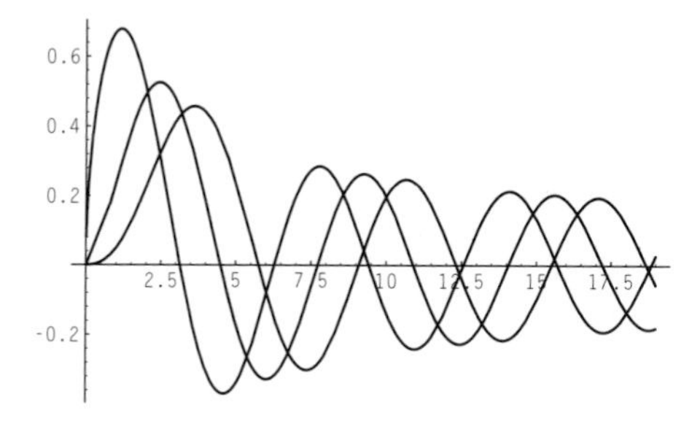

An inspection of the curves gives the first four eigenvalues for **el**=$\{0, 1, 2\}$. The eigenvalues for the S, P, and D states are approximately given by the three lists, respectively:

```
In[179]:= {guess[0], guess[1], guess[2]}=
           {  {3   , 6   , 9 , 13 }
             ,{4.49, 7.5, 11, 14 }
             ,{5.76, 9   , 12, 15 }
           };
```

The second way to find the roots for the Bessel function is to apply **FindRoot** to **BesselJ**. To find the energy eigenvalues for different choices of **el**, define the function **eigenE**:

```
In[180]:= eigenE[el_,guess_] :=
           ( guess
             //Map[FindRoot[ BesselJ[1/2+el,k]==0,{k,#}]& ,#]& )
```

where **guess** is the initial guess for **k**. Use the graphical result for the initial guess. It follows from **eigenE** that the allowable values of **k** are

```
In[181]:=  kvalues= Table[ eigenE[n,guess[n]] ,{n,0,2,1}];
```

Display the results in tabular form:

```
In[182]:= TableForm[ k/.kvalues
                ,TableSpacing -> {0, 2,2}
                ,TableHeadings->{{"S-Wave","P-Wave","D-Wave" }
                                ,{"first","second","third" ,"fourth" }}]
```

Out[182]=

	first	second	third	fourth
S-Wave	3.14159	6.28319	9.42478	12.5664
P-Wave	4.49341	7.72525	10.9041	14.0662
D-Wave	5.76346	9.09501	12.3229	15.5146

The results can also be displayed in the form of an energy-level graph:

```
In[183]:= list= k/.kvalues;

          levels=
          Table[ {{2i,list[[i,j]]},{2i+1,list[[i,j]]}}
              ,{i,1,3}
              ,{j,1,Length[list[[1]] ]}];

          Show[   {Graphics[ Map[Line,levels,{2}]     ]
                  ,Graphics[ Text["S state",{2.5,1}] ]
                  ,Graphics[ Text["P state",{4.5,1}] ]
                  ,Graphics[ Text["D state",{6.5,1}] ] }
                  ,Frame->True ];
```

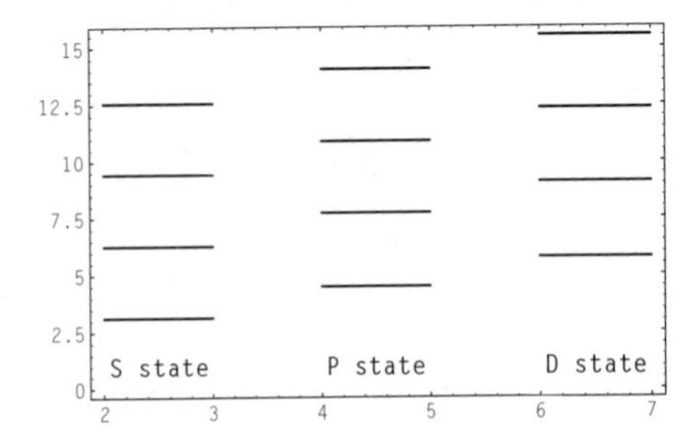

Part (c) Plot the first four eigenfunctions for the S, P, and D states. The graphics for the waves follow from the plot function

```
In[184]:= Clear[wavePlot];
          wavePlot[n_] :=
          Plot[  Evaluate[ rwave[n-1,k,1,r]
                       /.{k->(k/.kvalues[[n]])}]
            ,{r,.001,1}
            ,PlotStyle->{Dashing[{.01,.00}]
                        ,Dashing[{.05,.02}]
                        ,Dashing[{.03,.02}]
                        ,Dashing[{.01,.02}] }
            ,PlotLabel->FontForm[y,{"Symbol",10}]
            ,DisplayFunction->Identity
          ];
```

where **n**=$\{1,2,3\}$ are the S, P, and D states, respectively. The plots for the **el=0** wave functions that correspond to the four lowest eigenvalues are

```
In[185]:= Show[{wavePlot[1]},DisplayFunction->$DisplayFunction];
```

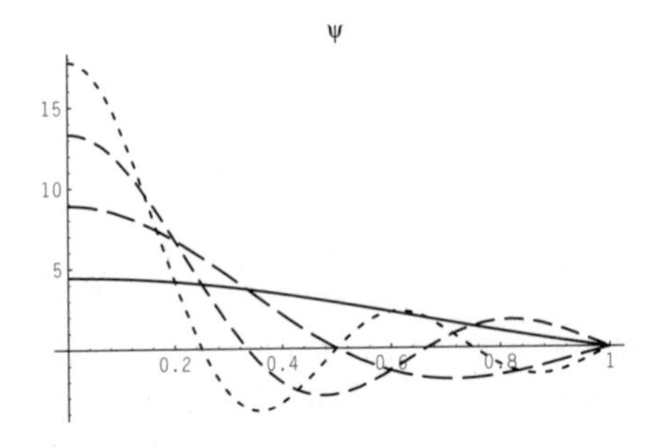

Likewise, the plots for the P and D waves are, respectively,

```
In[186]:= Show@GraphicsArray[{wavePlot[2],wavePlot[3]}];
```

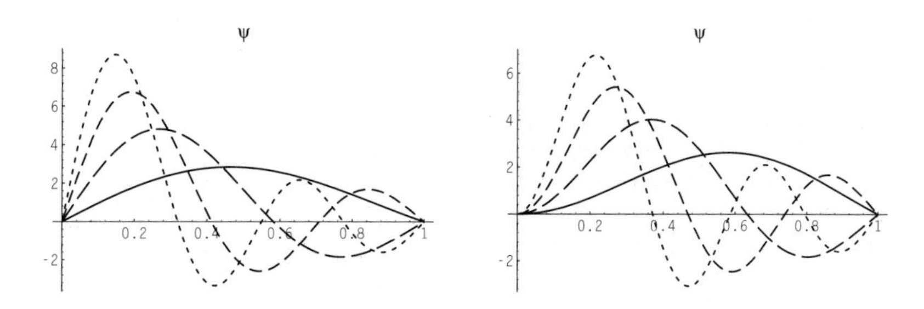

Problem 4: Particle with negative energy in a finite, spherical box

Consider a three-dimensional finite square well of radius a defined by

$$V(x) = \begin{cases} -V_0 & \text{if } r < a; \\ 0 & \text{if } a < r. \end{cases}$$

Assume the wave function is of the form $R[r]Th[\theta]Ph[\phi]$. The radial function $R[r]$ satisfies the equation

$$k^2 r^2 R[r] - \ell(\ell+1)R[r] + 2rR'[r] + r^2 R''[r] = 0,$$

where $E_n = ((\hbar k)^2/(2m)) + V$.

Part (a) Use the condition that the wave function must be finite at the origin and vanish as $r \to \infty$. Write the general form of the wave inside and outside the well.

Part (b) Use the boundary condition at the wall to derive an expression for the eigenvalues.

Part (c) Find the energy eigenvalues for $\ell = 0$ and plot the radial solution.

Part (d) Find the energy eigenvalues for $\ell = 1$ and plot the radial solution.

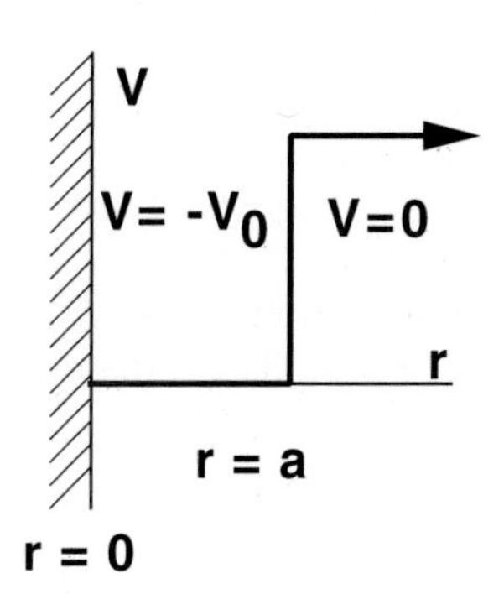

Remarks and outline This is the finite generalization of the infinite well in Problem 3. The wave function follows from the solution of the radial Schrodinger's equation having a constant potential. The regions $0 < r < a$ and $a < r$ must be treated separately. Again the radial functions are Bessel functions; however, the boundary conditions lead to more complex solutions. Solve the eigenvalue equation both graphically and via the **FindRoot** command for the S and P wave functions.

Required packages

```
In[187]:= Needs["Algebra`Trigonometry`"]
          Needs["Calculus`VectorAnalysis`"]
          Needs["Graphics`ParametricPlot3D`"]
```

Solution

```
In[188]:= Clear["Global`*"];
```

Part (a) The radial Schrodinger's equation is

```
In[189]:= rEq= 0 ==  (- el R[r] - el^2 R[r] + k^2 r^2 R[r]
                       + 2 r R'[r] + r^2 R''[r])  ;
```

Solving **rEq**, you get

```
In[190]:=   rRule= ( DSolve[rEq,{R[r]},r][[1]]
                     //PowerExpand
                     //Simplify
                   )

Out[190]=
                       1                             1
           BesselJ[- + el, k r] C[1] + BesselY[- + el, k r] C[2]
                       2                             2
{R[r] -> ---------------------------------------------------------}
                                 Sqrt[r]
```

The values of **k** inside and outside the well are, respectively,

```
In[191]:= kRule  ={k1 ->Sqrt[2 m (V0 - Wn)]/hbar,
                   k2 ->Sqrt[2 m (    - Wn)]/hbar} //PowerExpand

Out[191]=
          Sqrt[2] Sqrt[m] Sqrt[V0 - Wn]
{k1 -> -----------------------------,
                    hbar

           I Sqrt[2] Sqrt[m] Sqrt[Wn]
  k2 -> ----------------------------}
                    hbar
```

where **0<Wn<V0**.

Consider the region inside the well. The constant in front of **BesselY[el+1/2,r]** must be set equal to zero, since **BesselY** diverges as **r->0**. The unrenormalized wave function for **r<a** is

```
In[192]:= wave1[k_,el_,r_]= A BesselJ[1/2+el,k r]/Sqrt[r];
```

Outside the well, the wave function must decay as **r**$\rightarrow \infty$. To examine the behavior of the Bessel functions for large **r**, consider the combinations **(BesselJ+ I BesselY)** and **(BesselJ- I BesslY)**:

```
In[193]:=  symBes[k_, el_,r_]:=
             B (BesselJ[1/2+el,k r]+I BesselY[1/2+el,k r])/Sqrt[r];

           asymBes[k_, el_,r_]:=
             B (BesselJ[1/2+el,k r]-I BesselY[1/2+el,k r])/Sqrt[r];
```

To explore the asymptotic behavior, consider an explicit value of **el**, for example **el=2**. Using the fact that **k** is complex, set **(k=I q)** to find

```
In[194]:= ( (symBes[I q,2,r] )
              //HyperbolicToComplex
              //ExpandAll
              //Factor
           )

Out[194]=
                 -2 I                     2  2
          I B Sqrt[------] (3 + 3 q r + q  r )
                  Pi q r
          ------------------------------------
                      q r  2  5/2
                     E    q  r
```

and

```
In[195]:= ( (asymBes[I q,2,r] )
              //HyperbolicToComplex
              //ExpandAll
              //Factor
           )

Out[195]=
                q r        -2 I                      2  2
          -I B E    Sqrt[------] (3 - 3 q r + q  r )
                           Pi q r
          --------------------------------------------
                             2  5/2
                            q  r
```

symBes is exponentially decreasing, and **asymBes** is exponentially increasing as **r** goes to infinity. Therefore the wave function outside the well must be of the form

```
In[196]:=  wave2[k_,el_,r_]= symBes[k , el ,r ]

Out[196]=
                     1                          1
          B (BesselJ[- + el, k r] + I BesselY[- + el, k r])
                     2                          2
          --------------------------------------------------
                                Sqrt[r]
```

Part (b) The boundary conditions require the wave function and its derivative to be continuous at **r=a**. The continuity condition is described by **bc1**:

```
In[197]:= bc1= (wave1[k1,el,r]==wave2[k2,el,r]) //.{r->a};
```

and the derivative condition by **bc2**:

```
In[198]:= bc2= (D[wave1[k1,el,r],r] == D[wave2[k2,el,r],r]) //.{r->a};
```

Equation **bc1** gives a relation between the amplitudes, and **bc2** gives an equation for the eigenvalues. Solving **bc1** for **B**, you get

```
In[199]:= bc1Sol= Solve[bc1,B][[1]]
```

```
Out[199]=
                                1
                  A BesselJ[- + el, a k1]
                                2
  {B -> -(---------------------------------------------------)}
                 1                                 1
     -BesselJ[- + el, a k2] - I BesselY[- + el, a k2]
                 2                                 2
```

A is an overall normalization, and it can be arbitrarily set equal to one without loss of generality. Using **bc1Sol** to eliminate **B**, the derivative boundary condition becomes

```
In[200]:= bc2Sol= (bc2 //.bc1Sol/.A->1);
```

The energy eigenstates follow from choosing **el** and solving **bc2Sol**.

Part (c) Find the energy eigenvalues for **el=0** and plot the radial solution. Let the parameters have the values

```
In[201]:= allRules=Union[kRule,{hbar->1, a->1, m->1, V0->100}] ;
```

The eigenstate condition for **el=0** follows from **bc2Sol** with **el=0** (S state):

```
In[202]:= eq3= bc2Sol //.{el->0} //PowerExpand //Map[Together,#]& //Simplify

Out[202]=      2
         Sqrt[--] (a k1 Cos[a k1] - Sin[a k1])
              Pi                                   2 I (I + a k2) Sin[a k1]
         ------------------------------------- == ------------------------
                     2                             2
                    a  Sqrt[k1]                   a  Sqrt[k1] Sqrt[2 Pi]
```

The eigenvalues follow from graphing the left-hand and right-hand sides of **eq3** and observing where the curves intersect. Plotting the two sides of **eq3**, you get

```
In[203]:= Plot[{eq3[[1]], eq3[[2]]} //.allRules //Evaluate
            ,{Wn,0,99}
            ,PlotStyle->{ Dashing[{0.05,0.0}],
                          Dashing[{0.04,0.02}] }
          ];
```

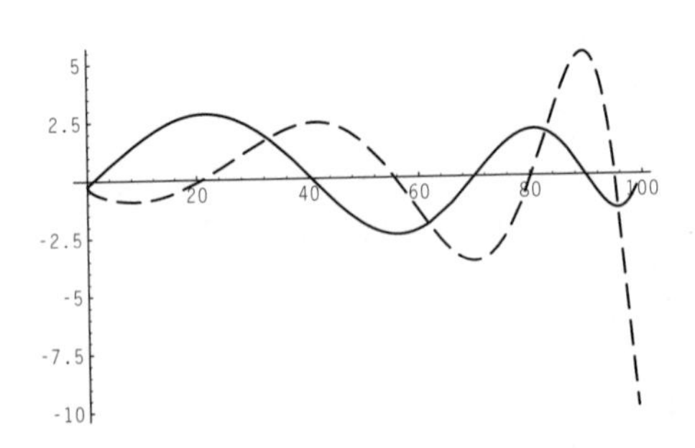

The eigenvalues occur at the intersection of the curves. The values of the first four eigenvalues are approximately $\{32, 61, 80, 95\}$. The numerical values for the four S-eigenstates follow (more accurately) from **FindRoot**:

```
In[204]:= eigen[0]= ( {32,61,80,95}
              //Map[FindRoot[Evaluate[eq3//.allRules ],{Wn,#}]&,#]&
              //Chop[#,10-8]&
              )

Out[204]= {{Wn -> 32.6732}, {Wn -> 61.6572}, {Wn -> 82.8523}, {Wn -> 95.6996}}
```

Evaluate **bc2Sol** to verify that these eigenvalues satisfy the derivative boundary conditions:

```
In[205]:= ( (bc2Sol[[1]]-bc2Sol[[2]])
                /.el->0
                //.allRules
                //.eigen[0]
                //N
                //Chop[#,10^-3]&
          )

Out[205]= {0, 0, 0, 0}
```

To plot the **el=0** wave, numerically evaluate the wave function inside and outside the well:

```
In[206]:= region1s= ( wave1[k1,0,r] /.A->1
                                    //.allRules
                                    //.eigen[0]
                                    //N           );

          region2s= ( wave2[k2,0,r] //.bc1Sol
                                    /.el->0
                                    /.A->1
                                    //.allRules
                                    //.eigen[0]
                                    //N           );
```

Define the plot operator:

```
In[207]:= plotOp[region_,range_List] :=
             Plot[ region //Evaluate
                  ,range
                  ,PlotStyle->{   Dashing[{0.05,0.0}],
                                  Dashing[{0.04,0.02}],
                                  Dashing[{0.03,0.03}],
                                  Dashing[{0.02,0.04}]}
                   ,DisplayFunction->Identity
                ];
```

and apply it to **region1s** and **region2s**:

```
In[208]:= Show[ plotOp[region1s,{r,0.001,1}]
               ,plotOp[region2s,{r,1,1.2}]
               ,Graphics[Line[{{1,-1},{1,3}}]]
               ,Graphics[
                  Text[FontForm["Boundary",{"Courier-Bold",10}],{1,2}]]
               ,DisplayFunction->$DisplayFunction
             ];
```

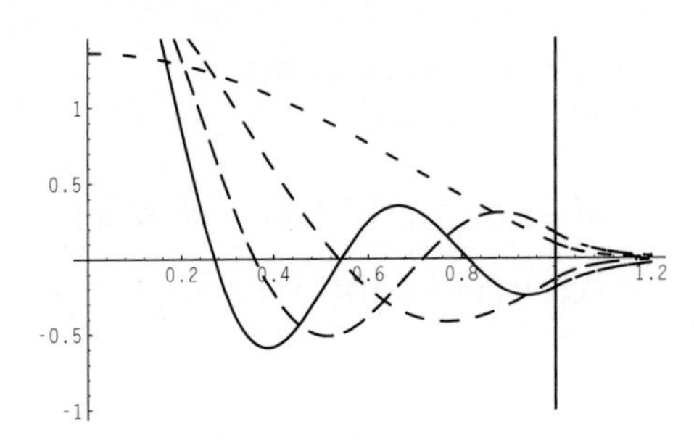

Part (d) Find the energy eigenvalues for **el=1** and plot the radial solution. The eigenstate condition for the P-wave is

```
In[209]:=  eq4= ( bc2Sol //.{el->1}
                   //Map[Together,#]&
                   //PowerExpand
                   //Simplify
            )
```

```
Out[209]=       2                                          2   2
          Sqrt[—] (2 a k1 Cos[a k1] - 2 Sin[a k1] + a  k1  Sin[a k1])
               Pi
          ---------------------------------------------------------- ==
                                3   3/2
                               a  k1

                          2   2        2
            (2 I + 2 a k2 - I a  k2 ) Sqrt[—] (a k1 Cos[a k1] - Sin[a k1])
                                          Pi
            ----------------------------------------------------------------
                                  3   3/2
                                 a  k1    (I + a k2)
```

Plotting the left-hand and right-hand sides of **eq4**, you get

```
In[210]:= Plot[{Re[eq4[[1]]], Re[eq4[[2]]]} //.allRules //Evaluate
                ,{Wn,1,99}
                ,PlotStyle->{Dashing[{0.05,0.0}],
                             Dashing[{0.04,0.02}] }
             ];
```

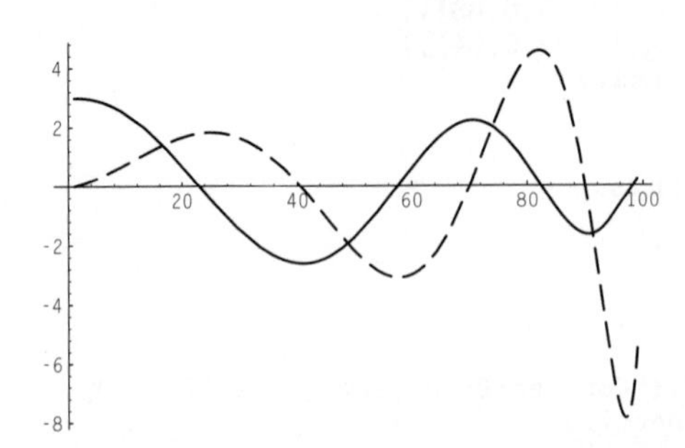

Again the eigenvalues occur at the intersection of the curves. The eigenvalues are approximately $\{16, 50, 77, 90\}$. The numerical values of the P-eigenstates follow from **FindRoot**:

```
In[211]:= eigen[1]=
          ( {16.544,50,77,90}
             //Map[FindRoot[Evaluate[eq4//.allRules ],{Wn,#}]&,#]&
             //Chop[#,10-8]&
           )

Out[211]= {{Wn -> 16.5443}, {Wn -> 48.8983}, {Wn -> 74.1329}, {Wn -> 91.2099}}
```

Evaluate the left-hand and right-hand sides of **bc2Sol** to verify the numerical solutions:

```
In[212]:= ( (bc2Sol[[1]]-bc2Sol[[2]])
                  /.el->1
                  //.allRules
                  //.eigen[1]
                  //N
                  //Chop[#,10^-4]&
           )

Out[212]= {0, 0, 0, 0}
```

Numerically evaluate the P-wave inside and outside the well to plot the wave functions:

```
In[213]:= region1p= ( wave1[k1,1,r] /.A->1
                                     //.allRules
                                     //.eigen[1]
                                     //N            );

          region2p= ( wave2[k2,1,r] //.bc1Sol
                                     /.el->1
                                     /.A->1
                                     //.allRules
                                     //.eigen[1]
                                     //N            );
```

The plots of the P-wave functions are

```
In[214]:= Show[ plotOp[Re[region1p],{r,0.001,1}]
               ,plotOp[Re[region2p] ,{r,1,1.2}]
               ,Graphics[Line[{{1,-1},{1,1.5}}]]
               ,Graphics[
                 Text[FontForm["Boundary",{"Courier-Bold",10}],{1,1}]]
               ,DisplayFunction->$DisplayFunction
              ];
```

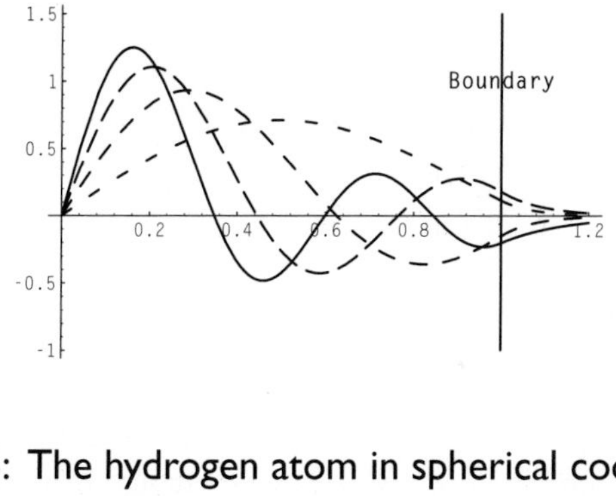

Problem 5: The hydrogen atom in spherical coordinates

Consider Schrodinger's equation with the coulomb potential $V = -Ze^2/r$. The angular dependence is given by $Y[\ell, m, \theta, \phi]$. The radial function satisfies the equation

$$0 = (-\ell R[r] - \ell^2 R[r] + k^2 r^2 R[r] + 2r R'[r] + r^2 R''[r]),$$

where $k^2 = (E_n - V[r])/(\hbar^2/(2m))$. Because E_n is negative, set $W_n = -E_n$.

Part (a) Find the normalized radial wave equation in terms of the orbital quantum number n and the angular quantum number ℓ. Normalize the solution using the criterion that $\int_0^\infty R[r]^2\, r^2\, dr = 1$.

Part (b) Graph the radial probability density $\rho[r] = r^2 R[r]^2$ for the states $n = \{1, 2, 3\}$.

Part (c) Plot the absolute value of the wave function for $\ell = 0$, 1, and 2 for all the allowable values of m and for fixed r.

Remarks and outline Use **DSolve** for the radial wave equation. Requiring the solution to vanish at the origin and infinity reduces the solution to a Laguerre polynomial.

Required packages

```
In[215]:= Needs["Algebra`Trigonometry`"]
          Needs["Calculus`VectorAnalysis`"]
          Needs["Graphics`ParametricPlot3D`"]
```

Solution

```
In[216]:= Clear["Global`*"];
```

Part (a) For a spherically symmetric potential, the angular dependence is given by spherical harmonics. The radial equation is

```
In[217]:= rEq= 0 == (- el R[r] - el^2 R[r] + k^2 r^2 R[r]
                 + 2 r R'[r] + r^2 R''[r])

Out[217]=
                          2          2 2                       2
          0 == -(el R[r]) - el  R[r] + k  r  R[r] + 2 r R'[r] + r  R''[r]
```

Substituting for the values of **k** and **V**,

```
In[218]:= rule1= {k   -> Sqrt[2 m (En-V[r])]/hbar
                ,V[r]-> -Z e^2/r
                ,En  -> -Wn};
```

rEq becomes

```
In[219]:= rEq2=rEq //.rule1 //ExpandAll

Out[219]=
                                           2                2
                          2         2 m r  Wn R[r]    2 e  m r Z R[r]
          0 == -(el R[r]) - el  R[r] - -------------- + --------------- +
                                              2                 2
                                          hbar              hbar

                      2
            2 r R'[r] + r  R''[r]
```

We have used the fact that the bound states have negative energy and defined **Wn** by **-En=+Wn>0**. The solution for **R** follows from applying **DSolve** to **rEq2**:

```
In[220]:= rSol= ( DSolve[rEq2 ,{R[r]},r][[1]]
                //PowerExpand //ExpandAll //Simplify )

Out[220]=
                                                   2
                el                                e  Sqrt[m] Z
        {R[r] -> (r   (C[2] HypergeometricU[1 + el - ------------------------,
                                                  Sqrt[2] hbar Sqrt[Wn]

                         2 Sqrt[2] Sqrt[m] r Sqrt[Wn]
              2 + 2 el, ------------------------------] +
                                    hbar

                                               2
                                              e  Sqrt[m] Z
           C[1] Hypergeometric1F1[1 + el - ------------------------,
                                           Sqrt[2] hbar Sqrt[Wn]

                         2 Sqrt[2] Sqrt[m] r Sqrt[Wn]
              2 + 2 el, ------------------------------])) /
                                    hbar

            (Sqrt[2] Sqrt[m] r Sqrt[Wn])/hbar
           E                                  }
```

The notation simplifies if you replace

```
In[221]:= rule2={ Wn-> e^4 m Z^2/(2 n^2 hbar^2)
              ,e -> hbar/Sqrt[m a0] };

          rSol2=(R[r]/.rSol) //.rule2 //PowerExpand

Out[221]=  el                                              2 r Z
          (r   (C[2] HypergeometricU[1 + el - n, 2 + 2 el, -----] +
                                                            a0 n

                                                             2 r Z
               C[1] Hypergeometric1F1[1 + el - n, 2 + 2 el, -----])) /
                                                              a0 n

            (r Z)/(a0 n)
           E
```

For **(1+el-n)<0**, the solution converges for large **r** if the coefficient in front of **HypergeometricU** is zero:

```
In[222]:= Series[ HypergeometricU[-a,b,r] ,{r,Infinity,0}]

Out[222]=  a
          r
```

Therefore the radial wave equation is proportional to

```
In[223]:= rSol3= ( r^el Hypergeometric1F1[ (1+el-n),2+2 el,(2 r Z)/(a0 n)]
                  )/  E^((r Z)/(a0 n))

Out[223]=  el                                            2 r Z
          r   Hypergeometric1F1[1 + el - n, 2 + 2 el, -----]
                                                       a0 n
          ----------------------------------------------------
                              (r Z)/(a0 n)
                             E
```

For **(1+el-n)** a negative integer, **rSol3** is a **LaguerreL** polynomial times a decreasing exponential and a power of **r** given by **r^el**:

```
In[224]:= int  /: IntegerQ[int]=True;

          rSol4= ((( rSol3
                       /.{(1+el-n) -> -int}
                     ) //Simplify
                     ) //.{int -> -(1+el-n)}
                     ) //Simplify

Out[224]=   el
          (r   Gamma[2 + 2 el] Gamma[-el + n]

                                               2 r Z
             LaguerreL[-1 - el + n, 1 + 2 el, -----]) /
                                               a0 n

             (r Z)/(a0 n)
           (E             Gamma[1 + el + n])
```

Define a normalizing function:

```
In[225]:= norm[n_,el_] :=
           Integrate[ (r^2 (r^el Exp[- r Z/(a0 n)]*
                LaguerreL[-1-el+n,1 +2 el ,2 r Z/(a0 n)])^2
                //ExpandAll )
            ,{r,0,Infinity}]
```

The normalized radial wave function is given by **rwave**:

```
In[226]:= rwave[n_,el_,r_] := rwave[n,el,r] =
           (  r^el Exp[- r Z/(a0 n)]*
              LaguerreL[-1-el+n,1 +2 el ,2 r Z/(a0 n)]/Sqrt[norm[n,el]]
             //PowerExpand  )
```

The principle quantum number is **n**=$\{1, 2, 3, \ldots\}$ and the angular momentum quantum number is **el**=$\{0, 1, \ldots, n-1\}$. Verify that **rwave** is normalized for **n**=$\{1, 2, 3\}$:

```
In[227]:= Table[ Integrate[ rwave[n,el,r]^2 r^2 //ExpandAll
                           ,{r,0,Infinity}]
           ,{n,1,3},{el,0,n-1}]

Out[227]= {{1}, {1, 1}, {1, 1, 1}}
```

The explicit wave function for **n=1** is

```
In[228]:= rwave[1,0,r]

Out[228]=        3/2
             2 Z
          ---------------
           3/2  (r Z)/a0
          a0     E
```

The two wave functions for **n=2** and **el**=$\{0, 1\}$ are

```
In[229]:=   rwave[2,#,r]& /@ {0,1}

Out[229]=           3/2        r Z
                   Z    (2 - ---)
                               a0                                 5/2
                                                                r Z
          {-------------------------------------, ------------------------------------}
                         3/2  (r Z)/(2 a0)                         5/2  (r Z)/(2 a0)
           2 Sqrt[2] a0     E                     2 Sqrt[6] a0     E
```

The three wave functions for **n=3** and **el=**$\{0,1,2\}$ are

```
In[230]:=    rwave[3,#,r]& /@ {0,1,2}

Out[230]=
                                                              2      5/2        2 r Z
       3/2        2                         2  2     Sqrt[-] r Z       (4 - -----)
   2 Z     (27 a0   - 18 a0 r Z + 2 r  Z )               3                   3 a0
  {---------------------------------------------, --------------------------------,
                      7/2   (r Z)/(3 a0)                  5/2  (r Z)/(3 a0)
      81 Sqrt[3] a0      E                          27 a0     E

              2    2  7/2
     2 Sqrt[--] r  Z
             15
   -----------------------}
          7/2   (r Z)/(3 a0)
     81 a0     E
```

Part (b) The plots of the probability densities **r^2 rwave^2** for the states **n=**$\{1,2,3\}$ are, respectively,

```
In[231]:= value={Z->1,a0->1};

          plot[n_,el_,rmax_]:=
            Plot[ r^2 rwave[n,el,r]^2 /.value //Evaluate
                  ,{r,0.01,rmax}
                  ,Ticks->None
                  ,DisplayFunction->Identity  ];

In[232]:= Show[
             GraphicsArray[
               Evaluate[
                   Table[plot[n,el, n*7],{n,1,3},{el,0,n-1}]
                 ]
               ]
            ];
```

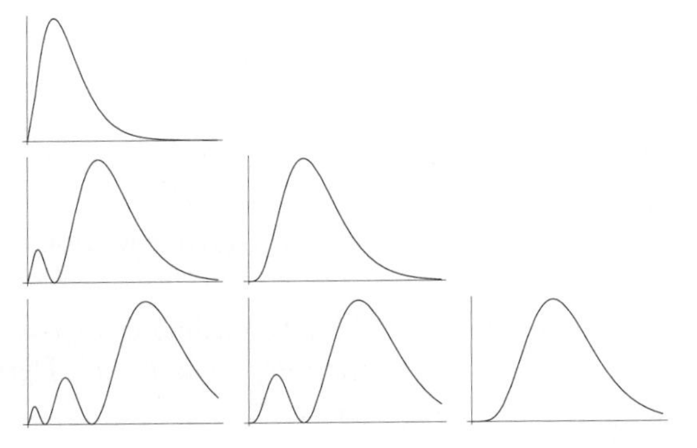

Part (c) The complete wave function is a product of the radial wave function and spherical harmonics. To plot the angular part of the wave function for fixed **r**, you need to consider only the plot of the spherical harmonics. Plot the absolute value of the spherical harmonics for **el**$=0$, 1, and 2 with the allowable values of **m={-el,el}**. Define

```
In[233]:= plot[l_,m_ ]:=
          SphericalPlot3D[
            (  Abs[SphericalHarmonicY[l, m, theta, phi]]
                    //Evaluate  )
              ,{theta,0,Pi}
              ,{phi,0,2 Pi}
              ,Boxed->False
              ,Axes->False
              ,DisplayFunction->Identity ];
```

The graphs follow from

```
In[234]:= pt1=Table[plot[el,m] ,{el,0,2} ,{m,-el,el}];
          Show[GraphicsArray[pt1]];
```

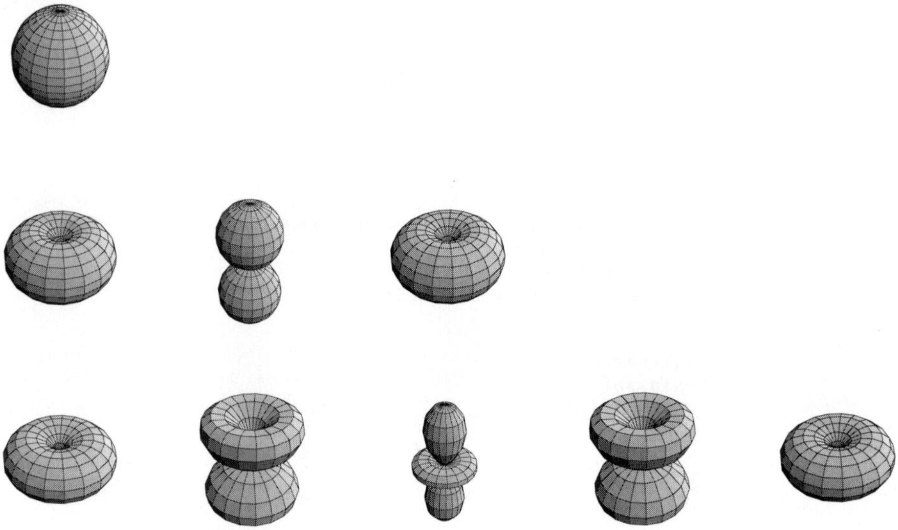

Problem 6: Separation in cylindrical and paraboloidal coordinates

Consider the following problems in cylindrical and paraboloidal coordinates:

Part (a) Express Schrodinger's equation in cylindrical coordinates r, θ, z. Assume a potential $V = V_r[r] + V_z[z]$ and wave function $f[r]Th[\theta]h[z]$. Derive equations for $f[r]$, $Th[\theta]$, and $h[z]$.

Part (b) Express Schrodinger's equation in paraboloidal coordinates $\{u, v, \phi\}$. Assume the potential $V = (V_u[u] + V_v[v])/(u^2 + v^2)$ and wave function $f[u]g[v]Phi[\phi]$. Derive equations for $f[u]$, $g[v]$, and $Phi[\phi]$.

Remarks and outline Assume the solution is separable. Express the user-defined operator **`schrodinger`** in **`Cylindrical`** coordinates. Apply it to the wave function and group terms. For Part (b), assume the wave function is of the form $f[u]g[v]Phi[\phi]$. Express the user-defined operator **`schrodinger`** in **`Paraboloidal`** coordinates, apply it to the wave function, and group the terms.

Required packages

```
In[235]:= Needs["Algebra`Trigonometry`"]
          Needs["Calculus`VectorAnalysis`"]
          Needs["Graphics`ParametricPlot3D`"]
```

Solution

```
In[236]:= Clear["Global`*"];
```

Part (a) Let the wave function be of the form

```
In[237]:= psi[r_,theta_,z_]=  f[r] Th[theta] h[z];
```

Applying the user-defined function **schrodinger** to **psi** with the potential **Vr[r]+Vz[z]**, you get

```
In[238]:= SetCoordinates[Cylindrical];

          eq1= ( -schrodinger[Vr[r]+Vz[z],En] @ psi[r,theta,z]
                  //ExpandAll )

Out[238]= En f[r] h[z] Th[theta] - f[r] h[z] Th[theta] Vr[r] -

                                                 2
                                             hbar  h[z] Th[theta] f'[r]
            f[r] h[z] Th[theta] Vz[z] + ---------------------------- +
                                                    2 m r

                2                                2
            hbar  h[z] Th[theta] f''[r]     hbar  f[r] Th[theta] h''[z]
            --------------------------- + --------------------------- +
                      2 m                            2 m

                2
            hbar  f[r] h[z] Th''[theta]
            ---------------------------
                          2
                     2 m r
```

Schrodinger's equation follows from setting **eq1** equal to zero. The **theta** variable separates if you multiply **eq1** by $r^2/(\psi\hbar^2/(2m))$:

```
In[239]:= eq2= ( eq1 r^2 /(hbar^2/(2 m)  psi[r,theta,z])
                 //ExpandAll  )

Out[239]=
                  2          2             2                       2
          2 En m r    2 m r  Vr[r]   2 m r  Vz[z]   r f'[r]    r  f''[r]
          --------- - ----------- - ------------ + ------- + --------- +
               2           2              2          f[r]       f[r]
           hbar        hbar           hbar

             2
            r  h''[z]   Th''[theta]
            --------- + -----------
              h[z]       Th[theta]
```

To separate terms having the same variable, define the **select** operator:

```
In[240]:= select[eq_,var_]:= Select[eq,(!FreeQ[#,var])&]
```

You get the equation for **Th** by setting the **theta** terms equal to a separation constant **-m^2**:

```
In[241]:= thetaEq1= select[eq2,theta] == -m^2

Out[241]= Th''[theta]        2
          ------------ == -m
          Th[theta]
```

Simplifying this equation, you get

```
In[242]:= thetaEq= Solve[thetaEq1,{Th''[theta]}][[1,1]] /.Rule->Equal

Out[242]=                       2
          Th''[theta] == -(m  Th[theta])
```

The solution of **thetaEq** is

```
In[243]:= thetaRule= DSolve[thetaEq,{Th},theta][[1]]

Out[243]=                           -I m theta          I m theta
          {Th -> Function[theta, E            C[1] + E          C[2]]}
```

An equation for **r** and **z** follows from using **thetaRule** to eliminate the **theta** variable in **eq2**:

```
In[244]:= eq3= eq2/r^2 //.thetaRule //Cancel //ExpandAll

Out[244]=
                    2
          2 En m    m     2 m Vr[r]    2 m Vz[z]    f'[r]     f''[r]    h''[z]
          ------- - -- - ---------- - ---------- + ------- + ------- + -------
              2      2        2            2       r f[r]     f[r]      h[z]
          hbar      r     hbar         hbar
```

Setting the **z** and **r** functions equal to a separation constant, you get the following equation for **h**:

```
In[245]:= zTerm=( select[eq3,z] == -mz^2
                  // Thread[h[z] #,Equal]&
                  //ExpandAll
                )

Out[245]= -2 m h[z] Vz[z]                        2
          --------------- + h''[z] == -(mz  h[z])
                   2
               hbar
```

and for **f**:

```
In[246]:= rTerm=( select[eq3,r] == mz^2
                  //Thread[r^2 f[r] #,Equal]&
                  //ExpandAll
                )

Out[246]=                        2
             2             2 m r  f[r] Vr[r]                 2              2  2
          -(m  f[r]) - ------------------------ + r f'[r] + r  f''[r] == mz  r  f[r]
                                   2
                               hbar
```

Part (b) To separate variables in **Paraboloidal** coordinates, assume the wave function is of the form

```
In[247]:= Clear[psi];
          psi[u_,v_,phi_]= f[u] g[v] Phi[phi];
```

Expressing Schrodinger's operator in **Paraboloidal** coordinates and applying it to **psi**, you get

```
In[248]:= SetCoordinates[Paraboloidal]

Out[248]= Paraboloidal[u, v, phi]

In[249]:= eq4= ( -schrodinger[V[u,v], En] @ psi[u,v,phi]
                   //ExpandAll );
          Short[eq4,3]

Out[249]= En f[u] g[v] Phi[phi] - f[u] g[v] Phi[phi] V[u, v] +

               2                                          2
           hbar  v g[v] Phi[phi] f'[u]                hbar  u f[u] g[v] Phi''[phi]
          ------------------------------- + <<4>> + ------------------------------
                   3            3                           3  2           4
              2 m u  v + 2 m u v                       2 m u  v  + 2 m u v
```

The **phi** variable separates if you multiply **eq4** by $(2mv^2u^2)/(\hbar^2\psi[u,v,\phi])$:

```
In[250]:= eq5= ( eq4 (2m u^2 v^2/hbar^2)/ psi[u,v,phi]
                   //Simplify
                   //Apart
                   //Expand
              );
          Short[eq5,3]

Out[250]=
                4  2              2  4            4  2
          2 En m u  v       2 En m u  v      2 m u  v  V[u, v]
          -------------- + -------------- - ------------------ -
              2   2    2       2   2    2        2   2    2
          hbar  (u  + v )  hbar  (u  + v )   hbar  (u  + v )

              2  4                2                        2  2
           2 m u  v  V[u, v]    u v  f'[u]              u  v  g''[v]
           ----------------- + ------------- + <<2>> + -------------- +
               2   2    2        2    2                  2    2
           hbar  (u  + v )     (u  + v ) f[u]          (u  + v ) g[v]

           Phi''[phi]
           ----------
            Phi[phi]
```

Equating the **phi** functions to a separation constant **-mphi^2**, you get

```
In[251]:= phiEq= ( select[eq5,phi]  == -mphi^2
                  //Thread[Phi[phi] #,Equal]&
                )    //Simplify

Out[251]=
                                    2
          Phi''[phi] == -(mphi  Phi[phi])
```

The solution of **phiEq** is

```
In[252]:= phiRule=DSolve[phiEq,{Phi},phi][[1]]

Out[252]=
                                 -I mphi phi           I mphi phi
          {Phi -> Function[phi, E             C[1] + E           C[2]]}
```

Eliminating **Phi** from **eq5** and multiplying by $(u^2+v^2)/(u^2v^2)$, the equation for **f[u]** and **g[v]** follows after some simplification:

```
In[253]:= eq6=(( eq5 (u^2+v^2)/(u^2 v^2)
                /.{u^2+v^2->temp}
                //Expand
              ) /.{temp-> u^2+v^2}
                /.phiRule
                //Simplify
              )
```

```
Out[253]=
               2                   2           2                    2             2
           mphi       2 En m u      mphi      2 En m v      2 m u  V[u, v]
         -(-----) + ---------- - ----- + ---------- - --------------- -
              2            2          2          2              2
             u         hbar          v       hbar           hbar

             2
         2 m v  V[u, v]     f'[u]      g'[v]      f''[u]    g''[v]
         --------------- + ------ + ------ + ------ + ------
                2           u f[u]    v g[v]     f[u]      g[v]
            hbar
```

The final separation of the **u** and **v** variables follows after replacing the potental in **eq6** by **V[u,v]**= $(V_u[u]+V_v[v])/(u^2+v^2)$ and simplifying:

```
In[254]:= eq7= ( eq6  /.{V[u,v]-> (Vu[u]+Vv[v])/( u^2+v^2)}
                    //Collect[#,{Vu[u],Vv[v]}]&
                    //Cancel
                )
```

```
Out[254]=
               2                   2           2                    2
           mphi       2 En m u      mphi      2 En m v      2 m Vu[u]     2 m Vv[v]
         -(-----) + ---------- - ----- + ---------- - --------- - --------- +
              2            2          2          2             2             2
             u         hbar          v       hbar         hbar          hbar

          f'[u]      g'[v]      f''[u]    g''[v]
         ------ + ------ + ------ + ------
         u f[u]    v g[v]     f[u]      g[v]
```

Equating the **u** and **v** functions to separation constants, you get

```
In[255]:= uEq= (  select[eq7,u]== -mu^2
                      //Thread[u^2 f[u]#,Equal]&
                      //ExpandAll
                  )
```

```
Out[255]=
                                   4                  2
                  2        2 En m u  f[u]     2 m u  f[u] Vu[u]
         -(mphi  f[u]) + --------------- - ------------------ + u f'[u] +
                                  2                    2
                              hbar                 hbar

             2                  2  2
            u  f''[u] == -(mu  u  f[u])
```

and

```
In[256]:= vEq= ( select[eq7,v] == -mv^2
                        //Thread[v^2  g[v]#,Equal]&
                        //ExpandAll
                   )
```

```
Out[256]=
                           4                2
          2         2 En m v  g[v]   2 m v  g[v] Vv[v]
-(mphi  g[v]) + ------------- - ------------------ + v g'[v] +
                          2                 2
                      hbar               hbar

   2              2  2
  v  g''[v] == -(mv  v  g[v])
```

6.3 Unsolved Problems

Exercise 6.1: Infinite potential well with rectangular perturbation

Consider the symmetric double potential barrier:

$$V(x) = \begin{cases} \infty & \text{if } x < 0; \\ 0 & \text{if } 0 < x < a_1; \\ V_1 & \text{if } a_1 < x < a_2; \\ 0 & \text{if } a_2 < x < a_3; \\ \infty & \text{if } a_3 < x. \end{cases}$$

Part (a) Solve the one-dimensional Schrodinger's equation for the wave function.

Part (b) Make a table of eigenvalues. Graph the eigenvalues as a function of V_1.

Part (c) Graph the first few wave functions for some fixed value of V_1. Use **Plot3D** to compare how the wave functions change with V_1.

Part (d) Develop a perturbation procedure that will solve this problem.

Exercise 6.2: Tilted square well

Consider the tilted potential well:

$$V(x) = \begin{cases} V_1 & \text{if } x < 0; \\ V_2(x/a) & \text{if } 0 < x < a; \\ V_3 & \text{if } a < x. \end{cases}$$

Part (a) Consider a particle bound in the well. Solve Schrodinger's one-dimensional equation for the eigenfunctions.

Part (b) Make a table of the eigenvalues and graph the first four wave functions.

Part (c) Find the transmission and reflection coefficients for a particle with energy greater than V_1. Graph the coefficients and show how the results depend on V_2.

Exercise 6.3: The Wentzel-Kramers-Brillouin approximation

Consider the one-dimensional Schrodinger's equation with a potential $V[x]$. Assume the wave function is of the form $u = R[x]Exp[iS[x]/\hbar]$.

Part (a) If $R[x]$ is a slowly varying function, derive an equation for $S[x]$ in terms of the potential $V[x]$.

Part (b) Derive a procedure that will return the probability of tunneling through a potential barrier.

Exercise 6.4: Plot of the electron probability density

Derive two operators using **`ScatterPlot`** that will plot the probability density of a wave function. Make the density of dots proportional to the probability density given by the wave function. Construct a density operator that will plot the results in two dimensions and a second operator that will plot the results in three dimensions. Apply the two operators to the hydrogen wave functions.

Exercise 6.5: Perturbed harmonic oscillator

Consider a particle that is bound by a potential $V = (1/2)kx^2 + px^4$. Treat the last term px^4 as a perturbation.

Part (a) Find the perturbed solutions for the eigenvalues and eigenfunctions.

Part (b) Compare the perturbed and unperturbed solutions by graphing the results as a function of p.

Part (c) Construct a perturbative function using **`Module`** for the one-dimensional Schrodinger's equation having an arbitrary potential.

Exercise 6.6: Perturbation theory

Consider the hydrogen atom in an electric field. The potential for the system is $V(r) = -e^2/r - eE_0z$.

Part (a) Express Schrodinger's equation in **`Paraboloidal`** coordinates.

Part (b) Solve for hydrogen eigenvalues and eigenfunctions when $E_0 = 0$.

Part (c) Treat E_0 as a perturbation and solve for the perturbed eigenvalues and eigenfunctions.

Part (d) Graph the results and compare the perturbed and unperturbed results.

Exercise 6.7: Separation of variables

Consider the coordinate systems *Mathematica* supports. Assume a three-dimensional wave function is separable and can be represented by a product of three functions, where each function depends on only one variable. Derive a procedure that will separate the variables for any coordinate system (where possible) and return differential equations for the three functions. Allow the user to choose the potential and coordinate system.

CHAPTER SEVEN

Relativity and Cosmology

7.1 Introduction

In this chapter, we solve problems taken from topics in special relativity, general relativity, and cosmology. The special relativity problems are similar to those found in *Classical Electrodynamics*, by John D. Jackson, John Wiley and Sons, Inc. and *Classical Mechanics*, by Herbert Goldstein, Addison-Wesley. The general relativity and cosmological problems were taken from topics covered in *Gravitation and Cosmology*, by Steven Weinberg, John Wiley & Sons, Inc., and *Gravitation*, by Misner, Thorne, and Wheeler, W. H. Freeman and Company. This chapter is divided into three sections:

1. Introduction
2. Problems
3. Unsolved Problems

The Introduction is divided into three subsections: (1) Special Relativity, (2) General Relativity and Cosmology, and (3) *Mathematica* Commands. The first two subsections include a short review of the foundations of special relativity, general relativity, and cosmology and explain those concepts and definitions needed to understand the problems.

The third subsection lists the *Mathematica* packages, user-defined rules, and user-defined procedures for boosts, relativistic velocity parameters, Christoffel symbols, curvature tensor, Ricci tensor, Killing equations, Einstein tensor, and geodesic equations. We have also included rules and functions that define the metrics and Christoffel symbols for the Schwarzschild and Kerr metrics. These user-defined procedures are not the most efficient; we were primarily interested in clarity, not speed, so the algebraic symmetries of the geometric operators were not considered. Several sophisticated packages are available to handle such geometric manipulations efficiently. Also, the use of differential forms was omitted because of space limitations. You are encouraged to take advantage of these symmetries by modifying these procedures and to use differential forms in the calculations.

The Problems section is divided into subsections on special relativity problems and on general relativity and cosmological problems. The methods used to solve these problems are not unique; you are encouraged to find procedures that will illuminate the physics and to find other approaches that make the calculations faster.

7.1.1 Special Relativity

The two basic postulates of special relativity

Einstein proposed in 1905 two postulates for special relativity. One relates to the constancy of the speed of light, the other to the formal invariance of physical laws:

1. **Principle of the Constancy of the Speed of Light:** The speed of light in free space has the same value in all inertial systems. The velocity of light is the same in all directions and is independent of the relative uniform motion of the inertial observer and inertial source.
2. **Principle of Relativity:** Physical laws and principles have the same form in all inertial systems. There does not exist any preferred inertial system.

The first postulate is the prescription used to find the transformation between inertial frames. The relation between inertial frames in classical mechanics is simple due to the absolute nature of time. To illustrate this classical transformation, consider two inertial frames of reference S and S'. Let the coordinates of a certain event in frame S be $\{t, x, y, z\}$ and in S', the coordinates of the same event are $\{t', x', y', z'\}$. The absolute nature of time means $t = t'$. Let the coordinate systems be coincident at $t = 0$ and assume the relative motion of the two inertial frames is along the x-direction. If the velocity of S' relative to S is V, then the classical transformation between the inertial frames is $x' = x - Vt$, $y' = y$, $z' = z$, $t' = t$. This classical transformation is called a Galilean transformation. Galilean transformations do not satisfy the first postulate of special relativity because they do not preserve the velocity of light.

Lorentz transformations

The transformations between inertial frames that keep the velocity of light constant are known as **Lorentz transformations.** A Lorentz transformation relates the coordinates $\{t, x, y, z\}$ of an event in S to the coordinates $\{t', x', y', z'\}$ of the same event in another inertial system S'. To construct the Lorentz transformation, consider the infinitesimal distance between two nearby points. Let $\{dt, dx, dy, dz\}$ be the infinitesimal distance between two nearby points in an inertial frame S. The space and time coordinates of a light ray between these nearby points satisfy the equation $0 = dt^2 - dx^2 - dy^2 - dz^2$. (We choose natural units and set the speed of light equal to one ($c = 1$) in this chapter.) The equation for light suggests we define a function ds^2 by the relation $ds^2 = dt^2 - dx^2 - dy^2 - dz^2$. We call ds^2 the line element. The propagation of light is described by $ds^2 = 0$. Similarly in s' the line element is $ds'^2 = dt'^2 - dx'^2 - dy'^2 - dz'^2$ and light is described by $ds'^2 = 0$. The first postulate of relativity is satisfied if the transformation between the frames leaves the line element invariant, $ds'^2 = ds^2$.

The line element ds^2 can be written as $ds^2 = \eta_{ij}\, dx^i dx^j$, where $\eta =$ `DiagonalMatrix[1, -1, -1, -1]` and the coordinates $dx^i = \{dt, dx, dy, dz\}$ are Cartesian. We assume the Einstein convention and sum over repeated indices. The matrix η is the metric of the relativistic spacetime. The geometry for special relativity is pseudo-Euclidean. In S', the line element is $ds'^2 = \eta_{ij}\, dx'^i dx'^j$. To find the transformation between the two inertial frames, we assume a linear transformation $x'^i = a^i{}_j\, x^j$. For $a^i{}_j$ to preserve the speed of light, we require

$$\eta^i{}_j = \sum_{m=1,4} \sum_{n=1,4} \eta^m{}_n \, a^i{}_m \, a^n{}_j.$$

A Lorentz transformation that relates coordinate frames at different velocities is called a **boost**. Consider the boost that connects an inertial frame S with another inertial frame S' moving with relative speed v along the x-axis. The transformation between the two inertial frames is $x' = \gamma(x - vt)$, $t' = \gamma(t - vx)$, $y' = y$, $z' = z$, where $\gamma = 1/\sqrt{1 - v^2}$. (Recall, $c = 1$.) The expression for $a^i{}_j$ is

$$a^i{}_j = \begin{pmatrix} \gamma & 0 & 0 & -v\gamma \\ 0 & 1 & 0 & 0 \\ 0 & 0 & 1 & 0 \\ -v\gamma & 0 & 0 & \gamma \end{pmatrix}.$$

Covariant equations and tensors

The second postulate of relativity requires the equations of motion to have the same form in all inertial coordinate systems. An equation written in a form-independent manner is said to be **covariant**. The construction of covariant equations leads to the study of tensors; **tensors** are those objects that are invariant under coordinate transformations.

The simplest of all tensors is a **scalar**. A scalar is trivially invariant under coordinate transformations. If f is a scalar in coordinates $\{x\}$, then in the coordinates $\{x'\}$ the scalar is f', where $f' = f$. For example, the line element ds^2 is a scalar, since $ds'^2 = ds^2$. A scalar is called a tensor of rank zero.

The next most elementary tensor is of rank one, or a **vector**. In four-dimensional spacetime, a vector has four components and can be either contravariant or covariant. If the components V^i $(i = 1, 2, 3, 4)$ transform like the coordinate differentials dx^i, the vector is contravariant, $V'^i = (dx'^i/dx^j)\, V^j$. The components of contravariant vectors are denoted with superscripts. If the components of a vector, V_i $(i = 0, 1, 2, 3)$, transform like $V'_i = (dx^j/dx'^i)\, V_j$, the vector is covariant. We can generalize these transformations to define tensors of arbitrary rank. An arbitrary tensor can be covariant, contravariant, or mixed.

Cartesian coordinates and "flat" spacetime

It is not always convenient to use Cartesian coordinates. For example, the line element in spherical coordinates $\{t, r, \theta, \phi\}$ is

$$ds^2 = dt^2 - dr^2 - r^2 d\theta^2 - r^2 \mathrm{Sin}[\theta]^2 d\phi^2.$$

The transformation between spherical and Cartesian coordinates is well known in this case. The existence of Cartesian coordinates is a special property of pseudo-Euclidean, or "flat," spacetime. Only certain metrics describe flat spacetimes. For an arbitrary metric, a transformation to a Cartesian coordinate system does not generally exist, except at a point. Given a metric in an arbitrary coordinate system, the necessary and sufficient condition that there must exist a coordinate transformation to Cartesian coordinates is that the components of the curvature tensor $R^i{}_{jk\ell}$ be zero. The curvature tensor is defined in terms of Christoffel symbols $\Gamma^i jk$ by

$$R^i{}_{jk\ell} = \frac{\partial \Gamma^i{}_{jk}}{\partial x^\ell} - \frac{\partial \Gamma^i{}_{j\ell}}{\partial x^k} + \Gamma^s{}_{jk}\Gamma^i{}_{\ell s} - \Gamma^s{}_{j\ell}\Gamma^i{}_{ks}.$$

The Christoffel symbols are related to the metric tensor g_{ij} by

$$\Gamma^s{}_{jk} = \frac{1}{2} g^{s\ell} \left[\frac{\partial g_{k\ell}}{\partial x^j} + \frac{\partial g_{j\ell}}{\partial x^k} - \frac{\partial g_{jk}}{\partial x^\ell} \right].$$

7.1.2 General Relativity and Cosmology

Spacetime metric

The theory of general relativity was formulated by Einstein in 1916 and is based on the principle that motion in noninertial systems is equivalent to certain gravitational fields. The axiom of indistinguishability between gravity and inertia is a consequence of the equivalence of inertial and gravitational masses. The equivalence principle is the basis of general relativity. Geometry enters the theory through the metric structure of spacetime. In contrast to special relativity, general relativity abandons the pseudo-Euclidean metric of special relativity, and the general form of the line element becomes $ds^2 = g_{ij}\, dx^i\, dx^j$, where g_{ij} is a function of the coordinates $\{x\}$. Einstein's field equations relate the metric g_{ij} to matter-energy in spacetime.

Field equations

General relativity uses Riemann metrics characterized by nonvanishing curvature components $R^i{}_{jk\ell}$. The pseudo-Euclidean metric is distorted by the presence of matter and energy, which is described by a stress-energy tensor T_{ij}. This distortion of spacetime is the manifestation of a gravitational field. The equations that relate the metric with the matter-energy follow from the curvature tensor. We define the Ricci tensor $R_{j\ell} = R^i{}_{ji\ell}$ and the Ricci scalar $R = R^i{}_i$. Einstein's field equations are $G_{ij} = -8\pi G T_{ij}$, where G_{ij} is the Einstein tensor and is defined by $G_{ij} = R_{ij} - (1/2) g_{ij} R$. In the vacuum case, the field equations reduce to the simpler form $G_{ij} = 0$ or, equivalently, $R_{ij} = 0$.

Free-falling test particles and light trajectories

Geodesics play the same role in a curved spacetime as "straight lines" do in flat spacetime. A timelike geodesic represents the motion of a particle in a gravitational field, $d^2x^i/ds^2 + \Gamma^i{}_{jk} v^j v^k = 0$, where $v^i = dx^i/ds$, $v^i v_i = +1$ and $\Gamma^i{}_{jk}$ are the Christoffel symbols defined by

$$\Gamma^i{}_{jk} = (1/2) g^{is} \left[\frac{\partial g_{si}}{\partial x^k} + \frac{\partial g_{sk}}{\partial x^j} - \frac{\partial g_{jk}}{\partial x^s} \right].$$

Null geodesics represent the motion of light, where the four-velocity v^i satisfies the relation $v^i v_i = 0$.

Robertson-Walker cosmology

The standard cosmological models are based on the **Cosmology principle**, that is, all positions in space are essentially equivalent. This cosmology principle implies the universe is spatially homogeneous and isotropic. The line element for such a universe can be written in the form

$$ds^2 = dt^2 - R^2[t]\left[\frac{dr^2}{(1-\kappa r^2)} + r^2 d\theta^2 + r^2 \text{Sin}[\theta]^2 d\phi^2\right],$$

where $R[t]$ is an unknown function of time. The signature of the model is denoted by $\kappa = +1,\ 0$, and -1, and the corresponding models are called closed, flat, and open, respectively. This metric is called the **Robertson-Walker metric** and assumes that matter can be described as a perfect fluid characterized by only a density and pressure, which are functions of time but not of space.

7.1.3 *Mathematica* Commands

We have constructed user-defined procedures for doing common repetitious calculations. To understand the procedures, you should execute the steps one at a time until you understand each. Try to make the procedures more time efficient. Many of them will fail in particular cases and you are encouraged to write conditional statements to handle these cases.

Packages

You may wish to turn off the spell checker before starting this chapter:

```
In[1]:= Off[General::spell ];
        Off[General::spell1];
```

In this chapter, you will use the following packages:

```
In[2]:= Needs["Algebra`Trigonometry`"];
        Needs["Graphics`ParametricPlot3D`"];
```

User-defined metric, boost, and velocity parameters

Metric

Choose the metric η:

```
In[3]:= eta=DiagonalMatrix[1,-1,-1,-1];
```

Note that *Mathematica* does not distinguish between the order of the operations:

```
In[4]:= p={E,0,0,pz};
        p.eta.p==eta.p.p

Out[4]= True
```

However, **p.p.eta** is not acceptable (try it). So we will typically write a scalar product in the order **p.eta.p** to suggest either a vector inner product $\langle p|\eta|p\rangle$ or the matrix multiplication $p_i\eta_{ij}p_j$.

Rule for relativistic velocity parameters

Several rules are useful in simplifying the algebra in special relativistic calculations:

```
In[5]:=  gam2v[gam_, v_]= {gam ->1/Sqrt[1-v^2]};
         v2gam[v_,gam_ ]= {v    ->  Sqrt[1-1/gam^2]};
         E2gam[En_,m_,gam_]=   {En-> gam  m};
         p2gam[p_,m_,v_,gam_]= {p -> gam  m  v};
```

These rules are straightforward, and we illustrate their use in the problems.

Boost along the x-axis

```
In[6]:=  boostx::usage= " The matrix boostx[gam,v] is the transformation matrix
         between two  coordinate systems moving with relative speed v
         along the x-axis. The symbol gam = 1/Sqrt[1-v^2], where the
         speed of light, c=1.";

In[7]:=  boostx[gam_,v_]:={{ gam   ,-gam v,0,0}
                           ,{-gam v, gam  ,0,0}
                           ,{     0,      0,1,0}
                           ,{     0,      0,0,1}};
```

Note that the inner product of a vector is invariant under boosts:

```
In[8]:=  p1={E,0,0,pz};
         p2=boostx[gam,v].p1;
         p1.eta.p1

Out[8]=   2      2
         E   - pz

In[9]:=  p1.eta.p1 == p2.eta.p2 /.gam2v[gam,v] //Simplify

Out[9]=  True
```

User-defined geometric procedures

We define procedures that calculate useful tensor properties. These procedures are constructed for clarity rather than efficiency.

Christoffel symbols

The Christoffel symbols are defined in terms of derivatives of the metric:

$$\Gamma^{i}{}_{jk} = \frac{1}{2} g^{is} \left[\frac{\partial g_{js}}{\partial x^k} + \frac{\partial g_{ks}}{\partial x^j} - \frac{\partial g_{jk}}{\partial x^s} \right].$$

```
In[10]:= christ::usage= "When christ[metric,variables] is applied
         to a metric, it returns the Christoffel
         symbols in the coordinate system given by variables. The
         Christoffel symbols are defined in terms of the
         derivatives of the metric.
         We have chosen spherical coordinates to be the
         default coordinates.";
```

```
In[11]:= christ[met_,var_:{t,r,theta,phi}]:=
          Module[{imet,temp1,temp2,i,s,j,k},
            temp1= Table[D[met[[i ,j ]],#] ,{i,4},{j,4}]& /@ var;
            imet = Inverse[met]//Simplify;
            temp2=(Table[ (1/2)
                    *Sum[ imet[[i,s]] (temp1[[j,s,k]]+temp1[[k,s,j]]
                                                   -temp1[[s,j,k]])
                       ,{s,4}]
                    ,{i,4},{j,4},{k,4}]
                  //ExpandAll //Simplify );
            Return[temp2]
          ];
```

Example: Christoffel symbols for a pseudo-Euclidean metric

Consider the pseudo-Euclidean metric expressed in **Spherical** coordinates:

```
In[12]:= sphMet=DiagonalMatrix[1,-1,-r^2,-r^2 Sin[theta]^2];
```

The Christoffel symbols are

```
In[13]:= sphChrist= christ[sphMet]

Out[13]= {{{0, 0, 0, 0}, {0, 0, 0, 0}, {0, 0, 0, 0}, {0, 0, 0, 0}},

          {{0, 0, 0, 0}, {0, 0, 0, 0}, {0, 0, -r, 0},

                                  2
           {0, 0, 0, -(r Sin[theta] )}},

                                1           1
          {{0, 0, 0, 0}, {0, 0, -, 0}, {0, -, 0, 0},
                                r           r

                      -Sin[2 theta]
           {0, 0, 0, -------------}},
                            2

                                       1
          {{0, 0, 0, 0}, {0, 0, 0, -}, {0, 0, 0, Cot[theta]},
                                       r

               1
           {0, -, Cot[theta], 0}}}
               r
```

Curvature tensor

The curvature tensor is defined by

$$R^i{}_{jk\ell} = \frac{\partial \Gamma^i{}_{jk}}{\partial x^\ell} - \frac{\partial \Gamma^i{}_{j\ell}}{\partial x^k} + \Gamma^s{}_{jk}\Gamma^i{}_{\ell s} - \Gamma^s{}_{j\ell}\Gamma^i{}_{ks}.$$

```
In[14]:= curvR::usage= "The curvature tensor components are returned
         when curvR[christ ,var] is applied to the Christoffel
         symbols christ with variables given by var.
         We have chosen Spherical variables as the defaults.";
```

```
In[15]:=  curvR[christ_,var_:{t,r,theta,phi}]:=
          Module[{temp1,temp2,i,j,k,s,l},
            temp1= Table[D[christ[[i,j,k]],#],{i,4},{j,4},{k,4}]& /@ var;
            temp2[i_,j_,k_,l_]:= ( temp1[[l,i,k,j]]- temp1[[k,i,j,l]]
                                +Sum[(christ[[i,s,l]] christ[[s,k,j]]
                                     -christ[[i,s,k]] christ[[s,l,j]]),{s,4}]);
            Return[ Table[ temp2[i,j,k,l],{i,4},{j,4},{k,4},{l,4}] ]
          ];
```

Example: Curvature tensor for pseudo-Euclidean metric

Consider the pseudo-Euclidean metric expressed in spherical coordinates. The metric and its Christoffel symbols are

```
In[16]:= sphMet=DiagonalMatrix[1,-1,-r^2,-r^2 Sin[theta]^2];
         sphChrist= christ[sphMet];
```

Applying the curvature operator to the Christoffel symbols, you get

```
In[17]:= eq1= curvR[sphChrist] //Simplify;
         eq1  //Short[#,4]&

Out[17]= {{{{0, 0, 0, 0}, {0, 0, 0, 0}, {0, 0, 0, 0}, {0, 0, 0, 0}},

           {{0, 0, 0, 0}, {0, 0, 0, 0}, {0, 0, 0, 0}, {0, 0, 0, 0}}, <<1>>,

           {{0, 0, 0, 0}, {0, 0, 0, 0}, {0, 0, 0, 0}, {0, 0, 0, 0}}}, <<2>>,

          {<<4>>}}
```

(Only the **Short** form of the solution is displayed.) All components of the curvature vanish as expected, since the spacetime is flat. Another way to verify that the curvature vanishes is to equate **eq1** to a **Table** of zeros:

```
In[18]:= eq1== Table[0,{i,4},{j,4},{k,4},{l,4}]

Out[18]= True
```

Ricci tensor

The Ricci tensor is defined by $R_{ij} = R^{k}{}_{ikj}$ and the Ricci scalar, by $R = R^{i}{}_{i}$:

```
In[19]:= ricciR::usage= "The Ricci tensor is returned when
         ricciR[curv ] is applied to the curvature tensor
         called curv. The Christoffel symbols and curvature
         tensor must be computed first.";

In[20]:= ricciR[curv_] := Table[
                             Sum[curv[[k,i,k,j]],{k,4}]
                           ,{i,4},{j,4}] //ExpandAll //Simplify
```

Example: Ricci tensor for Gödel metric

Consider the so-called Gödel metric

```
In[21]:= godelMet= {{1          , 0,    Exp[a x1], 0}
                   ,{0          ,-1,            0, 0}
                   ,{Exp[a x1], 0,Exp[2 a x1]/2, 0}
                   ,{0          , 0,            0,-1}};
```

where the coordinates are $\{x_0, x_1, x_2, x_3\}$. First, compute the Christoffel symbols and curvature tensor:

```
In[22]:= godelChrist= christ[godelMet,{x0,x1,x2,x3}];
         godelCurvature= curvR[godelChrist,{x0,x1,x2,x3}];
```

The covariant Ricci tensor follows from

```
In[23]:= godelRicci= ricciR[godelCurvature ] //Simplify

Out[23]=     2            2  a x1
         {{-a , 0, -(a  E      ), 0}, {0, 0, 0, 0},

              2  a x1          2  2 a x1
           {-(a  E     ), 0, -(a  E       ), 0}, {0, 0, 0, 0}}
```

The contravariant Ricci tensor is

```
In[24]:= Inverse[godelMet].godelRicci.Inverse[godelMet] //ExpandAll

Out[24]=     2
         {{-a , 0, 0, 0}, {0, 0, 0, 0}, {0, 0, 0, 0}, {0, 0, 0, 0}}
```

Killing's equations

Killing's equations are

$$\zeta_{\mu;\nu} + \zeta_{\nu;\mu} = 0,$$

where the semicolon ";" means covariant derivative and ζ_μ is the covariant Killing vector:

```
In[25]:= killingEq::usage="Killing's equations are computed
         when killingEq[ContraV,metric,christ,var] is applied
         to the contravariant vector ContraV. The operator returns
         only the nontrivial equations. The Christoffel symbols
         must be computed first.";

In[26]:= killingEq[ContraV_,metric_,christ_,var_:{t,r,theta,phi}]:=
         Module[{temp1,temp2,i,j,s},
              temp1=metric.ContraV;
              temp2=( Table[ D[temp1[[i]],var[[j]]],{i,4},{j,4}]
                     -Sum[christ[[s]]temp1[[s]],{s,4}]);
              table=  Table[If[ i>j ,True,temp2[[i,j]]+temp2[[j,i]]==0]
                         ,{i,1,4},{j,1,4}]
                        //Flatten
                        //Simplify;
              Return[ Select[ table, FreeQ[#,True]&] ]
         ];
```

Example: Killing vector equations in pseudo-Euclidean space

Consider the contravariant vector

```
In[27]:= vector= {0,0, f[r ]Sin[phi], f[r] Cos[phi] Cot[theta]};
```

expressed in spherical coordinates $\{t, r, \theta, \phi\}$. Assume a pseudo-Euclidean metric:

```
In[28]:= sphMet=DiagonalMatrix[1,-1,-r^2,-r^2 Sin[theta]^2];
         sphChrist= christ[sphMet];
```

Applying **killingEq** to the vector, you get

```
In[29]:= killingEq[vector,sphMet,sphChrist]

Out[29]=
                                                2
           2                              -(r  Cos[phi] Sin[2 theta] f'[r])
{-(r  Sin[phi] f'[r]) == 0, ---------------------------------------- == 0}
                                                      2
```

There are two nontrivial Killing equations. The contravariant vector is obviously a Killing vector if $f[r]$ is a constant.

Einstein tensor

The Einstein tensor is defined by

$$G_{ij} = R_{ij} - \frac{1}{2}\, g_{ij}\, R.$$

```
In[30]:= einstein::usage=" The Einstein tensor is returned
         when einstein[curvature,metric] is applied to the curvature
         tensor and metric.";

In[31]:= einstein[curv_,met_] :=
         Module[{temp1,temp2,i,j,s},
           temp1= Table[ Sum[curv[[s,i,s,j]],{s,4}]
                   ,{i,4},{j,4}] //ExpandAll //Simplify;
           temp2= Sum[ Inverse[met][[j,i]] curv[[s,i,s,j]]
                   ,{s,4},{i,4},{j,4}] //ExpandAll //Simplify;
           temp3=  Table[ temp1[[i,j]]- (1/2) met[[i,j]] temp2
                   ,{i,4},{j,4}] //ExpandAll //Simplify;
           Return[temp3]
          ];
```

Example: Einstein tensor for wave metric

Consider the metric for a "plane" gravity wave

```
In[32]:= waveMet= {{   0, -1/2,                  0,                 0 }
                  ,{-1/2,    0,                  0,                 0 }
                  ,{   0,    0, -L[u]^2 Exp[f[u]],                  0 }
                  ,{   0,    0,                  0, -L[u]^2 Exp[-f[u]] }};
```

where the coordinates are $\{u, v, x, y\}$. This metric corresponds to a gravity wave traveling in a direction orthogonal to the $\{x, y\}$ plane. First, compute the Christoffel symbols and curvature tensor:

```
In[33]:= Clear[u,v,x,y];
         waveChrist=christ[waveMet,{u,v,x,y}];
         waveCurvature= curvR[waveChrist,{u,v,x,y}];
```

The Einstein tensor follows from

```
In[34]:= waveEinstein=  einstein[waveCurvature,waveMet] //Simplify

Out[34]=
                2
          f'[u]     2 L''[u]
         {{------ + --------, 0, 0, 0}, {0, 0, 0, 0}, {0, 0, 0, 0},
             2        L[u]

           {0, 0, 0, 0}}
```

Geodesic equations

The Geodesic equations are defined by

$$\frac{d(\,g_{ij}\,dx^j/ds)}{ds} - (1/2)\frac{d\,g_{\ell m}}{dx^i}\,\frac{dx^\ell}{ds}\,\frac{dx^m}{ds} = 0:$$

```
In[35]:= geodesic::usage="Four geodesic equations are returned
         for the variables when  geodesic[metric,variables,parameter:s]
         is applied to the metric.";

In[36]:= geodesic[met_,var_,parameter_:s] :=
         Module[{met1,temp1,vel1,var1},
           vel1= D[ Through[var[parameter]] ,parameter];
           var1= Through[var[parameter]];
           met1= met /.Thread[var-> Through[var[parameter]]];
           temp1= ( D[ met1 .vel1,parameter]
                   -(1/2 D[met1,#].vel1.vel1& /@ var1)
                   == {0,0,0,0} )//Thread;
           Return[temp1]
          ]
```

Example: Geodesics for a pseudo-Euclidean metric

Consider a pseudo-Euclidean metric expressed in cylindrical coordinates $\{t, r, \phi, z\}$. The metric is

```
In[37]:= cylMet=DiagonalMatrix[1,-1,-r^2,-1];
```

Applying **geodesic** to **cylMet**, you get

```
In[38]:= geodesic[cylMet,{t,r,phi,z}] //ColumnForm

Out[38]= t''[s] == 0
                       2
         r[s] phi'[s]   - r''[s] == 0
                                    2
         -2 r[s] phi'[s] r'[s] - r[s]  phi''[s] == 0
         -z''[s] == 0
```

User-defined metrics and Christoffel symbols

The Schwarzschild and Kerr metrics are used in several problems in this chapter. It is useful to create functions for these metrics. We also give rules that will yield the explicit values for their components and Christoffel symbols.

Schwarzschild metric

The Schwarzschild metric is of the form

```
In[39]:= ssMet::usage="Gives the Schwarzschild metric";

         ssMet=DiagonalMatrix[g00[r],g11[r],-r^2,-r^2 Sin[theta]^2];
```

where the two components are given by the rule

In[40]:=
```
ssComp::usage="Gives a rule to evaluate the components of
the Schwarzschild metric";

ssComp={ g00->((    1- 2 m/# )&)
       , g11->((-1/(1- 2 m/#))&)};
```

The components are expressed in their pure form so the substitutions will work for derivatives of the components as well as the components. The Christoffel symbols for the Schwarzshild metric are

In[41]:=
```
ssChrist::usage="Gives the Christoffel symbols for
the Schwarzschild metric";

ssChrist=christ[ssMet] //.ssComp //Simplify
```

Out[41]=
```
              m                          m
{{{0, ──────────────, 0, 0}, {──────────────, 0, 0, 0}, {0, 0, 0, 0},
                    2                          2
        -2 m r + r                 -2 m r + r

                   m (-2 m + r)                                m
   {0, 0, 0, 0}}, {{────────────, 0, 0, 0}, {0, ──────────────, 0, 0},
                         3                                   2
                        r                            2 m r - r

                                                        2
   {0, 0, 2 m - r, 0}, {0, 0, 0, (2 m - r) Sin[theta] }},

                               1              1
  {{0, 0, 0, 0}, {0, 0, -, 0}, {0, -, 0, 0},
                               r              r

              -Sin[2 theta]
   {0, 0, 0, ──────────────}},
                    2

                                  1
  {{0, 0, 0, 0}, {0, 0, 0, -}, {0, 0, 0, Cot[theta]},
                                  r

       1
   {0, -, Cot[theta], 0}}}
       r
```

The calculations simplify if the explicit form of the metric is not evaluated until the final step; for this purpose, use **ssComp** to evaluate the nontrivial **g00** and **g11** components.

Kerr metric

The general form of the Kerr metric is

In[42]:=
```
kMet::usage="Gives the Kerr metric";

kMet ={{g00[r,theta],0              ,0              ,g03[r,theta]}
      ,{0           ,g11[r,theta],0              ,0           }
      ,{0           ,0              ,g22[r,theta],0           }
      ,{g03[r,theta],0              ,0              ,g33[r,theta]}};
```

The components are given by the rule

```
In[43]:= kComp::usage="Gives a rule to evaluate the components of
         the Kerr metric";

         kComp={

          g00->(( (#1^2-2 m #1 +a^2)/( #1^2+ a^2 Cos[#2]^2)-
                  a^2 Sin[#2]^2/( #1^2+ a^2 Cos[#2]^2))&)

         ,g03 -> ((-(#1^2-2 m #1 +a^2)/
                  (#1^2+ a^2 Cos[#2]^2) a Sin[#2]^2+
                  Sin[#2]^2/( #1^2+ a^2 Cos[#2]^2)(#1^2+a^2) a)&)

         ,g11 ->((-( #1^2+ a^2 Cos[#2]^2)/(#1^2-2 m #1 +a^2))&)

         ,g22 ->  ((-( #1^2+ a^2 Cos[#2]^2))&)

         ,g33 ->(((#1^2-2 m #1 +a^2)/
                ( #1^2+ a^2 Cos[#2]^2) a^2 Sin[#2]^4-
                  Sin[#2]^2/( #1^2+ a^2 Cos[#2]^2) (#1^2+a^2)^2)&)
         };
```

The Christoffel symbols for the Kerr metric are

```
In[44]:= kChrist::usage="Gives the Christoffel symbols for
         the Kerr metric";

         kChrist=christ[kMet];
         kChrist[[1,1]]

Out[44]=
                              (1,0)
             g33[r, theta] g00     [r, theta] -

                                 (1,0)
               g03[r, theta] g03     [r, theta]
         {0, -------------------------------------------------------

                                  2
                   2 (-g03[r, theta]  + g00[r, theta] g33[r, theta])

                              (0,1)
              g33[r, theta] g00     [r, theta] -

                                 (0,1)
                g03[r, theta] g03     [r, theta]
            , -------------------------------------------------------

                                   2
                    2 (-g03[r, theta]  + g00[r, theta] g33[r, theta])

            , 0}
```

Only the **kChrist[[1,1]]** element is displayed due to its length. Using **kComp** to look at a portion of **kChrist**, you get

```
In[45]:= kChrist[[1,1]] //.kComp //Together //Simplify
```

```
Out[45]=
                  2    2     2      2     2
           2 m (a  + r ) (a  - 2 r  + a  Cos[2 theta])
      {0, ------------------------------------------------,
              2            2    2      2    2             2
           (-a  + 2 m r - r ) (a  + 2 r  + a  Cos[2 theta])

                2
           -4 a  m r Sin[2 theta]
         ---------------------------, 0}
            2      2    2            2
          (a  + 2 r  + a  Cos[2 theta])
```

Protect user-defined operators

Next, protect the user-defined objects so that you can use **Clear["Global`*"]** to start a clean *Mathematica* session without erasing these functions:

```
In[46]:= Protect[boostx,christ,ContraV,curvR,einstein,eta,E2gam,gam2v,geodesic
         ,kComp,kMet,kChrist,killingEq,p2gam,ricciR,ssComp,ssMet,ssChrist,v2gam];
```

7.2 Problems

7.2.1 Special Relativity Problems

Problem 1: Decay of a particle

A particle of mass m_1 at rest decays into two particles of mass m_2 and mass m_3. Use the scalar invariant $m_3^2 = p_3 \cdot \eta \cdot p_3$ to find the kinetic energy of the particle with mass m_2. The four-momenta for the particle of mass m_3 is p_3.

Remarks and outline Write the four-momenta vectors for the three particles. Use the conservation of energy and momentum $p_3 = p_1 - p_2$ and write the scalar equation $m_3^2 = p_3 \cdot \eta \cdot p_3$ as $m_3^2 = (p_1 - p_2) \cdot \eta \cdot (p_1 - p_2)$. Solve for γ_2 and then construct the expression for the kinetic energy, $m_2\gamma_2 - m_2$.

Solution

```
In[1]:= Clear["Global`*"];
```

Part (**a**) In the rest frame of particle one, the four-momenta vectors are defined by the rule

```
In[2]:= momRule={ p1->m1      { 1, 0, 0,   0},
                  p2->m2 gam2{ 1, 0, 0, -v2},
                  p3->m3 gam3{ 1, 0, 0,  v3}};
```

Here, the fact that in the rest frame, **p2** and **p3** are back to back has been used; without loss of generality, you can align these particles along the z-axis. The conservation of the four-momenta gives the additional rule

```
In[3]:= conRule= {p3 -> p1-p2};
```

It follows from the scalar invariant $m_3^2 = p_3 \cdot \eta \cdot p_3$ and **conRule** that

```
In[4]:=   eq1= ( m3^2 == eta.p3.p3
                      /.conRule
                      /.momRule
                      /.v2gam[v2,gam2]
                      //Simplify
              )

Out[4]=     2        2                          2
          m3  == m1  - 2 gam2 m1 m2 + m2
```

The user-defined rule **v2gam** has been used to eliminate the speed **v2**. Solving for **gam2**, you get

```
In[5]:=   sol= Solve[eq1,gam2][[1]] //Simplify

Out[5]=                2      2     2
                     m1  + m2  - m3
          {gam2 -> ----------------}
                       2 m1 m2
```

The kinetic energy of **m2** is the total energy (**m2 gam2**) minus the rest mass (**m2**):

```
In[6]:=   eq3=  m2 gam2-m2  /.sol //Simplify

Out[6]=     2                  2     2
          m1  - 2 m1 m2 + m2  - m3
          ------------------------
                   2 m1
```

Problem 2: Two-particle collision

A particle of rest mass m_1 collides with a stationary particle of mass m_2. The collision produces two particles with masses m_3 and m_4. The threshold kinetic energy in the laboratory is the kinetic energy just sufficient to make the center of mass energy equal to the sum of the rest masses of the particles in the final state.

Part (a) Write the total four-momenta vectors for the initial particles in the laboratory frame and in the center-of-mass frame.

Part (b) Consider the scalar invariant $p \cdot \eta \cdot p$, where p is the total momentum. Evaluate the invariant in the laboratory and center-of-mass frames and equate the expressions to solve for the threshold energy. Express the results in terms of the kinetic energy of the particle with mass m_1.

Remarks and outline Write the two initial four-momenta for the two particles in the laboratory frame and form the total initial four-momentum, p_{LT}. Do the same for the total four-momentum as measured in the center-of-mass frame, p_{CMT}. Form the scalar equation $p_{LT} \cdot \eta \cdot p_{LT} = p_{CMT} \cdot \eta \cdot p_{CMT}$. Express the energy E_1 and the three-momentum p_1 in terms of kinetic energy. Use the definition of threshold energy and solve for the kinetic energy.

Solution

```
In[7]:=   Clear["Global`*"];
```

Part (a) Let the initial four-momenta of the particle with mass **m1** be **pL1** and the four-momenta of the stationary particle be **pL2**. The initial four-momenta of the two particles in the laboratory frame are

```
In[8]:=  pL1={E1,p1,0,0};
         pL2={m2, 0,0,0};
```

The total initial four-momentum is

```
In[9]:=  pLT=pL1+pL2;
```

Next, write the initial four-momentum in the center-of-mass frame:

```
In[10]:= pCM1={ECM1, kCM1,0,0};
         pCM2={ECM2,-kCM1,0,0};
```

Note that the fact that the sum of the three momenta is zero has been used. The total four-momentum is

```
In[11]:= pCMT=pCM1+pCM2;
```

Part (b) Consider the scalar invariant $p \cdot \eta \cdot p$, where p is the total momentum. Evaluating the scalar invariant in both frames and equating the expressions, you get

```
In[12]:= eq1= pLT.eta.pLT == pCMT.eta.pCMT

Out[12]=           2     2                  2
         (E1 + m2)  - p1  == (ECM1 + ECM2)
```

The threshold energy is obtained when

```
In[13]:= eq2=    ECM1+ECM2 == m3+m4;
```

where **m3** and **m4** are the rest masses of the final particles. To find the threshold energy in terms of the kinetic energy of particle one, you must express the energy **E1** and three-momentum **p1** in terms of kinetic energy. The kinetic energy **T1** is related to **E1** and **p1** by the equations

```
In[14]:= {eq3,eq4}= {E1==T1+m1, p1==Sqrt[E1^2-m1^2]}

Out[14]=                                    2     2
         {E1 == m1 + T1, p1 == Sqrt[E1  - m1 ]}
```

Solving for **T1** in terms of the rest masses, you get

```
In[15]:= sol= (Solve[{eq1,eq2,eq3,eq4},{T1},{E1,p1,ECM1}][[1]]
             //Simplify
           )

Out[15]=               2                2     2                 2
                 -m1  - 2 m1 m2 - m2  + m3  + 2 m3 m4 + m4
         {T1 -> ------------------------------------------}
                                  2 m2
```

Problem 3: Compton scattering

A photon with wavelength λ_0 collides with an electron of mass m at rest and scatters with a new wavelength λ_1. Let the incident photon move to the right along the x-axis and let the scattered photon make an angle θ_1 with the x-axis. The scattered electron moves along a direction making an angle $-\theta_2$ with the x-axis. Let p_1 and p_2 be the initial four-momentums of the photon and electron, respectively. The final four-momentums of the photon and electron are p_3 and p_4, respectively.

Part (a) Write the equations that follow from the conservation of four-momentum.

Part (b) Consider the scalar invariant $m^2 = p_4 \cdot \eta \cdot p_4$ and find the change in wavelength as a function of the photons scattering angle θ_1.

Part (c) Find the kinetic energy of the recoil electron.

Part (d) Find a relation between θ_1, θ_2, and λ_0.

Remarks and outline Write the four-momentum vectors for the initial photon and electron, p_1 and p_2. Write the final photon and electron four-momentums, p_3 and p_4. The equations for the conservation of four-momentum follow from $p_1 + p_2 = p_3 + p_4$. Write the scalar equation $m^2 = (p_1 + p_2 - p_3) \cdot \eta \cdot (p_1 + p_2 - p_3)$ and solve for the wavelength of the final photon. The kinetic energy of the recoil electron follows from $m\gamma_f - m$. The function γ_f follows from the conservation of four-momentum. The relation between θ_1, θ_2, and λ_0 follows from the equations for the conservation of four-momentum.

Solution

```
In[16]:= Clear["Global`*"];
```

Part (a) The initial photon and electron four-momentums are

```
In[17]:= p1 = h/lam0{1,0,0,1};
         p2 = m     {1,0,0,0};
```

where **lam0** is the initial photon wavelength and **h** is Planck's constant. The final four-momentum vectors are

```
In[18]:= p3= h/lam1{1,      Sin[o1], 0,    Cos[o1]};
         p4= m gamf{1, -vf Sin[o2], 0, vf Cos[o2]};
```

where **lam1** is the scattered wavelength of the photon and **vf** is the recoil velocity of the electron. The four equations that follow from the conservation of four-momentum are

```
In[19]:= eq1= p1+p2 == p3+p4 //Thread;
         eq1 //ColumnForm

Out[19]= h               h
         ---- + m == ---- + gamf m
         lam0          lam1
              h Sin[o1]
         0 == --------- - gamf m vf Sin[o2]
                lam1
         True
          h       h Cos[o1]
         ---- == --------- + gamf m vf Cos[o2]
         lam0      lam1
```

Part (b) Consider the scalar invariant $m^2 = p_4 \cdot \eta \cdot p_4$. It follows from the conservation of energy-momentum that $p_4 = p_1 + p_2 - p_3$. Expressing the scalar invariant in terms of p_1, p_2, and p_3, it follows that

```
In[20]:= eq2= ( m^2 == (p1+p2-p3).eta.(p1+p2-p3)
                //Simplify
                //ExpandAll
             )

Out[20]=
                 2                                    2
 2            -2 h        2 h m    2 h m     2     2 h  Cos[o1]
m   ==   ------------ + ------- - ------- + m  + --------------
          lam0 lam1       lam0     lam1            lam0 lam1
```

Solve **eq2** for **lam1**:

```
In[21]:= lamRule= Solve[eq2,lam1][[1]] //ExpandAll

Out[21]=
                       h    h Cos[o1]
         {lam1 -> lam0 + - - ---------}
                       m        m
```

Part (c) Use **lamRule** to eliminate **lam1** in **eq1[[1]]**:

```
In[22]:= eq3= eq1[[1]] /.lamRule

Out[22]=  h                                 h
         ---- + m == gamf m + --------------------------
         lam0                          h    h Cos[o1]
                               lam0 + - - ---------
                                       m       m
```

Solving **eq3** for **gamf**, you get

```
In[23]:= gamfRule= Solve[eq3,gamf][[1]] //Simplify

Out[23]=
                2                      2  2     2
               h  + h lam0 m + lam0  m  - h  Cos[o1] - h lam0 m Cos[o1]
         {gamf -> ------------------------------------------------------------}
                          lam0 m (h + lam0 m - h Cos[o1])
```

The kinetic energy is the total energy (**m gamf**) minus the rest mass energy (**m**):

```
In[24]:= (m gamf-m) /.gamfRule //Simplify

Out[24]=
                 2     o1 2
              2 h  Sin[--]
                       2
         ------------------------------
         lam0 (h + lam0 m - h Cos[o1])
```

Part (d) Use the second and fourth equations in **eq1** to get a relation between the angles **o1** and **o2**:

```
In[25]:= eq4= eq1[[{2,4}]] //Simplify

Out[25]=
              h Sin[o1]
         {0 == --------- - gamf m vf Sin[o2],
                lam1

           h       h Cos[o1]
          ---- == --------- + gamf m vf Cos[o2]}
          lam0       lam1
```

Solve for **gamf** and **Sin[o1]**:

```
In[26]:= sol1= Solve[eq4,{gamf,Sin[o1]}][[1]]
Out[26]=
                      (h lam1 - h lam0 Cos[o1]) Tan[o2]
         {Sin[o1] -> ---------------------------------,
                                   h lam0

                    (h lam1 - h lam0 Cos[o1]) Sec[o2]
          gamf -> ---------------------------------}
                           lam0 lam1 m vf
```

Next, obtain an equation for **Sin[o1]** and solve this for **Tan[o2]**:

```
In[27]:= eq5= Sin[o1] == (Sin[o1] /.sol1);

         sol2= Solve[eq5,{Tan[o2]}][[1]] /.lamRule //Simplify
Out[27]=
                                o1
                     lam0 m Cot[--]
                                2
         {Tan[o2] -> --------------}
                       h + lam0 m
```

where **lamRule** has been used to eliminate **lam1**.

Problem 4: Moving mirror and generalized Snell's law

Consider a mirror at rest in the reference frame S'. As observed in S', the mirror's rest frame, an incident beam of light makes an angle θ_m relative to the normal and, according to Snell's law, reflects at the same angle. The frequency of light in S' is f_M and is the same for the reflected and incident beam. The laboratory frame S is moving with velocity v along the normal. In the laboratory frame, the incident light has frequency f_i and makes an angle θ_i with the normal to the mirror. The reflected beam in S has frequency f_r and makes an angle θ_r relative to the normal.

Part (a) Show that the generalized Doppler formula is

$$f_i = f_r \frac{(1-v^2)}{(1+v^2+2v\text{Cos}[\theta_i])}.$$

Part (b) Show that the generalized Snell's law is

$$\text{Cot}[\theta_i/2]\ \text{Tan}[\theta_r/2] = \frac{(1-v)}{(1+v)}.$$

Remarks and outline Let $\{\theta_i, \theta_r\}$ be the incident and reflected angles of the beam and $\{f_i, f_r\}$ the incident and reflected frequencies as measured in the laboratory frame. Write the four-momentum vectors for the incident and reflected momentums in the laboratory and mirror reference frames. Relate these vectors by the user-defined Lorentz boost, **boostx**. Solve the resulting equations for the relevant parameters.

Solution

```
In[28]:= Clear["Global`*"];
```

Part (a) The momentum four-vectors in the laboratory frame for the incident and reflected waves are

```
In[29]:= Ki= fi{1, Cos[oi],Sin[oi],0};
         Kr= fr{1,-Cos[or],Sin[or],0};
```

where **{oi,or}** are the incident and reflected angles of the beam and **{fi,fr}** are the incident and reflected frequencies as measured in the laboratory frame. In the mirror's frame, the reflected and incident frequencies **fm** are equal and, by Snell's law, the angle of incident and reflection **om** are the same. The momentum vectors are

```
In[30]:= Kmi= fm{1, Cos[om],Sin[om],0};
         Kmr= fm{1,-Cos[om],Sin[om],0};
```

The four-momentums in the laboratory and mirror frames are related by Lorentz transformations. The incident four-momentums in the two frames are related by

```
In[31]:= eq1= (Ki ==  boostx[gam,v].Kmi) //Thread;
         eq1 //ColumnForm

Out[31]= fi == fm gam - fm gam v Cos[om]
         fi Cos[oi] == -(fm gam v) + fm gam Cos[om]
         fi Sin[oi] == fm Sin[om]
         True
```

The incident four-momentums give four equations. Another four equations follow from relating the reflected four-momentums:

```
In[32]:= eq2= (Kr == boostx[gam,v].Kmr) //Thread;
         eq2 //ColumnForm

Out[32]= fr == fm gam + fm gam v Cos[om]
         -(fr Cos[or]) == -(fm gam v) - fm gam Cos[om]
         fr Sin[or] == fm Sin[om]
         True
```

Use **Eliminate** to eliminate the variables **{Cos[or],Sin[or],Cos[oim],Sin[oim]}** in **eq1** and **eq2**:

```
In[33]:= eq3= {eq1,eq2,{fr !=0, fi !=0, gam !=0, v!=0}}//Flatten;

         eq4= Eliminate[eq3,{Cos[or],Sin[or],Cos[om],Sin[om]}]

Out[33]= fm != 0 && fr - 2 fm gam != 0 && fr != 0 && gam != 0 && v != 0 &&

                                             1    fm gam    fm gam v
           fi == -fr + 2 fm gam && Cos[oi] == -(-) + ------ - --------
                                             v    fi v        fi
```

Solving for **fi** and **fm**, you get the generalized Doppler shift:

```
In[34]:= sol1= Solve[eq4,{fi,fm}][[1]] //Simplify

Out[34]=
                           2
                   fr (1 - v )                       fr (1 + v Cos[oi])
         {fi -> -------------------, fm -> ---------------------------------}
                       2                                2
               1 + v   + 2 v Cos[oi]        gam (1 + v   + 2 v Cos[oi])
```

Part (b) To get a relation between the laboratory angles **oi**, **or**, and **v**, eliminate **{Cos[om],Sin[o** terms in **eq1** and **eq2**:

```
In[35]:= eq5=Eliminate[{eq1,eq2,{fr !=0, fi !=0,gam !=0, v!=0 }}//Flatten,
                  { Cos[om],Sin[om] }]
```

```
Out[35]= fm != 0 && fi != 0 && fi - 2 fm gam != 0 && gam != 0 && v != 0 &&

                                                 2 fm gam
          fr == -fi + 2 fm gam && Sin[oi] == (-1 + --------) Sin[or] &&
                                                    fi

                     1   fm gam   fm gam v
          Cos[or] == - - ------ + -------- &&
                     v    fr v       fr

                      1    fm gam   fm gam v
          Cos[oi] == -(-) + ------ - --------
                      v     fi v       fi
```

Next, use **sol1** to eliminate the **fi** and **fm** terms in **eq5**:

```
In[36]:= eq6= eq5 /.sol1 //Simplify
Out[36]=
                                                            2
             fr (1 + v Cos[oi])                     fr (1 - v )
        ---------------------------- != 0 && ---------------------- != 0 &&
                    2                             2
        gam (1 + v  + 2 v Cos[oi])           1 + v  + 2 v Cos[oi]

          -fr != 0 && gam != 0 && v != 0 &&

                              2
                       (1 + v  + 2 v Cos[oi]) Sin[or]
          Sin[oi] == ------------------------------- &&
                                     2
                                1 - v

                                             2
                       2 v + Cos[oi] + v  Cos[oi]
          Cos[or] == ---------------------------
                               2
                          1 + v  + 2 v Cos[oi]
```

The last two equations are two forms of the generalized Snell's law:

```
In[37]:= sol2= Solve[eq6,{Sin[oi],Cos[or]}][[1]] //Simplify
Out[37]=
                              2
                       (1 + v  + 2 v Cos[oi]) Sin[or]
          {Sin[oi] -> -------------------------------,
                                     2
                                1 - v

                                           2
                      2 v + Cos[oi] + v  Cos[oi]
          Cos[or] -> ---------------------------}
                              2
                         1 + v  + 2 v Cos[oi]
```

In particular, the **Cos[or]** equation can be simplified further by solving for **v**:

```
In[38]:= sol3= ( Solve[ Cos[or] == (Cos[or] /.sol2) ,v][[1]]
                   //Simplify
                   //PowerExpand
                   //Simplify
               )
```

```
Out[38]=
                    oi + or        oi - or
         {v -> Csc[———————] Sin[———————]}
                       2              2
```

which can be written in the form,

```
In[39]:= temp = (1-v)/(1+v);
         temp == (temp /.sol3) //Simplify
```

```
Out[39]= 1 - v          oi       or
         ————— == Cot[——] Tan[——]
         1 + v           2        2
```

Caution: You may have to modify the above step because different implementations of *Mathematica* will give slightly different algebraic results!

Problem 5: One-dimensional motion of a relativistic particle with constant acceleration

Consider a relativistic particle accelerated by a constant force directed along the x-axis. Assume the particle is moving along the x-axis and solve for its distance as a function of coordinate time t. The motion is assumed to be described by the relativistic equation

$$\frac{d}{dt}\left(\frac{v[t]}{\sqrt{1-v[t]^2}}\right) = g,$$

where $v[t] = x'[t]$, t is the coordinate time, and g is the acceleration constant due to the force.

Part (a) Solve for $v[t]$ and $x[t]$ and assume $x[0] = 0$ and $v[0] = v_0$. Show that the solution has the expected nonrelativistic limit.

Part (b) Plot $x[t]$ for the relativistic and nonrelativistic solutions.

Remarks and outline Use **DSolve** to solve for v. Let $x'[t] = v[t]$ and use **DSolve** again to find $x[t]$.

Solution

```
In[40]:= Clear["Global`*"];
```

Part (a) The equation of motion for constant acceleration is

```
In[41]:= eqMotion=  D[ v[t]/Sqrt[1-v[t]^2] ,t] == g
```

```
Out[41]=
              2
          v[t]  v'[t]                 v'[t]
         ——————————————— + ———————————————————— == g
                   2 3/2                   2
         (1 - v[t] )        Sqrt[1 - v[t] ]
```

where **t** is the coordinate time. Solving **eqMotion** for **v[t]**, you get

```
In[42]:= eq1= ( DSolve[{eqMotion},v[t],t][[2]]
                 /.{Rule->Equal}
                 //Simplify
                 //PowerExpand
              )
```

```
42]=                        g t + C[1]
      {v[t] == ─────────────────────────────────────────────}
                      2  2                          2
              Sqrt[1 + g  t  + 2 g t C[1] + C[1] ]
```

Then let **v[t]=x'[t]** and solve for **x[t]**:

```
43]:= eq2= eq1 /.{v[t]->x'[t]};

      dsol2= ( DSolve[{eq2} //Flatten,{x},t
                 ,{DSolveConstants -> CC}][[1]]
               //PowerExpand
               //Simplify
             )
```

```
t[43]=                             2  2                          2
                          Sqrt[1 + g  t  + 2 g t C[1] + C[1] ]
       {x -> Function[t, ───────────────────────────────────── + CC[1]]}
                                         g
```

Use these initial conditions:

```
[44]:= initialConditions=
         {(x[0] /.dsol2)==x0, (x'[0] /.dsol2)==v0}
```

```
t[44]=              2
         Sqrt[1 + C[1] ]                          C[1]
       {──────────────── + CC[1] == x0, ──────────────── == v0}
                g                                     2
                                        Sqrt[1 + C[1] ]
```

to eliminate the constants of integration:

```
[45]:= cRule=Solve[initialConditions,{CC[1],C[1]}][[2]] //Simplify
```

```
:[45]=                   1
                Sqrt[───────]
                           2
                     1 - v0                             I v0
       {CC[1] -> -(─────────────) + x0, C[1] -> ───────────────}
                         g                                   2
                                                Sqrt[-1 + v0 ]
```

and obtain the relativistic result:

```
46]:= relResult= ( x[t] /.dsol2
                        /.cRule
                        //Simplify
                 )
```

```
[46]=                                                2
             1                       2  2         v0         2 I g t v0
       Sqrt[───────]        Sqrt[1 + g  t  + ─────── + ───────────────]
                  2                                2                2
            1 - v0                           1 - v0    Sqrt[-1 + v0 ]
       -(─────────────) + ─────────────────────────────────────────────── + x0
               g                                 g
```

You can simplify this with the rule

```
7]:= sRule= {   Sqrt[a_] Sqrt[b_] :>  Sqrt[a b]
             ,1/(Sqrt[a_] Sqrt[b_]):>1/Sqrt[a b]
             };
```

and by grouping factors of **(1-v0^2)** as **1/gam^2**:

```
In[48]:= gamRule= {(1-v0^2) -> 1/gam^2, (-1+v0^2) -> -1/gam^2};

         relResult2= ( relResult
                        //.sRule
                        /.gamRule
                        //PowerExpand
                       )
```

```
Out[48]=
                        2  2                           2   2
     gam     Sqrt[1 + g   t   + 2 g gam t v0 + gam   v0 ]
   -(---) + ------------------------------------------------ + x0
      g                             g
```

To find the nonrelativistic solution, expand **relResult2** in powers of **{t,v[0]}** and keep only leading-order terms:

```
In[49]:= nonRelResult= (( Series[relResult,{t,0,2},{v0,0,2}]
                           //Normal
                           //Simplify
                          ) //.{v0^2->0}
                          )
```

```
Out[49]=
       2
    g t
    ---- + t v0 + x0
     2
```

You will obtain the standard result. You should also verify the initial conditions for the relativistic result:

```
In[50]:= (( {relResult,D[relResult,t]}
              /.{t->0}
              //Simplify
            ) //.gamRule
              //PowerExpand
          )
```

```
Out[50]= {x0, v0}
```

and for the nonrelativistic result:

```
In[51]:= {nonRelResult,D[nonRelResult,t]} /.{t->0}
```

```
Out[51]= {x0, v0}
```

Part (b) To plot the relativistic and nonrelativistic motions, assume the values

```
In[52]:= values={ x0->0, v0->1/2, g-> -0.1};
```

The comparison of the relativistic and nonrelativistic solutions is

```
In[53]:= Plot[{relResult,nonRelResult} /.values //Evaluate
           ,{t,0,12 }
           ,GridLines->Automatic
           ,AxesLabel->{"t","x"}
           ,PlotStyle->{{Thickness[0.01]}
                        ,{Dashing[{0.06,0.1}],Thickness[0.001]}}
           ,Epilog->Graphics[
             {Text[FontForm["Relativistic",{"Courier-Bold",10}],{10,1.3}]
             ,Text[FontForm[nonRelResult   ,{"Courier-Bold",10}],{ 8,0.3}]
             }][[1]]
          ];
```

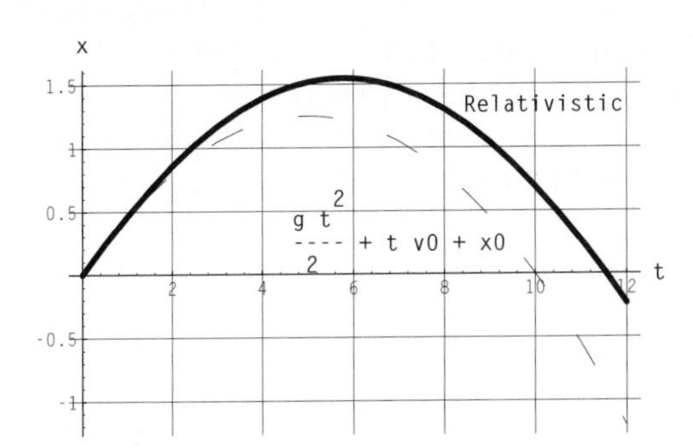

Problem 6: Two-dimensional motion of a relativistic particle in a uniform electric field

A particle with mass m and charge e moves in a uniform static electric field E_0. The electric field is assumed to be along the y-axis. The equations of motion for the particle are $p_x'[t] = 0$, $p_y'[t] = eE_0$, $p_z'[t] = 0$, where $\{p_x, p_y, p_z\}$ are the three space components of the four-momentum vector.

Part (a) Solve for the velocity and position of the particle as an explicit function of time. Assume the initial momentum is perpendicular to the electric field, $\{p_x[0] = p_{x0}, p_y[0] = 0, p_z[0] = 0\}$. Find the nonrelativistic limit.

Part (b) Compare the plots for the nonrelativistic limit with the relativistic plots.

Remarks and outline Use **DSolve** to solve the three equations $p_x'[t] = 0$, $p_y'[t] = eE_0$, $p_z'[t] = 0$ for $\{p_x, p_y, p_z\}$. Express the momentum in terms of the derivatives of the coordinates. For example, $p_x[t]$ is related to $x[t]$ by $p_x[t] = m\gamma dx/dt$, where $\gamma = 1/\sqrt{1-v^2}$. An expression for γ follows from the energy and momentum relation $(m\gamma)^2 = \vec{P}\cdot\vec{P} + m^2$, where $\vec{P} = \{p_x, p_y, p_z\}$. The resultant equations for the coordinates can then be solved using **DSolve**.

Solution

```
In[54]:= Clear["Global`*"];
```

Part (a) The equations of motion for the particle are

```
In[55]:= eqMotion={px'[t]==0, py'[t]==e E0, pz'[t]==0};
```

and the conditions at $t = 0$ are

```
In[56]:= initial= {px[0]==p0x, py[0]==0, pz[0]==0};
```

The equations of motion and initial conditions can be integrated to give expressions for the momentum components:

```
In[57]:= momRule=DSolve[eqMotion~Join~initial
                ,{px[t],py[t],pz[t]},t][[1]]

Out[57]= {px[t] -> p0x, py[t] -> e E0 t, pz[t] -> 0}
```

Then separately solve each equation that follows from **momRule**. The solution for **z[t]** is obviously zero. Consider the equation for **px[t]**. The momentum components are related to the position variables by $m\gamma dx/dt$, where $\gamma = 1/\sqrt{1-v^2}$. An expression for γ follows from the energy and momentum relation $(m\gamma)^2 = \vec{P}\cdot\vec{P} + m^2$, where $\vec{P} = \{p_x, p_y, p_z\}$:

```
In[58]:= pVec[t_]={px[t],py[t],pz[t]};

In[59]:= eq1= m gamma ==  Sqrt[pVec[t].pVec[t]+m^2 /.momRule]

Out[59]=                  2      2    2   2 2
         gamma m == Sqrt[m  + p0x  + e  E0  t ]
```

Solving **eq1** for **gamma**, it follows that

```
In[60]:= gammaRule=  Solve[eq1 ,gamma][[1,1]]

Out[60]=                  2      2    2   2 2
                  Sqrt[m  + p0x  + e  E0  t ]
         gamma -> ---------------------------
                               m
```

Then use **gammaRule** and **momRule** to express **px[t]** in terms of **x'[t]**:

```
In[61]:= eq2= (m gamma x'[t]==px[t]) /.momRule /.gammaRule

Out[61]=      2      2    2   2 2
         Sqrt[m  + p0x  + e  E0  t ] x'[t] == p0x
```

The expression for **x[t]** follows from applying **DSolve** to **eq2**:

```
In[62]:= xRule= ( DSolve[{eq2,x[0]==0},x[t],t][[1]]
                     //PowerExpand
                     //ExpandAll
                 )

Out[62]=                                                     2      2
                    p0x Log[e]     p0x Log[E0]    p0x Log[m  + p0x ]
         {x[t] -> -(----------) - ----------- - ------------------ +
                      e E0           e E0             2 e E0

                    2   2                  2      2    2   2 2
           p0x Log[e  E0  t + e E0 Sqrt[m  + p0x  + e  E0  t ]]
           ---------------------------------------------------}
                                 e E0
```

We assumed **x[0]=0**. The solution for **y[t]** follows similarly:

```
In[63]:= yEq  = (m gamma y'[t] == py[t]) /.momRule /.gammaRule;

         yRule= DSolve[{yEq, y[0]==0},y[t],t][[1]]

Out[63]=                  2      2           2      2    2   2 2
                   Sqrt[m  + p0x ]    Sqrt[m  + p0x  + e  E0  t ]
         {y[t] -> -(---------------) + ---------------------------}
                        e E0                     e E0
```

The nonrelativistic limit follows from expanding **x[t]** and **y[t]** in a **t** power series and keeping the leading-order term and then assuming $m^2 > p_{x0}^2$. It follows from **xRule** and **yRule** that the nonrelativistic equation for **{x,y}** is

```
In[64]:= relXY= {x[t],y[t]} /.xRule /.yRule;

         nonRelXY= ( Map[Series[#,{t,0,2}]&,relXY]
                         /.{p0x^2->0}
                         //PowerExpand
                         //Normal
                      )
Out[64]=
                    2
          p0x t   e E0 t
         {-----, -------}
            m      2 m
```

The motion in the x-direction moves with uniform speed and the motion in the y-direction moves with constant Newtonian acceleration. This motion is to be compared with the relativistic values for **{x,y}**:

```
In[65]:= relXY
Out[65]=
                                                       2      2
           p0x Log[e]     p0x Log[E0]    p0x Log[m  + p0x ]
         {-(----------) - ----------- - ------------------- +
             e E0           e E0            2 e E0

                      2   2                   2      2     2   2  2
           p0x Log[e  E0  t + e E0 Sqrt[m  + p0x  + e  E0  t ]]
           ------------------------------------------------------,
                                  e E0

                   2      2              2      2     2   2  2
            Sqrt[m  + p0x ]        Sqrt[m  + p0x  + e  E0  t ]
          -(---------------) + ----------------------------}
                e E0                     e E0
```

Part (b) To plot the motion, assume these values:

```
In[66]:= values={m->1, e->1, E0->0.1, p0x->0.1};
```

Then plot the positions as a function of time and compare the relativistic results with the nonrelativistic results. Define a plot operator **plotOp** that will plot a particular coordinate for the relativistic and nonrelativistic curves:

```
In[67]:= Clear[plotOp];
         plotOp[eqs_List,axis_,tRange_] :=
         Plot[ eqs //.values //Evaluate
           ,{t,0,tRange}
           ,GridLines->Automatic
           ,PlotStyle->{{Thickness[0.01]}
                       ,{Dashing[{0.06,0.1}],Thickness[0.001]}}
           ,AxesLabel->{"t", axis}
           ,DisplayFunction->Identity];
```

where **eqs** are the equations for the relativistic and nonrelativistic curves, **axis** labels the space-axis, and **tRange** is the range of the time parameter. Applying **plotOp** to the **y** and **x** components, you get

```
In[68]:= Show[GraphicsArray[
           {plotOp[{relXY[[1]],nonRelXY[[1]]},"x",30],
            plotOp[{relXY[[2]],nonRelXY[[2]]},"y",30]}
          ] ];
```

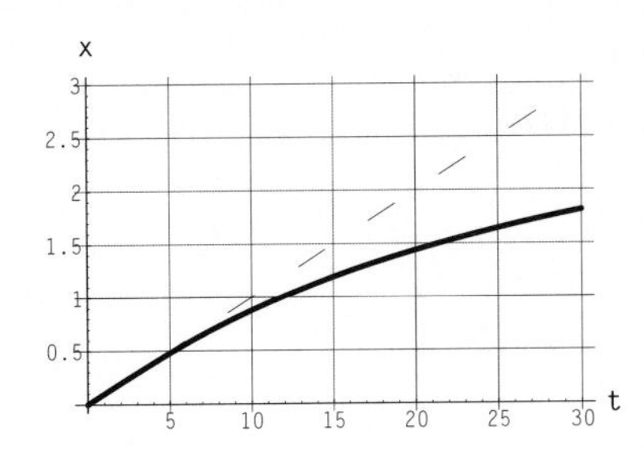

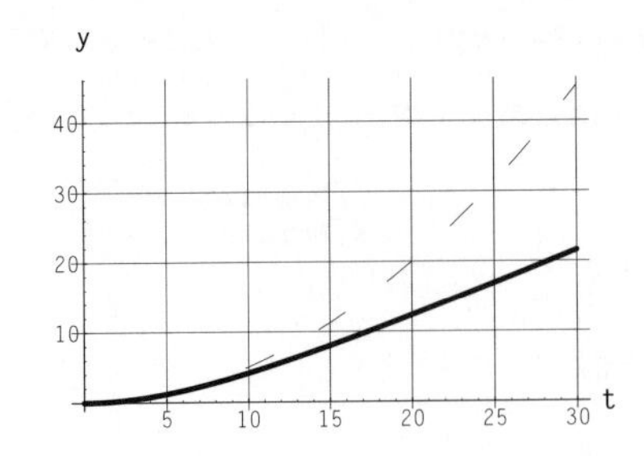

The dashed curve is for the nonrelativistic motion; the solid curve describes the relativistic motion. The curves on the right are for the **y** components, which are being accelerated by an electric field along the y-direction. There is no force in the x-direction, so the nonrelativistic **x** component moves with constant speed. This is not the case for the relativistic solution.

7.2.2 General Relativity and Cosmology

Problem 1: Schwarzschild solution in null coordinates

Consider a spherically symmetric metric expressed in null coordinates $\{u, r, \theta, \phi\}$. The metric is nondiagonal and is of the form

$$\eta_{ij} = \begin{pmatrix} g_{00}[r] & -1 & 0 & 0 \\ -1 & 0 & 0 & 0 \\ 0 & 0 & -r^2 & 0 \\ 0 & 0 & 0 & -r^2\mathrm{Sin}^2[\theta] \end{pmatrix}.$$

Find the solution of Einstein's vacuum equations with the boundary conditions that $g_{00}[r] \to (1 - 2m/r)$ as r goes to infinity.

Remarks and outline This metric is the Schwarzschild solution expressed in null coordinates. The Christoffel symbols, curvature tensor, and Ricci tensor follow from user-defined procedures. Set the Ricci tensor to zero and use **DSolve** to solve for g_{00}.

Solution

```
In[69]:= Clear["Global`*"];
```

Assume a coordinate system:

```
In[70]:=  varnull= {u,r,theta,phi};
```

and a metric of the form

```
In[71]:=   nullmet= {{g00[r],-1,0,0},{-1,0,0,0},
                {0,0, -r^2,0},{0,0,0,-r^2 Sin[theta]^2}};
```

The vacuum equations follow from setting the components of the Ricci tensor to zero. To evaluate the tensor, first find the Christoffel symbols:

```
In[72]:= nullchrist=christ[nullmet,varnull];
           Short[nullchrist,3]
```

```
Out[72]=     g00'[r]
         {{{-------, 0, 0, 0}, {0, 0, 0, 0}, {0, 0, -r, 0},
               2

                                  2
           {0, 0, 0, -(r Sin[theta] )}}, <<2>>,

                                    1
          {{0, 0, 0, 0}, {0, 0, 0, -}, {0, 0, 0, Cot[theta]},
                                    r

              1
           {0, -, Cot[theta], 0}}}
              r
```

and the curvature tensor:

```
In[73]:= nullcur= curvR[nullchrist , varnull]//Simplify ;
                  Short[nullcur,6]

Out[73]=       g00''[r]           -g00''[r]
         {{{{0, --------, 0, 0}, {---------, 0, 0, 0}, {0, 0, 0, 0},
                   2                  2

           {0, 0, 0, 0}}, {{0, 0, 0, 0}, {0, 0, 0, 0}, {0, 0, 0, 0},

                                  r g00'[r]
           {0, 0, 0, 0}}, {{0, 0, ---------, 0}, {0, 0, 0, 0},
                                      2

            -(r g00'[r])
           {------------, 0, 0, 0}, {0, 0, 0, 0}},
                 2

                                      2
                       r Sin[theta]  g00'[r]
           {{0, 0, 0, ----------------------}, {0, 0, 0, 0}, {0, 0, 0, 0},
                                2

                               2
            -(r Sin[theta]  g00'[r])
           {------------------------, 0, 0, 0}}}, <<3>>}
                       2
```

The Ricci tensor follows from the user-defined function **ricciR**:

```
In[74]:=  nullricci=  ricciR[nullcur ]//Simplify

Out[74]=   -(g00[r] (2 g00'[r] + r g00''[r]))   g00'[r]   g00''[r]
         {{-----------------------------------, ------- + --------, 0, 0},
                         2 r                       r         2

           g00'[r]   g00''[r]
          {------- + --------, 0, 0, 0}, {0, 0, -1 + g00[r] + r g00'[r], 0},
              r         2

                                2
          {0, 0, 0, Sin[theta]  (-1 + g00[r] + r g00'[r])}}
```

The Einstein vacuum equations follow from setting **nullricci** to zero. There are five nontrivial vacuum equations:

```
In[75]:= vacuumeq= Select[(nullricci//Flatten)==0//Thread,
                    FreeQ[#,True]&]//Together//Numerator

Out[75]=  -(g00[r] (2 g00'[r] + r g00''[r]))           g00'[r]    g00''[r]
         {------------------------------------ == 0, ------- + -------- == 0,
                          2 r                            r          2

          g00'[r]   g00''[r]
          ------- + -------- == 0, -1 + g00[r] + r g00'[r] == 0,
             r         2

                      2
          Sin[theta]  (-1 + g00[r] + r g00'[r]) == 0}
```

We have not listed the trivial (**0==0**) or **True** equations. It is sufficient to apply **DSolve** to the fourth equation in the list to get an expression for **g00**:

```
In[76]:= nullrule=DSolve[vacuumeq[[4]],g00 ,r]//Flatten

Out[76]=                               C[1]
         {g00 -> Function[r, 1 + ----]}
                                       r
```

The boundary condition requires **C[1]**$=-2m$. Substituting **g00** given by **nullrule** into **vacuumeq** satisfies all the equations:

```
In[77]:= vacuumeq/.nullrule

Out[77]= {True, True, True, True, True}
```

The Schwarzschild solution expressed in null coordinates is

```
In[78]:= nullmet/.nullrule/.C[1]->-2 m

Out[78]=       2 m                                            2
         {{1 - ---, -1, 0, 0}, {-1, 0, 0, 0}, {0, 0, -r , 0},
                r

                      2           2
          {0, 0, 0, -(r  Sin[theta] )}}
```

Problem 2: The horizons and surfaces of infinite redshift

The Kerr metric **kMet** is defined in the Introduction, and its components follow from the user-defined rule **kComp**. The surfaces of infinite redshift are defined by the equation $g_{00} = 0$, and the event horizons are defined by the inverse metric component equation $g^{11} = 0$. Assume $m^2 > a^2$ and find the surfaces of infinite redshifts and horizons. Plot the surfaces.

Remarks and outline The Kerr metric is a case in which the surfaces of infinite redshift and the horizons differ. The equations for the surfaces follow from solving $g_{00} = 0$ and $g^{11} = 0$ for r.

Required packages

```
In[79]:= Needs["Algebra`Trigonometry`"];
         Needs["Graphics`ParametricPlot3D`"];
```

Solution

```
In[80]:= Clear["Global`*"];
```

Part (a) Consider the Kerr metric defined in the Introduction. Setting the **g00** component to zero, you get the equation for the surfaces of infinite redshift:

```
In[81]:= eq1=(g00[r,theta]/.kComp)==0//Simplify

Out[81]=  2                 2    2
         a  - 4 m r + 2 r  + a  Cos[2 theta]
         ------------------------------------ == 0
             2      2    2
            a  + 2 r  + a  Cos[2 theta]
```

Solving for r, you can see that

```
In[82]:=   redshift= Solve[eq1,r]//TrigReduce//ExpandAll

Out[82]=
                          2   2            2
         {{r -> m - Sqrt[m  - a  Cos[theta] ]},

                            2   2            2
           {r -> m + Sqrt[m  - a  Cos[theta] ]}}
```

There are two surfaces of infinite redshift: inner and outer. Assume $m^2 > a^2$. The horizons follow from setting the inverse component $g^{11} = 0$ to zero:

```
In[83]:= eq2= Inverse[kMet][[2,2]] ==0  //Simplify

Out[83]=       1
         -------------- == 0
         g11[r, theta]
```

The radial positions of the event horizons follow from evaluating the metric components in **eq2** with the rule **kComp** and solving for **r**:

```
In[84]:= horizon= Solve[eq2 /.kComp,r] //Simplify

Out[84]=
                          2    2                         2    2
         {{r -> m - Sqrt[-a  + m ]}, {r -> m + Sqrt[-a  + m ]}}
```

There are two horizons. The four surfaces follow from **redshift** and **horizon**. The graphics for the surfaces follow from applying the command **SphericalPlot3D** to **horizon** and **redshift**:

```
In[85]:= list= {{r /. redshift},{r /.horizon}} //Flatten;

         pt= SphericalPlot3D[Evaluate[#/.m->1/.a->.995],
                   {theta,0,  Pi},{phi,-Pi/2,Pi/2 },
                   AspectRatio->Automatic,
                   Boxed->False, Axes->False,
                   DisplayFunction->Identity ]& /@ list;
```

where we let $m = 1$ and $a = 0.995$. Displaying the results, you get

```
In[86]:= Show[pt, ViewPoint->{-1, -.6,.4},
         DisplayFunction->$DisplayFunction];
```

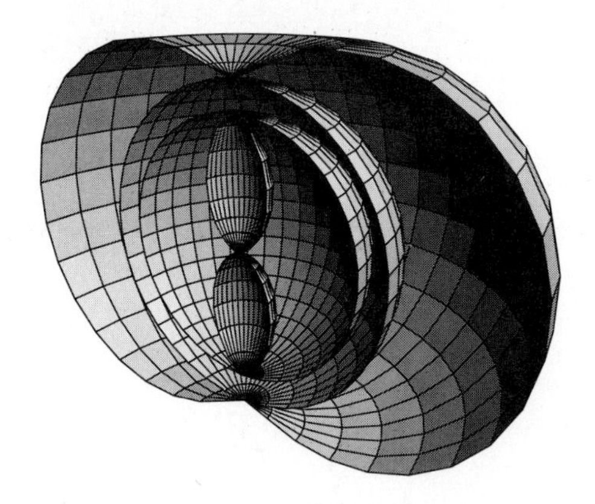

Only one half of the surfaces have been graphed so that the inner surfaces can be seen. The inner and outer surfaces are the surfaces of infinite redshift, and the two middle surfaces are the horizons.

Problem 3: Killing vectors and constants of motion

Consider the following problems for the Schwarzschild and Kerr metrics on Killing vectors.

Part (a) Consider the following four contravariant vectors:

$$k_t = \{1, 0, 0, 0\}$$
$$k_z = \{0, 0, 0, -1\}$$
$$k_x = \{0, 0, -\mathrm{Sin}[\phi], -\mathrm{Cos}[\phi]\mathrm{Cot}[\theta]\}$$
$$k_y = \{0, 0, \mathrm{Cos}[\phi], -\mathrm{Sin}[\phi]\mathrm{Cot}[\theta]\}$$

Verify that they satisfy Killing's equation for the Schwarzschild metric.

Part (b) Let $V = \{t'[s], r'[s], \theta'[s], \phi'[s]\}$ be a geodesic four-velocity vector. Constants of motion follow from equating the scalar dot product $V \cdot \eta \cdot K$ to a constant, where K is a Killing vector and η is the metric. Express the components of the four-velocity V in terms of the Killing constants.

Part (c) Verify that the two contravariant vectors $k_t = \{1, 0, 0, 0\}$, $k_z = \{0, 0, 0, -1\}$ satisfy Killing's equation for the Kerr metric.

Part (d) Express the two components of the four-velocity $\{t'[s], r'[s]\}$ for the Kerr metric in terms of Killing constants. Write the explicit expressions when $\theta = \pi/2$.

Remarks and outline The metric (**ssMet**) and Christoffel symbols (**ssChrist**) for the Schwarzschild and Kerr solutions were calculated in the Introduction. Apply the user-defined operator **killingEq** to the contravariant vectors to verify that they are Killing vectors.

Solution

```
In[87]:= Clear["Global`*"];
```

Part (**a**) Consider the contravariant vectors:

In[88]:=
```
kt={1,0,0,0};
    kz={0,0,0,-1};
      kx={0,0,-Sin[phi],-Cos[phi]Cot[theta]};
       ky={0,0,Cos[phi],-Sin[phi]Cot[theta]};
```

Apply the user-defined operator **killingEq** to the four contravariant vectors to find that

In[89]:=
```
( killingEq[#,ssMet,ssChrist]& /@ {kt,kz,kx,ky}
      /.ssComp
      //Simplify
  )
```

Out[89]=
```
{{True}, {}, {}, {}}
```

There are no nontrivial equations, so the four contravariant vectors **{kt,kz,kx,ky}** are Killing vectors.

Part (**b**) The constants of motion follow from the scalar dot product $V \cdot \eta \cdot K$, where V is the geodesic four-velocity vector, K is the Killing vector, and η is the metric. Let the four-velocity be

In[90]:=
```
velocity={t'[s],r'[s],theta'[s],phi'[s]};
```

Define the four Killing constants **{en,jx,jy,jz}** by the equations:

In[91]:=
```
 kconst= { en==ssMet.velocity.kt,
           jx==ssMet.velocity.kx,
           jy==ssMet.velocity.ky,
           jz==ssMet.velocity.kz};
kconst//ColumnForm
```

Out[91]=
```
en == g00[r] t'[s]
        2
jx == r  Cos[phi] Cos[theta] Sin[theta] phi'[s] +

     2
    r  Sin[phi] theta'[s]
        2
jy == r  Cos[theta] Sin[phi] Sin[theta] phi'[s] -

     2
    r  Cos[phi] theta'[s]
        2            2
jz == r  Sin[theta]  phi'[s]
```

Solving for the components of the four-velocity in terms of the Killing constants, you get the four first integrals:

In[92]:=
```
trule= Solve[kconst[[1]],t'[s]][[1,1]]/. ssComp
phirule= Solve[kconst[[4]],phi'[s] ][[1,1]]
jrule=  Solve[kconst[[{2,3}]],{phi'[s],theta'[s]}][[1]]//
                                                    Simplify
```

Out[92]=
```
                en
t'[s] -> ---------
              2 m
         1 - ---
               r
```

```
Out[92]=                             2
                        jz Csc[theta]
          phi'[s] -> ----------------
                              2
                             r
```

```
Out[92]=              Csc[theta] Sec[theta] (jx Cos[phi] + jy Sin[phi])
          {phi'[s] -> ------------------------------------------------,
                                              2
                                             r

                         -(jy Cos[phi]) + jx Sin[phi]
            theta'[s] -> ----------------------------}
                                      2
                                     r
```

Part (c) Verify that the two contravariant vectors

```
In[93]:=  kt={1,0,0,0};
              kz={0,0,0,-1};
```

satisfy Killing's equations for the Kerr metric. Applying the user-defined operator **killingEq** to the two contravariant vectors, you can see that

```
In[94]:= killingEq[#,kMet,kChrist]&/@{kt,kz}

Out[94]= {{}, {}}
```

All components of Killing's equation are satisfied, so **kt** and **kz** are Killing vectors for the Kerr metric.

Part (d) The two Killing constants **{en,jz}** are defined by

```
In[95]:= kconst= { en== kMet.velocity.kt,
                   jz== kMet.velocity.kz}

Out[95]= {en == g03[r, theta] phi'[s] + g00[r, theta] t'[s],

           jz == -(g33[r, theta] phi'[s]) - g03[r, theta] t'[s]}
```

Solving for **t'[s]** and **phi'[s]**, you get the following first integrals:

```
In[96]:=
         eq1=( Solve[kconst ,{t'[s],phi'[s]}][[1]]  //Simplify)

Out[96]=                jz g03[r, theta] + en g33[r, theta]
         {t'[s] -> ------------------------------------------------,
                                    2
                   -g03[r, theta]  + g00[r, theta] g33[r, theta]

                           jz g00[r, theta] + en g03[r, theta]
           phi'[s] -> ------------------------------------------------}
                                     2
                      g03[r, theta]  - g00[r, theta] g33[r, theta]
```

Explicitly evaluate the Kerr components and specialize the results to the equatorial plane **theta=Pi/2**. **eq1** then becomes

```
In[97]:=  eq1/.kComp/.theta->Pi/2//Simplify
```

Out[97]=

```
                  2                2              3
         -2 a  en m + 2 a jz m - a  en r - en r
{t'[s] -> ---------------------------------------,
                     2            2    3
                -(a  r) + 2 m r  - r

             2 a en m - 2 jz m + jz r
 phi'[s] -> -------------------------}
               2          2    3
              a  r - 2 m r  + r
```

Problem 4: Potential analysis for timelike geodesics

Consider the Schwarzschild metric and assume the timelike geodesics are confined to the plane $\theta = \pi/2$. Use the two first integrals that follow from Killing vectors: $E = t'[s](1 - 2m/r)$ and $\ell = r^2\phi'[s]$.

Part (a) Consider the timelike condition $1 = v \cdot \eta \cdot v$, where v is the geodesic four-velocity and η is the metric. If the effective potential V is defined by the equation $r'[s]^2 == -(1 - E^2) - V$, show that $V = (-2\ell^2 m)/r^3 + \ell^2/r^2 - (2m)/r$. The first term is the relativistic correction, the second is the centrifugal barrier, and the third is the gravity potential.

Part (b) Show that the potential has two extrema if $\ell^2 > 12m^2$.

Part (c) Graph the three parts of the effective potential for $\ell^2 > 12m^2$. Also graph the effective potential for the three cases, $\ell^2 > 12m^2$, $\ell^2 = 12m^2$, and $\ell^2 < 12m^2$.

Remarks and outline The potential follows from the timelike criteria $1 = v \cdot \eta \cdot v$, where η is the Schwarzschild metric and v is the four-velocity vector. The extrema follow from taking the derivative of V with respect to r and setting the results to zero. Let **en** represent the energy E.

Solution

In[98]:=
```
Clear["Global`*"];
```

Part (a) The restriction of the orbit to the **theta=Pi/2** plane is accomplished with the rule,

In[99]:=
```
planerule= {theta ->((Pi/2 )&) };
```

The rules for the two first integrals along with rules for their derivatives are

In[100]:=
```
SSconst = { t'[s]->en/ (1-2 m/r[s] ),
            t''[s]->D[en/ (1-2 m/r[s]) ,s] ,
              phi'[s]->l/r[s]^2,
                phi''[s]->D[l/r[s]^2,s] } ;
```

The potential for the orbit follows from the timelike criteria $1 = v \cdot \eta \cdot v$, where η is the Schwarzschild metric and v is the four-velocity vector. Define the four-velocity:

In[101]:=
```
vel={t'[s],r'[s],theta'[s],phi'[s]};
```

Eliminate **t'[s]** and **phi'[s]** with the rule **SSconst** and restrict the motion to the **theta=Pi/2** plane. The timelike condition becomes

```
In[102]:=   eq1=   1== ((ssMet/.ssComp/.r->r[s]/.theta->theta[s] ).
                        vel.vel )/.planerule/.SSconst

Out[102]=
                 2          2               2
               en          l            r'[s]
     1 == ----------- - ------- - -----------
                2 m           2           2 m
           1 - ----       r[s]      1 - ----
               r[s]                     r[s]
```

Defining the potential by

```
In[103]:= eq2=    r'[s]^2==    -(1-en^2 ) -    V;
```

you find that by eliminating **r'[s]** in **eq1** and **eq2**, **V** is

```
In[104]:= eq3=Solve[{eq1,eq2},V,r'[s]][[1,1]]/.r[s]->r //
                                      Simplify//ExpandAll

Out[104]=
                2       2
            -2 l  m    l    2 m
     V -> -------- + -- - ---
               3      2     r
              r      r
```

The first term is the relativistic correction, the second is the centrifugal barrier, and the third is the gravity potential.

Part (b) The maximum and minimum values follow from taking the derivative of V and setting the results to zero. These points are located at

```
In[105]:=    {rmax,rmin} =r/.Solve[D[V/.eq3,r]==0,r] //Simplify

Out[105]=  2           4        2 2     2            4        2 2
          l  - Sqrt[l  - 12 l  m ]   l  + Sqrt[l  - 12 l  m ]
         {------------------------, ------------------------}
                   2 m                        2 m
```

The potential has one maximum and one minimum point for $\mathtt{l}^2 > 12m^2$. The values of the potential at these points are

```
In[106]:= {Vmax,Vmin}=  {V/.eq3/.r->rmin, V/.eq3/.r->rmax} //
                                            Together//Simplify

Out[106]=      2 2     2      2            4       2 2
          4 l  m  (-l  + 8 m  - Sqrt[l  - 12 l  m ])
         {---------------------------------------------,
                   2         4       2 2  3
                 (l  + Sqrt[l  - 12 l  m ])

             2 2     2      2            4       2 2
          4 l  m  (-l  + 8 m  + Sqrt[l  - 12 l  m ])
          ---------------------------------------------}
                   2         4       2 2  3
                 (l  - Sqrt[l  - 12 l  m ])
```

Part (c) The behavior of the potential depends on the value of **l**. Consider values of **l^2 > 12m^2**. The graphics for the potential for different values of **l** follow from

```
In[107]:= list= {eq3[[2]],List @@ eq3[[2]] } //Flatten

Out[107]=       2       2             2    2
          -2 l  m    l     2 m  -2 l  m  l     -2 m
        {-------- + --- - ---, --------, ---, ----}
              3       2     r       3      2    r
             r       r             r      r

In[108]:=   potplot[L_] :=
            Plot[  list //.{m->1,l->L}  //Evaluate,
                      {r,2.1,50 },
                 Axes->False,Frame-> True,
            PlotLabel-> eq3[[2]],
            FrameLabel ->{"radius","potential","radius","potential"},
            FrameTicks ->Automatic  ,
            PlotStyle->  {Thickness[.006]  ,
                           Dashing[{0.01}],
                           Dashing[{0.02}],
                          Dashing[{0.03}]       },
            Epilog->  Graphics[
            {PointSize[0.02],Map[Point,{{rmax, Vmin},{rmin, Vmax} }]/.
                                             m->1/.l->L}][[1]],
               DisplayFunction->Identity
                     ];
```

m=1 and markings for the extrema points have been introduced via the command **Epilog**. Add to these graphics the text:

```
In[109]:= label=
           Graphics[{
            Text[FontForm[" Unstable ",{"Times-Roman",10}],{ 3,.16}],
            Text[FontForm[" Stable    ",{"Times-Roman",10}],{ 16,-.07}],
            Text[FontForm[  eq3[[2,1]],{"Times-Roman",10}],{ 40, -.1}],
            Text[FontForm[  eq3[[2,2]],{"Times-Roman",10}],{ 20,.1}],
            Text[FontForm[  eq3[[2,3]],{"Times-Roman",10}],{ 13,0}]
          }];
```

Combining the potential and text graphics and letting **L=1.2 Sqrt[12]**, you get

```
In[110]:= Show[potplot[1.3 Sqrt[12]],label,
                PlotRange->{.2,-.2},
                  DisplayFunction->$DisplayFunction];
```

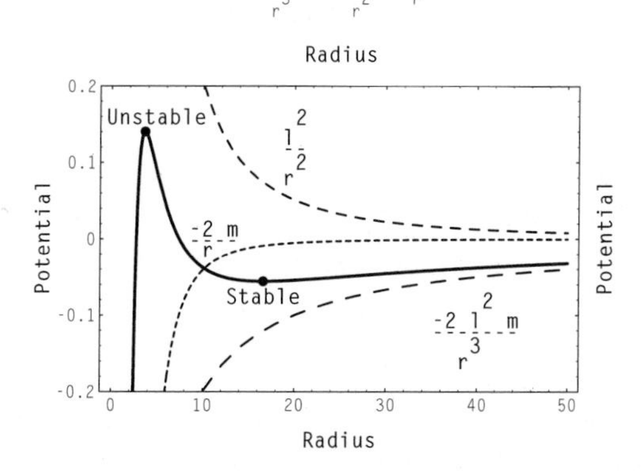

There are two circular orbits: one at the maximum point, the other at the mini point. The circular orbit closest to the origin is unstable; the other is stable. Bound o exist for orbits between the bottom of the well and zero potential. The orbits bet zero potential and the top of the potential curve marked "Unstable" turn around at s finite radius and return toward infinite **r**. The orbits above the "Unstable" mark s to the origin. The potential changes as **l** approaches **Sqrt[12]/m**. At **l=Sqrt[12** the two extrema points merge and there are no extrema points for **l< Sqrt[12]/m**. example, the shape of the potential for **l={0.9 Sqrt[12],Sqrt[12],1.1Sqrt[12]**

```
In[111]:= Plot[  Evaluate[  eq3[[2]] /.m->1/.
                l->{ .9 Sqrt[12],  Sqrt[12],  1.1 Sqrt[12] }  ],
                {r,2.1,15 },
            Axes->False,Frame-> True,
          PlotLabel->" Potential with Different l",
          FrameLabel ->{"Radius","Potential","Radius","Potential"},
          FrameTicks ->Automatic ,
          PlotStyle-> {Dashing[{0.01}]
                      ,Dashing[{0.02}]
                      ,Dashing[{0.03}]},
          PlotRange->{0,-.2 },
          Epilog->
           Graphics[
            {Text[FontForm[l^2>12  m^2//OutputForm,{"Times-Roman",10}]
                      ,{4.5, -.02}],
             Text[FontForm[l^2==12 m^2//OutputForm,{"Times-Roman",10}]
                      ,{5, -.09}],
             Text[FontForm[l^2<12  m^2//OutputForm,{"Times-Roman",10}]
                      ,{5, -.14}]
            }][[1]]
                                                                    ];
```

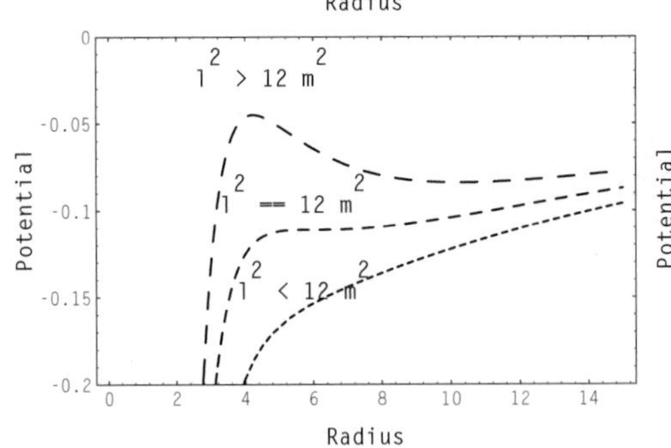

Problem 5: Time it takes to fall into a black hole

Consider the radial Schwartzschild geodesics. Assume the geodesic four-velocity is g by $v = \{t'[s], r'[s], 0, 0\}$, where $E = t'[s](1 - 2m/r)$. Let the initial conditions be $r'[0$ and $r[0] = r_0$.

Part (a) Derive an equation for $r'[s]$ from the timelike condition $1 = v \cdot \eta \cdot v$.

Part (b) Verify from the user-defined function **geodesic** that $\{t'[s], r'[s], 0, 0\}$ satisfi geodesic equations, where $t'[s]$ is given by the first integral and $r'[s]$ is given by the timelike condition derived in Part (a).

Part (c) Derive an expression for the proper time as a function of the radial position.

Part (d) Derive an expression for the coordinate time as a function of the radial position.

Part (e) Plot the proper and coordinate times as a function of r.

Remarks and outline The simplest geodesics are the radial trajectories. The radial solution follows from the Killing constant $E = t'[s](1 - 2m/r)$ and the timelike condition $1 = v \cdot \eta \cdot v$. You should use the user-defined operator **geodesic**, with the condition that the angular variables are constants, to verify that the solution is a geodesic. The calculation for the proper time as a function of the radial coordinate simplifies if you consider the change of variables $r = (r_0 \mathrm{Cos}[x[s]^2)$ and then solve for $x[s]$. The expression for the coordinate time follows from integrating $t'[s]/r'[s]$ over r. Let **en** represent the energy E.

Caution: Not all *Mathematica* versions will select the same order for the roots, so adjust this problem according to your specific *Mathematica* version.

Solution

```
In[112]:= Clear["Global`*"];
```

Part (a) The expressions for **t'[s]** and **t''[s]** are assumed to be given by the following rules:

```
In[113]:= construle={t'[s]-> en/(1-2 m/r[s]),
                          t''[s]->D[en/(1-2 m/r[s]),s]};
```

Define the radial four-velocity and metric:

```
In[114]:= vel = {t'[s],r'[s],0,0};
          met=     (ssMet/.ssComp/.r->r[s] );
```

An equation for **r'[s]** follows from the timelike condition **1==met.vel.vel** and **construle**:

```
In[115]:= eq1= 1==met.vel.vel/.construle

Out[115]=
               2               2
             en             r'[s]
    1 == ----------- - -----------
               2 m            2 m
          1 - ----       1 - ----
              r[s]           r[s]
```

Solving **eq1** for **r'[s]**, you get

```
In[116]:= eq2= Solve[eq1,r'[s]][[1,1]]/.Rule->Equal

Out[116]=
                                        2
                 Sqrt[2 m - r[s] + en  r[s]]
    r'[s] == -(-----------------------------)
                        Sqrt[r[s]]
```

You can express the constant **en** in terms of the initial velocity **r'[0]=0** and initial position **r[0]=r0** by evaluating **eq2** at **s=0**:

```
In[117]:=     eq3=   0== eq2[[2]]/.s->0/.r[0]->r0

Out[117]=                             2
                  Sqrt[2 m - r0 + en  r0]
          0 == -(-----------------------)
                        Sqrt[r0]
```

Solving **eq2** and **eq3** for **r'[s]** and **en**, you get

```
In[118]:= rprimerule=MapAt[Factor,(Solve[{eq2,eq3},{r'[s],en}][[2]]),
                                    {1,2,2,1 }]//PowerExpand

Out[118]=               Sqrt[2] Sqrt[m r0 - m r[s]]                   2 m
          {r'[s] -> -(-----------------------------), en -> Sqrt[1 - ---]}
                          Sqrt[r0] Sqrt[r[s]]                         r0
```

Only those relations that correspond to inward motion have been kept. The radial equation follows

```
In[119]:= radialeq= rprimerule[[1]]/.Rule->Equal

Out[119]=              Sqrt[2] Sqrt[m r0 - m r[s]]
          r'[s] == -(-----------------------------)
                          Sqrt[r0] Sqrt[r[s]]
```

Part (b) Verify that **{t'[s],r'[s]}** satisfy the geodesic equations, where **t'[s]** and **r'[s]** are given by **construle** and **radialeq**. The four-velocity for the timelike radial geodesic follows from **radialrule** and **construle**:

```
In[120]:= geovel= vel/.(radialeq/.Equal->Rule)/.construle

Out[120]=     en          Sqrt[2] Sqrt[m r0 - m r[s]]
          {--------, -(-----------------------------), 0, 0}
                2 m          Sqrt[r0] Sqrt[r[s]]
           1 - ----
               r[s]
```

The geodesic equations follow from the user-defined operator **geodesic**. Applying the **geodesic** operator to the Schwarzschild metric:

```
In[121]:=  metric= ssMet/.ssComp

Out[121]=       2 m                        1                         2
          {{1 - ---, 0, 0, 0}, {0, -(-------), 0, 0}, {0, 0, -r , 0},
                 r                        2 m
                                      1 - ---
                                           r

                          2          2
            {0, 0, 0, -(r  Sin[theta] )}}
```

you get four geodesic equations:

```
In[122]:= geoeq=   geodesic[metric,{t,r,theta,phi}  ];

                 Short[geoeq,4]
```

```
Out[122]= 2 m r'[s] t'[s]          2 m
          {──────────────── + (1 - ────) t''[s] == 0,
                   2               r[s]
               r[s]

                    2
         2 m r'[s]
        ────────────────── + <<2>> == 0, <<2>>,
             2 m  2      2
        (1 - ────)  r[s]
             r[s]

                                   2
        -2 r[s] Sin[theta[s]]  phi'[s] r'[s] -

                                     2
          2 Cos[theta[s]] r[s]  Sin[theta[s]] phi'[s] theta'[s] -

              2                  2
          r[s]  Sin[theta[s]]  phi''[s] == 0}
```

Set the angular variables equal to a constant in the geodesic equations:

```
In[123]:= eq4=   geoeq/. {phi ->(((phi0))&),theta->(((theta0))&)}//
                                                 Simplify
```

```
Out[123]= 2 m r'[s] t'[s]          2 m
          {──────────────── + (1 - ────) t''[s] == 0,
                   2               r[s]
               r[s]

                  2            2
         m r'[s]        m t'[s]     r[s] r''[s]
        ────────────── - ──────── + ──────────── == 0, True, True}
                     2        2     2 m - r[s]
        (-2 m + r[s])     r[s]
```

The final step in the verification follows from eliminating **r'[s]**, **r''[s]**, and **t'[s]** with **construle**, **rprimerule**, and the derivative of **rprimerule**, respectively:

```
In[124]:= eq4/. r''[s]->D[(r'[s]/.rprimerule),s]/.
                          construle/.rprimerule//Simplify
```

```
Out[124]= {True, True, True, True}
```

Part (c) An equation for the proper time as a function of the radial position follows from integrating **1/r'[s]** over **r**. However, the final form of the solution is simpler if you change the variables and then integrate. Define a function **x[s]** by

```
In[125]:= rchange= r->((r0 Cos[x[#]]^2  )&) ;
```

Notice that **x=0** if **r[s]=r0** and **x=Pi/2** if **r[s]=0**. If you express **radialeq** in terms of **x[s]**, you get an equation for **x'[s]**:

```
In[126]:= radialeq /.rchange
```

```
Out[126]= -2 r0 Cos[x[s]] Sin[x[s]] x'[s] ==

                                                 2
            Sqrt[2] Sqrt[m r0 - m r0 Cos[x[s]] ]
          -(------------------------------------)
                                          2
                  Sqrt[r0] Sqrt[r0 Cos[x[s]] ]
```

```
In[127]:=    eq5= ( Solve[radialeq /.rchange,x'[s]][[1,1]]
                     //Simplify
                     //PowerExpand
                  )
```

```
Out[127]=                              2
                  Sqrt[m] Sec[x[s]]
          x'[s] -> -----------------
                              3/2
                     Sqrt[2] r0
```

The proper time in terms of the function of **x** follows from integrating **1/x'[s]** over the variable **x[s]**:

```
In[128]:= Stimex= Integrate[( 1/x'[s]/.eq5),{x[s],0,x} ]//Simplify
```

```
Out[128]=   3/2
          r0    (2 x + Sin[2 x])
          ----------------------
            2 Sqrt[2] Sqrt[m]
```

To express **Stimex** in terms of **r**, you must eliminate **x**. To do this, invert the expression for **r** in **rchange** to get an expression for **x** as a function of **r**:

```
In[129]:=   xrule=     Solve[ r ==(r[s]/.rchange),x[s]][[2]]/.
                                                  x[s]->x
```

```
Out[129]=                  Sqrt[r]
          {x -> ArcCos[--------]}
                       Sqrt[r0]
```

The equation for the proper time as a function of **r** follows from using **xrule** to eliminate **x** in **Stimex**:

```
In[130]:= Stime=Stimex/.xrule
```

```
Out[130]=   3/2              Sqrt[r]                        Sqrt[r]
          r0    (2 ArcCos[--------] + Sin[2 ArcCos[--------]])
                             Sqrt[r0]                       Sqrt[r0]
          ------------------------------------------------------------
                                2 Sqrt[2] Sqrt[m]
```

Part (d) The expression for coordinate time follows from integrating the ratio **t'[s]/r'[s]** over **r[t]**. The expression **t'[s]/r'[s]** follows from **construle** and **rprimerule**:

```
In[131]:= eq6= ( - t'[s]/r'[s] /.construle/.
                            rprimerule/.r[s]->r[t] )//ExpandAll//
                                                 PowerExpand//Simplify
```

```
Out[131]=                                  3/2
                  Sqrt[-2 m + r0] r[t]
          -------------------------------------------
          Sqrt[2] Sqrt[m (r0 - r[t])] (-2 m + r[t])
```

The coordinate time as a function of **r** follows from integrating **eq6** over **r[t]**. It is faster to find the indefinite integral and then evaluate the expression at the limits **{r0,r}**. Integrating **eq6** over **r[t]**, you find that

```
In[132]:= eq7= Integrate[ eq6 , r[t] ]//Simplify;
          eq7 //Short[#,4]&

Out[132]=      3/2
          -2 m     Sqrt[2 m - r0] Log[Sqrt[2] Sqrt[m] - Sqrt[r[t]]]
          ------------------------------------------------------- +
                            Sqrt[m (2 m - r0)]

             3/2
           2 m     Sqrt[2 m - r0] Log[Sqrt[2] Sqrt[m] + Sqrt[r[t]]]
           ------------------------------------------------------- + <<3>> -
                            Sqrt[m (2 m - r0)]

           Sqrt[-2 m + r0] Sqrt[m (r0 - r[t])] Sqrt[r[t]]
           ----------------------------------------------
                            Sqrt[2] m
```

Evaluating the integral at the end points, you get the coordinate time:

```
In[133]:= Ttime=(eq7/.r[t]->r0)-(eq7/.r[t]->r )//Simplify//PowerExpand;
                Short[Ttime,3]

Out[133]= Sqrt[r] Sqrt[-2 m + r0] Sqrt[-r + r0]
          ------------------------------------ + 2 <<2>> + <<8>> -
                   Sqrt[2] Sqrt[m]

                Log[m]                5/2
           2 m (------ + Log[-4 m        Sqrt[r] + <<3>> -
                  2

                Sqrt[2] m (2 m - r0) Sqrt[-2 m + r0] Sqrt[-r + r0]])
```

Only the short form of the solution is shown.

Part (e) Plot the coordinate time and proper time for the values:

```
In[134]:=  value={m->1, r0->8 } ;
```

The graphics for the plots are

```
In[135]:= pt1= Plot[
           Evaluate[(Ttime/.value//Chop[#,10^-5]&)],{r,2.009,7.99},
                                  DisplayFunction->Identity];

           pt2= Plot[Evaluate[Stime/.value ],{r,0,8},
                                  DisplayFunction->Identity];
```

After you combine these two curves and add labels, the coordinate time and proper time graphs are

```
In[136]:=  Show[ {pt1,pt2, Graphics[{Line[{{2,0},{2,40}}],
                 Text[FontForm["Proper Time",
                        {"Times-Roman",10}],{1.2,22}],
                 Text[FontForm["Coordinate Time",
                        {"Times-Roman",10}],{4.5,30}],
                 Text[FontForm["r=2 m ",
                        {"Times-Roman",10}],{2.6,10}]}]},
                 AxesLabel->{"Distance","Time"},
                 DisplayFunction->$DisplayFunction,
                                      PlotRange->{0,40}   ] ;
```

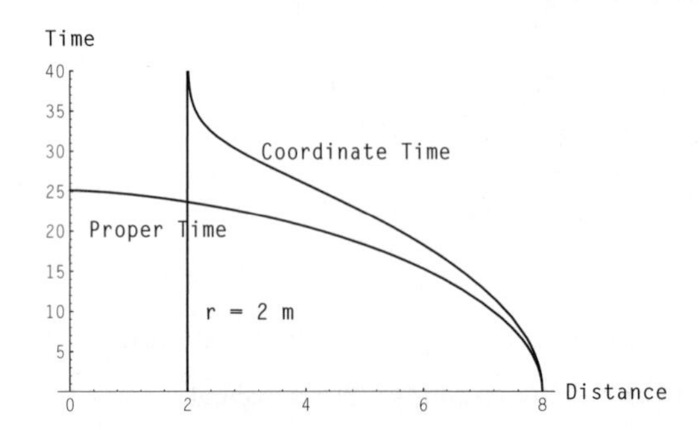

Problem 6: Circular geodesics for the Schwarzschild metric

Consider the timelike circular geodesics for the Schwarzschild metric with four-velocity given by $v = \{t'[s], r'[s], \theta'[s], \phi'[s]\}$. Assume the geodesics are in the plane $\theta[s] = \pi/2$ and the radius is a constant $r[s] = r_0$.

Part (a) Solve for $t[s]$ and $\phi[s]$ with the initial conditions $t[0] = 0$ and $\phi[0] = 0$.

Part (b) Find the proper and coordinate periods for the circular orbit.

Remarks and outline Apply the user-defined operator **geodesic** to the Schwarzschild metric and then assume $r = r_0$ (constant) and $\theta = \pi/2$. The Schwarzschild metric and its components were defined in the Introduction. Solve the resulting equations for $t[s]$ and $\phi[s]$ using the initial conditions $t[0] = 0$ and $r[0] = 0$. Eliminate the variable s and write $t[\phi]$.

Solution

```
In[137]:= Clear["Global`*"];
```

Part (a) For circular orbits in a plane, assume **r[s]** and **theta[s]** are constants:

```
In[138]:= circlerule=  {r ->((( r0))&),theta ->((( Pi/2) )&)} ;
```

Four equations follow from the geodesic equation. Applying the user-defined operator **geodesic** to the **ssMet**, you get

```
In[139]:=  eq1= geodesic[ssMet/.ssComp ,{t,r,theta,phi} ]/.
                                     circlerule//Simplify
```

```
Out[139]=
                                                    2
        2 m                          2     m t'[s]
 {(1 - ----) t''[s] == 0, r0 phi'[s]  -  ---------- == 0, True,
        r0                                    2
                                            r0

      2
  -(r0  phi''[s]) == 0}
```

Solving **eq1[[1]]** for **t[s]** with the initial conditions **t[0]=0**, you get

```
In[140]:= trule=  DSolve[{eq1[[1]],t[0]==0},t,s ]/.C[2]->c1

Out[140]= {{t -> Function[s, s c1]}}
```

Likewise, you solve **eq1[[4]]** for **phi[s]** with the initial conditions **phi[0]=0**:

```
In[141]:= phirule=  DSolve[{eq1[[4]],phi[0]==0},phi,s ]/.C[2]->c2

Out[141]= {{phi -> Function[s, s c2]}}
```

The constants **c1** and **c2** are assumed to be positive. One equation between **c1** and **c2** follows from **eq1[[2]]**:

```
In[142]:= eq2= eq1[[2]] /.trule/.phirule//Flatten
Out[142]=
            2
          c1  m       2
 {-(---------) + c2  r0 == 0}
           2
         r0
```

A second equation follows from the timelike condition **1==met.vel.vel**:

```
In[143]:=
     vel= {t'[s],r'[s], theta'[s],phi'[s]};
     met=    (ssMet/.ssComp/.{theta->theta[s],r->r[s]}  ) ;
    eq3= 1==met.vel.vel/.circlerule/.trule/.phirule//Flatten
Out[143]=
              2      2 m       2   2
 {1 == c1  (1 - ----) - c2  r0 }
                    r0
```

Solving **eq2** and **eq3** for **c1** and **c2**, you get

```
In[144]:= construle=(Solve[{eq3,eq2}/.m->-temp r0//Flatten,
                                   {c1,c2}][[3]]//ExpandAll)/.
                                     temp->- m/r0//PowerExpand
Out[144]=
                     1                          Sqrt[m]
 {c1 -> ----------------, c2 -> -----------------------}
                   3 m                      3 m     3/2
         Sqrt[1 - ----]        Sqrt[1 - ----] r0
                   r0                       r0
```

Only those solutions in which **c1** and **c2** were positive were kept, and **m->-temp r0** was used to reduce the relation to a convenient form. The equations for the coordinate time and angle as a function of **s** are

```
In[145]:= eq4=   CoordT  ==( t[s]/.trule/.construle )[[1]]
          eq5=   Phi ==( phi[s]/.phirule/.construle)[[1]]
```

```
Out[145]=                s
          CoordT == ---------------
                             3 m
                   Sqrt[1 - ---]
                             r0
Out[145]=             Sqrt[m] s
          Phi == -------------------
                          3 m     3/2
                 Sqrt[1 - ---] r0
                          r0
```

Part (b) The coordinate period follows from expressing the coordinate time in terms of the angle **Phi** and setting **Phi=2 Pi**:

```
In[146]:= coordperiod=CoordT/.Solve[{eq4,eq5},CoordT,s][[1]]/.Phi->2Pi
Out[146]=         3/2
          2 Pi r0
          ---------
          Sqrt[m]
```

The proper time follows from setting **Phi= 2 Pi** in **eq5** and solving for **s**:

```
In[147]:= properperiod= MapAt[Apart,s/.Solve[eq5, s][[1]] /.
                                             Phi->2 Pi,{5,1}]
Out[147]=                    3 m    3/2
          2 Pi Sqrt[1 - ---] r0
                         r0
          ---------------------------
                 Sqrt[m]
```

MapAt was used to express the proper period in a convenient form.

Problem 7: Field equations for Robertson-Walker cosmology

Consider the Robertson-Walker metric defined by the diagonal metric

$$\eta_{RW} = \begin{pmatrix} 1 & 0 & 0 & 0 \\ 0 & \frac{-R[t]^2}{(1-\kappa_{RW} r^2)} & 0 & 0 \\ 0 & 0 & -R[t]^2 r^2 & 0 \\ 0 & 0 & 0 & -R[t]^2 r^2 \mathrm{Sin}^2[\theta] \end{pmatrix}.$$

Assume the model is described by the contravariant energy-momentum tensor $T^{ij} = (\rho + P)V^i V^j - g^{ij}(P)$, where $V^i = \{1,0,0,0\}$.

Part (a) From Einstein's equation with a cosmological constant λ, show that $R[t]$ obeys the following two equations:

$$\begin{aligned} R''[t] &= R[t](\lambda - 12G\pi P[t] - 4G\pi\rho[t]))/3 \\ R[t]'^2 &= (-3\kappa_{RW} + \lambda R[t]^2 + 8G\pi R[t]^2 \rho[t])/3 \end{aligned}.$$

Part (b) Assume the pressure and density are related by the equation of state $P[t] = \alpha\rho[t]$. Show from the conservation of energy-momentum $T^{ij}{}_{;j} = 0$ that

$$\rho[t] = \rho[0]\, \frac{R_0^{3(1+\alpha)}}{R[t]^{3(1+\alpha)}}.$$

The symbol ";" stands for the covariant derivative.

Part (c) Let $\alpha = 0$ and show that the equations for $R'[t]^2$ and $R''[t]$ can be written as

$$R''[t] = \frac{\lambda R[t]}{3} - \frac{4G\pi R_0^3 \rho[0]}{3R[t]^2}$$

$$R'[t]^2 = -\kappa_{RW} + \frac{\lambda R[t]^2}{3} + \frac{8G\pi R_0^3 \rho[0]}{3R[t]}.$$

Remarks and outline To derive Einstein's equations, calculate the Christoffel symbols, the curvature tensor, and the Einstein tensor with user-defined procedures. The Einstein tensor gives the left-hand side of the field equations. To get the right-hand side, form the contravariant energy-momentum tensor $T_{\mu\nu} = (\rho[t] + P[t])\, U_\mu U_\nu - \eta_{\mu\nu}^{RW} P[t]$, where u is the covariant four-velocity and η_{RW} is the metric. The Einstein equations with cosmological constant follow from $G_{\mu\nu} = -8\pi G T_{\mu\nu} - \lambda\, \eta_{\mu\nu}^{RW}$, where $G_{\mu\nu}$ is the Einstein tensor. To construct the equations $T^{ij}{}_{;j} = 0$, create a user-defined function that will take the covariant derivative of a contravariant tensor. The remaining calculations follow from using **DSolve**.

Solution

```
In[148]:= Clear["Global`*"];
```

Part (a) The Robertson-Walker metric is

```
In[149]:= RWmet= DiagonalMatrix[
           {1 ,-R[t]^2 / (1-(krw) r^2) ,
                  -R[t]^2 r^2,
                        -R[t]^2 r^2 Sin[theta]^2}];
```

To derive Einstein's equations, first calculate the Christoffel symbols and curvature tensor. The Christoffel symbols follow from the user-defined function **christ** and the curvature tensor follows from the user-defined function **curvR**:

```
In[150]:= RWchrist= christ[RWmet];
          RWcurv  =  curvR[RWchrist ]//Simplify;
           Short[RWcurv,8]

Out[150]= {{{{0, 0, 0, 0}, {0, 0, 0, 0}, {0, 0, 0, 0}, {0, 0, 0, 0}},

               R[t] R''[t]                R[t] R''[t]
      {{0, ---------------, 0, 0}, {---------------, 0, 0, 0}, {0, 0, 0, 0},
                       2                          2
           -1 + krw r                 1 - krw r

                                     2
       {0, 0, 0, 0}}, {{0, 0, -(r  R[t] R''[t]), 0}, {0, 0, 0, 0},

         2
       {r  R[t] R''[t], 0, 0, 0}, {0, 0, 0, 0}},

                     2                2
      {{0, 0, 0, -(r  R[t] Sin[theta]  R''[t])}, {0, 0, 0, 0},

                        2                2
       {0, 0, 0, 0}, {r  R[t] Sin[theta]  R''[t], 0, 0, 0}}}, <<3>>}
```

Only the short form of the curvature components have been displayed. The Einstein tensor follows from the user-defined procedure **einstein**:

```
In[151]:= einsteinT= einstein[RWcurv,RWmet]

Out[151]=
                          2
       -3 (krw + R'[t] )
     {{------------------, 0, 0, 0},
                 2
             R[t]

                      2
        krw + R'[t]  + 2 R[t] R''[t]
      {0, ------------------------------, 0, 0},
                          2
                 1 - krw r

              2               2
      {0, 0, r  (krw + R'[t]  + 2 R[t] R''[t]), 0},

                 2            2                 2
      {0, 0, 0, r  Sin[theta]  (krw + R'[t]  + 2 R[t] R''[t])}}
```

The only nonzero elements of the Einstein tensor are the diagonal components. To get an expression for the stress-energy tensor, first define the contravariant and covariant four-velocity:

```
In[152]:=  vel={1,0,0,0};           (* Contravariant four-velocity*)

          covel= RWmet.{1,0,0,0}; (* Covariant four-velocity*)
```

The expression for the covariant stress-energy tensor is

```
In[153]:=  T=   Table[
           (rho[t]+ press[t])covel[[i]]covel[[j]]-
                                  RWmet[[i,j]]press[t],
                                          {i,4},{j,4}];
```

The four nontrivial Einstein equations with a cosmological constant follow from

```
In[154]:= eq1= Map[Flatten,
                    einsteinT == -8 Pi G T -lambda RWmet]//
                                Thread//Select[#1,FreeQ[#,True]&]&//
                                                 Simplify

Out[154]=
                          2
       -3 (krw + R'[t] )
     {------------------ == -lambda - 8 G Pi rho[t],
                  2
              R[t]

                   2                                                   2
      krw + R'[t]  + 2 R[t] R''[t]      (-lambda + 8 G Pi press[t]) R[t]
      ------------------------------ == ----------------------------------,
                      2                                    2
             1 - krw r                           -1 + krw r
```

```
  2           2
r  (krw + R'[t]  + 2 R[t] R''[t]) ==

  2                                  2
 r  (lambda - 8 G Pi press[t]) R[t]  ,

 2              2            2
r  Sin[theta]  (krw + R'[t]  + 2 R[t] R''[t]) ==

  2                                 2            2
 r  (lambda - 8 G Pi press[t]) R[t]  Sin[theta]  }
```

Solving **eq1** for **R'[t]^2** and **R''[t]**, you get the desired solutions:

```
In[155]:= fieldeq=Solve[eq1/.R'[t]->Sqrt[x],{R''[t],x}][[1]]/.
                        x->R'[t]^2/.  Rule->Equal//Simplify

Out[155]=
                R[t] (lambda - 12 G Pi press[t] - 4 G Pi rho[t])
    {R''[t] == ------------------------------------------------,
                                      3

                                      2                2
           2    -3 krw + lambda R[t]  + 8 G Pi R[t]  rho[t]
     R'[t]  == ---------------------------------------------}
                                      3
```

Part (b) Consider the equations that follow from the conservation equation $T^{ij}{}_{;j} = 0$. The contravariant stress energy tensor is

```
In[156]:=  conT=Table[ (rho[t] + press[t])vel[[i]] vel[[j]] -
                     Inverse[RWmet][[i,j]] press[t] ,
                                 {i,4},{j,4}]//Simplify

Out[156]=
                                              2
                                    (1 - krw r ) press[t]
    {{rho[t], 0, 0, 0}, {0, ---------------------, 0, 0},
                                              2
                                          R[t]

                                                      2
           press[t]                         Csc[theta]  press[t]
     {0, 0, ---------, 0}, {0, 0, 0, ---------------------}}
              2    2                              2    2
             r  R[t]                             r  R[t]
```

Next, create a user-defined function that will return the covariant derivative of a contravariant tensor of rank two:

```
In[157]:=  dconT[conT_,christ_,var_:{t,r,theta,phi}] :=
             Simplify[ExpandAll[Table[
               D[conT[[i,j]],var[[k]]]+
                      Sum[christ[[i,s,k]]conT[[s,j]]+
                                 christ[[j,s,k]]conT[[i,s]],
                                        {s,4}],{i,4},{j,4},{k,4}]]]
```

When **dconT[conT,christ,var]** is applied to a contravariant tensor of rank two given by **conT**, it returns its covariant derivative. The Christoffel symbols are given by **christ** and the coordinates are given by **var**. Spherical coordinates are the default coordinates. Applying **dconT** to **conT**, you get

```
In[158]:= eq2= dconT[conT,RWchrist ] ;
                Short[eq2 ,4]

Out[158]=
                                     (press[t] + rho[t]) R'[t]
          {{{rho'[t], 0, 0, 0}, {0, -------------------------, 0, 0},
                                             R[t]

                  (press[t] + rho[t]) R'[t]
             {0, 0, -------------------------, 0},
                          R[t]

                     (press[t] + rho[t]) R'[t]
             {0, 0, 0, -------------------------}}, <<2>>,
                              R[t]

                      (press[t] + rho[t]) R'[t]
             {{0, 0, 0, -------------------------}, <<2>>,
                              R[t]

                          2
               Csc[theta]  press'[t]
             {-----------------------, 0, 0, 0}}}
                      2     2
                     r  R[t]
```

Only the short form of **eq2** has been shown. Summing over the last two indices, you get $T^{ij}{}_{;j}$,

```
In[159]:= eq3= Table[Sum[eq2[[i,s,s]],  {s,1,4} ] ,{i,1,4}]

Out[159]=  3 (press[t] + rho[t]) R'[t]
          {--------------------------- + rho'[t], 0, 0, 0}
                     R[t]
```

The conservation equations follow from setting the four equations in **eq3** to zero. The only nontrivial equation follows from the first element in the list:

```
In[160]:=    eq4=eq3[[1]]==0

Out[160]= 3 (press[t] + rho[t]) R'[t]
          --------------------------- + rho'[t] == 0
                     R[t]
```

Consider the equation of state defined by the rule

```
In[161]:= eqstate={press[t]->alpha rho[t]};
```

Using **eqstate** to eliminate the pressure in **eq4** and solving for **R[t]**, you can see that

```
In[162]:= eq5= DSolve[{eq4/.eqstate,R[0]==R0 } ,R[t],t][[1,1]]/.
                          Rule->Equal//
                Simplify//PowerExpand//ExpandAll

Out[162]=           3/(3 + 3 alpha) + (3 alpha)/(3 + 3 alpha)
          R[t] == (R0

                                          2                           2
                    3/(3 + 3 alpha)  + (3 alpha)/(3 + 3 alpha)
             rho[0]                                                    ) /

                  1/(3 + 3 alpha)
            rho[t]
```

If you solve **eq5** for **rho[t]**, you get a rule for the behavior of **rho[t]**:

```
In[163]:= rhorule=  Solve[eq5/.rho[t]->x,x][[1,1]]/.x->rho[t] //
                     PowerExpand//Simplify

Out[163]=            3 (1 + alpha)
                  R0                rho[0]
          rho[t] -> ---------------------------
                        3 (1 + alpha)
                     R[t]
```

Part (c) Eliminating the pressure and **rho** in the field equations (**fieldeq**) and setting **alpha** to zero, you get

```
In[164]:= eq7=fieldeq/.rhorule /.eqstate/.alpha->0//
                                       ExpandAll//Simplify

Out[164]=                                    3
                  lambda R[t]   4 G Pi R0  rho[0]
          {R''[t] == ----------- - -----------------,
                       3                  2
                                      3 R[t]

                                 2           3
            2           lambda R[t]   8 G Pi R0  rho[0]
       R'[t]  == -krw + ------------ + ----------------}
                             3            3 R[t]
```

Problem 8: Zero-pressure cosmological models

The equations for zero-pressure Robertson-Walker models can be written as

$$R''[t] = (\lambda R[t])/3 - (4G\pi R[0]^3 \rho[0])/(3R[t]^2)$$

or

$$R'[t]^2 = -\kappa_{RW} + (\lambda R[t]^2)/3 + (8G\pi R[0]^3 \rho[0])/(3R[t]).$$

Parameters more directly related to observations are

1. the Hubble expansion rate, $H_0 = R'[0]/R[0]$;
2. the dimensionless deceleration parameter, $q_0 = -R''[0]/(H_0^2 R[0])$;
3. the dimensionless density parameter, $\sigma = (4G\pi\rho[0])/(3H_0^2)$; and
4. the dimensionless cosmological parameter, $L = \lambda/(3H_0^2)$.

Part (a) Show that q_0 and κ_{RW} satisfy the following equations:

$$q_0 = -L + \sigma$$
$$\kappa_{RW} = -(H_0 R[0])(1 - L - 2\sigma)$$

Part (b) Derive an analytical expression for the cosmological age as a function of q_0 when the cosmological parameter is zero. Plot the results.

Part (c) Consider models with a nonzero cosmological parameter and make a table of ages for different values of σ and q_0.

Remarks and outline Relations for $\{q_0, \kappa_{RW}\}$ follow from evaluating the two field equations at $t = 0$ and expressing the results in terms of q_0, H_0, σ, L. The age of the model follows from integrating $1/R'[t]$ from $R = 0$ (beginning) to $R = R_0$ (now). The integration can be analytically solved for the special case of zero cosmological constant. For the nonzero cosmological constant, use **NIntegrate** to evaluate the ages.

Caution: Not all *Mathematica* versions will select the same roots; adjust this problem according to your *Mathematica* version.

Solution

```
In[165]:= Clear["Global`*"];
```

Part (**a**) The field equations for zero-pressure and nonzero cosmological models are

```
In[166]:= eq1=R''[t] ==
            (lambda R[t])/3 -(4 G Pi R[0]^3 rho[0])/(3 R[t]^2);
          eq2=R'[t]^2 ==
            -krw +(lambda R[t]^2)/3 +(8 G Pi R[0]^3 rho[0])/(3 R[t]);
```

Define the Hubble constant (**Ho**), deceleration parameter (**qo**), dimensionless cosmological parameter (**L**), and dimensionless density (**sigma**) by these rules:

```
In[167]:= parameterrule={R'[0]->Ho R[0],
                         R''[0]->- Ho^2 qo R[0],
                         rho[0]->3 Ho^2 sigma/(4 G Pi),
                         lambda->L 3 Ho^2                 };
```

Relations for **{qo,krw}** follow from evaluating the field equations at **t=0**. Evaluate **eq1** and **eq2** at **t=0** and express the results in terms of **{qo,Ho,sigma,L}**. The relations for **{qo,krw}** follow from the command **Solve**:

```
In[168]:= qrule= Solve[{eq1,eq2}/.t->0/.parameterrule,
                                 {qo,krw}]//Flatten//Simplify

Out[168]=
                                    2                         2
          {qo -> -L + sigma, krw -> Ho  (-1 + L + 2 sigma) R[0] }
```

Part (**b**) The ages follow from **eq2**. Simplify the notation in **eq2** by defining **X[t]** with the rule,

```
In[169]:= change=    {R'[t]->X'[t] R[0],R[t]->X[t] R[0]};
```

The variable **X[t]** is more convenient than **R[t]** because the present time occurs at **X=1**. Expressing **eq2** in terms of **X**, you get

```
In[170]:= (eq2/.parameterrule/.qrule/.change//
                         Solve[#,X'[t]]&)

Out[170]=
                                                2 sigma          2
          {{X'[t] -> -(Ho Sqrt[1 - L - 2 sigma + ------- + L X[t] ])},
                                                 X[t]

                                              2 sigma          2
            {X'[t] -> Ho Sqrt[1 - L - 2 sigma + ------- + L X[t] ]}}
                                               X[t]
```

Caution: At this step you want the positive root; different *Mathematica* versions order the roots differently.

```
In[171]:= eq4= {X'[t]->
              Ho (1 - L - 2 sigma + (2 sigma)/X[t] + L X[t]^2)^(1/2)};
```

The ages follow from integrating **1/X'[t]** from **X=0** (beginning) to **X=1** (now). If you set **Ho=1**, then the time is in units of **1/Ho** (Hubble time units). The integration can be solved analytically for the special case of zero cosmological constant. Setting **L=0** and **Ho=1** and integrating **1/X'[t]** from **X=0** to **X=1**, the age in Hubble units follows from **eq4**:

```
In[172]:= integrand=( 1/X'[t] /.eq4
                               /.{L->0, Ho->1, sigma->qo}
                               //Together
                     )

Out[172]=                 1
          ----------------------------------
               2 qo + X[t] - 2 qo X[t]
          Sqrt[-----------------------]
                        X[t]

In[173]:= ageo=Integrate[ integrand,{X[t],0,1}] //Together

Out[173]= -Sqrt[1 - 2 qo] + qo Log[1 + Sqrt[1 - 2 qo] - qo] - qo Log[qo]
          --------------------------------------------------------------
                          Sqrt[1 - 2 qo] (-1 + 2 qo)
```

where the fact that **sigma=qo** when **L=0** is used. Display the results by plotting the age as a function of **qo**. The labels for the curves are given by

```
In[174]:= label=  Graphics[
           {PointSize[.03],Point[{.5,Evaluate[Limit[ageo,qo->.5]]}],
           Text[FontForm["k=0,Flat",{"Times-Roman",10}],{.73,.67}],
           Text[FontForm["k=1,Closed",{"Times-Roman",10}],{1.4,.58}],
           Text[FontForm["k=-1,Open",{"Times-Roman",10}],{.36,.82}]
                                                                    }];
```

Define a plot function to graph the age as a function of **qo** for a given Hubble constant:

```
In[175]:= ageplot[Ho_] :=Plot[ ageo ,{qo,.00001,2},
               Frame-> True,
            PlotLabel-> "Age for zero cosmological constant",
            FrameLabel ->{"10^-29 gm/cc","Age",
                               "Density","Billion Years"},
            FrameTicks ->
            { {0,{1/2, .000196 Ho^2},
                {1,2* .000196 Ho^2   }, {1.5,3*.000196 Ho^2 } },
            { {.5,.5*9.78*(100/Ho) }, {.6,.6*9.78*(100/Ho)},
                {.7,.7*9.78*(100/Ho) }, {.8,.8*9.78*(100/Ho)},
                {.9,.9*9.78*(100/Ho) }, {1,1*9.78*(100/Ho)  }  }},
            GridLines->Automatic,Epilog->label[[1]]              ]
```

The labels were included with the option **Epilog**. The input **Ho** converts the time scale to units of billions of years. Displaying the results for a particular choice of **Ho**, you get

```
In[176]:= ageplot[50];
```

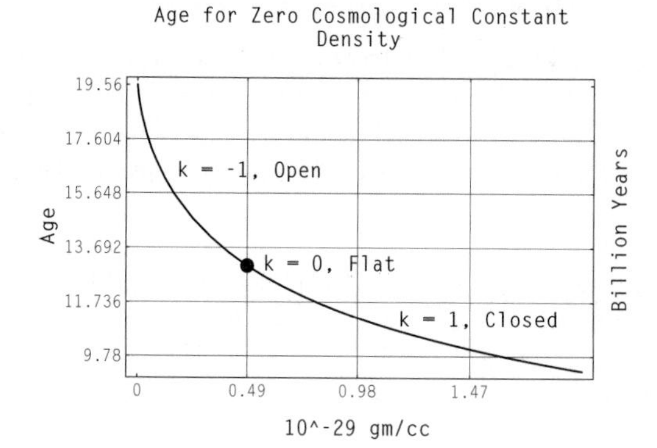

Part (c) Consider the cosmological age for models with nonzero **lambda**. The age can no longer be expressed in terms of elementary functions, so numerically evaluate the ages for different **sigma** and **qo**. The age follows from integrating **1/X'[t]** from **X=0** to **X=1**. Use **NIntegrate** to evaluate the results that follow from **eq4**:

```
In[177]:= ageL[q_,dimmass_] :=  NIntegrate[1/X'[t]/.eq4/.Ho->1/.
     L-> dimmass-q/.qo->q/.sigma->dimmass/.
                               X[t]->x,{x ,0,1} ]//N[#,2]&
```

The age is a function of the deceleration parameter called **q** and the dimensionless density (**sigma**) called **dimmass**. Instead of just evaluating the age for one set of parameters, form a list of ages for different decelerations and dimensionless densities. Also, convert the age from Hubble units to billions of years. The table of age values for different Hubble constants follows from

```
In[178]:= agetable[Ho_]:=Table[  N[ 9.78*(100/Ho) ageL[q,sigma],3],
                                  {q,0,2 ,.5},{sigma,0,2.5,.5}];
```

Displaying the result with **TableForm** and a particular value of **Ho**, you get (this takes a while)

```
In[179]:=  TableForm[ agetable[50 ]   ,TableSpacing -> {0, 3,3}   ,
  TableHeadings->{{"qo=0","qo=.5","qo=1",
                            "qo=1.5","qo=2"  },
  {"sigma=0"," =.5"," =1",   " =1.5   "," =2 ",  " =2.5"} }]
```

Out[179]=

	sigma=0	=.5	=1	=1.5	=2	=2.5
qo=0	19.6	14.	12.5	11.5	10.8	10.3
qo=.5	17.	13.	11.8	10.9	10.3	9.88
qo=1	15.4	12.3	11.2	10.5	9.93	9.51
qo=1.5	14.2	11.6	10.7	10.	9.57	9.19
qo=2	13.2	11.1	10.2	9.68	9.25	8.9

Problem 9: The expansion and age of the standard model

The standard cosmological models are Robertson-Walker models with zero pressure and zero cosmological parameter. The scale factor obeys the equation $R'[t]^2 = -\kappa_{RW} + 2q_0 H_0^2 R[0]^3/(R[t])$, where $R[0] = 1/H_0\sqrt{\kappa_{RW}/(2q_0 - 1)}$ for $\kappa_{RW} \neq 0$.

Part (a) Find the solution for **R[t]** and age for the model when $\kappa_{RW} = 0$. Graph the results.

Part (b) Find the solution for **R[t]** and age for the model when $\kappa_{RW} = 1$. Graph the results.

Part (c) Find the solution for **R[t]** and age for the model when $\kappa_{RW} = -1$. Graph the results.

Remarks and outline The solution for $R[t]$ follows from applying **DSolve** to the scale factor equation. For $\kappa_{RW} = 1$ or -1, the solution is long. It is more transparent to change the variables and define the solution in parametric form. The graphics follow from **Plot** for the case $\kappa_{RW} = 0$ and from **ParametricPlot** for the cases $\kappa_{RW} = 1$ and -1.

Solution

```
In[180]:= Clear["Global`*"];
```

Part (a) The equation for the scale factor **R[t]** when **krw=0** is

```
In[181]:=  eq1=R'[t]^2 ==  Ho^2 R[0]^3/( R[t])

Out[181]=
                      2    3
               2    Ho  R[0]
          R'[t]  == ---------
                      R[t]
```

A solution for **t** follows directly from integrating **1/R'[t]** over **R**:

```
In[182]:= eq2=  t== Integrate[ Sqrt[ 1/eq1[[2]] ],R[t]]//
                                          PowerExpand

Out[182]=
                      3/2
                2 R[t]
          t == -----------
                        3/2
               3 Ho R[0]
```

The age for the model follows from setting **R[t]=R[0]** in **eq2**:

```
In[183]:=          eq2[[2]]/.R[t]->R[0]

Out[183]=  2
          ----
          3 Ho
```

The expression for **R[t]** follows from solving **eq2** for **R[t]**:

```
In[184]:= scale0=   Solve[ eq2 , R[t]][[1,1]]/.Rule->Equal

Out[184]=
                     3 2/3    2/3  2/3
          R[t] == (-)      Ho     t     R[0]
                     2
```

Use **Plot** to graph **R[t]**:

```
In[185]:= Plot[Evaluate[scale0[[2]]/.{Ho->1,R[0]->1}],{t,0,  1},

          Frame-> True,
            PlotLabel-> scale0,
             FrameLabel ->{"Time (Hubble units)","R[t]"," "," "},
             FrameTicks -> { {{0, "Big Bang"},{1/3,"1/(3Ho)"},
                              {2/3,"Now "},{1,"1/Ho"}}   ,
                              {0,{1/2," R[0]/2 "},{1,"R[0]"}}},
             Epilog->Graphics[{PointSize[.03],
                                          Point[{2/3,1}]}][[1]]];
```

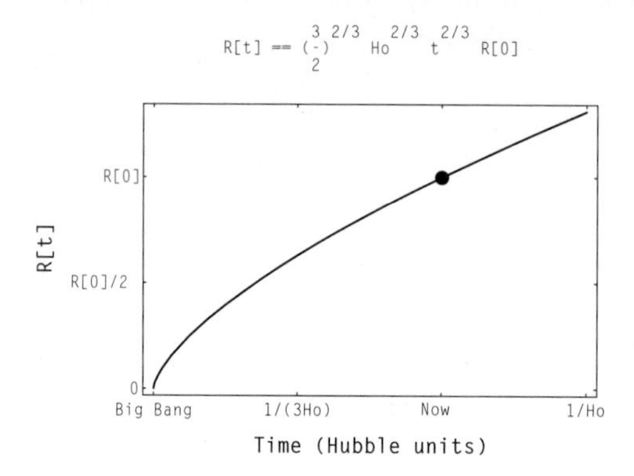

Part (b) Consider the closed model with **krw=1** (**qo>1/2**). The equation for the scale factor **R[t]** is

```
In[186]:= eq3=R'[t]^2 == -1+ 2 qo Ho^2 R[0]^3/(  R[t])    ;
```

For **krw=1**, the value of **R[0]** expressed in terms of **qo** is

```
In[187]:= R0rule=R[0] -> 1/Ho Sqrt[ 1/ ( 2 qo-1)  ];
```

Eliminating **R[0]** in **eq3**, you get

```
In[188]:= eq4=eq3/.R0rule //PowerExpand
Out[188]=
             2                    2 qo
          R'[t]  == -1 + ---------------------------
                                          3/2
                         Ho (-1 + 2 qo)      R[t]
```

A solution for **t** follows directly from **eq4** by integrating **1/R'[t]** over **R[t]**. However, do not solve **eq4** this way; the solution is simpler to obtain if you express **R[t]** in parametric form. Define **R** in terms of the parameter **x** with the rule

```
In[189]:= change=R->((2 qo/(Ho(2qo-1)^(3/2))Sin[x[#]]^2)&);
```

The parametric representation for **R[x]** is

```
In[190]:= scale1=R[x] /.change/.x[x]->x//PowerExpand
Out[190]=
                       2
            2 qo Sin[x]
          ----------------
                         3/2
          Ho (-1 + 2 qo)
```

An equation for **x'[t]** follows from using **change** to eliminate **R** in **eq4**:

```
In[191]:= eq5= Solve[eq4/.change,x'[t]][[2]]//
                                   Simplify//PowerExpand

Out[191]=
                            3/2          2
                 Ho (-1 + 2 qo)    Csc[x[t]]
        {x'[t] -> ----------------------------}
                            4 qo
```

The expression **t[x]** follows from integrating **1/x'[t]** over **x[t]**:

```
In[192]:= time1= Integrate[(1/x'[t]/.eq5),{x[t],0,x}]//
                                   PowerExpand//Simplify

Out[192]= qo (2 x - Sin[2 x])
          -------------------
                          3/2
           Ho (-1 + 2 qo)
```

The model starts at **x=0** (**t=0**) and completes the cycle at the Big Crunch, when **x=Pi** (**R=0**). To find the age, first find the value of **x** when **scale1=R[0]**:

```
In[193]:=  Solve[R[0]==scale1,x] /.R0rule//ExpandAll//PowerExpand

Out[193]=
                      Sqrt[-1 + 2 qo]                   Sqrt[-1 + 2 qo]
        {{x -> -ArcSin[-----------------]}, {x -> ArcSin[-----------------]}}
                      Sqrt[2] Sqrt[qo]                  Sqrt[2] Sqrt[qo]
```

Caution: Take the positive root; different *Mathematica* versions order the roots differently.

Picking the positive root, you get

```
In[194]:= xage= {x -> ArcSin[( (-1 + 2*qo)^(1/2))/
                             (2^(1/2)*qo^(1/2))]}

Out[194]=
                       Sqrt[-1 + 2 qo]
        {x -> ArcSin[-----------------]}
                       Sqrt[2] Sqrt[qo]
```

When **x=xage**, you get the current values of **R[x]** and **t[x]**. The age for the model follows from evaluating **time1** at **xage**:

```
In[195]:=   age1=time1/.xage//ExpandAll//Simplify

Out[195]=
                       Sqrt[-1 + 2 qo]                       Sqrt[-1 + 2 qo]
        qo (2 ArcSin[-----------------] - Sin[2 ArcSin[-----------------]])
                       Sqrt[2] Sqrt[qo]                      Sqrt[2] Sqrt[qo]
        ------------------------------------------------------------------------
                                                    3/2
                                     Ho (-1 + 2 qo)
```

Plot **R[t]** for the particular values of **qo** and **Ho** given by

```
In[196]:= value={ qo->1, Ho->1};
```

The plot for **R** follows from applying **ParametricPlot** to **{t[x],R[x]}**. Add to these graphics a point for the current value of **{t[x],R[x]}** and a point at maximum expansion. The graphics for these points along with the text are given by **label**:

```
In[197]:= label= Graphics[{PointSize[.03],
            Text[FontForm[
             (time1/.xage/.value//N[#,2]&)"/Ho==age",
                          {"Times-Roman",10}],{2,1}],
            Point[{time1/.x->Pi/2/.value,
                              scale1/.x->Pi/2/.value}],
            Point[{time1/.xage/.value,
                              scale1/.xage/.value}] }] ;
```

Applying **ParametricPlot** to **{time1,scale1}**, you get the plot for the closed model:

```
In[198]:= ParametricPlot[Evaluate[{time1,scale1}/.value],{x,0,  Pi},
           Frame-> True,
            PlotLabel->  "Closed Model, qo=1 " ,
            FrameLabel ->{" t " ,"R[t]"," "," "},
              FrameTicks ->
                {{{0,"Bang"},{age1/.value//N,"age"},
                  {3 age1/.value//N,"3* age"},
                  {time1/.value/.x->Pi/2,"Max"},
                  {9 age1/.value//N,"9* age"},
                  {time1/.value/.x->Pi,"Crunch"}}   ,
                {0,1/2,{1, "R[0]"},1.5,2 }},
                                  Epilog->label[[1]]]   ;
```

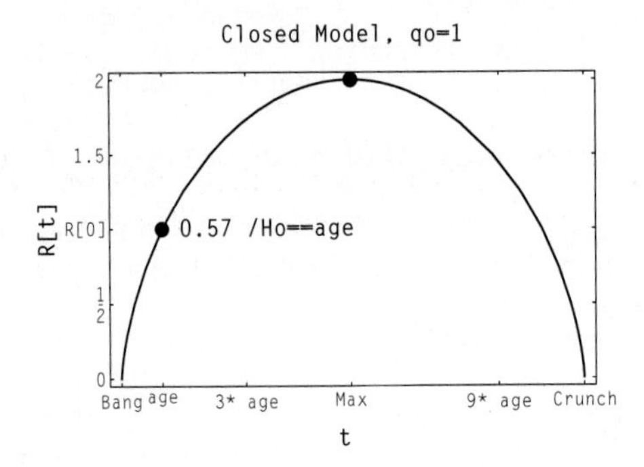

The time has been expressed in units of the current age, and the current age is expressed in units of the Hubble age.

Part (c) Consider the closed model with **krw=-1** (**qo<1/2**). The equation for the scale factor **R[t]** is

```
In[199]:=  eq6=R'[t]^2 ==  1+ 2 qo Ho^2 R[0]^3/(  R[t]);
```

For **krw=-1**, the value of **R[0]** in terms of **qo** and **Ho** is

```
In[200]:=  R0rule=R[0] -> 1/Ho Sqrt[ 1/ (1- 2 qo )];
```

Eliminating **R[0]** in **eq6**, you get

```
In[201]:= eq7=eq6/.R0rule//PowerExpand
Out[201]=       2                      2 qo
          R'[t]   == 1 + --------------------------------
                                       3/2
                         Ho (1 - 2 qo)      R[t]
```

Solve **eq7** in the same way that you solved for **R[t]** in Part (b). Express **R** in terms of the parameter **x**:

```
In[202]:=  change=R->((2 qo/(Ho(1-2 qo)^(3/2))Sinh[x[#]]^2)&);
```

The parametric representation for **R[x]** is

```
In[203]:= scale2=R[x] /.change/.x[x]->x//PowerExpand

Out[203]=
                   2
  2 qo Sinh[x]
 --------------
            3/2
 Ho (1 - 2 qo)
```

Next, change variables in **eq7** and get an equation for **x'[t]**:

```
In[204]:= eq8= Solve[eq7/.change,x'[t]][[2]]//
                                         Simplify//PowerExpand

Out[204]=
                             3/2            2
               Ho (1 - 2 qo)     Csch[x[t]]
 {x'[t] -> ----------------------------------}
                           4 qo
```

The expression for **t[x]** follows from integrating **1/x'[t]** over **x[t]**:

```
In[205]:= time2= Integrate[(1/x'[t]/.eq8),{x[t],0,x}]//
                                         PowerExpand//Simplify

Out[205]=
        qo (2 x - Sinh[2 x])
 ---------------------------------
 Sqrt[1 - 2 qo] (-Ho + 2 Ho qo)
```

The parametric expression for **t** is given by **time2** and the parametric expression for **R** follows from **scale2**. The model starts at **x=0** (**t=0**), evolves to the present at **R[x]=R[0]**, and then keeps expanding. To find the age, first find the value of **x** when **scale2=R[0]**:

```
In[206]:= (  Solve[R[0]==scale2,x]
              /.R0rule
              //PowerExpand
              //ExpandAll
              //PowerExpand
           )

Out[206]=
                        Sqrt[1 - 2 qo]                            Sqrt[1 - 2 qo]
 {{x -> -ArcSinh[---------------------]}, {x -> ArcSinh[---------------------]}}
                     Sqrt[2] Sqrt[qo]                          Sqrt[2] Sqrt[qo]
```

Caution: Take the positive root; different *Mathematica* versions order the roots differently.

Picking the positive root, you get

```
In[207]:= xage= {x -> ArcSinh[(1 - 2 qo)^(1/2)/(2^(1/2) qo^(1/2))]} ;
```

The age for the model follows from

```
In[208]:=  age2= time2/.xage

Out[208]=
                  Sqrt[1 - 2 qo]                              Sqrt[1 - 2 qo]
 qo (2 ArcSinh[---------------------] - Sinh[2 ArcSinh[---------------------]])
                Sqrt[2] Sqrt[qo]                           Sqrt[2] Sqrt[qo]
 -----------------------------------------------------------------------------
                      Sqrt[1 - 2 qo] (-Ho + 2 Ho qo)
```

Plot **R[t]** for the particular values of **qo** and **Ho** given by

```
In[209]:= value={ qo->.02 , Ho->1};
```

The plot for **R** follows from applying **ParametricPlot** to **{t[x],R[x]}**. Add to these graphics a point for the current value of **{t[x],R[x]}**. The graphics for this point along with the text are given by **label**:

```
In[210]:= label= Graphics[{PointSize[.03],
            Text[FontForm[
                  (time2/.xage/.value//N[#,2]& )"/Ho==age",
                                                 {"Times-Roman",10}],
             {1.5 time2/.xage/.value,scale2/.xage/.value}],
            Point[{time2/.xage/.value,scale2/.xage/.value}]}];
```

Applying **ParametricPlot** to **{time2,scale2}**, you get the plot for the open model:

```
In[211]:= ParametricPlot[Evaluate[{time2,scale2}/.value],
                                                      {x,0,3 },
           Frame-> True,
            PlotLabel-> "Open Model, qo=.02" ,
            FrameLabel ->{" t " ,"R[t]"," "," "},
              FrameTicks ->
           {{{0, " Bang"},{.5 age1/.value//N ,"age/2"},
             {age1/.value//N ,"age"},{2age1/.value//N,"2age"},
             {3/2 age1/.value//N ,"3/2 age"}},
                                  { 0,1/2,{1, "R[0]"},1.5,2 }},
                                                 Epilog->label[[1]]];
```

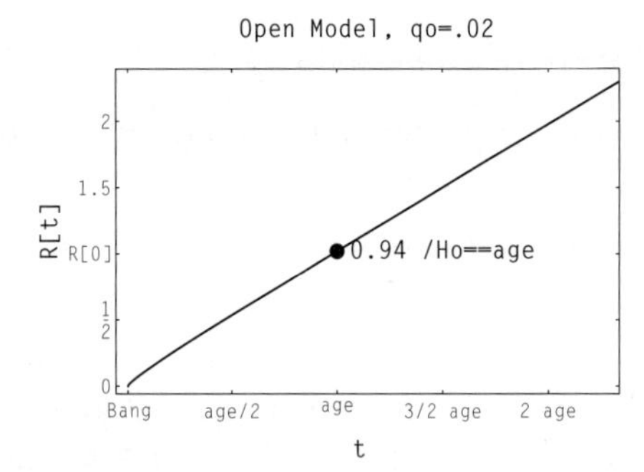

The time has been expressed in units of the current age, which is expressed in units of the Hubble age.

7.3 Unsolved Problems

Exercise 7.1: User-defined covariant derivatives

Construct user-defined procedures for taking the covariant derivatives of the following tensors:

1. Covariant tensor of rank 1
2. Contravariant tensor of rank 1
3. Covariant tensor of rank 2
4. Contravariant tensor of rank 2

5. Mixed tensor of rank 2
6. A mixed tensor of arbitrary rank

Exercice 7.2: User-defined geometric procedures

Construct user-defined procedures for calculating the Christoffel symbols and curvature tensor. Use their algebraic symmetries and make the programs as efficient as possible. Use the command **`Timing`** to compare the speeds of your programs with those of the user-defined programs listed in the Introduction.

Exercise 7.3: Lorentz boosts

Consider the problems on Lorentz transformations.

Part (a) Show that two successive Lorentz boosts in the same direction are equivalent to a single Lorentz boost. Find the formula for the addition of velocities that follows from the product of two Lorentz boosts.

Part (b) The electromagnetic three-vector fields E and B are elements of a second-rank contravariant tensor F^{ij}. In frame S, the second-rank tensor is defined as

$$F^{ij} = \begin{pmatrix} 0 & -E_x & -E_y & -E_z \\ E_x & 0 & -B_z & B_y \\ E_y & B_z & 0 & -B_x \\ E_z & -B_y & B_x & 0 \end{pmatrix}.$$

In a boosted reference frame, the three-vector fields are E' and B'. Find the relation between the electric and magnetic fields in frames S and S'.

Exercise 7.4: Schwarzschild solution

Consider spherical coordinates $\{t, r, \theta, \phi\}$ and assume the metric is diagonal with components given by $\{g_{00}[r], g_{11}[r], -r^2, -r^2\text{Sin}^2[\theta]\}$. Find the vacuum solutions of Einstein's field equations. Assume the boundary conditions $g_{11} \to -1 - 2m/r$ and $g_{00}[r] \to 1 - 2m/r$ as r goes to infinity.

Part (a) Set the components of Einstein's tensor to zero and solve for the metric components.

Part (b) Set the components of the Ricci tensor to zero and solve for the metric components.

Part (c) Verify that all components of the Ricci and Einstein tensors are zero.

Exercise 7.5: Null geodesics for the Schwarzschild metric

Consider null geodesics for the Schwarzschild metric.

Part (a) Find the effective potential and extremum points of the potential.

Part (b) Plot the potential and discuss the possible orbits. Also make a phase plot around the stable and unstable equilibrium points.

Part (**c**) Solve the geodesic equations for each type of orbit.

Exercise 7.6: Elliptical timelike orbits for the Schwarzschild metric

Consider the timelike geodesics for the Schwarzschild metric. Assume the orbit is in the $\theta = \pi/2$ plane. Consider the bound orbits.

Part (**a**) Use perturbation theory to show that the bound orbits are precessing ellipses.

Part (**b**) Make a table for the angles of precession for all the planets.

Exercise 7.7: Circular geodesics for the Kerr metric

Consider the timelike circular geodesics for the Kerr metric. Assume the geodesics are in the plane $\theta = \pi/2$ and r_0 is the radius of the orbit. Solve for the proper and coordinate periods of the circular orbit. Be sure to consider both clockwise and counterclockwise motions.

Index

/., 4, 19
//., 4, 19
/:, 9
/;, 8, 55
:>, 38
?, 10
#, 2, 14
$DisplayFunction, 235, 319
&, 2, 14
^:=, 9
!!, 11

Abs, 71, 363
Algebra`SymbolicSum`, 265
Algebra`Trigonometry`, 10
And, 286
animate, 21, 62, 93, 134, 170, 328, 335
animate, 15
animation, 114, 120, 138
animation for linear motion, 93
Apart, 184
Append, 13
Array, 126
Attributes, 12
Automatic, 29
Axes, 56, 76, 96
AxesLabel, 46, 51, 56, 60, 68, 74, 96
AxesOrigin, 326

BesselJ, 7, 19, 261, 350, 354
BesselY, 261, 350, 354
boost, 373
boost along the x-axis, 376
boostx, 376, 390
BoxRatios, 45, 46

canonical momentum, 211
Canonical transformations, 145
Cartesian, 54, 309, 311, 324, 331, 336, 341
center-of-mass frame, 385
centrifugal, 42
centripetal forces, 24
change of variables procedure, 305
Chop, 357, 359
christ, 376, 398, 417
Christoffel symbols, 373, 376, 417
Clear, 11, 12
Coefficient, 338
CoefficientList, 44, 86, 156, 280
Collect, 3, 39, 44, 73, 160, 332, 337, 348
Complex, 306
complex conjugate, 5
complex conjugate rule, 306
ComplexToHyperbolic, 147, 163, 306
ComplexToTrig, 13, 330
Conjugate, 5, 8, 9
conjugate, 306
conjugateRule, 5
conservative force, 24
Context, 11, 61
context, 16, 17
Contexts, 11
ContourPlot, 57, 60, 96, 111, 235, 244, 251, 253, 273, 298
Contours, 60
ContourShading, 57, 96
ContourSmoothing, 57
Convert, 27
CoordinatesFromCartesian, 253, 282, 287, 291, 298
CoordinatesToCartesian, 7, 62, 185, 221, 246
coriolis, 24, 42
Courier-Bold, 313, 319, 325, 332, 357, 359, 394
CourierB, 33, 67, 73
covariant derivative, 419
covariant equations, 373
CrossProduct, 43, 49
Curl, 63–65
curvature tensor, 373, 377
curvR, 377, 399, 417
Cylindrical, 65, 364, 365
CylindricalPlot3D, 278, 290, 293–295, 297

Dashing, 14, 70, 74
dconT, 419
declare, 9
DensityPlot, 96
derivative, 13, 29
Det, 126
DiagonalMatrix, 78, 417
differential equation, 20
diffSeriesOne, 85, 86, 98, 153, 191, 216, 337
dimensionless cosmological parameter, 421
dimensionless deceleration, 421
dimensionless density, 421
dipole, 26, 60, 233
DisplayFunction, 235, 319, 322
Do, 15, 62, 138, 327, 328, 335
Doppler shift, 390
dot product, 78

double brackets, 6
double pendulum, 135
doublePlot, 89, 103, 108, 113, 116, 118, 120, 123
Drop, 65
DSolve, 13, 14, 28, 29, 32, 38, 51, 66, 69, 98, 100, 105, 259, 261
Duffing equation, 109
E, 2
EarthMass, 27
EarthRadius, 27
eigenfrequencies, 131, 136, 140
eigenfrequency, 126
Eigensystem, 141
Eigenvalues, 141
eigenvalues, 90, 140
eigenvalues and eigenvectors for small oscillating systems, 90
eigenvector, 126
Eigenvectors, 141
eigenvectors, 90, 131, 136
einstein, 380, 418
Einstein tensor, 374, 380, 418
electric field, 24
electric field lines, 234
electric potential, 24
ElectronCharge, 76
ElectronMass, 76
Eliminate, 20, 390
Epilog, 74, 319, 325, 340, 394, 407, 408, 423, 426, 430
Equal, 13, 50, 337, 348
equilibrium, 83
equipotential, 60
equipotential lines, 234
Evaluate, 14, 313, 332, 333, 335, 340, 351, 352, 357, 359, 363, 397, 408
Expand, 3
FaceGrids, 46, 68, 74
Factor, 3, 4, 410
fftSpectrum, 89, 117, 123
FilledPlot, 95
FindMinimum, 248
FindRoot, 20, 320, 351, 356–359
First, 8
first-order perturbation solution, 155
firstDiffSeries, 153, 208, 213
firstOrderPert, 155, 194, 205, 209
Fit, 81
Fit, 25, 35
flat spacetime, 373
Flatten, 13
flux, 304, 309
flux, 309, 324, 331
FontForm, 33, 46, 67, 73, 313, 319, 325, 326, 332, 352, 357, 359, 394, 407, 408, 414, 428, 430
forced duffing oscillator, 114
four-vector, 78
Fourier spectrum, 117, 123
fourier spectrum of a one-dimensional oscillating system, 89
Fourier transform, 70, 81, 89
Frame, 30, 33, 70, 74
FrameLabel, 33, 34, 70, 74
FrameTicks, 318, 319, 423, 426, 428, 430
FreeQ, 260, 317, 342, 346, 365, 399, 418
FullForm, 5
Function, 38
Galilean transformation, 372
Gamma, 362
Gauss's theorem, 244
GaussianIntegers, 3
general physics, 23
general relativity, 374
generalized coordinates, 143
generalized Doppler formula, 389
generalized force, 144
generalized momentum, 144
generalized Snell's law, 389
geodesic, 374
geodesic, 381, 410, 414
Geodesic equations, 381
Global, 12
Global`, 11
Gödel metric, 378
Grad, 54, 62, 244, 251, 255, 282, 287
gradient, 54, 62
Graphics, 33, 39, 334, 352, 359
Graphics3D, 343
Graphics`PlotField`, 16
Graphics`PlotField3D` ScaleFunction, 61
Graphics`PlotField` ScaleFunction, 16, 62
GraphicsArray, 89, 111, 112, 115, 119, 128, 133, 322, 326, 363, 364, 397
Grapics, 34
GravitationalConstant, 27
gravity wave, 380
GrayLevel, 55, 248
GridLines, 29, 33, 34, 40, 70, 74
Hamilton, 148, 167, 207, 208, 215, 219
Hamilton characteristic functions, 222
Hamilton's canonical equations, 145
Hamilton's characteristic functions, 146
Hamilton's equations, 148, 207, 211
Hamilton's principal function, 146, 219
Hamilton-Jacobi, 143, 145
Hamilton-Jacobi equation, 146, 219, 222
Hamilton-Jacobi equations, 151
Hamiltonian, 144
hamiltonian, 309, 312
Hamiltonian operator, 304
Hamiltonians, 143
HamiltonJacobi, 151, 222, 227
harmonic oscillator, 13, 140
Hermite polynomials, 307, 338
Hermite solution, 307
HermiteH, 19, 339, 343
Hold, 220, 223
holonomic, 143
Hubble expansion rate, 421
Hue, 170
HyperbolicToComplex, 147, 306, 355
Hypergeometric1F1, 360, 361
HypergeometricU, 360, 361
Identity, 235
If, 57, 283
Im, 326
induced charge, 251
induced charge density, 244, 247
Infinity, 32, 54, 62
IntegerQ, 9, 307, 339, 361
Integrate, 12, 56, 64
integrate, 12, 20
integration, 20
InterpolatingFunction, 14, 112, 115, 119, 122, 169
Inverse, 401

InverseLaplaceTransform, 72, 99

Join, 14

kChrist, 383
kComp, 382
Kerr metric, 382, 404
Killing's equation, 402
Killing's equations, 379, 404
killingEq, 379, 403, 404
kMet, 382

laboratory frame, 385
Lag, 147, 159, 160, 162, 166, 172, 176, 180, 186, 200
Lagrange multipliers, 144
Lagrange's equations, 144, 147
Lagrangian, 144
Lagrangian multiplier, 160, 172
Lagrangian multipliers, 177
Lagrangians, 143
LaguerreL, 19, 361, 362
Laplace transform, 72
Laplace transforms, 140
Laplace's equation, 230
LaplaceTransform, 99
Laplacian, 250, 258
least-squares fit, 25
leastSquares, 25, 36, 81
Legendre polynomial, 241
Legendre polynomials, 232, 238, 244, 307
Legendre solution, 307
LegendreP, 19, 307, 308, 348
level, 4
Limit, 39, 47, 56, 66, 177
limit, 32, 54, 64
limiting cycle, 118
Line, 359, 414
line element, 372, 373
ListPlot, 29, 30, 35, 70, 71, 134
Log, 18
LogicalExpand, 285
LogicExpand, 286
Lorentz transformations, 372, 390

magnetic induction, 24
magnetic vector potential, 24
Map, 3, 44, 73, 160, 330, 332, 337, 348, 351, 352, 356–359, 396, 407
Map[Series, 122
MapAll, 184
MapAt, 160, 410, 416
MatrixForm, 125
MaxSteps, 115, 119, 122
method of images, 243, 246
metric, 375
metric tensor, 78
Mod, 9
Module, 10, 91, 231, 232, 236
Monopole, 233, 239, 243, 246
monopole, 26
monopole, 60
MultipoleP, 237, 238, 254
MultipoleSH, 236–238, 254

NDSolve, 14, 112, 115, 118, 119, 122, 169, 197
Needs, 10
Newton's second law, 23
Newtonian motion, 23
non-inertial frame of reference, 24
nonholonomic, 143
nonlinear pendulum, 121
normal frequencies, 84
normal mode, 129
normal modes, 84, 128, 132, 136, 137, 140
NRoots, 20
NSolve, 20
null coordinates, 398
numerical value, 2
numOrbit, 198

O[t], 41
Options, 3
oscillations, 83
oscMovie, 93, 114, 118, 120
Outer, 92, 125
OutputForm, 408

Packages, 10
parabolic coordinates, 227
Paraboloidal, 364, 367
ParametricPlot, 15, 33, 34, 39, 51, 87, 128, 173, 292, 298, 428
ParametricPlot3D, 18, 45, 59, 164, 248
Partition, 321
perturbative solution, 139
phase diagram, 87
phase plane, 96, 140
phase planes, 83
phase plot, 89
phase plot for one-dimensional system, 87
phase plots and time evolution for a one-dimensional system, 89
phase trajectories, 120
phasePlot, 89, 104, 108, 116
Pi, 2
PlanckConstantReduced, 76
Plot, 14, 15
Plot3D, 18, 68, 74, 101, 107, 110, 168, 263, 268, 273, 326, 333, 343
PlotGradientField, 17, 57, 62, 235, 283, 291, 299
PlotGradientField3D, 59, 61, 241
PlotJoined, 70
PlotLabel, 46, 51, 67, 96, 322
PlotPoints, 18, 59, 60, 62, 68, 74, 326
PlotRange, 15, 59, 76
PlotStyle, 14, 29, 70, 74, 128, 325, 326, 334, 340
PlotVectorField, 16, 251
PlotVectorField3D, 64
Point, 40, 170, 407, 423, 426, 430
PointSize, 29, 35, 40, 164, 170, 407, 423, 426, 428, 430
PolarPlot, 190, 194, 197
Polygon, 55
postfix form, 2
potential diagrams, 83
PotentialExpansion, 234, 240, 244, 247
PowerExpand, 330, 349, 356, 358, 360, 361
probability current, 304
Product, 10
Protect, 9, 12
pseudo-Euclidean, 372, 373
pseudo-Euclidean metric, 378, 381
pure forms, 13, 184
pure functions, 2, 20, 159

Random, 19, 35, 70, 81
Rationalize, 35
Re, 326, 335
Reduce, 6, 20, 126, 316, 317
reflection, 324
Ricci scalar, 374
Ricci tensor, 374, 378
ricciR, 378, 399
Riemann metrics, 374
Robertson-Walker metric, 375

Roots, 6
RotateLabel, 34, 40
rotating coordinate system, 42
Rule->Equal, 38
Rules, 4, 9

scalar potential, 24
ScaleFunction, 16, 17, 252
schrodinger, 308, 311, 336, 341, 365
Schrodinger's equation, 304
Schwarzschild metric, 381
FreeQ, 259
Select, 260, 317, 342, 346, 365, 399, 418
Series, 43, 62, 286, 396
series expansion for second-order equation, 85
series expansion solution for second-order equation, 153, 305
SetCoordinates, 258, 260, 282, 309–311, 324, 331, 336, 341, 345, 365, 367
shadowing, 16
ShadowPlot3D, 96
ShadowPosition, 96
Short, 49, 316
Show, 33, 34
Sign, 8, 334
Simplify, 3, 348
small oscillations, 90, 140
smallOsc, 90, 127, 131, 136, 141
Snell's law, 391
Solve, 5, 20
Sort, 321
Spherical, 62, 345
spherical harmonics, 231
SphericalHarmonicY, 348, 363
SphericalPlot3D, 363, 401
ssChrist, 382
ssComp, 381
ssMet, 381
step function, 8
stress-energy tensor, 374
Sum, 97, 99, 191, 231, 232, 236, 267, 276, 280, 286, 289, 293, 296
sum, 41
superposition of point charges, 238
Symbol, 313, 326, 352
Table, 9, 268, 340, 348, 351, 352, 363, 424
table, 9, 29, 33
TableForm, 313, 321, 351, 424
TableHeadings, 313, 321, 351, 424
TableSpacing, 321, 351, 424
Text, 33, 56, 73, 313, 319, 325, 352, 394, 407, 408, 414, 423, 428, 430
Thickness, 70, 74, 128, 327, 340
Thread, 7, 43, 44, 49, 50, 79, 86, 126, 132, 137, 234, 246, 253, 282, 287, 291, 317, 318, 342, 348, 366, 367, 418
three coupled harmonic oscillators, 130
time behavior of phase plot for a one-dimensional system, 88
time-evolved phase diagram, 88
time-independent quantum mechanics, 304
timePhasePlot, 88, 89, 104, 108, 116
Times-Roman, 407, 408, 414, 423, 428, 430
Timing, 8, 83
Together, 330, 358
transmission, 324
Transpose, 137
TrigToP, 232, 241, 244
TrigToY, 231, 236
two coupled harmonic oscillators, 124
two postulates for special relativity, 372

Union, 356
Unprotect, 9
user-defined geometric procedures, 376
user-defined metric, boost, and velocity parameters, 375
user-defined solutions of differential equations, 307

van der Pole equation, 118
van der Pole oscillator, 118
vChange, 305, 337, 347
vector potential, 64
VectorHeads, 59, 61, 62
VectorQ, 185
VEPlot, 235, 239, 247, 250, 256, 263, 264, 268, 274, 284
ViewPoint, 46, 65, 68, 74

Mathematica® FOR PHYSICS

With a foreword by Stephen Wolfram

Robert L. Zimmerman, University of Oregon
Fredrick I. Olness, Southern Methodist University

Mathematica® is a powerful mathematical software system for researchers, students, and anyone seeking an effective tool for mathematical analysis. Tools such as *Mathematica* have begun to revolutionize the way science is taught. Now there is a book specifically for teachers and students of physics who wish to use *Mathematica* to visualize and display physics concepts and to generate numerical and graphical solutions to physics problems.

Throughout the book, the complexity of both the physics and *Mathematica* is systematically extended to broaden the tools the reader has at his or her disposal, and to broaden the range of problems that can be solved. This text is an appropriate supplement for any of the core advanced undergraduate and graduate courses.

Highlights

- Provides *Mathematica* solutions for the canonical problems in the physics curriculum.
- Uses the power of *Mathematica* to develop the reader's use of visualization in problem solving.
- Emphasizes the graphical capability of *Mathematica* to help develop the reader's intuition for the physics behind a given problem.

About the Authors

Robert L. Zimmerman is a Professor of Physics and research associate in the Institute of Theoretical Science at the University of Oregon. He has taught graduate courses in mathematical physics, theoretical mechanics, electrodynamics, quantum mechanics, general relativity, and cosmology. He received his Ph.D. from the University of Washington.

Fredrick I. Olness is an Assistant Professor of Physics at Southern Methodist University in Dallas, Texas. He received his Ph.D. from the University of Wisconsin, an SSC Fellowship in 1993, and is a member of the CTEQ collaboration — a novel collaboration of theorists and experimentalists.

ISBN 0-201-53796-6

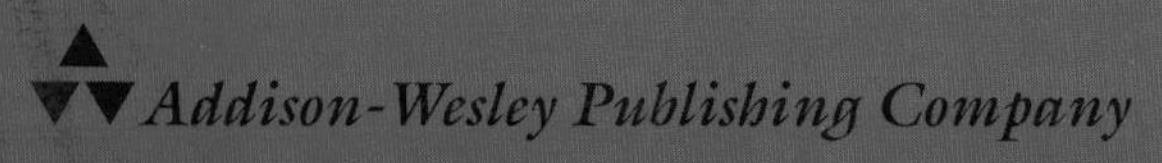